Christian E.W. Steinberg

Ecology of Humic Substances in Freshwaters

Springer
Berlin
Heidelberg
New York
Hong Kong
London
Milan
Paris
Tokyo

Christian E.W. Steinberg

Ecology of Humic Substances in Freshwaters

Determinants from Geochemistry to Ecological Niches

With 213 Figures and 19 Tables

Springer

Professor Dr. Christian E.W. Steinberg
Humboldt-Universität zu Berlin und
Leibniz-Institut für Gewässerökologie und Binnenfischerei
Müggelseedamm 310
12587 Berlin

ISBN 978-3-642-07873-6

Library of Congress Cataloging-in-Publication Data
A catalog record for this book is available from Library of Congress.

Bibliographic information published by Die Deutsche Bibliothek.
Die Deutsche Bibliothek lists this publication in the Deutsche Nationalbibliografie;
detailed bibliographic data is available in the Internet at http://dnb.ddb.de

Springer-Verlag Berlin Heidelberg New York
a member of Springer Science+Business Media

http://www.springer.de

Printed in Germany

Cover Design: Erich Kirchner, Heidelberg, Germany

Printed on acid-free paper

Dedication

I dedicate this book to my wife *Anette* for all the borrowed common leisure time during weekends and vacations. *Anette*, I thank you very much for your patience and understanding. That must be love.

Acknowledgments

I gratefully acknowledge the help of so many friends and colleagues: Gudrun Abbt-Braun, Valeria Amé, George R. Aiken, Jacqui Aitkenhead, Jarkko Akkanen, David Bastviken, Nanna Buesing, Kent B. Burnison, Yona Chen, Helmut Fischer, Fritz H. Frimmel, Egil Gjessing, Elisabeth Gross, Markus Haitzer, Sebastian Höss, Karsten Kalbitz, Maris Klavinš, Renate Klöcking, Edgar Klose, Pirkko Kortelainen, Jussi Kukkonen, Bernd Marschner, Eurico Melo, Uwe Münster, E. Mike Perdue, Irina V. Perminova, Hans Rudolf Schulten, Ruben Sommaruga, Maxim Timofejew, Lars Tranvik, Rolf Vogt, and Quinlan Wu. I am also deeply indebted to the members of my working group: Jörg Gelbrecht, Thomas Meinelt, Andreas Nicklisch, Andrea Paul, Stephan Pflugmacher, Thomas Rossoll, Anke Sachse, Angelika Stüber, Claudia Wiegand, and Elke Zwirnmann, and to our enthusiastic students, Kerstin Greulich, Philipp Hillmeister, Nanke Meems, Constanze Pietsch, Torsten Preuer, Stefanie Pütz, Wiete Rieger, and Dominik Zak.

Furthermore, I thank the staff of the Leibniz Institute of Freshwater Ecology and Inland Fisheries, especially Brigitte Spieler, Karin Römer, Magdalena Sieber, and Vera Henke, for providing office, library, and technical support. I wish to express my gratitude to all colleagues and publishing houses, who gave copyright permission for figures and chapters.

I am very grateful to Andrew Fyson for working hard on the earlier version of the manuscript, and Sarah L. Poynton for language polishing and particularly for very rigorous, yet very supportive, comments on the manuscript.

Many thanks are also due to the staff of Springer-Verlag, particularly Janet Sterritt-Brunner and Christian Witschel, for their understanding, and their help in preparing my book.

Contents

Glossary

ANC	acid neutralizing capacity
BCF	bioconcentration factor of chemicals (metals, xenobiotics)
bioconcentration	undirected uptake of metals and xenobiotic chemicals via epithelia or membranes
bioaccumulation	bioconcentration plus uptake of metals and xenobiotic chemicals via food
biomagnification	enrichment of metals and xenobiotic chemicals within the food web
BP	before present, term common in paleo-sciences
cDOC	chromophoric dissolved organic carbon
DOC, TOC	dissolved organic carbon, total organic carbon
DON	dissolved organic nitrogen
DOP	dissolved organic phosphorus
DOS	dissolved organic sulfur

enzymes, affected by humic substances
DNAases, esterase, glutathione peroxidase, glutathione-*S* transferase, glycogen phosphorylase, guaiacol peroxidase, indole-3-acetic acid oxidase, metalloprotease, mixed function oxygenase, phosphatase, phosphorylase, protease

enzymes, involved in humification
aminopeptidase, cellobiose oxidase, cellulase, laccase, lignin peroxidase (syn. ligninase), peroxidase, phenoloxidase, phenolperoxidases, polyphenoloxidase

enzymes, involved in reduction of the oxidative stress
catalase, glutathione peroxidase, guaiacol peroxidase, superoxide dismutase

enzymes, involved in organohalogen production
bromoperoxidase, chloroperoxidase, haloperoxidase

fluorescence quenching	decreasing fluorescence of a fluorophore, such as a PAH, by binding to a non-fluorescent compound
FA	fulvic acid
HA	humic acid
HAP	humic acid precursors

HS	humic substances
HiA	hydrophilic acids
HiN	hydrophilic neutrals
HiB	hydrophilic bases
HoA	hydrophobic acids
HoN	hydrophobic neutrals
K_{DOC}	partition coefficient between water and DOC
K_{OC}	partition coefficient between water and colloidal and particulate OC (organic carbon)
K_{OW}	partition coefficient between water and *n*-octanol
molecular weight	more stringently called molecular mass, in Da. However, the book follows the less stringent terminology 'molecular weight', common in limnological literature
NOM	natural organic matter, isolated by reverse osmosis or ultrafiltration
OC	organic carbon
PAH	polycyclic aromatic hydrocarbon, such as anthracene, benzo[*a*]pyrene, fluoranthene, pyrene
anthracene	linear aromatic hydrocarbon with three fused rings
fluoranthene	angular aromatic hydrocarbon with four fused rings
pyrene	angular aromatic hydrocarbon with four fused rings
B*a*P	benzo[*a*]pyrene, aromatic hydrocarbon with five fused rings
PAR	photosynthetically active radiation (400–700 nm)
QSAR	quantitative structure-activity-relationship
ROS	reactive oxygen species (singlet oxygen, hydrogen super oxide)
SEC	size exclusion chromatography
SRP	soluble (molybdenum blue) reactive phosphorus
UV	ultra violet radiation
VIS (vis)	visible light
xenobiotic chemical	organic, often halogenated, chemicals produced by man; consideration of the natural halogen chemistry (Chap. 6) suggests that the separation between natural and xenobiotic potentially toxic chemicals is fictitious

1 Introduction

Ecology of dead organic matter in freshwater? Or the ecological control by dead organic matter? The link of 'ecology' and 'dead' looks to be a strange alliance, perhaps even a contradiction. The most important objects of limnological studies are living organisms, and their interactions with the surrounding biotic and abiotic world. Organisms build the central dogma in ecology with the predominance of food-webs on the one hand and nutrient and energy cycling on the other hand. Which organisms consumes which and why? What are the strategies to acquire food? How much energy is transferred from one trophic level to the next? How does this determine the Darwinian fitness of organisms, particularly under the influence of various natural and man-made stress factors? Is the description of more or less unidirectional trophic relationships of biocoenoses (guilds) superior to biomass spectra that reflect an energetic continuum? What does biodiversity mean with respect to ecosystem functioning and evolution? These are some of the current questions of ecology and limnology. These items are, however, only one part of the life-bound processes in nature.

Eventually, all organisms are eaten or die, and thereby release organic substances which are finally substrates for various humification processes. Humification processes are quantitatively the second most important biogeochemical processes on earth after photosynthesis, and ensure the stability of the global cycling of energy and materials. However, in terms of scientific research on both biogeochemical processes, it is conspicuous that until very recently, the humification process is as good as unnoticed or unstudied. Humic substances (HS) were, and still are, difficult to define from a chemical point of view in comparison with, for example, proteins, carbohydrates and lipids. One may casually say: they are wastes, even 'dirt', and in the case of environmental samples interfere with the chemical analysis of compounds of interest.

In contrast, the source materials of humification contain functional groups or structures with much biological information. There is no plausible reason why a portion of this biological information cannot be retained following the humification processes. It is a general feature that aquatic HS contain a certain, relatively high proportion of peptides (Steinberg 1977; Watt 1996; Box 2.1). For example, it is well known that peptides are signal molecules, acting not only within organisms but also have regulative functions in ecosystems. The behavior of animals can be influenced by such signal chemicals (Browne et al. 1998; Rittschof 1990). The association of HS and peptides offers protection against microbial degradation (Münster et al. 1999c,d; Steinberg 1977; Wetzel 1991, 1993). Hence, peptides can persist in the aquatic environment, and thus the function of peptides as signals may not be completely lost as they may subsequently be released from HS masking, and may reach the target organisms.

From this general reflection, it is hardly surprising that the HS have diverse ecological and physiological roles in aquatic systems. Some functions of the original substances can be expected, but other functions are not so easily predicted. One can clearly say that freshwater HS are not only accountable for physical characteristics such as light absorption, and chemical characteristics such as acid status or photolytic release of microbial substrates, but that they also interact indirectly and even directly with living organisms. The fact that these matters are little studied is certainly due to the traditional view that HS are (with the exception of photolytic cleavage) inert, refractory, or in some other way passive in ecosystems. This idea is at best outdated, if not false. In this book, I shall discuss several of the recently discovered control functions of HS that indirectly or directly affect aquatic organisms.

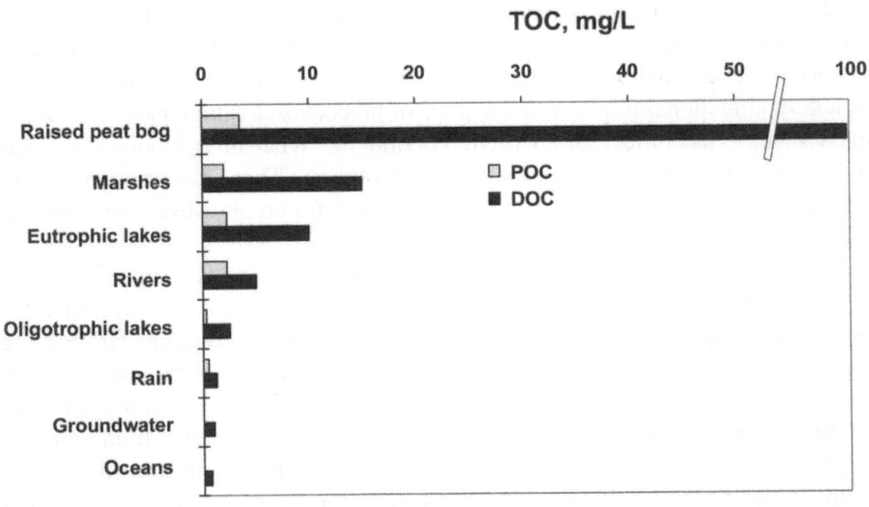

Fig. 1.1. Concentrations of particulate (POC) and dissolved (DOC) carbon compounds in natural waters (modified from Thurman 1985, with kind permission of Kluwer Academic Publishers)

In freshwaters, there are differences in autochthonous and allochthonous humification pathways, which clearly differ in terms of reaction products. HS are an important component of the total organic carbon (TOC), dominated by dissolved organic carbon (DOC). If one considers the relationship of carbon in living organisms, to that in the total particulate and dissolved dead fractions, then one finds that the proportion in the dead fractions exceeds that in living organisms by approximately one order of magnitude (Table 1.1).

The concentration of organic carbon compounds in most natural freshwaters is in the range 0.5 to 50 mg/L C as TOC or DOC. With DOC concentrations in excess of 100 mg/L, inland saline lakes in semiarid regions of the Canadian prairies contain some of the highest known DOC concentrations (Arts et al. 2000; Waiser and Robarts 2000). In a few peat bogs or in soil waters of wetlands, values can reach more than 100 mg/L DOC (Chen et al. 1994; Cronan 1990; Kortelainen 1999a). In the interstitial water of minerotrophic fens, DOC concentrations as high as 300 mg/L are reported (Sachse et al. 2001a). A schematic overview is shown in Fig. 1.1.

Table 1.1. Estimates of carbon in organisms in relation to dead organic carbon in the ecosystems of the world (after various authors from Killops and Killops 1993)

	Ratios of carbon in living organisms to dead organic carbon	Dead organic carbon, % of total organic carbon
Terrestrial Systems	1:3	75
Freshwater	1:10	90
Oceans	1:400	> 99

The variability of freshwater DOC concentrations is markedly less than the water quality parameters, such as oxygen, nutrients, or biochemical oxygen demand. There is some kinetic buffering, both in the release from catchment areas and in freshwaters themselves, as described in detail in Chap. 3.

In contrast to the quantitative dominance of dead organic matter in the environment, the knowledge of its ecological role is minor compared to that about the organisms themselves. This statement is even more applicable to HS. In older publications, HS are referred to as 'refractory' or 'inert', which implies that they have little involvement in ecosystem processes, or are persistent with low degradation by microorganisms. This view is apparently supported by data on the turnover of HS in soil (Fig. 1.2). From this schematic diagram, it is clear that the complete turnover time for HS is determined by soil depth. Near the surface, in the root-zone, the complete turnover takes 'only' a few years to a few decades since in this layer, microbial activity is highest due to easily biodegradable organic substrates. With increasing depth, the turnover increases to several thousand years. Thus poorly or totally non-degradable substances persist, and it is this material, which after passage in groundwater, finally appears in surface waters. These materials do indeed have an age of decades to thousands of years (Chap. 3.1.3.3).

Previous estimates of DOC turnover in freshwater, as shown for example with peptides and proteins, also give figures of several thousand years (Steinberg 1977). Marine studies, too, using ^{14}C dating, indicate that POC and DOC are essentially inert, as an age of 1–6 10^3 years is commonly found (Bauer et al. 1992; Druffel and Williams 1990; Williams and Druffel 1987)[1]. If these age estimates are applicable to the total pool of organic substances, they strongly suggest a pic-

[1] The DOC pool contains fractions with long and short turnover times, both in the HMW and LMW fractions (Santschi et al. 1995). This applies to marine as well as freshwater DOC.

ture of high persistence for these substances, or at least a particular fraction of them. This is misleading and distracts from the dynamic role of organic matter, including HS, in freshwater systems. The dynamic role of HS in various metabolic ecosystem pathways, as well as interactions of HS with organisms, will be the subjects of the following chapters of the book.

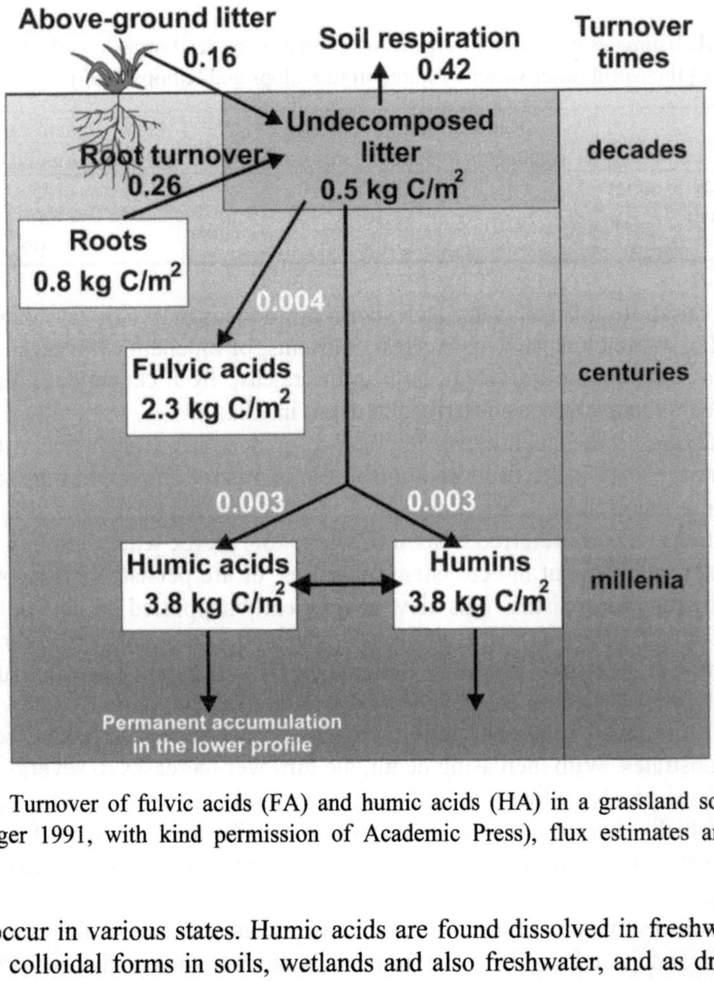

Fig. 1.2. Turnover of fulvic acids (FA) and humic acids (HA) in a grassland soil (from Schlesinger 1991, with kind permission of Academic Press), flux estimates are in kg $C/m^2/a$

HS occur in various states. Humic acids are found dissolved in freshwater, in solid or colloidal forms in soils, wetlands and also freshwater, and as dry, solid forms in lignite and coals (Killops and Killops 1993). In surface waters, these supposedly inert organic carbon components and HS have decisive functions through various physical, chemical, and biological processes. These substances regulate many metabolic processes in freshwater ecosystems. An understanding of these functions of HS will give new insights into the functioning of aquatic ecosystems. For the formation of biocoenoses, this biogeochemical matrix is definitely very important. For example, HS have a decisive role in determining nutrient content and bioavailability, and also the underwater light conditions. In addition, in waters rich in HS, the classical food-web starting with algae or macrophytes, is reversed with the detritus food chain quantitatively more important than autochthonous primary production. Additionally, direct interactions of HS with organisms will favor the less sensitive forms, and will structure guilds (biocoenoses). Comparing different humic sources and their effects on different aquatic species, there appears not be a most sensitive species. That means that different humic sources from different catchments will produce guilds with different species compositions. This applies at least to the quantitative compositions.

1.1 Definitions

The term 'humic substances' comprises three groups of substances: **fulvic acids**, **humic acids** and **humin**. This distinction is based on the traditional fractionation of soil HS. The treatment of total HS with dilute alkali, dissolves the fulvic and humic acids, leaving the undissolved humins behind. If one acidifies this alkaline extract, the humic acids (HA) precipitate, leaving the fulvic acids (FA) in solution. This operational definition is shown in Fig. 1.3 below.

HA can be further fractionated into hymatomelin, grey and brown humic acids, through extraction in alcohol or redissolving in alkaline solution. This differentiation, however, is not applied in aquatic ecology. The fractionation of aquatic HS is based on their differing solubility in acids or alkalis. Here one makes use only of the pH dependent absorption of organic solutions at the hydrophobic surfaces. Under acidic conditions, the carboxyl groups of organic acids are not further dissociated (for example –COOH). These organic molecules can now be adsorbed on a fixed hydrophobic phase (hydrophobic resins such as XAD-resins) and build a coating on their surface (Fig. 1.4). Under acidic conditions, the organic bases cannot be adsorbed on the surface of adsorbers, since the basic groups, such as $–NH_2$ now are protonated as $–NH_3$. Organic bases are first adsorbed under non-protonating conditions such as basic and neutral pH; the organic acids dissociate, and are desorbed from the adsorber. According to the definition of the International Humic Substances Society (IHSS), the adsorbed material is described as HS.[2] These principles for definition of HS components are summarized in Fig. 1.5.

Other schemes are developed according to process or object. Another scheme for the further separation of HS and non-HS comes from the work of Leenheer

---

2 Criticisms of this procedure has occasionally been expressed, for example, by Shuman (1990) in studies on the role of the hydrophilic acids. In more recent studies (Aiken and Cotsaris 1995; Egeberg et al. 1999, 2002; Fettig 1999; Gjessing et al. 1998, 1999; Rozan et al. 1999), the role of natural organic components in water is studied in an ecological context or in relation to drinking water processing. These studies do not separate DOM into hydrophobic and hydrophilic components, but look for DOM as so-called natural organic matter (NOM) isolated by reverse-osmosis.

(1981) (Fig. 1.6). Only the adsorbed components of HS – the sum of the hydrophobic components – are considered, as in the previous scheme (Fig. 1.5).

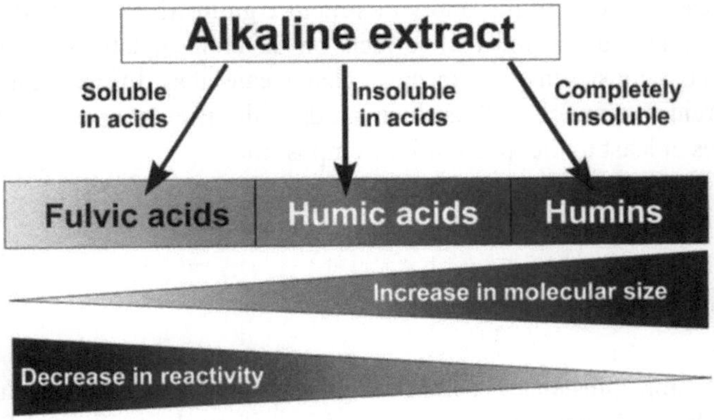

Fig. 1.3. Scheme for the operational definition of humic substances

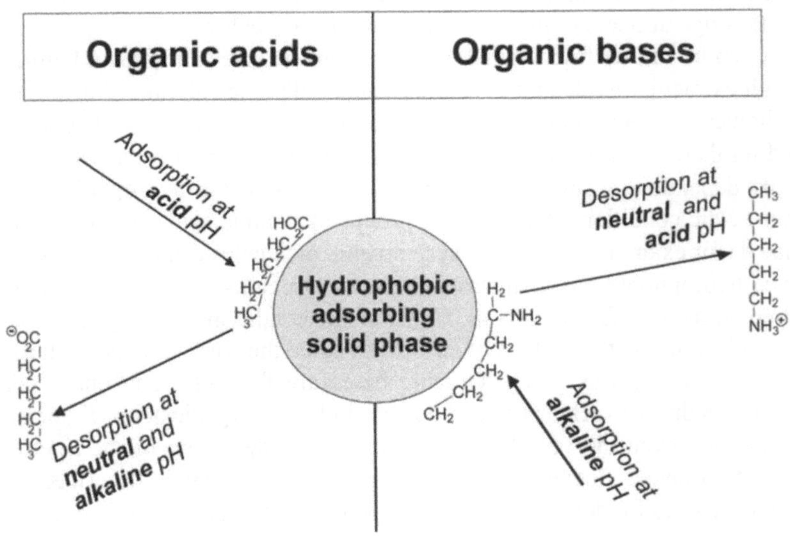

Fig. 1.4. The effect of acidity (pH value) on the ionization and solubility of adsorbable organic acids and bases (from Peuravuori and Pihlaja 1999, with kind permission of Backhuys Publishers). Per definition, the adsorbable compounds are HS. Hydrophilic compounds that do not belong to HS are lost by this separation.

An overview of possible acidic, basic, and neutral functional groups present in aquatic HS is given in Table 1.2. This classification is based on the capacity of the functional groups to donate or accept protons, as FA and HA do. Neutral functional groups can neither donate nor accept protons. FA contain smaller and more strongly oxidized units than do HA. This is shown also in the higher content of oxygen-containing functional groups (Table 1.3). One therefore concludes that the FA are probably the oxidized breakdown products of humic acids (Chap. 2).

Of the major chemical elements, dead organic matter contains 0.03–0.2% P, and according to pioneering results of Öberg (1998) 0.01–0.5% Cl. The Cl content (in soils) is therefore similar to that of P! This means that one of the most important components of soil organic substances has, until recently, been overlooked. The turnover of organic-bound Cl in soils is still little understood.

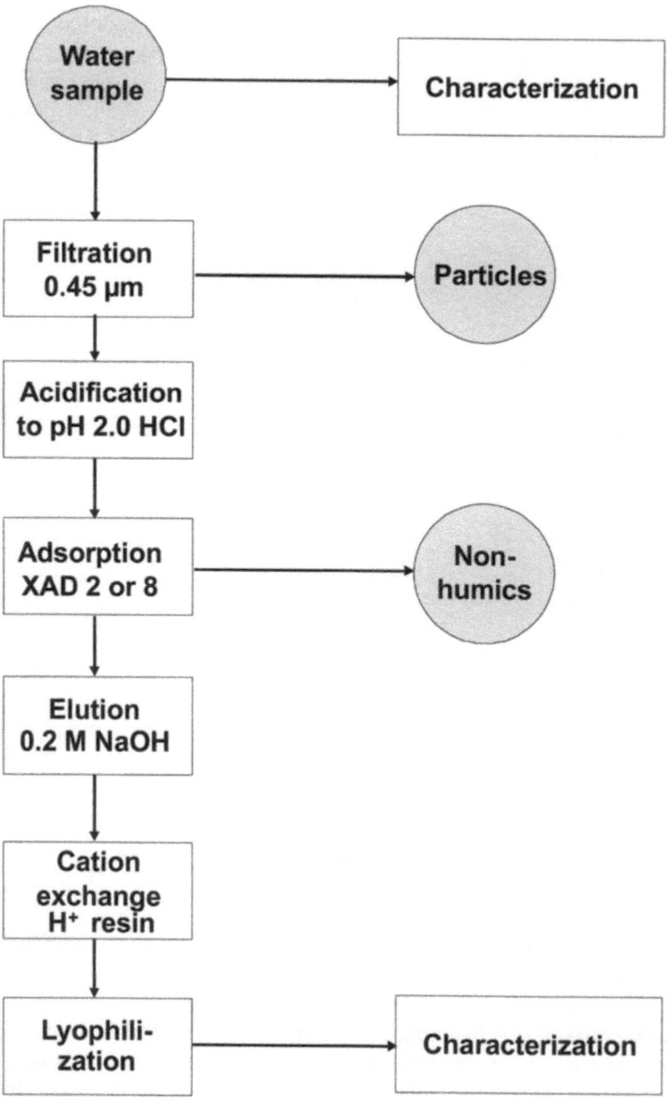

Fig. 1.5. Scheme of the isolation procedure for aquatic HS (after Frimmel 1990, with kind permission of Wiley & Sons)

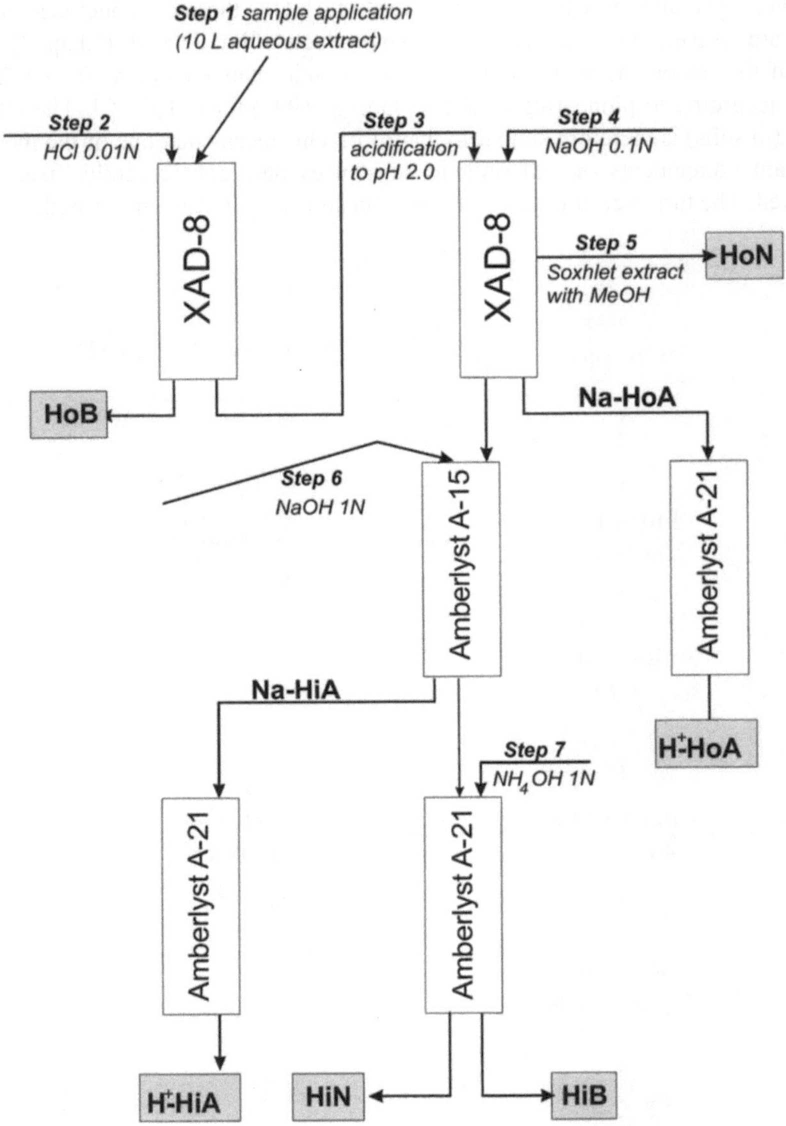

Fig. 1.6. Scheme for the isolation of HS and other organic carbon components (based on Leenheer 1981, after Chefetz et al. 1998, with kind permission of Wiley-VCH). Only hydrophobic components are considered as humic substances (HS). HiA hydrophilic acids; HiB hydrophilic bases; HiN hydrophobic neutral substances; HoA hydrophobic acids; HoN hydrophobic neutral substances. Several authors successfully apply this or a similar fractionation protocol to NOM in percolating and surface waters, for instances Dunnivant et al. (1992); Geyer (1994); Guggenberger and Zech (1993a); Hongve (1999, 2000); Kaiser et al. (2001c); Malcolm (1991); Mattsson et al. (1998); McCarthy et al. (1993); Roila et al. (1994a). Compare Chap. 3, particularly 3.2

Table 1.2. Important functional groups of dissolved organic carbon (from Peuravuori and Pihlaja 1999, after Thurman 1985)

Functional group	Structure	Where found
	Acidic groups	
carboxylic acid	(Ar-)R-CO_2H	90% of all DOC
phenolic OH	Ar-OH	aquatic HS, phenols
enolic hydrogen	(Ar-)R-CH=CH-OH	aquatic HS
quinone	Ar=O	aquatic HS, quinones
	Basic groups	
amine	(Ar-)R-CH_2-NH_2	amino acids
amide	(Ar-)R-C=O(-NH-R)	peptides
imines	CH_2=NH	unstable
	Neutral groups	
alcoholic OH	(Ar-)R-CH_2-OH	aquatic HS, sugars
ether	(Ar-)R-CH_2-O-CH_2-R	aquatic HS
ketone	(Ar-)R-C=O(-R)	aquatic HS, volatiles, keto-acids
aldehyde	(Ar-)R-C=O(-H)	sugars
ester, lactone	(Ar-)R-C=O(-OR)	aquatic HS, hydroxy acids, tannins
cyclic imides	(R-)O=C-NH-C=O(-R)	aquatic HS

R aliphatic compounds and Ar aromatic ring. In addition to certain amino acids sulfur (S) may also occur as mercapto/thiol compounds (–SH) and sulfonic acid derivatives (Ar–)R–SO_3H. It is notable that the backbone may be R or Ar, but mostly it is a combination of both.

Table 1.3. Elemental composition (weight %) and functional group content (mequ/g) in HA and FA (from Schnitzer 1978)

Element	HA	FA
C	53.6–58.7	40.7–50.6
H	3.2–6.2	3.8–7.0
N	0.8–5.5	0.9–3.3
O	32.8–38.3	39.7–49.8
S	0.1–1.5	0.1–3.6
Functional groups	**HA**	**FA**
acid groups, total	5.6–8.9	6.4–14.2
carboxylic acids	1.5–5.7	5.2–11.2
phenolic OH	2.1–5.7	0.3–5.7
alcoholic OH	0.2–4.9	2.6–9.5
quinoide/keto C=O	0.1–5.6	0.3–3.1
methoxy OCH_3	0.3–0.8	0.3–1.2

1.2 Operational Definition of 'Dissolved'

In addition to the definition of HS, the separation of dissolved and particulate matter is operationally defined through filtration with membrane filters of given pore-size. Usually, filters with a pore-size of 0.2–0.45 µm are used. Occasionally finer or more rarely coarser pored filters are applied. The operational definition of dissolved organic matter is given in Fig. 1.7. Filters allow not only truly dissolved molecules to pass through, but also colloidal material which comprises up to nearly 50% of total DOC in marine waters (Chin et al. 1998). For example, the filtrate contains a part of the humin fraction and various complexes with FA and particularly HA.

Generally, HS comprise up 60–80% of the total DOM (Steinberg and Münster 1985) of which 30–40% is aromatic carbon (Thurman 1985). In dystrophic waters, the HS content can rise to 80–95% of the total DOC (Münster 1999a,b). HS have a molecular weight of a few hundred to several thousand Da. The latter exist as colloids with diameters of 1 nm to 1 µm, and can pass through the pores of most membrane filters. HS exist as particles if the diameter exceeds 1 µm.

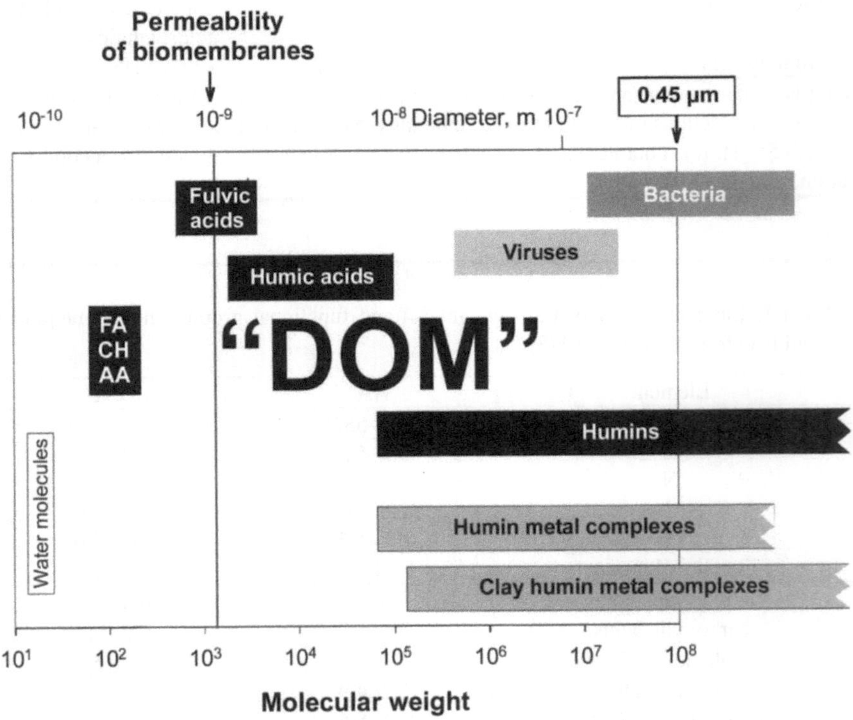

Fig. 1.7. Graphic definition of dissolved organic matter including small organic molecules and the three humic substance fractions. AA = amino acids, CH = carbohydrates, FA = fatty acids (modified from Thurman 1985, with kind permission of Kluwer Academic Publishers)

1.3 The Age of Humic Substances

Regarding the age of HS in the environment, there are large discrepancies depending on whether chemical or biological-biochemical methods have been employed. Chemical analyses and isotope studies yield an age for DOM in marine and freshwater systems of several hundred to around 6,000 years. Recently, Raymond and Bauer (2001) add another piece to the puzzle. They write that the input from rivers is, by itself, more than large enough to account for the apparent steady-state turnover times of up to 6000 years for oceanic dissolved organic carbon. But paradoxically, terrestrial organic matter, derived from land plants, is not detected in seawater and sediments in quantities that correspond to its input. Raymond and Bauer find that rivers are sources of old and young terrestrial dissolved organic carbon of predominantly old terrestrial particulate organic carbon. These findings contrast with earlier data that suggest terrestrial organic matter might be generally newly produced. The authors also find that much of the young dissolved organic carbon can be selectively degraded over the residence times of river and coastal waters, leaving an even older and more refractory component for oceanic export. Thus, pre-aging and degradation may significantly alter the structure, distributions, and quantities of terrestrial organic matter before its delivery to the oceans.

In contrast, microbial-biochemical methods discover an age of only days to weeks (Amon and Benner 1994). This demonstrates not only the lack of agreement between the various estimates, but also our poor understanding of the structure and function of DOM in aquatic ecosystems. The lack of knowledge is even more pronounced with dissolved HS (Münster 1999b, and references therein).

1.4 Aims of the Book

Since the dissolved HS concentrations exceed the total carbon in all living organisms, Wetzel (1995) states that the trophic structure in the pelagic zone is not only a particulate world. Wetzel continues by saying that '*population fluxes are not representative of the material and energy fluxes of either the composite pelagic region or the lake ecosystem. Metabolism of particulate and especially dissolved organic detritus from many pelagic and non-pelagic autochthonous and from allochthonous sources dominates both material and energy fluxes. Because of the very large magnitudes and relative chemical recalcitrance of these detrital sources, the large but slow metabolism of detritus provides an inherent ecosystem stability that energetically dampens the ephemeral, volatile fluctuations of higher trophic levels*'. This statement is equivalent to the general designation of HS as '*entropy buffer*' by Ziechmann (1994) (Chap. 6.4). Wetzel further explains that '*continued application of animal-orientated relationships to the integrated, process-driven couplings of the aquatic ecosystems impedes understanding of quantitative ecosystem pathways and control mechanisms.*'

In Wetzel's statements, there is already the beginning of a comprehensive and ecological understanding of dead organic matter in freshwater. However, it remains in some instances, still somewhat traditional particularly regarding direct interactions of HS with aquatic organisms and to controls of aquatic guilds and pathways by HS. The ecological role of dissolved HS in ecosystems is really more extensive. Dissolved HS not only dampen trophic fluctuations, but commonly also promote, regulate, and stimulate reactions of organisms, or provide some kind of information to aquatic biota. To date, only a few properties are known: with the UV-B increase related to global climate change, dissolved HS provide the best natural shield; in particular geological (non-calcareous) regions, dissolved HS determine freshwater chemistry, and provide the main buffer system.

If one orders the effects of dissolved HS in freshwater systems over space and time, the following sequence occurs, which is considered in this treatise:

- In the long-term, dissolved HS can determine or strongly change the chemistry of whole lakes and regions.
- In whole lakes, dissolved HS act through fast reactions as the most important natural purification system for xenobiotics and allelochemicals.
- In the mid- to long-term, dissolved HS influence or strongly impair processes of global climate change in entire regions.
- Dissolved HS can inhibit ('entropy buffer') or stimulate, short to mid-term metabolic processes in freshwater systems.
- Dissolved HS can act as inorganic or organic nutrients in single or multi-species systems.
- Dissolved HS rapidly change the bioconcentration and toxicity of metals and xenobiotics.
- Upon exposure to light, dissolved HS can rapidly release oxidized substances (reactive oxygen species, ROS), of which the toxic potential has not yet been fully realized.
- In the short-term, dissolved HS are, paradoxically, natural xenobiotic chemicals and act as an information source at the individual and single-species level:
    - They influence the productivity of the nematode *Caenorhabditis elegans*.
    - They affect the transformation system of aquatic macrophytes, invertebrates, and fishes.
    - They interfere within the photosynthetic electron chain of algae and macrophytes.
- This short-term interference determines ecological niches and, in turn, influence the structure and activity of guilds.

Box 1.1 Chemical Building-Blocks and Reactivity of Humic Substances

Random or Systematic Elemental Composition?

Owing to the heterogeneity of source materials that enter the humification processes, it is not possible to give a generally applicable structural formula for HS. The distribution of elements or building blocks of the HS molecule is therefore described statistically. However, the distribution is by no means random, as is shown graphically in Figs. B.1.1.1 and B.1.1.2, where elemental ratios and structural ratios are plotted.

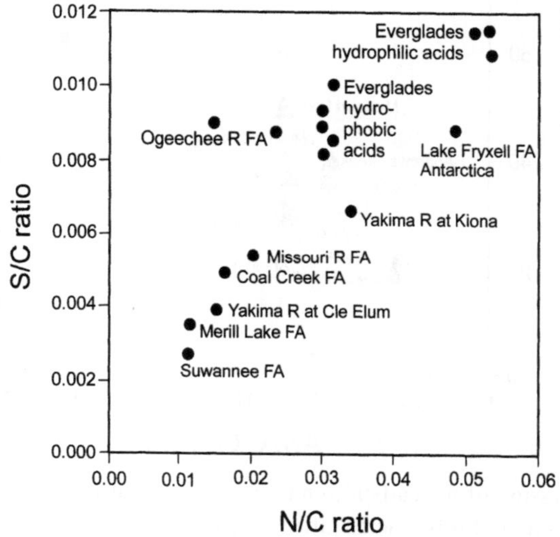

Fig. B.1.1.1. Regression of the sulfur to carbon ratio against the nitrogen to carbon ratio in fulvic acids (after Aiken, pers. comm.)

The predominant source materials can be identified from element composition, since particular source materials can only produce HS which have clearly predictable composition, structural characteristics, and therefore predictable properties. This also means that knowledge of the element composition, or the proportion of building blocks, and functional groups, can successfully be applied to the prediction of potential reactions in the aquatic ecosystem. In many instances, the quality of HS in the aquatic environment is the result of catchment processes. For instance, Möller et al. (2002) report that organic S is significantly correlated with total S, organic C, and total N, indicating that there is a close relationship between C, N, and S cycling in soils[1]. Consequently, a low N content in a aquatic HS is associated with a low S content. One extreme is the material from Lake Fryxell in Antarctica, for which, however it is known that no terrestrial plants, and therefore no lignin, is available as source material. Instead, algal and bacterial materials serve as educts of humification (McKnight et al. 1991). These materials are pigments, carbohydrates, and peptides. From the latter comes the relatively high N content of the HS. In contrast to Lake Fryxell, the FA from the Suwannee River is S rich and N poor. The catchment is characterized by wetlands with black water and negative redox conditions, and the source material is lignin rather than peptides or carbohydrates. That means that elemental ratios of HS are so fundamental that they are independent of their allochthonous or autochthonous origin.

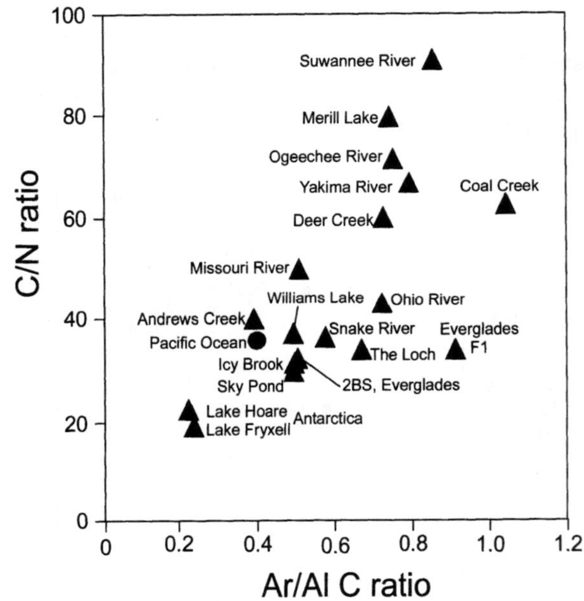

Fig. B.1.1.2. Regression of the carbon to nitrogen ratio against the ratio of aromatic fraction to aliphatic fraction of fulvic acids (after Aiken and Cotsaris 1995, supplemented with data from Aiken, pers. comm. and McKnight 1997). 2BS is a hydrophilic fraction from an oligotrophic, low-sulfide site, F1 is a eutrophic, sulfidic side in the Everglades (Benoit et al. 2001)

When the C-N ratio is plotted against the ratio of aromatic to aliphatic fractions

[1] The organic P cycling appears to be disconnected from the organic C cycling to some extent, since the retention of organic P seems to smaller than that of organic C (Kaiser 2001).

determined by ^{13}C-NMR (Fig. B.1.1.2), there is a clear linear relationship between the aromatic content and the C content, from which only one peat-dominated sample in the Everglades (F1), and the sample from Coal Creek deviate. It can be seen from this diagram that HS with low aromaticity also have a relatively low C content. This applies particularly to FA of Lake Fryxell which have a high proportion of aliphatics and a high N, but low C content. The products of the humification processes in this lake have similarities to those of the open ocean (Fig. B.1.1.2) (Stuermer and Harvey 1974). In contrast are the samples from rivers and lakes with lignin-rich sources such as the Suwannee River. Lignin is an important structural component of terrestrial vascular plants and of wood. For example, the HS from the Suwannee River have a very high proportion of aromatic structures and a very low N content. In the catchment of this river, peptides and proteins play minor roles in the humification process when compared to lignin. This means that lignin as the source of HS precursors leads automatically to a high aromatic content in the HS. The FA between the extremes of Lake Fryxell and the Suwannee River represent the high diversity of HS precursors and biogeochemical interactions, which the precursors and HS themselves underlie.

An overview of the elemental composition for the elements usually present in HS is shown in Table B.1.1.1. Interestingly, the elemental compositions follow a Gaussian distribution.

The mean proportion of aliphatic, aromatic, and excess C is reported by Perdue (1998) as: C_{al} 38 %, C_{ar} 27 %, C_{XS} 22 %, and COOH 13 %. C_{XS} is C which is not aliphatic or aromatic, and is not bound to carboxyl groups. Although the natural variability in composition can be considerable, the general distribution of the means of the non-carboxyl C in C_{al}, C_{ar} and C_{XS} is represented in Fig. B.1.1.3. This figure shows that the structural variability is probably mainly due to the variability in the aliphatic fraction (carbohydrates, sugar acids, and amino acids), which are the major components subject to easy biodegradation (Perdue 1998; see also Chap. 5.1.4).

Table B.1.1.1. Statistical data on the composition of freshwater fulvic acids (after Rice and MacCarthy 1991, from Perdue 1998)

Parameter	Mean	Standard deviation
Carbon(mmol/g)	38.9	3.6
Hydrogen (mmol/g)	42.0	7.0
Nitrogen (mmol/g)	1.6	1.5
Oxygen (mmol/g)	28.7	3.2
Sulfur (mmol/g)	0.4	0.3
Carboxyl (mmol/g)	5.0	0.5
mean molecular weight (g/mol)	1000	100

The various compositions of HS have wide-reaching consequences for their behavior in ecosystems. It is particularly the aromatic structures, and to a lesser degree the aliphatic structures, which are responsible for most of the properties of

HS in the aquatic environment.

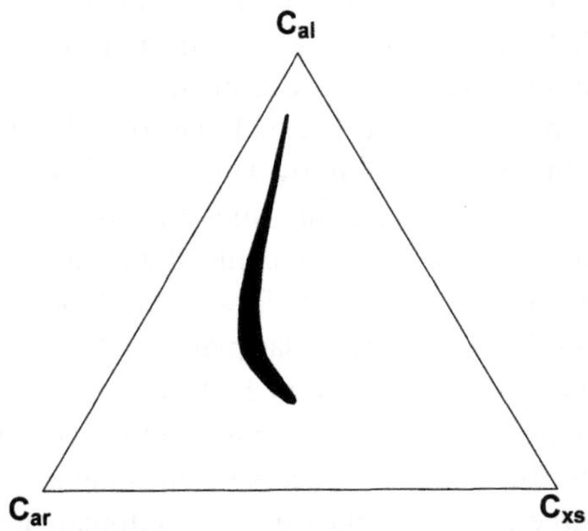

Fig. B.1.1.3. Structural distribution of the non-carboxylic DOC in freshwater fulvic acids based on average composition. The peaks in this diagram correspond to100 % aliphatic C (C_{al}), 100 % aromatic C (C_{ar}) and 100 % excess C (C_{ex}). The peak of the opposite edge accounts for 0 % of the particular C form (and modified from Perdue 1998, with permission of Springer Verlag)

Steps to Predict Environmental Behavior

Aromatic structures such as phenols, alkylaromatics, or quinones, are the essential carriers of π-electrons and determine the color, and:

- absorb UV radiation and radiation of the visible spectrum and can initiate a variety of photoreactions comprising photomineralization, photolytic release of inorganic and organic nutrients, indirect photolysis of allelochemical and xenobiotic substances;
- in some instances, the aromaticity is a strong predictor for the binding of heavy metals, for example, Cd (Kozuch and Pempkowiak 1997), Co (Steinberg et al. 2000); silver, lead, and metal mixtures (Playle et al. pers comm; Richards et al. 2001; Rose-Janes and Playle 2000);
- are responsible for adsorption onto clay particles (Specht et al. 2000). This applies particularly to the high-molecular weight fractions. This adsorption via aromatic structures also applies to co-precipitation with hydrolyzing Al hydroxide in soft-water lakes suffering from acid rain (Chap. 3.2.2.3). Furthermore, the aromatic structures control the change in HS quality during passage through soils (Kaiser and Zech 1997);
- aromatic structures are linked to aggregate formation when HS concentrations are high (see below);
- with quinone- and phenol-mediated dechlorination reactions, may potentially

contribute significantly to the fate of chlorinated contaminants (Barkovskii and Adriaens 1998; Chap. 6);

- more specifically, low molecular size lignin moieties may initiate aggregation of the aromatic rings to higher molecular weight humic compounds (Haiber et al. 2001b);
- the quinoide structures easily form radicals and can therefore function as electron acceptors and are consequently responsible for the role of HS as redox-catalysts (Scott et al. 1998), and as herbicides, reducing photosynthetic activity in aquatic plants (Chap. 7.2).

  Quinoide structures are probably also responsible for the interactions of HS with peptides and proteins which produce materials with significant healing properties in human medicine (Flaig 1997; Ziechmann 1996).

Ionic structures such as carboxyl, phenolic hydroxyl, and amino groups reduce the hydrophobicity and increase the water-solubility of HS. The carboxyl and phenolic hydroxyl groups, among others, are responsible for complexation of metal ions (Linnik 1996, 1998; Linnik and Nabivanets 1984). However, Mao et al. (2001) report about one third of the aromatic C–O groups is not phenolic C–OH, but C–OCH_3.

Aliphatic structures, particularly the carbohydrates and amino acids, are almost the structural antithesis of aromatic structures and decrease the strength of the properties of these. Further, the aliphatic structures can act as substrates for heterotrophic processes when accessible to exo-enzymes. Generally, the biodegradation appears to be slow (Hoitink and Grebus 1997, also see Box 8.1). Very recently, acid polysaccharides are described to be involved in metal binding in colloidal organic matter (Quigley et al. 2002). Some studies demonstrate that metal binding HA molecules of molecular weight >1 kDa occur in colloidal rather than dissolved state, and that the state of occurrence depends on both the properties of the surrounding medium and the chemical identity of the investigated metal species (Guentzel et al. 1996; Stordal et al. 1996). Fig. B.1.1.4 shows the linear relationship between partition coefficient (log K_c) of, for example, ^{234}Th as well as other heavy metals and the measured or amended carbohydrate content of marine organic C. The carbohydrate content explains approximately 80% of the binding variances. Future studies should determine whether or not this mechanism also applies to dissolved organic matter.

The **aliphatic fraction**, and clearly not the aromatic fraction, is involved in the formation of methylated metals, as Aiken (2000) shows for mercury in the Florida Everglades. The lower the aromaticity, the greater the methyl mercury formation potential. This is probably valid for all metals which can form organyls.

Aliphatic structures may also be involved in adorption of lipophilic compounds. It is known from terrestrial ecological chemistry that cuticular wax of leaves and needles are very good passive collectors of lipophilic environmental chemicals (Sabljic et al. 1990). Chefetz et al. (2000) describe similar accumulators for pyrene and particulate NOM, such as humins, humic acids, degraded lignin peat and brown coal.

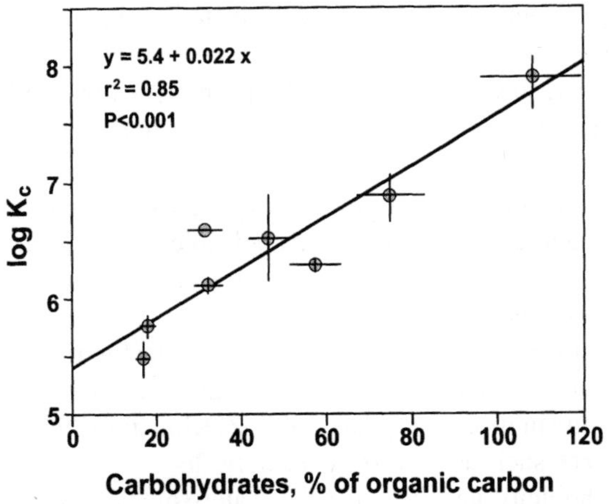

Fig. B.1.14. Log K_c values in >1 kDa permeate seawater plotted against the measured or amended fraction of carbohydrate C as percentage of total organic C (from Quigley et al. 2002, with permission of the American Society of Limnology and Oceanography)

Natural organic matter (NOM), including HS, can serve as precursors for by-product formation during chlorination of raw water in drinking water processing (for review see Nikolaou and Lekkas 2001). Trihalomethanes (THM) are the first category of chlorination by-products to be detected in drinking water, followed by haloacetic acids and haloacetonitriles, haloketones, chloropicrin, and MX [(Z)-2-chloro-3-dichloromethyl-4-oxobutenoic acid, a strong bacterial mutagen] at lower concentrations. Two of the main factors controlling the formation of chlorination by-products are the concentration and properties of NOM. Several different mechanisms must be involved in the formation of, for instance, trihalomethanes (THM), potentially cancer-causing agents. As an example, aromaticity of FA in the Suwannee River, which correlate with specific UV absorption, are an important factor in the THM formation during water treatment (Aiken 2000). Aiken reports a weak linear correlation between the UV absorption and the THM formation potential. Also Nikolaou and Lekkas (2001) state that the degree of aromaticity of NOM greatly affects yields of chlorination by-products.

As a contrasting example, freshly formed allochthonous materials in the Sacramento-San Joaquin River Delta, California, USA, appear to be the source materials for THM formation during the chlorination of HS containing water (Bergamaschi et al. 2000). The material is not lignin or its degradation products, and it remains an open question as to whether this 'fresh material' comprises the aliphatic compounds of HS. These constrasting statements mean that the currently known structural features can not yet be definitively attributed to the formation of these potentially mutagenic and cancer-causing agents.

Box 1.2 Short History of Aquatic Humic Substance Research[1]

Early Days

The history of humic substance (HS) research clearly reflects the difficulties of investigating these biogeochemical materials. Despite this, important progress has been made since the middle of the last century. Visionary ideas from the past are difficult to access today unfortunately, partly owing to the fact that publications are written in languages that are now unusual. Thus, not all that is published today is really new, particularly that pertaining to the interactions between HS and nutrients.

The history of HS research begins with the discoveries by the German chemist Achard in 1786. His German school develops the first isolation and classification schemes, and also uses the term 'Huminstoffe' (humic substances). It comes from the Latin word 'humus' which means soil or earth, and is subsequently used by Saussure (1804) to describe dark colored organic material in soils. In the 19th century, the most important contributions to the knowledge about the chemical properties of HS are made by the famous Swedish chemist Berzelius (1839) and his school. He pioneers the study of aquatic HS, and isolates HS from a spring in Porta, Sweden. As a result of his studies, it is evident that HS are not individual compounds but a mixture of similar but not identical molecules. This discovery is a discouragement for chemists and leads to a decrease in general scientific interest in these substances.

Scientific Schools

The major advance in investigations on aquatic HS begins in the 20th century with the work of Aschan (1907, 1908) who makes the first contributions on the overall role of these materials. Aschan isolates preparative quantities of HS via precipita-

[1] Based on the outline by Klavinš (1997) who made the previously hidden eastern European publication available (with kind permission of the author).

tion by metal salts, resulting in the first element analysis of comparatively clean humic substance samples. The composition of aquatic HS that Aschan describes, C 48.8%, H 4.7%, O 44.3%, N 2.2%, are similar to those achieved with modern methods.

Odén (1914, 1919, 1922) regards HS as colloids and recognizes their role in the transport of substances in aquatic ecosystems. Odén (1919) is the first to determine the presence of HS iron complexes and HS nutrient complexes in the aquatic environment. He also recognizes that HS are the main factor determining the water color. In addition, Odén develops the division of HS into humins, humic acids and fulvic acids, which is still valid today.

An important impetus to humic substance research in the 20th century is provided by soil science and soil chemistry. Aquatic HS are considered as seepage from soils and wetlands. In this period, substantial contributions on humification theory are made by Russian authors (Waksman 1936, Kononova 1966, Alexandrova 1980), German authors (Flaig 1955, Flaig et al. 1975; Ziechmann 1980, 1994, 1996; Scheffer and Ulrich 1960), and Canadian authors (Schnitzer and Khan 1972; Stevenson 1982).

For a long time, research on aquatic humic substances proceeds hand in hand with that on organic carbon. Birge and Juday (1926, 1934) consider organic material in water from the perspective of limnology and hydrobiology. Their studies lead to the integration of chemical analysis of dissolved organic matter (DOM) into standard hydrochemical practice. Further results of their studies are the definition of dissolved and particulate material. Ohle (1934, 1935, 1937) considers organic material as an important compartment of the aquatic ecosystem and discusses this material from a limnological perspective. Ohle (1935) treats the organic materials as colloids which co-precipitate nutrients, particularly phosphorus and iron.

After the second world war, humic substance research is dominated by Russian laboratories, but this work is not well known. Intensive studies of HS from southern Russian seas are carried out by Datsko (1939, 1940, 1956, 1959). In this period, the organic carbon is routinely recorded and provides the basis for the geochemistry of marine organic carbon. Overall budgets, turnover processes and both seasonal and diurnal changes in organic carbon are studied. Datsko (1959) describes the geochemical role of organic substances in the binding and transport of nutrients in water, particularly phosphorus.

Skopintsev (1934, 1947, 1979, 1982) is the first to emphasize the importance of aquatic HS in the cycling of organic matter. He analyzes HS in oceans, groundwater and freshwater and describes their formation. Skopintsev states that about 90 % of marine HS have a planktonic origin, which means that they are of autochthonous origin – a very insightful statement. A special structural feature of these substances is the predominance of aliphatic compounds. A further visionary opinion of Skopintsev (1950) is that aquatic HS are not only a reservoir of nutrients, but are also an energy source. Similar studies are carried by Maystrenko (1965), particularly in the Ukraine.

Russian scientists are also the first to isolate aquatic HS in gram quantities, and develop standardized isolation and analytical procedures (Ponomareva and Etlinger 1954). Shevchenko and Taran (1963) isolate HS from surface waters and discover the seasonal variation in elementary composition. Studies are also begun to determine the molecular size of HS (Fotijev 1964, 1970; Gonchareva et al. 1961; Semenov 1972).

The success of research on aquatic HS is closely linked to the development of isolation procedures (Malcolm 1985). The early isolation methods included vacuum evaporation (Barth and Acheson 1962; Ponomareva and Etlinger 1954; Shapiro 1957), freeze drying (Black and Christman 1963) and flow-extraction (Shapiro 1957). Certainly, from the perspective of our present knowledge, these early extracts are rather impure mixtures of various substances, including inorganic minerals and small non-humic organic molecules. The improvement of isolation techniques allow HS to be isolated from waters such as rivers, groundwater and rainwater with much lower concentrations than in wetlands or bogs.

The introduction of instrumental analysis considerably improves studies on aquatic HS. Aquatic HS are routinely studied using infrared spectroscopy, UV-vis spectroscopy, fluorescence spectroscopy, paper chromatography, and the elementary composition and functional group content are determined (Shapiro 1957, Semenov et al. 1964). These studies reveal clear differences between aquatic HS, and soil or peat HS, and serve as the basis for the understanding of the various humification processes in different environmental compartments. However, the value of the results is reduced due to the presence of contaminants in the samples.

Sephadex Period

The present phase of humic substance research is linked to the introduction of sorption methods for isolation, purification and fractionation. Improved isolation methods provide much cleaner (low ash content) samples for further study. The introduction of gel-filtration (with Sephadex gels) (Gjessing 1965, 1971) is an important step towards the quantitative determination of aquatic HS, as well as determination of their properties and ecological role. Malcolm (1985) describes the introduction of gel-filtration as the beginning of the 'Sephadex period'.

The success of humic substance research is connected to developments in (protein) biochemistry. However, the conventional methods which are developed for biopolymers need modification for HS (Lindquist 1967). Some specific methods for the isolation, fractionation and separation of these very complex molecules are necessary. The best examples of such methods are the isolation of aquatic HS as suggested by Thurman and Malcolm (1981) and the fractionation according to Leenheer (1981, 1985). At present, development and refinement of specific methods are still an important subject of humic substance research.

The discovery that HS are the source of trihalomethanes in drinking water processing is a breakthrough in humic substance research (Rook 1974). These studies stimulate further humic substance research, and encourage the use of complicated techniques such as ^{13}C-NMR. In addition, the general awareness of environmental problems stimulate humic substance research further.

Scientific Upsurge

The low point in humic substance research at the beginning of the 20th century due to the difficulty of studying these biogeopolymers, is evident not only in the small number of scientists studying this material, but also by the work being published only in low rated scientific journals. This situation is now changed and results on aquatic humic substance research can now be found in the most visible journals such as 'Nature' and 'Science'.

Various centers for aquatic humic substance research develop, for example, Gjessing's group in Norway (Gjessing 1976), Schnitzer's group in Canada, Christman's group in the USA (e.g. Christman and Gjessing 1983), the group of the late Malcolm of the US Geological Survey (Aiken et al. 1985, Thurman 1985), and the group of Frimmel in Karlsruhe (Frimmel et al. 2002). Malcolm is one of the founders in the early 1980's of the International Humic Substance Society which holds biannual scientific meetings and publishes proceedings. These activities give the international HS research new impetus and lead to an intensive international exchange of ideas and results not only between scientists of one discipline, but also between disciplines such as soil science, limnology, and marine science. Humic substance research becomes so well established that the American Chemical Society holds symposia on this subject and publishes proceedings (for instance, see Gaffney et al. 1996). Recently, the Northeastern University in Boston, MA, carries out HS Seminars and publishes the proceedings with the Royal Society of Chemistry. Chemical and environmental chemical questions form the core of these publications rather than basic ecological questions, a focus consistent since about 1980.

Chemical and ecological studies on humic substance in Germany are further stimulated by the establishment in 1992 of the Priority Program 'ROSIG' (Refraktäre organische Säuren in Gewässern – Refractory Organic Substances in the Environment = ROSE) of the Deutsche Forschungsgemeinschaft, initiated and coordinated by Frimmel (Technical University, Karlsruhe) (Frimmel et al. 2002). More ecologically orientated studies on HS are motivated by two books, Hessen and Tranvik (1998) and Keskitalo and Eloranta (1999). It is to be hoped that in the future, more high quality works will be published, and the paradigms of 'dead organic matter' (particulate and dissolved detritus) further revised.

2 Origin of Humic Substances in Freshwater: Biogeochemical Pathways

According to the traditional view of humification, there are two different pathways by which HS can form (Anonymous 2001; Hatcher and Spiker 1988). In the **degradative pathway** formed macromolecules are partially degraded to HS. Most macromolecules are plant biopolymers, such as lignin or cellulose. Humins are partly modified plant biopolymers. The other metabolic pathway comprises **condensation polymerization** reactions with three different models for abiotic condensation processes: the polyphenol model, the melanoidin model, and the polyunsaturated structure model.

While the degradative pathway in which plant polymers such as lignin or cellulose are decomposed, occurs primarily in the vegetated catchment or in a littoral zone with macrophytes, the condensation pathway can be more or less considered as autochthonous, taking place in extreme lakes with catchments without any higher plant cover, such as lakes in the Antarctica, or in open oceans. Terrestrial systems and lakes in Antarctica (and open oceans) represent two extremes: whereas in soils, the lignin-containing terrestrial plants dominate the formation of HS from lignin sources, the HS in extreme lakes and open oceans are dominated by products of lignin-free material degradation. Freshwater systems lie between these two extremes.

In many instances, both condensation and degradation may be occurring simultaneously because humification is a dynamic process with no unique unidirectional vector (Hatcher and Spiker 1988). Recent results from soil science show that the differentiation into two separate pathways is not so clear (Chap. 2.3).

2.1 Degradative Pathway

Vascular, non-vascular plants, microorganisms and animals are decomposed into biochemical compounds. During microbial degradation, the readily available labile macromolecules are degraded and lost. During humification, possibly some small molecules are linked into the polymer matrix. Refractory components or biopolymers such as lignin, cutin, suberin, N-containing aliphatic macromolecules, melanin, and other as yet unidentified biopolymers are selectively conserved. Some of these later become what is described as 'humin'. These humins are the source materials for humification via oxidative degradation. With increased degradation, the humins can be further oxidized and converted to macromolecules which have the same molecular weight as the humins but are already more 'functional', having a higher content of carboxyl, carbonyl and hydroxyl groups. The increased functionality leads to increased solubility in bases, and results in the formation of HA.

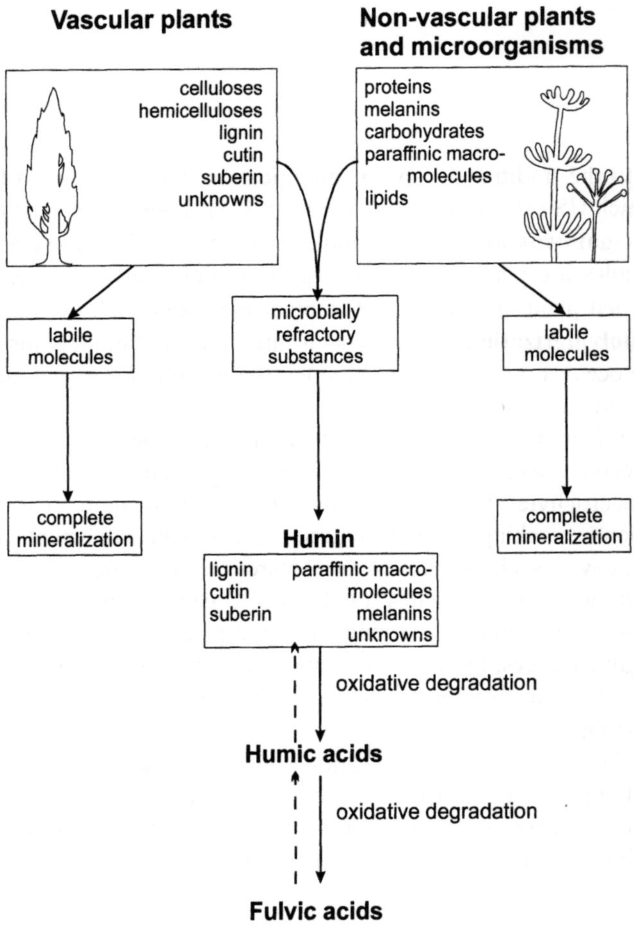

Fig. 2.1. Degradative pathway for formation of HS. This pathway occurs in the soils of a catchment area and is more or less allochthonous. Dotted arrows show that the transformation is possible but is of minor importance (from Hatcher and Spiker 1988, with kind permission of Wiley & Sons)

Further degradation produces FA, with reduced molecular weight, increased functionality, and increased solubility in acids and bases. As a consequence of the increased degradation, the structure of the molecules becomes more variable and they lose their similarity to the source materials. This degradative model of HS production is shown schematically in Fig. 2.1: the FA are diagenetically below the HA, which in turn lie below the humins.

2.2 Condensation Pathway

Compared to the allochthonous input of dissolved HS in freshwater systems, the condensation pathway appears quantitatively almost totally unimportant (Hatcher and Spiker 1988). However, in freshwater ecosystems, the individuals and populations develop simultaneously in close contact to their chemical environment. One important component of the chemical environment are the HS. In Chap. 8, we will see that HS, particularly allochthonous HS, are natural chemicals with similar properties to chemicals introduced by man, the xenobiotics. These 'natural xenobiotics' have several adverse effects. We hypothesize that autochthonous HS have a minor adverse impact than allochthonous HS. Hence, the autochthonous humification pathway deserves particular attention.

Indirect indications that the autochthonous pathway applies come from studies on extracellular organic substances in bacteria-free algal cultures, in which the filtrate develops a UV spectrum that is characteristic of HS, and distinct from that of plant pigments (Stabel 1977; Steinberg 1977). Direct evidence for autochthonous HS production comes from marine studies over a long time period. For example, Lara and Thomas (1995) describe the incorporation of radioactively labeled C into the hydrophobic DOC fractions of the cultured Arctic diatom *Thalassiosira tumida.* Most of the acid and neutral hydrophobic DOC fractions, indicators of dissolved HS, are produced in the logarithmic growth phase. The sum of all hydrophobic DOC fractions comprise 14% of the particulate C present, a significant proportion. Tranvik (1992a,b) presents evidence that freshwater microorganisms, too, can form compounds similar to dissolved HS from labile low-molecular weight DOC. A maximum of 19% of the produced DOC is hydrophobic, and about half of these substances have a molecular weight >0.7 kDa as determined by gel-filtration.

Three different modes of action for the abiotic condensation may apply: the polyphenol model, the melanoidin model, and the polyunsaturated structure model.

2.2.1 Polyphenol Model

The polyphenol model is based on the reaction of quinones with amines and ammonia as most important step of the polymerization. By this route N is incorporated into HS. One origin of the quinones is the microbial degradation of lignins, which leads to phenols, and in a second step to quinones. Yet, based on studies of the Lake Fryxell (a lake in Antarctica without any terrestrial vegetation in its catchment, compare Chap. 2.4), McKnight (pers. comm.) assumes that the role of bio-quinones from decaying microorganisms and algae in most of the metabolic pathways of aquatic ecosystems is strongly underestimated.

Transition metal oxides such as MnO_2, ZnO, or CuO, and cations such as Mn^{2+} and Fe^{3+}, can accelerate polymerization from dissolved catechol (1,2 dihydroxybenzene) and other catechol and pyrogallol derivatives (Larson and Hufnal 1980). Such colored catechol polymers form much faster in river water than in de-ionized water at the same pH. Sediments and clay also enhance the polymerization rate.

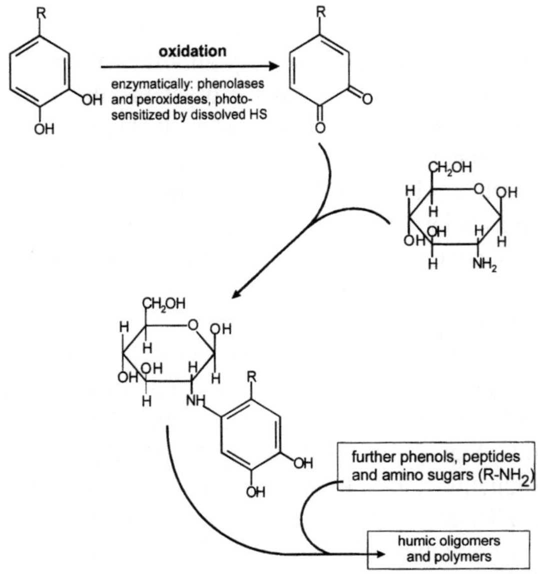

Fig. 2.2. Example of the polyphenol model: oxidative formation of C-N bonds between labile organic molecules and phenols in the autochthonous humification process (modified from Martin et al. 1975)

The autochthonous formation of aquatic HS can also occur via oxidation with enzyme catalysis or through indirect photolysis. Many enzymatic pathways include phenolase and peroxidase (Bollag and Myers 1992; de Haan et al. 1981; Martin et al. 1975; Steinberg 1977). A comparable autoxidation process can also occur. Oxidation applies to several dihydroxyphenols, such as catechol, but not all dihydroxyphenols. For instance, resorcinol, a *m*-dihydroxyphenol, does not undergo autoxidation (Sütfeld 1998). However, after adding a cryptomonad exoenzyme, Sütfeld (1998) observes a disappearance of resorcinol and an enhanced catechol conversion. Photosensitive dissolved HS reactions probably also take place in which photoactive species such as OH•-radicals, singlet oxygen or hydrogen peroxide mediate coupling reactions between labile and refractory organic components (see below). These reactions lead to the formation of bonds such as C–C, C–O, C–N, and N–N, between HS and labile organic molecules. An example of such a process is shown in Fig. 2.2.

Autoxidation of phenolics leads to HS. In alkaline solutions and in the presence of oxygen, phenolics are subject to little understood reactions which lead to brown, homogenous end products (Eller 1921 from Ziechmann 1994). As these products are remarkably similar to natural HS, they can be used as model substances. Brown coal (lignite) degradation generates HS-like substances which originate from phenolics through condensation reactions, as occurs in Lake Schwelvollert near Leipzig in north eastern Germany (Kopinke et al. 1995).

2.2.2 Melanoidin Model

The melanoidin model is based on the reaction between monomer reduced sugars and amino acids, which is first described by Maillard (Stuermer 1975; Ziechmann 1994). The amino group of peptides reacts with the carbonyl group of the sugar. The product (Schiff'sche base) undergoes further reactions to result in brown products (melanoidines). The melanoidines are comparable to HS.

Next steps are reorganization, cyclization, and decarboxylation leading to complex, brown-colored mixtures (melanoidines). Yet, this condensation of amino acids and sugars is not sufficient to explain the total quantity of aliphatic structures in HS. Ishiwatari (1985) and Yamamoto and Ishiwatari (1989, 1992) experimentally confirm the occurrence of this humification pathway by allowing protein (casein) to react with sugar (glucose) under relatively mild conditions (70°C for a few days), and hypothesize that humins, HA, and FA are formed almost simultaneously through the Maillard reaction between proteins – not amino acids! – with sugars from dead algae and bacteria in sediments. In the view of these authors, biolipids take also part in the reaction. Yamamoto and Ishiwatari summarize their results and hypothesis as the 'protein-based melanoidin model'. Through further studies with model amino acids in freshwater and marine HS, Yamamoto and Ishiwatari (1992) consider their hypothesis confirmed.

2.2.3 Polyunsaturated Structure Model

This model describes the formation of HS with polyunsaturated compounds, such as fatty acids, carotenoids, or alkenones as precursors. The reaction proceeds under UV light between single fatty acids so that the molecules become cross-linked (Harvey et al. 1983, 1984; Harvey and Boran 1985). The weakness of this model is that it does not explain the high N content of HS.

The polymerization pathways described above are more efficient with higher concentrations of reactants. Therefore, reactions at interfaces play an especially important role in the formation of aquatic HS. Some examples:

1. Decaying organisms contain high concentrations of precursors.
2. Organic substances in particles can be in a dynamic equilibrium with the surrounding liquid phase.
3. In aquatic systems, lipids, fatty acids, and triglycerides predominate in particulate or colloidal form.
4. Filter feeders concentrate organic substances in intestines and fecal material.
5. Surface-active molecules can exist as variously sized aggregates (a few molecules to micelles or colloids).
6. River and lake particles ('snow') can be synthesized from dissolved substances in the absence of microbial processing (Carlough 1994).
7. In all these examples, organic molecules come together within bonding distance so that intermolecular reactions are feasible (Stuermer 1975).

2.3 Genesis of Humic Substances

According to present knowledge, the strict separation of degradation and condensation pathways is no longer tenable – except for teaching purposes. HS formation appear to be more independent from any living process as described above. For instance, Ziechmann (1996) writes: *'According to new results, HS and related molecules are formed at a very early point in chemical evolution from CH_4, H_2O, NH_3, and H_2 as well as from their products. With this astonishing fact that HS can also be built from small inorganic building blocks and their derivatives, a corner stone of previous HS research has collapsed. Up to now, it is thought that HS are a typical component of active soils (and other media, e.g. or water)...'*

Humus or at least HS-like molecules become established in ecosystems independently from life and death events and play a definitive role in early evolution. Ziechmann (1994) and Bada and Lazcano (2000) point out that, in the pioneering studies of Stanley L. Miller (1955) on the origin of life in the primitive reducing atmosphere 4×10^9 years ago, colored tarry substances[1] were present, however, were not described or characterized in further detail since Miller focused on the amino acids to identify them as the missing biomolecules to prove the hypothesis that life on Earth has developed in a reducing atmosphere. Ziechmann (1994) and his co-workers repeat Miller's experiment using more energy. Dark brown substances, HS, or at least HS-like materials, also form in this experiment, as shown in UV and IR spectra, and by precipitation with hydrochloric acid. According to their element composition, these materials are somewhere between humic acid precursors (HAP) and HA.

Ziechmann (1994) sees the most important role of HS to be in chemical and biological evolution (Fig. 2.3): HS eliminate reactive radicals, which otherwise inhibit or prevent the synthesis of amino acids, carbohydrates, and nucleic acids. Thus, without HS serving as radical scavengers, in the primitive atmosphere, evolution would not be possible, since no biomolecules could otherwise form in the strong UV radiation.

The discovery of HS formation in the primitive atmosphere has further far reaching consequences for the understanding of ecosystem functioning. The HS are to be granted the role of an independent ecosystem component, such as atmosphere, water, or light, since they come into being simultaneously with early life. This means that living organisms have to adapt to humus or HS-like materials

---

[1] In their congratulations to Miller on the occasion of his 70th birthday, Bada and Lazcano (2000) write about the pioneering work of the laureate: *'He started the spark and began to heat the water flask gently. After two days, the water had turned pale yellow and a tarry residue coated the inside of the 'atmosphere' flask around the electrodes. Anxious and unable to wait any longer, he stopped the experiment at that point and analyzed the water for amino acids....Stanley repeated the experiment, this time sparking the apparatus for a week and boiling the water rather than heating it gently. At the end of the week, the inside of the sparking flask was coated with an oily scum and the water solution was yellow-brown. Now the glycine spot on the paper chromatogram was far more intense...'*

with which they come in contact from the very time they evolve. Because HS are natural environmental chemicals, they exert a chemical stress on organisms in many ways, and thus determine ecological niche, and structure biocoenoses and guilds. The chemical stress applies not only to terrestrial systems, but also to freshwater systems in which the humus content is relatively low.

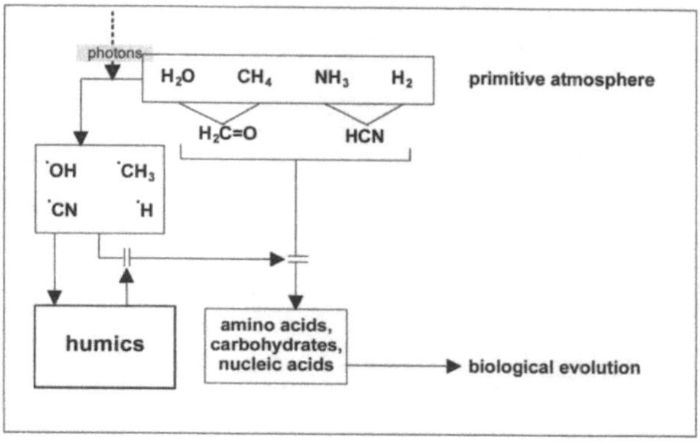

Fig. 2.3. The role of HS or HS-like materials in the ancient atmosphere as radical scavengers so that biomolecules such as amino acids, carbohydrates or nucleic acids can be formed (after Ziechmann 1994, with kind permission of Spektrum Akademischer Verlag)

Once organisms have evolved, and through the formation of humus from dead biomass, the quality of HS becomes more diverse and its quantity greater. For organisms, the adaptive pressure rises. Already from this, the general statement follows that organisms can only develop with HS in their environment. Since HS interfere with cell-internal metabolism, biological systems have to defend themselves against their protector against UV radiation and chemical radicals – the HS. We hypothesize that this chemical stress exerted on bacteria, fungi, plants, and animals since the primitive Earth, is likely one major reason for the fact that the 'defense' systems are very similar in all organisms. In fact, the transformation (**detoxification**) systems are very conservative, occurring in only slightly altered forms in organisms from bacteria to mammals.

There is abundant literature on the various concepts of chemical HS formation (Hatcher and Spiker 1988; Hedges and Oades 1997; Kononova 1966; Orlov 1995; Ziechmann 1994, 1996). To get a short insight, we will briefly follow Ziechmann's (1996) description. The HS formation, as a simultaneously synthesic and degradative process, is presented graphically in Fig. 2.4. After biosynthesis of aromatic and non-aromatic compounds, the building blocks of HS, a partial microbial degradation follows. This **metabolic phase** comprises basic biochemical processes, that are enzymatically controlled and dependent on environmental factors. Humic acid precursors (HAP) are formed from the aromatic compounds of these materials in the **radical phase**, via radical intermediates. With the HAP, non-HS are also channeled to the **conformation phase**. With the establishment of HS, the final state of this biogeochemical material is reached.

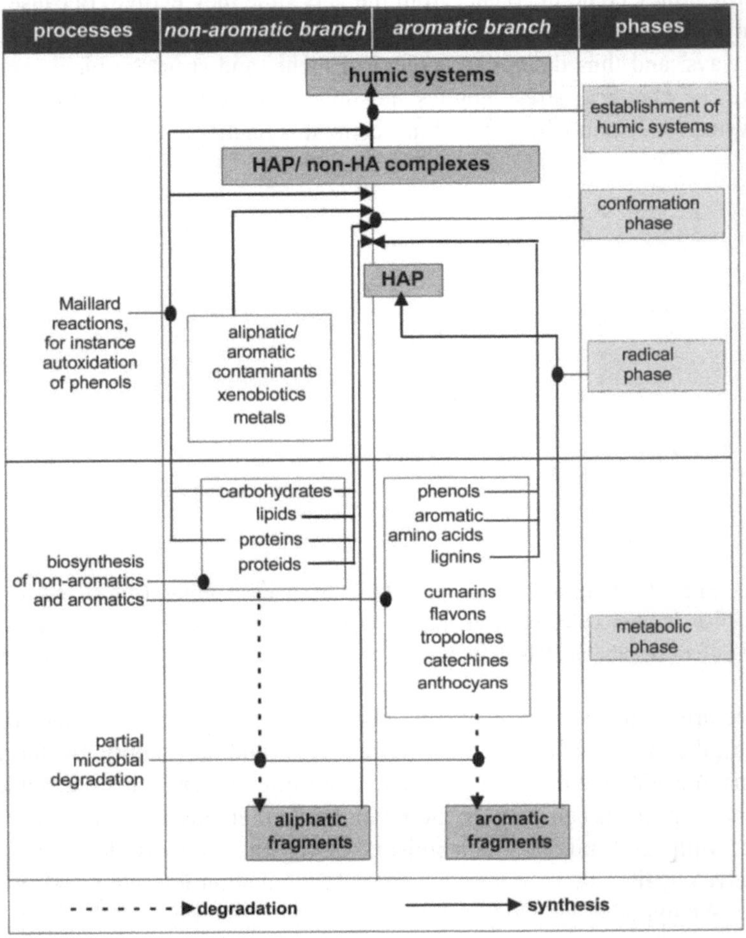

Fig. 2.4. Genesis of HS (after Ziechmann 1996, with kind permission of Spektrum Akademischer Verlag). HAP = humic acid precursors, HA = humic acids. Further details in text.

It must be stressed that a special feature of Ziechmann's (1996) model is that it considers the potential reactions of several intermediate compounds. This means that every part can react with every other part, so that as a result, chance events can result neither in a linear reaction, nor finally in a defined chemical constitution for HS. Thus, since HS are formed under naturally diverse conditions, no individual chemistry of formation can occur (Ziechmann 1996).

HS are relatively stable, but by no means inert ecosystem compounds. During humification the more or less stable end products are HS formed, as displayed in generalized form (Fig. 2.5).

The substances comprise:

1. **Precursors** (PC) which are fundamental for the formations of HS, but are considered non-humic and are different from the HS, such as through their color.

- Primary precursors (pPC): aromatic compounds which can easily be transformed into radicals; they start the humification.
- Secondary precursors (sPC) with non-fixed structure. They must be able to react with HS fractions to be incorporated into the conformation phase.

2. **Humic substances** (HS)
   - Humic acid precursors (HAP) which are converted to humic acids (HA) during the humification process.
   - Humic acids (HA): relatively stable acidic HS.
   - Humins (HM): end products of the humification process, very insoluble, poorly reactive.

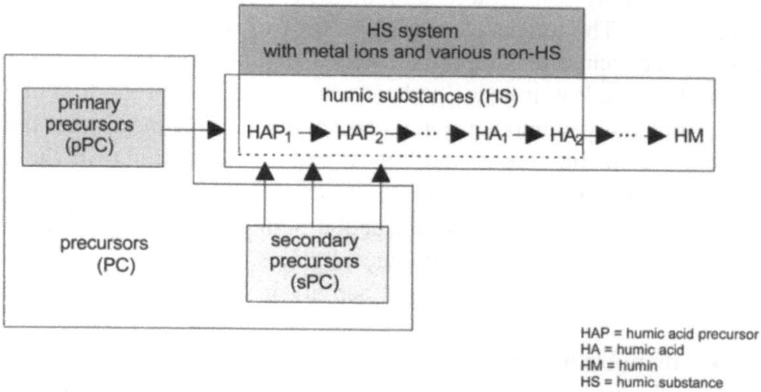

Fig. 2.5. Humic substance fractions and humification (from Ziechmann 1996, with kind permission of Spektrum Akademischer Verlag)

2.4 Recent Findings about Production and Diagenesis

HS are subject to many geochemical and biological diagenetic processes. For instance, Clair et al. (1989) show that microbial activity with 'fresh' HS is considerable, and decreases as the HS are increasingly altered. The changes in HS in nonsterile microcosms cause an increase in pH and a loss of strong organic acids, which only microbial activity can account for.

It is repeatedly stressed that plant material plays an important role as source material for HS in terrestrial and freshwater systems. This is confirmed experimentally by Filip and Alberts (1995), and Chefetz et al. (1998) among others. Filip and Alberts (1995) show that fresh and dead tissue from *Spartina alternifolia*, a plant which dominates salt marshes of the American Atlantic and Gulf of Mexico coasts, contains HS-like material which are released into seawater and sediments. Owing to spectroscopic properties, these materials resemble HS in sediments. The authors also describe that the sediment microflora of salt marshes is able to use these HS-like substances, under aerobic or semi-anaerobic conditions, as additional sources of nutrients and a single source of C and N. Depending on the particular culture conditions and type of HS, between 27 and 100% of the HS-like substances are utilized. It is particularly noteworthy that HA formed under aerobic conditions exhibit changes in elemental composition and structure, characteristic of diagenetic transformation of HS in sediments. These changes are increases in C content, C:H and C:N ratios, optical density, and infrared absorption typical of aromatics. Decreases occur in oxygen content and infrared absorption characteristic of proteins and carbohydrate-like structures.

Biotic transformation also proceeds in freshwater. Claus and Filip (1998) isolate the phenoloxidase forming fungus *Cladosporium cladosporioides* from the water of a raised peat bog lake. With high concentrations of C and N, under field conditions found only at interfaces, this fungus can form laccase and develop high lignolytic activity. The fungus cleaves about 60% of riverine and groundwater HS added at high concentrations (approximately 1 mg/ml). The HS are also structurally altered: they are less aromatic and contain an increased proportion of aliphatic structures, principally amino acids and carbohydrates. This means that the fungus can attack the aromatic core of the HS, in addition to splitting off the aliphatic side chains or peptides and carbohydrates, as carried out by other microorganisms. Therefore, *C. cladosporioides* has metabolic similarities to lignin-decomposing fungi, except that it requires high N concentrations, in contrast to *Phanerochaete chrysosporium* which exhibits maximal activity in N-poor and anaerobic conditions.

In a recently published study, Claus et al. (1999) calculate the proportion of C which is transformed to HS during the decay of readily decomposable substrates, such as starch and peptone and complex biomass. After six months incubation, the cultures are a dark color and HS like materials can be isolated. The organic C of the substrate is indeed mineralized, and only approximately 3% is transformed into HS. The total yield of HS differs according to substrate (peptone> yeast>starch), the origin of the inoculum (river water>lake water>groundwater), and incubation temperature (20 °C >10 °C). The microbially formed HS are similar to natural aquatic HS according to their spectroscopic and electrophoretic properties. They do, however, have a higher proportion of more aliphatic (carbohydrates, peptides) and less aromatic compounds.

The ^{13}C- and ^{15}N isotope content of the microbially formed HS varies according to microbial inoculum and organic substrate added to the culture, and gives evidence of the underlying processes. The biomass of the mixed microorganism community from the raised peat bog lake and from river water, and also the pure culture of *Bacillus sphaericus* and *C. cladosporioides* exhibited a ^{13}C enrichment of 1.2 to1.9 ‰. The mean isotopic composition of C in groundwater HS is -25.6 to -25.8 ‰ and in soil HS is -22.5 to -24 ‰. The microbially produced HS accord well with these values -22.2 to -23.7 ‰ (Claus et al. 1999). Thus, the isotopic composition is a mirror of the microbial biomass. This can mean that there are similarities between the synthesis of microbial biomass and of HS. This implies that the humification of the added organic substrates is anabolic, that is to say that it occurs through energy-requiring, enzymatic reactions.

Further insight into the short-term humification processes can be gained from the studies of composting processes (Chefetz et al. 1998). The operational fractionation scheme is shown in Fig. 1.6. The studies shows that during the composting process, the proportion of polysaccharide structures in the DOM declined. The ^{13}C-NMR- and FTIR spectra of the hydrophobic acid fraction indicate polyphenols and HS-like structures, while the spectra of the hydrophobic neutral fraction have characteristics of aliphatic structures. The hydrophilic neutral fraction spectra confirm the polysaccharide character of this material. In the last phase of composting – after about 100 days – a constant DOM concentration is attained in which the relative decrease in the hydrophilic neutral fraction (polysaccharide) is compensated for by an increase in the sum of the hydrophobic acid and neutral fractions (polyphenols and HS-like structures as well as aliphatics). From this it can be concluded that the DOM contains little biodegradable organic materials, and a larger proportion of HS-type macromolecules. These macromolecules from the mature compost are, as a rule, able to complex metals and thereby enhance plant growth better than younger compost (Box 8.1).

It can be concluded that both allochthonous and autochthonous pathways of HS formation occur, and the products of both pathways underlie diagenetic processes. The ratio of allochthonous production to autochthonous input varies from water body to water body and depends on factors such as the ratio of lake surface area to catchment area, the catchment area structure (vegetation, industrialization, settlements and their solid and liquid wastes), the hydrologic inputs to the water body, the autochthonous production of the water body, and relative size of the pelagic and littoral zones of the particular lake. In general, the higher the aromatic content of the HS, the more higher plants (which contain lignin) are available as source materials for the humification process. Where such materials are absent, as for example in Arctic and Antarctic lakes and in the open ocean, aliphatic structures dominate the HS structure. This is shown impressively in the studies of McKnight et al. (1991, 1994): in two Antarctic lakes (Lake Fryxell and Lake Hoare in the Dry Valley), the group describes the chemical nature of FA, known to be of autochthonous origin, arising from bacteria and algae which grow under a continuous ice cover. The two lakes differ in terms of the deep-water oxygen regime and salinity status. Nevertheless, the FA are chemically very similar in both lakes, indicating that the chemical conditions in the water column has no clear influence on the nature of these molecules. Compared to FA from other natural freshwater systems, those from the Antarctic lakes have a low C:N ratio (19–25) and low aromatic C content (5–7% of the total C, Box 1.1). These characteristics are also seen in two Antarctic coastal ponds, although the values for C:N ration and aromatic C content are not as low as in the Dry Valley lakes (McKnight et al. 1994). The FA of Arctic lakes and ponds are similar to those from marine environments (Harvey and Boran 1985).

These statements are strongly supported by a study of terrestrial HS and humification processes in an Antarctic coastal region (Beyer et al. 1997). In this region, histosols are the dominant soil type, formed through degradation of mosses and algae. In all horizons, carbohydrates and lipids dominate the soil organic material. Carboxyl groups are found in fatty acids and amino acids. Aliphatic compounds are also enriched, which is probably due to selective conservation. The alkyl components consist of lipids, fatty acids, and sterols. A proportion of the latter can originate from hopanoids, pentacyclic triterpenoids from microbial membranes (Ourisson and Rohmer 1992). Sterols comprise up to 15% of the soil organic material. Interestingly, aromatic structures are also found in the histosolic soils which originate from other sources than lignin (Fig. 2.6).

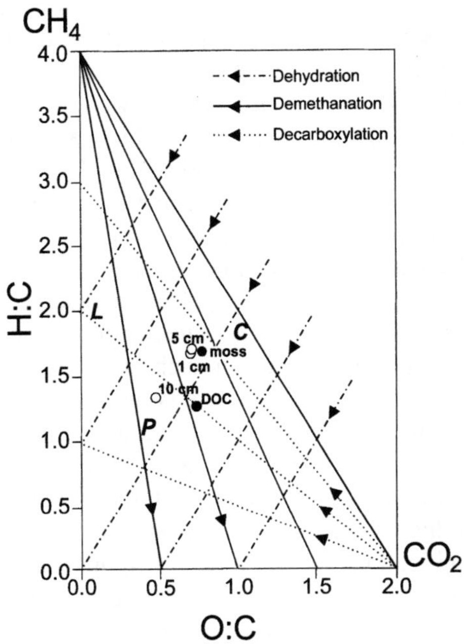

Fig. 2.6. Atomic ratio of O:C plotted against H:C of *Sphagnum* moss, of peat at three depths (open circles), and of DOC from Lake Hohlohsee in the northern Black Forest. The positions of important biotic components such as carbohydrates (**C**), lignin (**P**, phenylpropane) and lipids (**L**) are shown for comparison. The lines represent the various condensation reactions, such as dehydration, decarboxylation and demethylation in coal formation (after Kracht and Gleixner 2000, with kind permission of Elsevier Science)

An instructive study regarding the separation of autochthonnous and allochthonous humification pathways (Kracht and Gleixner 2000), shows how strongly both humification pathways are interconnected. Kracht and Gleixner describe the HS in the water and *Sphagnum* belt of a raised peat bog lake (Lake Hohlohsee, northern Black Forest, Germany, the HS and NOM isolates of this small lake are later referred to as 'raised peat bog'), and offer new data revising the previously held views. The prior view is that the HS in such waters are washed in from the *Sphagnum* belt. Kracht and Gleixner (2000) analyze the stable isotopes in the water column, in the living *Sphagnum* plants, and at various depths in the *Sphagnum* peat. Biogeochemical transformations are identified through isotope fractionation which are so characteristic that conclusions about particular metabolic processes can be drawn. The isotope signal of plant material, one source for humification, is primarily shaped by photosynthesis. The subsequent processes further discriminate between the various C isotopes. Lignin and other wood materials generally contain 3–6 ‰ less ^{13}C than do carbohydrates. Further isotope fractionation occurs during degradative processes. Following the isotope signals of individual compounds, one can identify the decisive processes which lead to the products. Therefore, changes in the isotope values indicate a transformation of source molecules, while constant values indicate conservation of these molecules (Lichtfouse et al. 1998; van Bergen et al. 1998). The conservation is considered a new pathway in the formation of soil organic matter. Kracht and Gleixner show that this new humification pathway applies to aquatic systems too.

For Lake Hohlohsee, element analysis provides information on processes in the peat and in the lake. The C content increases from the moss to the deepest peat layers, while the oxygen and hydrogen content decrease. In an element ratio diagram (O:C–H:C, Fig. 2.6), a dehydration process from moss to DOC also becomes evident, which leads to a loss of hydroxyl groups. Alternatively, this relative increase in C in the O:C and H:C ratios can be explained by various stages of microbial degradation of molecules such as cellulose. Overall, the atomic relationship indicates a replacement of carbohydrate dominated structures by phenolic polymers, which nevertheless cannot be derived from lignin since *Sphagnum,* as a moss, is lignin-free. The atomic relationships further shows that the DOC in the raised peat bog lake is neither identical to the *Sphagnum* moss nor the peat from various depths (Fig. 2.6).

Five different processes play a role in humification of organic matter in Lake Hohlohsee:

1. biological degradation of source material,
2. selective preservation of individual chemical compounds,
3. formation of microbial biomass (food-web effect),
4. introduction of microbially produced carbohydrates and
5. the mixing of original and altered material.

In order to demonstrate the corresponding processes, five pyrolysis products from the depth profile are selected (Fig. 2.7). The relative quantities of 1-dodecene and anhydropyranose increase with depth in the peat. This is most distinct for the anhydropyranose. Simultaneously, the isotopic content of the anhydropyranose tends to increase with profile depth. It is almost certain that this compound is an anabolic product from microorganisms. Anhydropyranose, but not 1-dodecene, increases simultaneously with the content of heavy stable C isotopes. The constancy of the δ value indicates that 1-dodecene, possibly a lipid pyrolysis product, is selectively conserved. The $\delta^{13}C$ values of 4-hydroxy-5,6-dihydro-2H-pyran-2-one decreases slightly from -21.7‰ in the moss, to relatively constant values around -24‰ in the peat. This pyrolysis product is most probably derived from a pentose. Since the source of this carbohydrate can be either plants or microorganisms, it is assumed that the isotope effect is caused by an introduction and/or substitution of carbohydrates from bacteria and fungi. The quantities of 2-furaldehyde and phenol clearly show a decrease from moss to the peat, but only the δvalue for 2-furaldehyde remains constant, which may be due to gradual degradation.

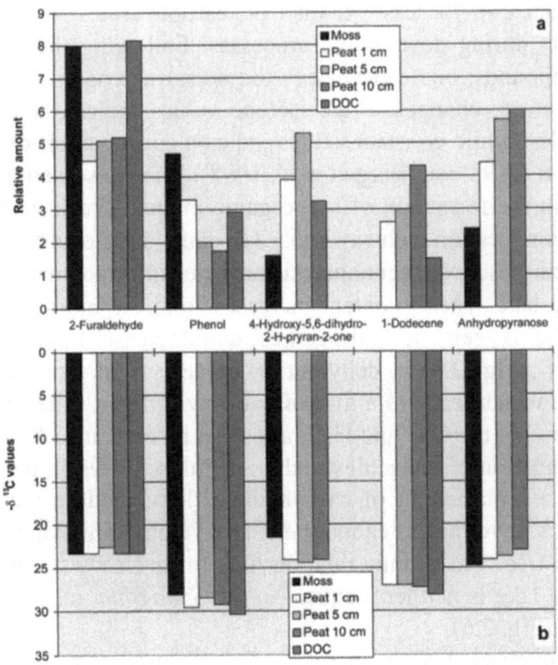

Fig. 2.7. Relative peak areas (**a**) and $\delta^{13}C$ values (**b**) of selected pyrolysis products: 2-furaldehyde, phenol, 4-hydroxy-5,6-dihydro-2H-pyran-2-one, 1-dodecene, and anhydropyranose from *Sphagnum* moss, peat, and DOC from Lake Hohlohsee (after Kracht and Gleixner 2000, with kind permission of Elsevier Science)

The isotope signatures of individual molecules shows that DOC in water is poor in ^{13}C in comparison to the same products in the peat. The most probable cause for this type of impoverishment is the inclusion of isotopically light C. Kracht and Gleixner (2000) assume therefore that the DOC in water is principally derived from microbial processes different from those occurring in peat. The inclusion of 'light' C, for example from bacterially respired $^{12}CO_2$, CO_2 from the oxidation of $^{12}CH_4$ or acetic acid, is sufficient to explain this effect.

What is surprising is that the HS in Lake Hohlohsee do not have an allochthonous origin from the *Sphagnum* belt, but appear to be predominantly from autochthonous sources, or at least a mixture of original and microbially altered material. These findings are probably applicable to many lakes in the boreal zone. Such a situation may also apply generally to non-eutrophic, so-called clear water lakes with a supposedly low allochthonous input (Chap. 9).

The findings of Kracht and Gleixner (2000) agree partly with findings of Ertel et al. (1993) who study bog FA and peat FA with a coarser biomarker technique. The fromer authors compare the lignin and carbohydrate. Although Ertel et al. (1993) find that bog FA and peat FA have similar biomarker compositions and that the peat provides most of the aquatic FA at the period of study, additional sources of lignin- and carbohydrate depleted FA to the bog are also needed to quantitatively match the relatively low levels of recognizable biochemicals in the aquatic FA from the bog. Probably, the pathway of methane oxidation, as discussed by Kracht and Gleixner (2000), contributes also to the bog studied by Ertel et al. (1993).

Although there are some key publications on chemical and microbial humification, as well as on diagenesis, our knowledge of this aspect still remains comparatively poor. However, one can finally conclude that existing HS in particular freshwater bodies is overall in equilibrium between allochthonous and autochthonous oxidative production. According to the most recent studies, the autochthonous process, and the coupling between allochthonous inputs and autochthonous photolytic and above all microbial transformations, have previously been underestimated. A graphical summary is shown in Fig. 2.8.

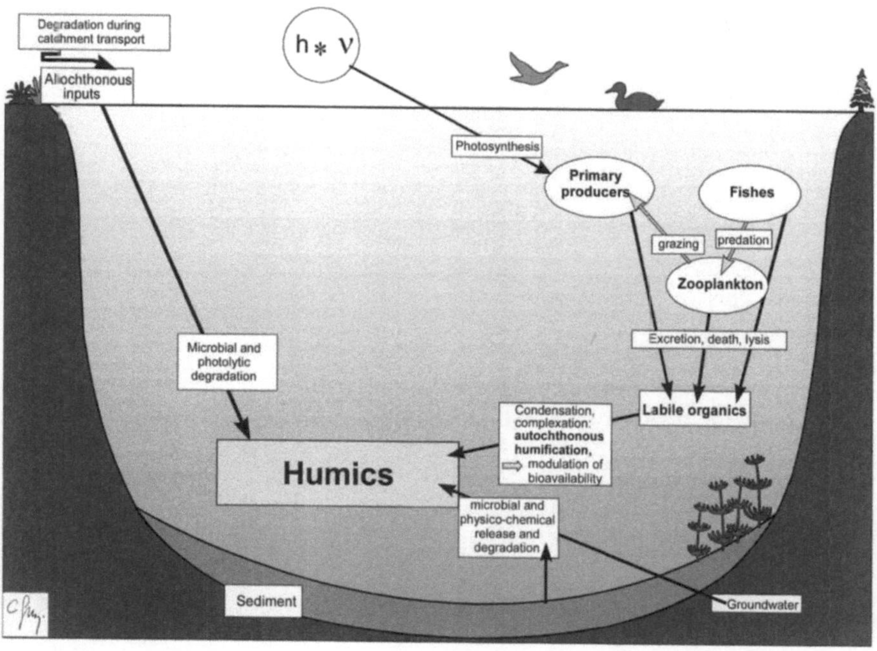

Fig. 2.8. Graphical summary of the allochthonous and autochthonous humification pathways in freshwater systems. The abiotic, spontaneous synthesis from existing components of the primitive atmosphere is not shown. According to new biogeochemical research, the autochthonous pathway has thus far been underestimated. Furthermore, the HS found in freshwaters are a mixture of original and modified material. h∗ν is solar radiation

Box 2.1 Structural Aspects of the Reactivity of Humic Substances in Ecosystems

Humic substances are polydispersed polyelectrolytes (Orlov 1995; Wagoner et al. 1997), and as such, they share a characteristically high degree of molecular irregularity and heterogeneity (Kleinhempel 1970). As a consequence, there are no two identical HS molecules in water or soil, and this implies that one cannot give a general structural formula for HS. Usually the HS molecule must be described by statistical features: ratios and variations of the various structural entities. The structure of HS is the subject of several congress proceedings (Davies and Ghabbour 1998; Hayes et al. 1989). When the structure of HS is considered in the following, it is only in terms of particular structural features responsible for certain geochemical or biochemical processes.

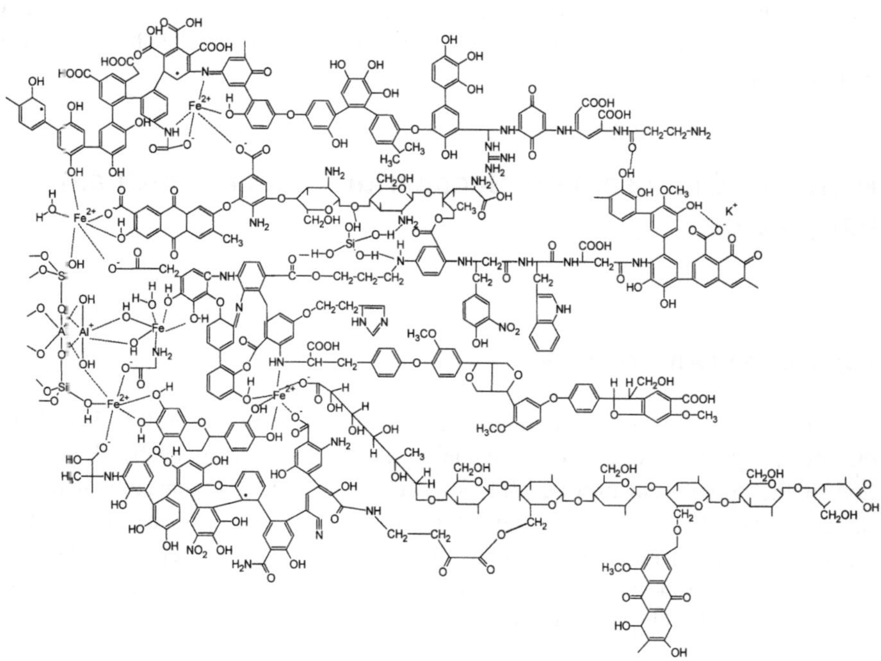

Fig. B.2.1.1. Proposed structure of soil HS (from Kleinhempel 1970).

Although no general structural formula for humic substance can be given, studies are published on the structural formulae of molecular fragments. It is obvious that there are as many structural formulae as authors, and every description is at least partly correct. Particular structural features are portrayed for particular possible ecological, eco-physiological or eco-chemical reactions. One of the many proposed structures is that of Kleinhempel (1970) shown in Fig. B.2.1.1. It shows a theoretical soil HS molecule bound to clay minerals and Fe hydroxide. The aromatic structures, aliphatic chains, carbohydrates, free radicals and peptide linkages, and chinoide structures are easily identified. The acidic nature of HS is attributable to the many –COOH- and phenolic OH-groups which are not balanced by the basic functional groups such as -NH_2.

Only recently, Mao et al. (2001) provide new structural information on a humic acid using a series of two-dimensional ^{1}H–^{13}C heteronuclear correlation solid-state NMR experiments. Based on their previous study of a series of peat HA from Europe and North America (Mao et al. 2000), the studied HA was found to be fairly typical. Mao et al. (2001) report that carbon-bound methyl groups (C–CH_3) are found to be near both aliphatic and O-alkyl, but not aromatic groups. Most OCH_3 groups are connected directly with the aromatic rings, as is typical in lignin. About one third of the aromatic C–O groups are not phenolic C–OH, but C–OCH_3. COO groups are found predominantly in OCH_n-COO environments, but some are also bound to aromatic rings and aliphatic groups. The authors state that all models of humic acids in the literature lack at least some of the features observed here. From an ecological perspective, the findings of Mao et al. (2001) signify that all properties related to lignin and its degradation products obviously play a more fundamental role in environmental and ecological behavior of HS than previously considered.

Peptides, Glucosamine, Saccharides, and Long-Chain Fatty Acids

Peptides and Amino Acids

The amino acid fraction of aquatic and terrestrial HS comprises 0.5–5% of the dissolved organic carbon (DOC) (Jahnel et al. 1998; Thurman 1985). A calculation of the amino acid nitrogen fractions from the total nitrogen content of soil samples reveal that 26–34% of the FA-fraction, and 28–35% of the HA-fraction are amino acids. From an elemental analysis of nitrogen, Jahnel et al. (1998) estimate a figure of 0.4–26.7% as the proportion of amino acid nitrogen in HS. Humic acids contain a higher proportion of hydrolyzable amino acids than do fulvic acids.

Enzymatic analysis is used to determine the bioavailability of the amino acids.

If one assumes that rather unspecific microbial proteinases are able to cleave all bioavailable amino acids (Jahnel et al. 1993), then 73–99% of the HS nitrogen is not bioavailable. However, it is more likely that the enzyme does not hydrolyze all bioavailable amino acids (Jahnel et al. 1994). Major amino acids that are released upon hydrolysis are glycine and L-enantiomers of alanine, serine, threonine, leucine, aspartic acid, and glutamic acid (Jahnel et al. 1994; Steinberg 1975). In many soils, HA contain relatively higher amounts of basic amino acids, such as arginine, histidine, and lysine than FA, but relatively lower amounts of acidic amino acids, such as aspartic and glutamic acids (Ding et al. 2001). The authors suggest that the percentage of basic amino acids in HA may increase with an increasing degree of humification.

Glucosamine

The intermediate between amino acids and carbohydrates, the nitrogen containing carbohydrate glucosamine, is analyzed by Jahnel and Frimmel (1996). Up to 8% of the total HS nitrogen is in this form. The authors show that there are HS building blocks, which possess a constant ratio of glucosamine to amino acids (as is known for the bacterial cell wall). Thus, the ratio of glucosamine to amino acids is so robust that the amino acid content can be calculated from the glucosamine content of the HS.

Carbohydrates

The second most important class of biosubstrates are carbohydrates, which appear as monosaccharides, oligosaccharides, polysaccharides, and a fraction of HS. These represent 5–25% of organic substance in soils and up to 20% in sediments. In surface freshwater, 5–10% of DOC is mono- and polysaccharides and up to 5% is HS bound carbohydrates. There are less carbohydrates in the FA than in the HA fractions (Jahnel and Frimmel 1995). The application of a combined technique of tangential flow multistage ultrafiltration and two-dimensional NMR spectroscopy reveal that carbohydrate moieties predominantly occur in higher molecular mass fractions (>10 kDa) of isolated HS (Haiber et al. 2001a). For instance, in the Suwannee River FA acid 29% of DOC is carbohydrate in the fraction 50–11 kDa, and in a raised peat bog water (Lake Hohlohsee, northern Black Forest, Germany) even 41%.

The carbohydrate content and patterns are dependent on various factors. First, certain biotic source materials build refractory organic materials with characteristic carbohydrate patterns. While arabinose, xylose, and glucose indicate plant origin, rhamnose, fucose, mannose and galactose indicate microbial origin. The ratio of these two groups of monosaccharides can therefore be used as biomarker of the origin of the material. Jahnel et al. (1998) show that HS in domestic waste water have a high proportion of monosaccharides of microbial origin, whereas in HS from soil water, brown-water and groundwater, glucose or xylose dominate, that points to the prevalence of plant origin. The biological availability, and thus the degree of biodegradation or 'age' of the organic substances also determines the carbohydrate patterns. For instance, glucose is very rapidly metabolized in comparison to other monosaccharides.

Watt et al. (1996) show data for the content of hydrolyzable amino acids and carbohydrates in ten humic and fulvic acid isolates from British and Irish freshwater sources. They find 0.5–5.1% dry weight as amino acid and 0.2–3.1% dry weight as neutral sugars. The authors observe a clear influence of land use on the qualitative composition, in particular that of the carbohydrates. A higher proportion of microbial carbohydrates is found in some isolates from forested catchments and in groundwater, while plant material carbohydrates (arabinose and xylose) are found in FA from pastures and peat bogs. Again, humic acids generally contain substantially more amino acids and carbohydrates than do fulvic acids. Interestingly, the amino acid and carbohydrate fractions are not distributed randomly in the HS isolates, but there is a significant correlation between them: the amino acid content is about 1.7 times higher than the carbohydrate content (Fig. B.2.1.2).

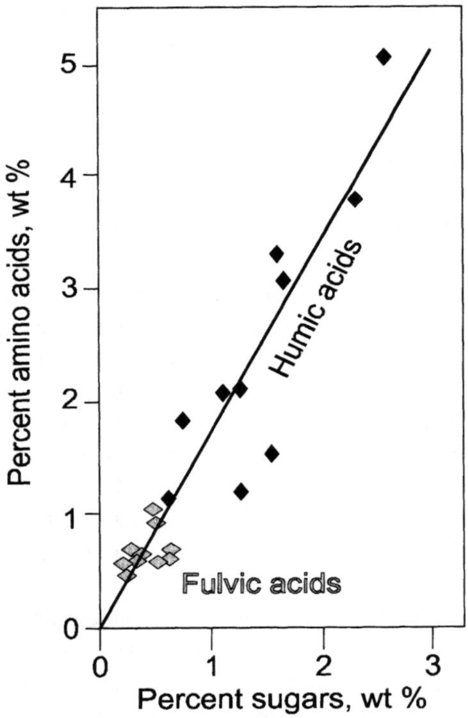

Fig. B.2.1.2. Biplot of the proportion of amino acids against the proportion of carbohydrates in aquatic humic substance on a weight basis ($r^2 = 0.875$) (after Watt et al. 1996, with kind permission of the International Humic Substances Society)

Long-chain Fatty Acids

Lipids in organic matter consist of a wide variety of organic compounds such as fatty acids, *n*-alkyl hydrocarbons, *n*-alkyl alcohols, sterols, terpenes, polycyclic hydrocarbons, chlorophyll, fats, waxes, and resins (Stevenson 1994). Fatty acids represent the most abundant class of lipids and probably the most thoroughly investigated (Jandl et al. 2002; Poltz 1972; Schnitzer et al. 1986; Weete 1976). The role of lipids in selective degradation and preservation and in the neogenesis of soil organic matter is reported by Lichtfouse et al. (1998) and van Bergen et al. (1998) (see Chap. 2). Schulten et al. (2002) show that clear differences in concentrations of *n*-alkyl fatty acids exist between DOC in river and raised peat bog water. More recently, Jandl et al. (2002) report total concentrations of $C_{14:0}$–$C_{28:0}$ *n*-alkyl fatty acids in Suwanne River NOM with 309 μg/g, raised peat bog NOM with 180 μg/g, raised peat bog HA 43.1 μg/g, and raised peat bog FA with 42.5 μg/g. Furthermore, differences among these samples are reflected in the concentration of individual fatty acids with a dominance of $C_{16:0}$ over $C_{18:0}$, $C_{14:0}$, $C_{15:0}$ and $C_{24:0}$. In contrast to aromatic building blocks, the long-chain fatty acids do not belong to the photostable compounds in HS (Schmitt-Kopplin et al. 1998). However, Jandl et al. do not find higher concentrations of fatty acids in soil organic matter than in surface water DOC – rather the contrast applies: the concentrations of long-chain fatty acids in aquatic DOC exceed those in soil dissolved organic matter by far. This contradiction remains to be solved, but nevertheless, long-chain fatty acids, in combination with carbohydrates and amino compounds, are the structural counterparts of the aromatic building blocks (see Box 1.1) and affect the environmental behavior of HS and their interactions with aquatic organisms (Chap. 8).

Secondary and Tertiary Structure

In addition to the previously described chemical building blocks, the macromolecular structure also determines the role of HS in biogeochemical processes.

At low HS concentrations, HS of aquatic origin exist primarily as dissolved ions. At higher concentrations, colloids and precipitates are formed. The same aggregation takes place when HS react with cations and protons. Aggregation and dis-aggregation have consequences for the biogeochemical behavior of HS and with biological activities such as (food) particle filtering.

Myneni et al. (1999) describe the effect of protons, divalent, and trivalent cations on fulvic acids from the Suwannee River, a standard HS from the International Humic Substances Society. Using a high-resolution X-ray microscope, Myneni et al. (1999) find no recognizable structures in concentrations of less than 1 g/L C. Above this concentration, fulvic acids build aggregates of various size and form. In acid solutions, FA form global aggregates and ring-like structures, whereas in alkaline solutions small aggregates are formed. Divalent cations cause thread-like structures, and trivalent cations cause globular as well as thread-like structures (Fig. B.2.1.3).

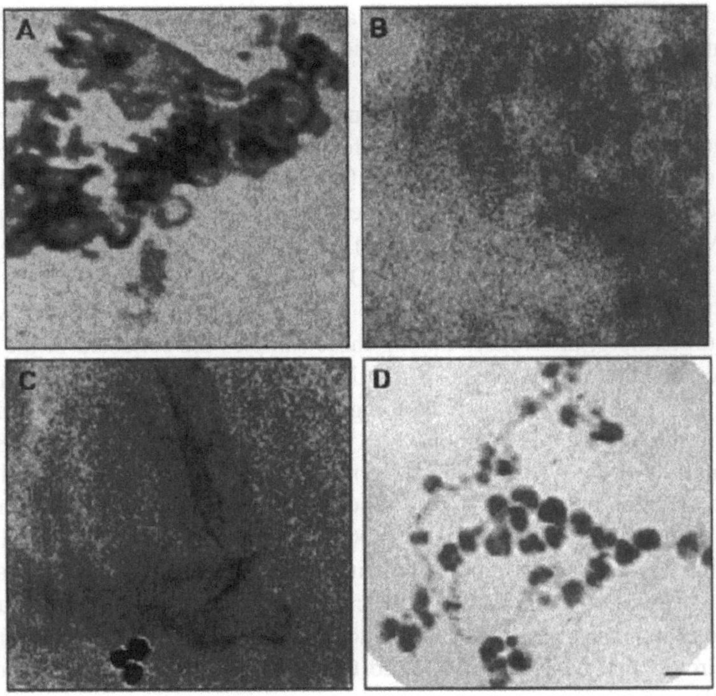

Fig. B.2.1.3. High resolution X-ray micrographs of HS. Scale bar in D applies to all figures and indicates 500 nm (from Myneni et al. 1999, with kind permission of Science). Effects of solution pH: **A**: globular and ring-like structures in acidic solution. **B**: uniform, small aggregates in alkaline solution. Effects of cations: **C**: thread-like structures with divalent cations. **D**: globular and thread-like structures with trivalent cations

Provided that the above displayed structures also occur under conditions of natural DOC concentrations, one can easily see that aggregates >1–2μm in diameter are formed in presence of divalent metal cations in hard-water lakes, and these can be taken up by zooplankton grazers as a major carbon source (see Chap. 9). An additional prerequisite for this ecological interpretation, is that these results obtained by X-ray microscopy are robust and can be confirmed by independent methods. For instance, in a study of the aggregate sizes of 18 Norwegian NOM isolates, several of which will be described with respect to their effects on aquatic organisms (Chap. 8), Lead et al. (1999) show that the use of transmission electron microscopy (TEM) alone will lead to the conclusion that the samples are composed primarily of large (up to several micrometer) aggregates. The concurrent use of additional methods, such as fluorescence correlation spectroscopy, capillary electrophoresis, and atomic force microscopy leads to the opposite conclusions, namely that the NOM is composed primarily of small, mobile colloids and that the large aggregates are quantitatively less important. However, the use of TEM allows the visualization of these aggregates, which otherwise are largely ignored by fluorescence correlation spectroscopy and atomic force microscopy. Nevertheless, a new picture of marine DOC showing that 10–40% is in the form of colloids which can aggregate (Chin et al. 1998), indirectly supports the finding of Myneni et al. (1999). Hence, the aggregation is feasible and is an important initial step in the formation of POC, which can serve as C source for heterotrophic particle feeders.

In acidic soils and sediments with a high carbon content, HS form dense structures with a low surface to volume ratio (Myneni et al. 1999). These structures restrict the openness of aggregate micropores for microorganisms, and reduce the diffusion of oxygen and other substances into the aggregates. Under alkaline conditions, the HS structures are more open, and therefore more amenable to microbial and chemical reactions. This means that the HS are better able to take part in biogeochemical processes in alkaline than in acidic soils and sediments.

The X-ray microscope technique is also used by Thieme et al. (2002) and confirms a flock-like formation of humic acids under the influence of cations. A dispersion of the soil humic acid (mixed forest, BS1 from the DFG Priority Program ROSE) is treated with di- and trivalent cations, and the valence of the introduced cation correlates with the degree of flocculation. The flocks reach a diameter of few µm. Concerning ecological implications, this means that the flocks become available for filter feeders.

Furthermore, Thieme et al. (2000) show which groups of the HS take part in the precipitation of FeOOH from Fe-containing groundwater. By precipitation, the free C=O groups are reduced in number. The authors demonstrate that reduced HS contain an open multiple thread-like structure; in contrast, oxidized HS form coils. In reduced HS, the bioavailability is decreased, as is the potential toxicity.

Using transmission electron microscopy, Schnitzer (1994) shows that with increasing pH or HS concentration, a fine woven net of fibrils forms (see also Fig. B.2.1.3), which then bind together to form leaf-like structures which are perforated by cavities of various sizes. These cavities can trap or fix organic and inorganic molecules. This observation is supported by molecular modeling as shown by Schulten in various studies (Schulten 1999, 2002; Schulten and Leinweber 2000) (see Fig. B.2.1.7), as well as by Bailey et al. (2001) in a slide show part.

Before X-ray microscopy or molecular modeling were available, there were some attempts made to determine or model the secondary structure of HS. It must be mentioned again that HS are heterogeneous and polydispersed, and thus it is difficult to give a general secondary structure for HS (Stevenson 1994). In order to describe the macromolecular structure of HS in terms of properties such as molecular size, molecular form, or molecular weight, a series of different concepts were developed, namely:

- rigid globular particles (Visser 1964)
- flexible ellipsoids (Orlov et al. 1972)

• closed coils (Cameron et al. 1972).

Four concepts, which give a special impetus or provide intensive discussions, will be described in detail here:

• protein-like macromolecules (Chen and Schnitzer 1976; Ghosh and Schnitzer 1980),
• micelles (Wershaw 1986),
• supermolecular associations (Cameron et al. 1972; Piccolo 1997; Piccolo et al. 2001)
• molecular modeling (Bailey et al. 2001; Schulten 1999).

Protein-like Macromolecules

Methods of protein biochemistry are used by Ghosh and Schnitzer (1980) to show that at high concentrations of HA and FA, and at low pH values or high concentrations of neutral salts, spherical colloids form. At higher pH values and lower concentrations of HS and neutral salts, linear colloids form. Fig. B.2.1.4 shows these findings in schematic form. With this model, one can resolve the contradiction in the literature between the concepts of the so-called rigid 'spheroid colloids' and the flexible 'linear molecules'.

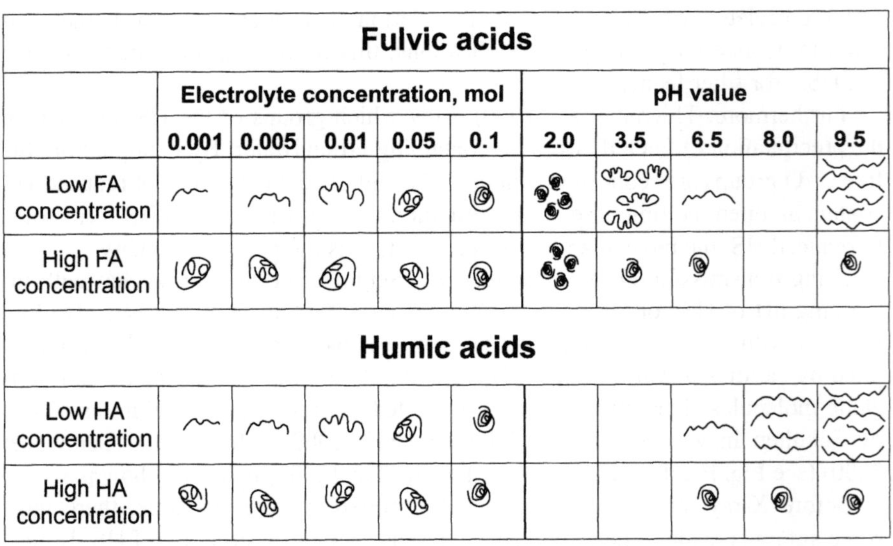

Fig. B.2.1.4. Classical model of the macromolecular structure of HS in relation to ionic strength and pH value (after Ghosh and Schnitzer 1980, with kind permission of Lippincott Williams & Wilkins)

This model also gives the first evidence that one single configuration does not apply for both HA and FA. The configurations vary with environmental conditions in both soils and freshwater. This picture is confirmed for various environmental conditions by using poteniometric titrations (Ephraim et al. 1995) or poly-acryl gel-electrophoresis (PAGE) (Münster 1982, 1985; Steinberg and Münster 1985). In an earlier study, Schnitzer (1978) postulates that HS are not individual molecules, but rather associations of molecules of various origins.

In several instances, particularly high-molecular weight or colloidal fractions with their carbohydrate moieties are most efficient in binding metals (Quigley et al. 2002). From recent transmission electron microscopy studies too, it is apparent that typical macromolecules and colloids in humic-rich surface waters consist of irregular networks of organic and inorganic entities in the size range from some nanometers up to the micron level (Buffle and Leppard 1995a,b; Wilkinson et al. 1997).

Micelles

For soils and sediments, Wershaw (1986) put forward the so-called micelle model, for conditions where HS concentrations are high. One of the most important features of this model is the interaction between HS and hydrophobic pollutants and metals. According to this model, HS comprise a set of oligomers and simple compounds which are formed in part through the decomposition of plant materials. These degradation products are stabilized through incorporation in HS aggregates, linked together by weak hydrogen bonds, π–π-bonds, or hydrophobic interactions. The resulting structures are like micelles or bio-membranes, with a hydrophilic exterior and hydrophobic interior (Fig. B.2.1.5). Hydrophilic compounds and ions such as metal ions interact with the hydrophilic HS surface, while hydrophobic molecules such as organic compounds migrate to the interior.

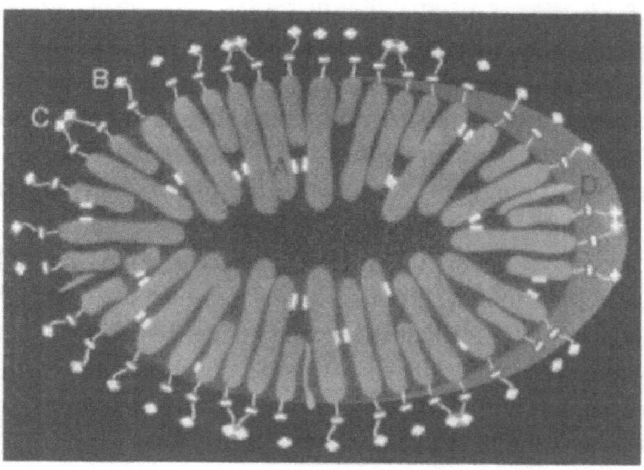

Fig. B.2.1.5. Schematic representation of a humin micelle. The amphiphilic molecules are indicated by **A**, monovalent cations by **B**, divalent cations by **C** and xenobiotic molecules by **D** (from Wershaw 1986 et al., with kind permission of Kluwer Academic Publishers)

Wershaw (1986) shows that some HS micelles are always dispersed in the water (soil water, sediment interstitial water, or standing water) and therefore are mobile. At higher pH values, a higher proportion of micelles are dispersed in the solution since more acidic groups dissociate. The resulting increasing load on the micelle surface leads to a break up of the larger structures, and the formation of smaller structures such as the micelles. Barak and Chen (1992) also support the micelle model of Wershaw, through studies of equivalent radii of the HS molecules.

In further studies, Wershaw et al. (1986, 1989) show that it is very likely that the model developed for soil and sediment solutions, also applies to water bodies. This model explains very clearly the interactions between HS and hydrophobic xenobiotics. In later studies, these insights are substantially proved. Using quantitative structural relationships, Haitzer et al. (1999a-c), Perminova et al. (1999), and Steinberg et al. (2000) show that HS, particularly the aromatic groups, bind xenobiotics.

Following Wershaw's studies, several groups try to confirm the existence of HS micelles. A so-called critical concentration of HS must be exceeded for aggregation into micelles to occur. Guetzloff and Rice (1994) using a commercial humic acid, prove that this critical concentration is as high as 7.4 g/L. Such high concentrations are not environmentally realistic.

Within the priority program 'Refractory Organic Substances in the Environment' (ROSE) of the Deutsche Forschungsgemeinschaft (DFG), various HS are studied for example with respect to their UV/vis spectroscopic properties. Langhals et al. (2000) emphasize that all HS fulfill the Lambert-Beer law perfectly up to a concentration of 100 mg/L DOC. This means that the chromophoric groups are totally **independent** of each other below this concentration. Only above this concentration, do the chromophores interact. One of the possible interactions could be the formation of micelles or other aggregates. Concentrations do not occur in the water column of lakes or rivers, but may occur in soil solutions and in the interstitial water of sediments. At concentrations found in open waters (few mg/L), interactions between the truly dissolved HS and xenobiotics do occur.

The micelle model can potentially explain another phenomenon, the increase in bioconcentration of lipophilic chemicals in the presence of HS. If the HS form micelle-like structures, or as Wershaw (1986) also emphasizes, membrane-like structures – and have bound organic substances in their hydrophobic interior, these structures could migrate through biomembranes. However, there is no empirical evidence for this mechanistic concept. Increases in bioconcentrations of lipophilic organic compounds appear with HS concentrations which do not lead to micelle formation, but with particularly low DOC concentrations, which for many waters are environmentally relevant (Haitzer et al. 1998).

Supramolecular Associations

Another model is put forward by Piccolo (1997) and his group (Piccolo et al.

1996, 2001) following a pioneering idea of Cameron et al. (1972). Piccolo's ideas are developed following experiments in which HS are fractionated by size exclusion chromatography (SEC) after adding monocarboxylic acid (methanoic, ethanoic, propanonic, and butanoic acids) to the HS. Building blocks and non-humic material are obtained. Piccolo (1997) proposes irregular micelles as the supermolecular structures, in contrast to the regular micelles of Wershaw's model. Piccolo suggests hydrophobic interactions as the dominant bonding mechanism between the individual micelles. These weak links explain the change in HS molecular size under alkaline and weakly alkaline pH conditions. In 1986, Wershaw already discusses this bonding type as an important, but not the only mechanism involved in aggregate building and the binding of hydrophobic chemicals.

Without doubt, Piccolo's model has fascinating aspects, since it explains both the splitting and aggregate formation of HS, and it could, for example, clearly explain the aging of fulvic acids. However, Piccolo's model is developed from the results of only one method, namely SEC in two modifications, gel chromatography and high performance size exclusion chromatography. In order to reduce and account for artifacts and biases which are inherent in all techniques, such far reaching conclusions as Piccolo's should have been proved by other independent methods, such as:

- dynamic light scattering, X-ray scattering (Tombácz et al. 1997; Wagoner et al. 1997)
- capillary electrophoresis (de Nobili et al. 1997; Schmitt et al. 1997)
- viscosimetry (Chen and Schnitzer 1976; Kawahigashi et al. 1997)
- Raman spectroscopy (Sanchez-Cortes et al. 1997)
- ultracentrifugation (Cameron et al. 1972; Wilkinson et al. 1993)
- atomic force microscopy or fluorescence correlation spectroscopy (Wilkinson et al. 2000)
- new developments in mass spectroscopy, such as time-of-flight secondary ion mass spectroscopy (Szymczak et al. 2000)
- on-line coupling of SEC with independent mass assessments (Persson et al. (2000).

Recent studies show that even aggregate formation during aging may be an artifact, since it does not always occur. Müller et al. (2000), for instance, describe the stability of SEC fractions of a raised peat bog NOM over time by re-injection of selected subfractions. Over 5 weeks, no significant changes in the UV chromatograms can be detected, suggesting a high stability of the fractions and negligible alterations of the molecular weight distribution during that time period.

Summarizing the controversy on size and shapes of HS, Clapp and Hayes (1999) conclude that HS can be macromolecular, but the extent of their macromolecularity can vary depending, possibly, on the substrate materials, on the extent of the biological degradation and synthesis processes, and on the environments in which they are formed and found. However, Clapp and Hayes believe that it is not possible at present to state the extent to which HS in solution are associations of polydispersed molecules, or polydispersed macromolecules ranging in size from the relatively small to the very large.

Recently, Haiber et al. (2001b) discover one mode of chemical action, by which high-molecular weight fractions are formed with increased aromaticity. Low-molecular size lignin moieties undergo demethylation of the methoxy groups. With increasing molecular size of the HS fraction, the amount of decomposition of phenylpropane side chains as well as the degree of aromatic substitution is enhanced. Hence, the decomposition of the phenylpropane side chains of the lignin building block is accompanied not only by demethylation of the methoxy groups, but also by an aggregation of aromatic moieties. This mechanism applies to FA as well as HA, as studied with Suwannee River isolates and displayed in Fig. B.2.1.6.

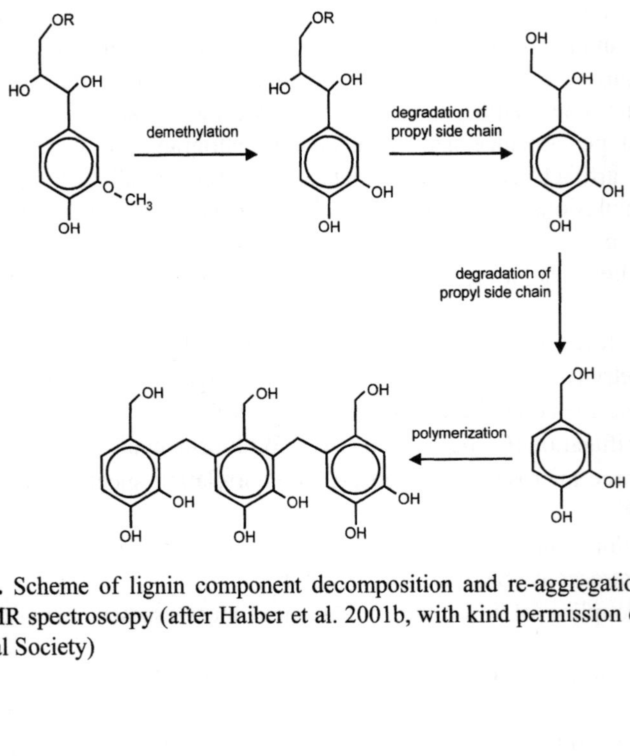

Fig. B.2.1.6. Scheme of lignin component decomposition and re-aggregation as derived from 2D NMR spectroscopy (after Haiber et al. 2001b, with kind permission of the American Chemical Society)

Molecular Modeling

A completely different, SEC independent, route to study HS conformation is taken by Schulten (for example 1999) when he 'creates' DOC molecules of several thousand atoms by molecular modeling. The DOC building blocks are identified through pyrolysis. Mechanistic molecular calculations are employed to confirm structural and three-dimensional models. These models include a geometrical optimization of DOC molecules according to the principle of energetic minimization, or in other words, the formation of structures with the lowest possible energy content. Schulten designs colloids with molecular weights of >40 kDa and an elemental composition of $C_{1788}H_{1958}O_{1224}N_{60}S_8$. Furthermore, he shows that the lengths and angles of covalent bonds are plastic, and voids and splits in the colloids are recognizable. These characteristics are especially important for the ecological and eco-chemical behavior of HS, such as inclusion and binding of (organic) nutrients and xenobiotic substances.

Inclusion is one of the possible mechanisms by which HS can mask non-HS. Further, lipophilic chemicals such as pentachlorophenol and DDT are bound to HS molecules mainly by van der Waals forces and hydrogen bonds which temporarily immobilize the xenobiotics (Schulten 1999). Also the degradation products of xenobiotics are bound to HS by hydrogen bonds.

Schulten (1999) describes, how HS capture, for instance, hydroxyatrazines, degradation products of atrazine. The hydroxyatrazines are immobilized by hydrogen bonds from hydrogen in the imino group (in the *s*-triazine ring), rather than from the hydroxyl group which replaces the chlorine atom at the *s*-triazine ring. The molecular modeling approach can also model the trapping of carbohydrates and amino acids into HS molecules (Schulten and Leinweber 2000) provided these are not covalently bound during humification. Fig. B.2.1.7 shows how the binding of hydroxyatrazine, peptides, and saccharides can occur. Hydrogen bonds are again the most important linkages for these biomolecules. On the basis of the high polarity of the HS and the low binding energy, the hydrogen bonds show an extraordinarily large flexibility and structural variability.

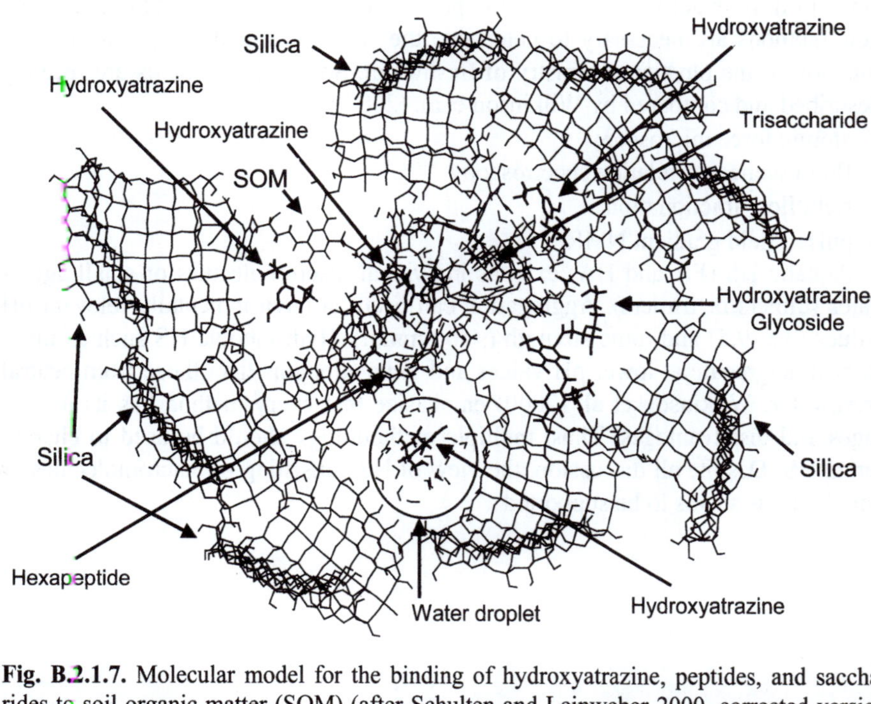

Fig. B.2.1.7. Molecular model for the binding of hydroxyatrazine, peptides, and saccharides to soil organic matter (SOM) (after Schulten and Leinweber 2000, corrected version from the authors with kind permission). The SOM contains eight peripheral silicic acid groups, five bound hydroxyatrazine as well as one hexapeptide, and one trisaccharide

Schulten (1999) emphasizes that all results of such molecular models are only acceptable when the predicted properties can be independently verified through analytical studies, for example ^{13}C NMR spectroscopy, pyrolysis-FIMS, or pyrolysis-GC/MS.

Outlook

Many properties and functions of HS in aquatic ecosystems can be explained by the models presented above. However, no model alone can explain everything. There is as yet no complete model of HS which can explain their properties and functions in ecosystems from the primary and secondary structure of these biogeochemical matrices. There are currently two different schools of thought. In one, poorly flexible, rigid molecules or aggregates are favored, while the other defines a supramolecular structure as the basis for HS in ecosystems.

Humic substances play a key role in the ecology and eco-chemistry of aquatic systems. They influence the bioavailability and transport of trace metals and organic chemicals, modify the stability of mineral colloids, and can bind directly to (Campbell et al. 1997; Meinelt et al. 2001) or be removed from, aquatic organisms (Pflugmacher et al. 1999; Wang et al. 1999). It is therefore of profound importance to have exact knowledge of the physico-chemical nature of HS. Therefore, new methods are necessary to determine the size, form, and charge of HS as a function of the physico-chemistry of freshwater. Some new methods are recently described and compared by Wilkinson et al. (2000):

- atomic force microscopy
- fluorescence correlation spectroscopy
- field flow fractionation
- pulsed field gradient NMR.

Aquatic HS (FA and HA) exist as individual macromolecules or small aggregates (dimers or trimers). Aggregation can occur at environmentally relevant pH values (3.5–9.5) and ionic strength (<500 mmol). Hydrophobic HS such as those in peat, aggregate at lower pH values and ionic strengths than those from neutral freshwater. Wilkinson et al. (2000) emphasize that every method has its advantages and disadvantages. Thus, multiple methodology should be used to characterize HS. Despite all discussion, the view of HS consisting of macromolecules or smaller units seems to be supported.

3 Humic Substances as Geochemical Determinants

Humic substances in freshwaters reflect, in many ways, the processes in the catchments. Under undisturbed, natural conditions, HS are tightly linked to soil processes and are connected to vegetation development. The size of the terrestrial catchment has an influence on the quality and quantity of materials washed into freshwater bodies. In the long-term, dissolved HS provide a clear reflection of climatic conditions and their development, as well as connected processes such as specific hydrological conditions. If man intervenes in the structures of the catchment, clear markers are left in the total biogeochemical cycles of a landscape. It is no surprise that, for example, wetland drainage is associated with an increased export of C compounds, and impacts water chemistry and biota in receiving lakes and rivers.

The role of dissolved HS in determining water chemistry, exemplified by recent limnological and paleolimnological studies, includes:

- the influence of the catchment on the quantity, quality and molecular weight of NOM inputs,
- the contribution of dissolved HS to present water acidity,
- paleolimnological studies relating to acidification processes in selected lakes in the Bavarian Forest (Germany) and in Lapland (Finland),
- Holocene climate trends in the boreal tree line of the Canadian Northwest Territories and in Finnish Lapland and
- effects of human settlement and related processes on the north eastern German landscape, as reflected in sediments of a river-lake area (River Schlaube system, Brandenburg State, Germany).

3.1 Influence of the Catchment

Despite a very slow calculated turnover rate in the various soil horizons (Fig. 1.2), turnover does occur in the soils of a lake or river catchment. Hence, one can establish, depending on the type and size of the catchment, specific quantities, compositions, and the molecular weight distributions of dissolved HS entering the freshwater bodies.

The influence of catchment areas and their diverse properties on the quantity of HS, or more general DOC, are summarized by Mulholland et al. (1990). The conclusions still remain valid and illustrative today, as described below. Major potential sources of dissolved HS for aquatic ecosystems are the soil horizons rich in organic material, detritus accumulated in aquatic ecosystems, interfaces between terrestrial and aquatic systems such as floodplains and wetlands, as well as aquatic primary producers themselves. Internal sources appear of minor importance to the bulk DOC, as shown by the relatively small temporal and spatial variation in DOC concentration and its main fractions in lakes dominated by autochthonous production(Steinberg and Münster 1985; Wetzel 2001).

In ecosystems with a relatively high hydrological turnover, temporal and spatial variability are not independent characteristics of DOC distribution. They are, in fact, closely coupled. For example, a spatial concentration gradient of DOC in a flowing stream is observed as a temporal change in DOC as measured at any given point. Observations of significant temporal variability in DOC concentration (>5 mg/L) appear in running waters in time spans of hours to days (for storm events in small streams), to weeks and months (for floods in major rivers). DOC concentration are typically positively correlated with discharge, and usually show clockwise hysteresis (higher concentrations at equivalent water discharge during the rising stage compared to the falling stage). Seasonal variation in DOC concentrations is also observed, particularly in systems with large seasonal changes in hydraulic residence times (Mulholland et al. 1990). The DOC fluctuations in lakes are small, in comparison to those in streams, perhaps with the exception of small lakes with high terrestrial or wetland inputs.

Although the DOC concentrations in surface waters can vary from 0.5 to >50 mg/L (Fig. 1.1), the fluctuation in large rivers (>100,000 km^2) is clearly smaller, ranging from 2–15 mg/L. The analysis of large data sets on spatial variability of DOC concentrations in the world's rivers, shows no clear relationship between mean concentration and climate (such as temperature), drainage, or the particular vegetation (such as net production, vegetation type) (Fig. 3.1). However, many other studies in Finland (see below), New Zealand (Moore 1987), and the south eastern USA (Mulholland and Kuenzler 1979) show a positive relationship between the DOC concentration and the surface area of wetlands.

Mulholland et al. (1990) conclude that there is apparently a lack of knowledge on the flow paths of waters, and therefore DOC input routes, from underground. Since this statement in 1990, only a few additional studies further our knowledge in this area. From the viewpoint of the microbial processes in hyporheïc sediments, Fiebig (1995) quantifies the input pathways of a small mountain stream, the Breitenbach in Schlitzerland (Hessen, Germany). He concludes that the surface input of organic C does not explain the high microbial production in this particular stream. The mean DOC concentrations in the groundwater (5.5 mg/L) markedly exceeds the mean concentration in the stream (1.4 mg/L). That means that the DOC concentration in the groundwater of the Breitenbach area is almost independent of the flow rate or season.[1] Mass balances in this stream show that

[1] This is, however, not a general phenomenon. Meyer and Tate (1983) find increased DOC concentrations in autumn in the run-off from fresh leaf litter. In contrast, Ford and Nai-

about 1 g/m/d DOC enters the stream via groundwater (annual mean). Of this, as much as two thirds is already immobilized in biofilms in the hyporheïc sediment. Evidently the varying loading into the hyporheïc sediment from groundwater is clearly compensated for by the increased immobilization in biofilms, which causes a more or less constant DOC concentration in the stream.

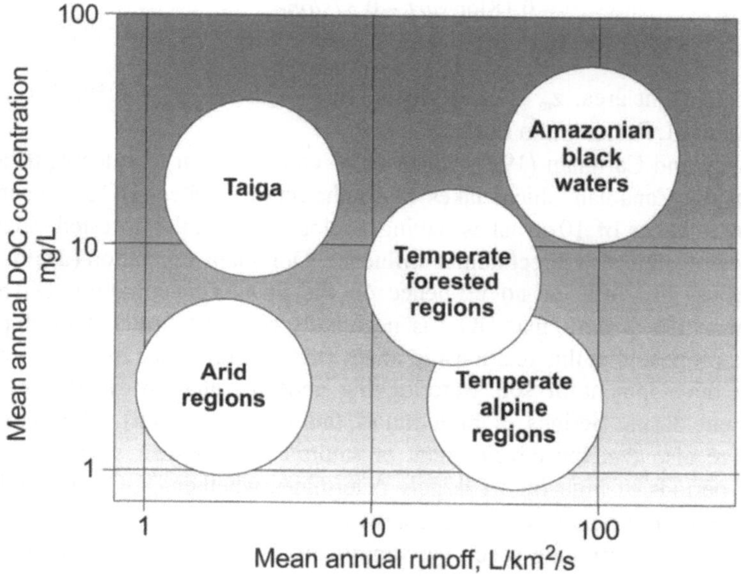

Fig. 3.1. Estimates of mean annual DOC concentrations as a function of the mean annual runoff for biomes in different climate zones (after Mulholland et al. 1990, with kind permission of Wiley & Sons)

3.1.1 Influence on Humic Substance Quantity

3.1.1.1 Water Color

There are a variety of studies to explain the geochemistry of lakes, or – to a lesser degree – rivers, according to size and structure of catchment areas and water basin morphometry. Gorham et al. (1986), Engstrom (1987), Rasmussen et al. (1989), and Schindler (1971) attempt to correlate water color, as a surrogate for HS content, with catchment area features and lake morphometry. The basic assumption of these studies is that water color originates mainly from allochthonous sources. However, slope gradient, soil development, or precipitation also impact water

man (1989) describe a summer maximum in DOC in groundwater, attributed to a concentration effect.

color. Furthermore, HS input is diluted by precipitation on the water surface.

For 287 pristine lakes in North America, Rasmussen et al. (1989) find the following multiple regression for water color (Eq. 3.1)

$$\log color(mg/LPt) = 1.6 + 3.4\log\frac{CA}{surf} - 0.59z_m - 0.59z_m - 0.18\log surf - 0.37 slope \tag{3.1}$$

CA= catchment area, z_m = mean depth, surf = surface, r^2 = 0.58. The color is given in mg/L Pt (platinum salt).

D'Arcy and Carignan (1997) relate DOC concentration to catchment topography. In the Canadian Shield lakes of southeastern Québec, DOC concentration vary by a factor of 10 in lakes having geologically similar forested catchments that are not subject to direct human influence. Dominant vegetation (deciduous vs. coniferous) has little or no influence on the DOC concentration. D'Arcy and Carignan (1997) show that DOC is negatively related to catchment slope. The authors propose that this relationship arises from the influence of catchment slope on the development of soil waterlogging and saturated overland flow in the catchment during periods of high runoff (autumn, snowmelt). The authors hypothesize also that the development of saturated zones near streams and lakes during periods of high runoff largely determines whether DOC produced in the forest floor will be exported to surface waters or intercepted in the soils.

Flow-path partitioning (deep vs. superficial) is a determining factor of DOC export to surface waters. The significance arises from the strong vertical stratification of biogeochemical processes in soils. The canopy and the upper organic soil horizons are major sites of DOC production in forests. However, DOC concentrations in the soil solution decrease markedly as water percolates down and reaches mineral soils. For example, Cronan and Aiken (1985) observe mean summer DOC concentrations of 19 mg/L in the forest floor and 5.5 mg/L in the B horizon of hardwood and conifer stands. McDowell and Wood (1984) and McDowell and Likens (1988) describe DOC concentrations of 30–50 mg/L just below the forest floor compared with 3–6 mg/l in the B horizon.

D'Arcy and Carignan (1997) explain their findings by applying Darcy's law (flow in porous media) to the inclined plane. If Q (m^3/s) is the flow through a saturated porous medium of width w (m), K_s (m/s) is the hydraulic conductivity, and (degree) is the inclination of the plane, then the height h (m) of the saturated zone above an impermeable boundary is given by:

$$h = \frac{Q_s}{wK_s \sin(\Theta)} \tag{3.2}$$

Thus, h is inversely proportional to K_s and to $\sin\Theta$. In other words, for a constant Q_s, constant K_s, and constant till thickness, the depth to the water-table will increase with increasing catchment slope. Eq. 3.2 can be applied to an idealized catchment segment where L_t is the total length (m) from ridge to stream or lake,

r is the inflow (m/s) as rainfall or snowmelt equally distributed over the entire surface, and d is the till thickness (m). For a vertically constant K_s, the length of the water-saturated zone (L_{sat}) is given by Fig. 3.2 (Eq. 3.3):

$$L_{sat} = L_t - \frac{K_s}{r} d \sin(\Theta) \tag{3.3}$$

At constant till thickness and constant rainwater or meltwater inflow, the likelihood that a catchment will generate saturated surface runoff increases with decreasing slope and with increasing distance between ridge and stream.

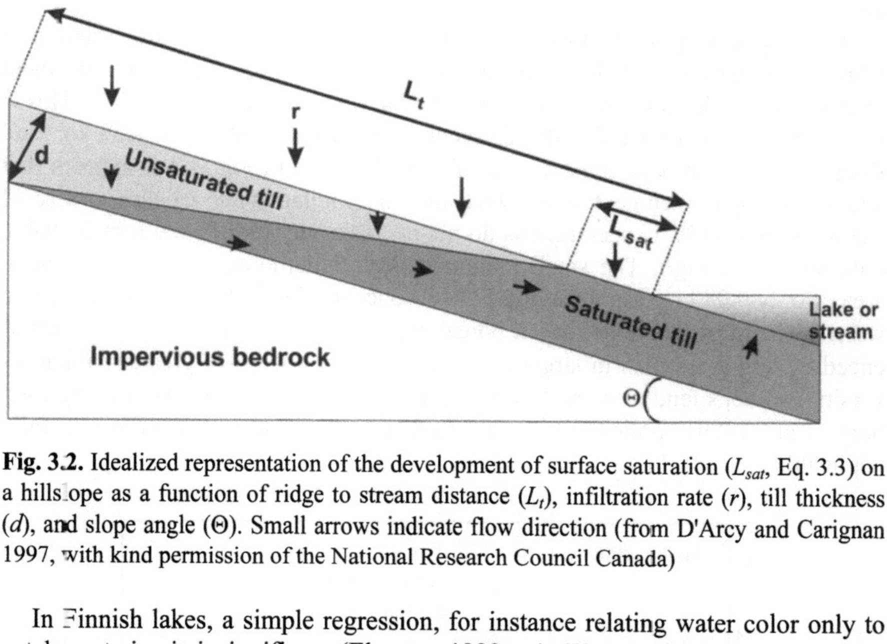

Fig. 3.2. Idealized representation of the development of surface saturation (L_{sat}, Eq. 3.3) on a hillslope as a function of ridge to stream distance (L_t), infiltration rate (r), till thickness (d), and slope angle (Θ). Small arrows indicate flow direction (from D'Arcy and Carignan 1997, with kind permission of the National Research Council Canada)

In Finnish lakes, a simple regression, for instance relating water color only to catchment size is insignificant (Eloranta 1999a–c). Water color can be related to HS content (TOC) according to the Eqs. (3.4, 3.5), based on work in Finland (Kortelainen 1993):

$$1 \text{ mg/L Pt} = 11.47\,([\text{TOC}] - 3.55) \tag{3.4}$$

or

$$1 \text{ mg/L TOC} = 0.0872 \text{ Pt} + 3.55;\ r^2 = 0.86;\ n = 976. \tag{3.5}$$

The prerequisite of Eqs. 3.4 and 3.5 is that there is a constant relationship between color and TOC. This applies if the vegetation is more or less homogeneous within a region or a climate zone. However, if there are changes in the vegetation, particularly from herbaceous to wooden dominance, or from coniferous to deciduous forest, and *vice versa*, the above equations may not apply. For instance in Lake Tsuolbmajavri in Finnish Lapland, between 6,000 and 3,000 years before present (BP), pine and birch retreat and pine finally disappears from the catchment. Reduced quantities of lignin and cellulose are deposited on the soil. As a consequence, water color decreases, but TOC remains more or less stable (Seppä and Weckström 1999; Fig. 3.37).

3.1.1.2 TOC/DOC Concentrations

The best models of the relationship between catchment features and organic C are from Scandinavia. The catchment size has a basic influence on DOC inputs into water bodies for catchments that are only sparsely inhabited. This applies both for quantity of DOC and its average molecular size, and for both running waters and lakes.

Egeberg et al. (1999) show for ten water bodies that 43% of the variation in DOC concentration could be accounted for by the size of the particular catchment area: the larger the catchment, the lower the DOC concentration (Fig. 3.2). This is particularly evident for Lake Maridalsvann with a catchment area of 252 km^2 and River Topdal with a catchment area of 1880 km^2. From Fig. 3.2 it appears that above a critical catchment size, which may be similar to the catchment size of Maridalsvann, DOC concentrations do not decrease any further and remain stable with some 3–4 mg/l. The small mountain lake Hellerudmyra with a catchment area of only 0.08 km^2 has the highest DOC concentration (19 mg/L) of the studied waters. The reason may be that in boreal areas, small catchments are more influenced by bog areas than in larger catchments, where there is a greater habitat diversity and bogs tend to be replaced by forests. The regression described by Egeberg et al. (1999) applies only to the studied region and is not more generally applicable.

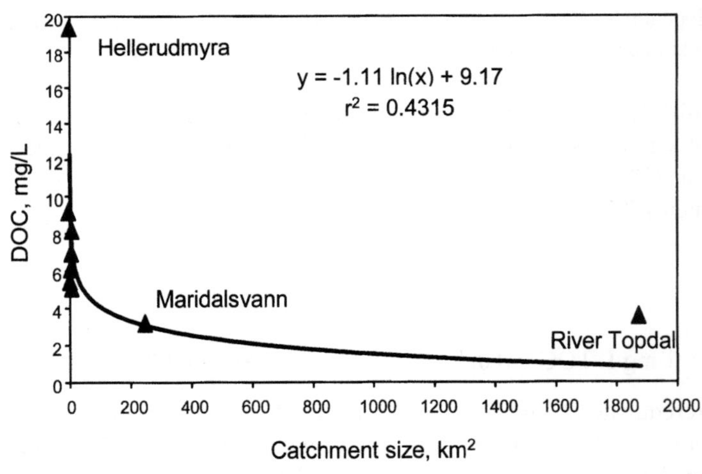

Fig. 3.3. Influence of catchment size on DOC concentration (non-fractionated = natural organic matter, NOM) in Norwegian freshwaters (modified from Egeberg et al. 1999, with kind permission of Elsevier Science)

A large Finnish data set in which about 1/3 of the catchments are covered by bogs and 2/3 are used by forests is studied by Kortelainen (1993). She concludes that:

- peatland and forested areas in Finland produce large quantities of organic materials which lead to high TOC concentrations in small, shallow lakes (Fig. 3.3). Lakes with the highest organic C concentrations are generally small with large catchments, and are mainly located in central and southern Finland;
- the proportion of the catchment covered by bogs and peatland is the most powerful predictor of TOC concentration (Fig. 3.4). This finding agrees well with the statement of Eckhardt and Moore (1990), that for Quebec streams, the proportion of peatland area in the catchment determines the TOC concentration. In general, the finding of Kortelainen (1993) agree also with Swedish findings, although the Swedish estimates are rougher. Meili (1992) estimates that in remote soft-water lakes of the Swedish forest region, autochthonous C is typically <5 mg/L. Most lakes in this region receive significant amounts of HS originating from coniferous forest soils or peatland in the catchment area. In most humic lakes with a water color of ≥50 mg/L Pt (8 mg/L DOC), more than half of the organic C in the surface water is of allochthonous origin, and in polyhumic lakes (>200 mg/L Pt, 21 mg/L DOC) the proportion can exceed 90%;

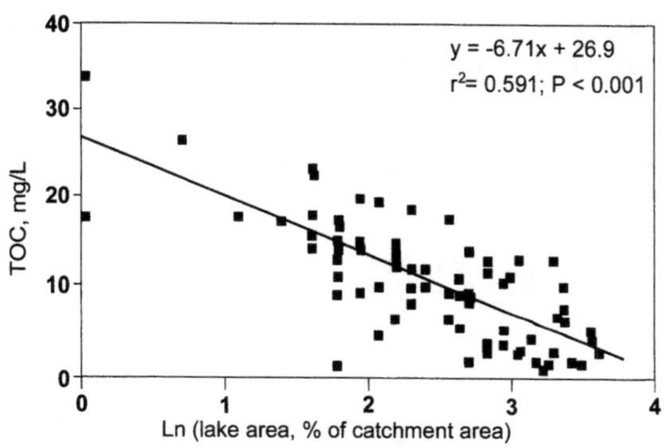

Fig. 3.4. Relationship between relative lake surface area and TOC concentrations in Finland (n=78) (from Kortelainen and Mannio 1988, with kind permission of Kluwer Academic Publishers)

- however, although the proportion of peatland in northern Finnish catchments is greater than in the south, the TOC concentrations are lower in the north. The colder climate, the longer frost period, the lower primary production, and lower decomposition rates in northern Finland, all account for the lower organic C outputs and the lower TOC concentrations in the northern lakes (see also paleolimnological studies in Lakes Hirvaslompolo and Tsuolbmajavri below);
- the mean annual output of TOC of Finnish catchments lies between 4.5 and >7 t/km^2 (Kortelainen and Saukkonen 1992, 1998). The annual output from forested areas is 2.6–8.8 g/km^2 with a clear tendency to higher values.

If the average charge density of DOC is 9.7 μeq/mg, the TOC exports produce an acidity loading to water bodies of 25–85 keq/km²/a. This value agrees well with the model value of Oliver et al. (1983), who assume a charge density of 10 μeq/mg C.

From his studies, Hongve (1999) concludes that Norwegian data show weak direct relationships between the occurrences of mires in the catchment of lakes, and water color. Where the catchment has a high proportion of surface runoff due to thin or impermeable soils or swamp areas, high concentrations of DOC may occur in lakes. That means that the impact of wetlands may be more indirect, via specific hydrological conditions in bogs, mires, and wetlands.

From the Scandinavian studies, it can be concluded that in boreal regions, peatland and bogs appear to be the main source of TOC/DOC quantity in rivers and lakes. Furthermore, the percentage surface area that peatland and bogs constitute, is a more or less strong predictor for the TOC/DOC concentrations in adjacent surface waters. Does this relationship also apply to non-boreal climates? In a regional study of north eastern Germany (selected rivers and streams in Brandenburg State), Behrendt et al. (1999) also describe a linear regression relating DOC concentration to the percentages of wetlands in the catchments (Eq. 3.6):

$$[\text{DOC, mg/L C}] = 3.1 + 0.68\ [\%\ \text{wetland area}] \qquad (3.6)$$

A similar relationship is described for Adirondack lakes, New York, USA (Eq. 3.6) (Driscoll et al. 1994) with a steeper slope, meaning that the impact of wetlands is stronger in the Adirondack region (Eq. 3.7) than in north German lowlands:

$$[\text{DOC, mg/L C}] = 0.56 + 1.42\ [\%\ \text{wetland area}];\ r^2 = 0.88. \qquad (3.7)$$

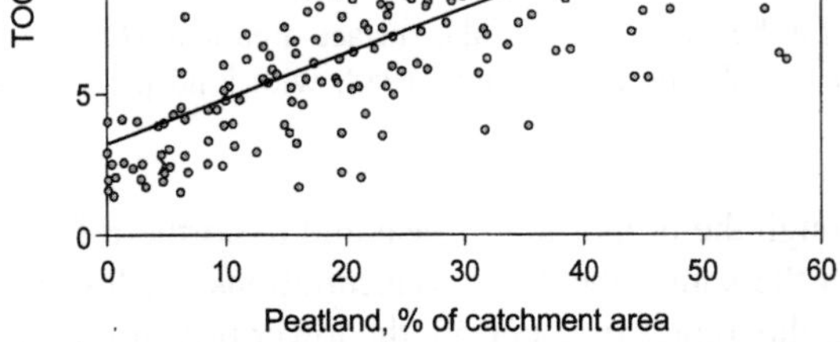

Fig. 3.5. Relationship between the proportion of peatland in the catchment area and the TOC concentrations in northern Finland (n=183) (from Kortelainen 1993, with kind permission of National Research Council Canada)

The equivalent equation (Eq. 3.8, see also Fig. 3.5) for the Finnish studies is for TOC, rather than DOC, and equals:

$$[TOC, mg/L] = 3.29 + 0.158\ [\%\ peatland];\ r^2=0.49. \quad (3.8)$$

The Finnish regression has the least steep slope, which indicates the coldest climate, the longest frost period, the lowest primary production, and the lowest decomposition rates of all described areas.

3.1.1.3 Molecular Weight

Another variable, believed to be of importance for geochemical and biological effects in freshwaters, is the average molecular weight of NOM and how it is determined by the catchment size. This relationship is only scarcely studied. From two Norwegian studies (Egeberg et al. 2002 and Gjessing et al. 1999), we can show, how the catchment size determines the mean molecular weight of NOM (Fig. 3.6). This Figure contains the same sampling location as Fig. 3.3, except that the River Topdal is omitted. In this graph, the molecular weights determined by diffusivimetry are plotted against catchment sizes. Egeberg et al. (2002) compare different methods for determination of molecular size of NOM so that the absolute amounts, but not the trend, may differ. Up to 90% of the variability in molecular weights can be explained by catchment sizes. This means that almost independent of the absolute DOC input, transport and microbial decomposition processes along the hydrologic pathways appear to play a definite role in determining the molecular weight and therefore also molecular size. The longer the time for the passage of the NOM into the lake or river, the greater the probability that molecules are reduced in size by hydraulic or microbial processes.

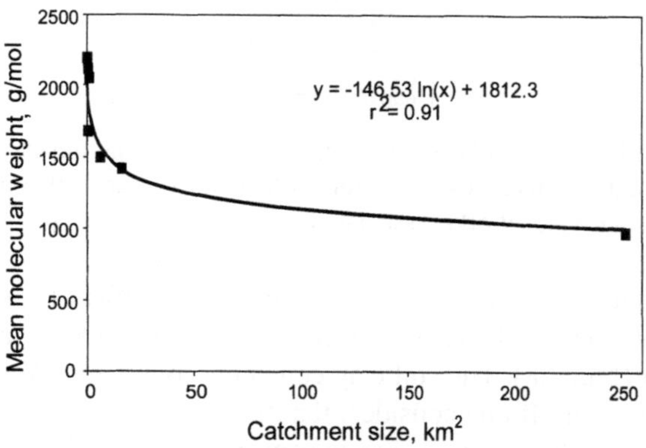

Fig. 3.6. Relationship between catchment size and mean molecular weight of NOM (natural organic matter) in Norwegian freshwaters as determined by diffusivimetry (combined from Egeberg et al. 2002 and Gjessing et al. 1999)

At present, there are no comparable studies available from other regions, or studies using other methods, so that it is unclear whether or not such a dependence of the molecular weight on the catchment size is a general phenomenon.

From the above, it is clear that most of the DOC in nearly all freshwater bodies comes from the catchment and not from the water body itself. The structure, vegetation, and microbial activity of the catchment, as well as the physical and chemical properties of the soil and the hydrological conditions, have a decisive influence on the quantity and quality of DOC in the water body. An illustrative example, how soil properties impact bulk variables such as riverine DOC concentrations, is the paper of Aiken and Cotsaris (1995). Two small South Australian catchments in the Lofty Mountain Range are studied, the Retreat valley (1.3 km^2) and the Lawless (3.0 km^2), catchments with similar land-use and precipitation, differing only in the clay content of the soil. It is obvious (Fig. 3.7) that these minerals can act as sinks for organic C: the higher the clay content, the higher the organic C storage capacity in the soil, and the greater the probability that the organic material is retained in the soil until degraded by microorganisms (Aiken and Cotsaris 1995).

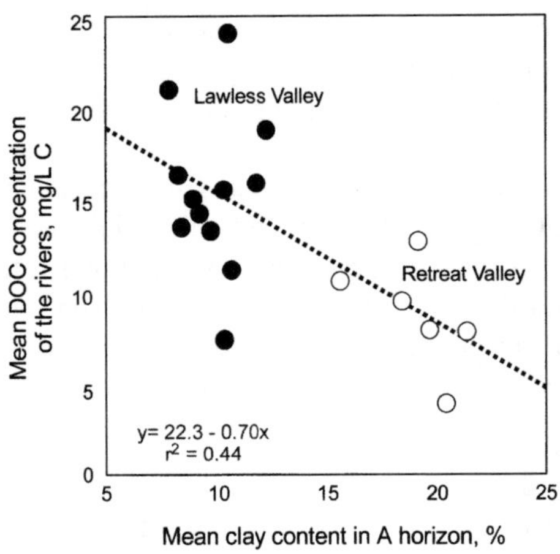

Fig. 3.7. Relationship between DOC concentrations in rivers and the clay content in soil A horizons in the Lofty Range, Australia (from Aiken and Cotsaris 1995, with kind permission of the American Water Works Association)

If one considers all the available data, there is a significant positive correlation between DOC content in the running waters of these two catchments and the clay mineral content (Fig. 3.8). The soil clay content accounts for 44% of the variation in DOC concentration. If one considers that the type of clay mineral is not further differentiated nor is the cation exchange capacity or proportion of cations considered, the bulk clay content is a comparably strong predictor for riverine DOC concentrations.

3.1.1.4 General Predictors of DOC Flux

In a more general approach, Aitkenhead and McDowell (2000) estimate the relationship between DOC flux and soil C:N ratio on a biome basis, to determine if this ratio can predict annual riverine DOC flux for local and global scales. The authors state that most of the studies on riverine DOC dynamics are small-scale and catchment-specific. Relatively few studies examine the factors that might be responsible for differences in DOC flux between catchment or biome types.

Aitkenhead and McDowell (2000) examine DOC fluxes of 164 rivers, subdivided into 15 biome types including tropical rain forest, coniferous forests, peatland, deciduous forests, mixed forests, and grasslands. At a global scale, mean soil C:N ratio of a biome accounts for 99.2% of the variance in annual riverine DOC flux among biomes. The regression is displayed in Fig. 3.8.

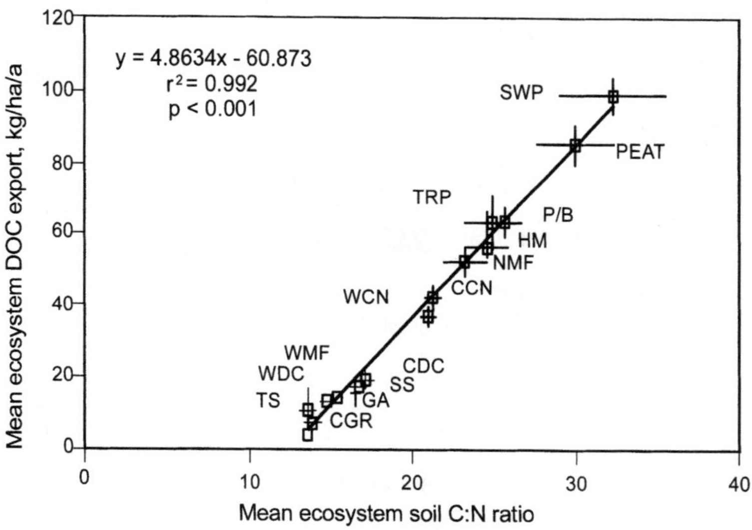

Fig. 3.8. The relationship between mean (± SE) ecosystem DOC export and mean (± SE) soil C:N for the 15 biome types used in model construction (after Aitkenhead and McDowell 2000, with kind permission of the American Geophysical Union). CCN cool conifer forests; CDC cool deciduous forests; CGR cool grassland; HM heath moorland; NMF northern mixed forests; P/B peat/boreal mix; PEAT peatland; SS Siberian steppe; SWP swamp forests. TGA taiga; TRP tropical forests; TS tropical savanna; WCD warm deciduous forests; WCN warm conifer forests; WMF warm mixed forests

A strong relationship is observed between soil solution DOC concentration and soil C:N across the nine coniferous sites included in the study (Fig. 3.9). Soil C:N ratios range from 11 to 55. Soil solution DOC concentration range from 23.9 in a Douglas fir forest in the Beaujolais Mountains of France to 105.0 mg/L in a red spruce forest in Howland, Maine, USA. Ninety-four percent of the variance in soil solution DOC concentration is explained by soil C:N ratios.

A high percentage of wetlands in a catchment can also lead to high DOC flux, and coniferous forests tend to have higher DOC flux than do deciduous forests (Aitkenhead and McDowell 2000 and references therein). Aitkenhead and McDowell (2000) show that catchment soil C:N is an effective integrator that incorporates the effects of all the other important variables, such as physical, chemical, biological, and hydrological variables, and is thus the best predictor of DOC flux.

The relationship between soil solution DOC concentration and soil C:N ratio (Fig. 3.9) suggests that riverine DOC flux may be much more intimately linked to biotically driven soil organic matter dynamics than previously suspected. For instance, for a hardwood forest, Aitkenhead and McDowell (in press) show that mean annual DOC and DON concentrations are positively related to fungal biomass ($r^2 = 0.84$, $r^2 = 0.62$, respectively), suggesting that fungal biomass (and activity?) may be responsible for DOC and DON production in the soil.

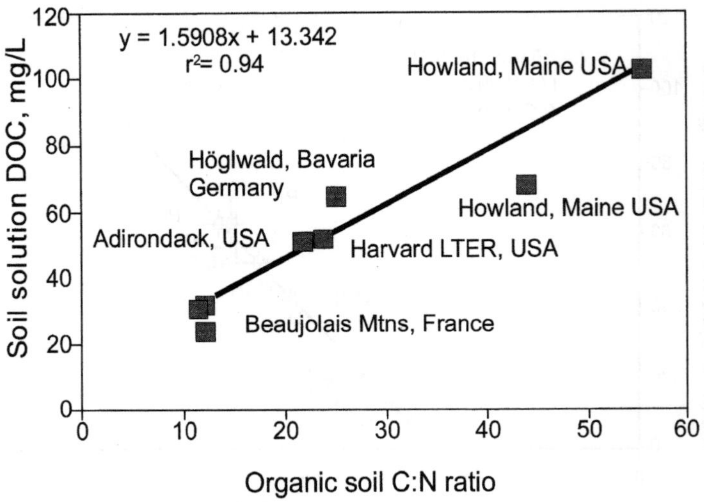

Fig. 3.9. The relationship between DOC concentration in soil solution from eight coniferous sites collected with zero tension lysimeters and soil C:N (from various authors, after Aitkenhead and McDowell 2000, with kind permission of the American Geophysical Union)

3.1.2. Hydrological Events

The transport of allochthonous organic C into lakes and streams largely depends on the hydrologic conditions in the catchment. For instance, Reemtsma et al. (1999) describe the dominance of hydrological processes for the specific case of wastewater infiltration sites (sewage farms). Organic matter is released in two phases, with an initial peak desorption of readily available organic matter, and a second phase of steady-state release that is governed by diffusion.

Under natural conditions, one would expect differences in DOC quality and quantity between base-flow and high-flow situations. The flow of water in the soil will determine the quality and quantity of DOC in the surface water. This is vividly described by Aiken and Cotsaris (1995). If the soil is water unsaturated, the precipitation can percolate through the top-most soil horizons and the unsaturated zones, finally entering the groundwater. Although the DOC concentration can be very high in the top-most soil horizons, during transport through the soil, the DOC is exposed to various processes such as adsorption and biodegradation. A classic example: during the summer growth period, DOC concentrations in the interstitial water of a podsol in the Adirondack Mountains (New York State, USA) vary between 21 and 32 mg/L in the O-A horizon, 5–7 mg/L in the B horizon, and 2–4 mg/L in the groundwater (Cronan and Aiken 1985). During base-flow periods, groundwater is the main source of DOC and thus determines riverine DOC quantity and quality.

Although the DOC concentrations in lower soil layers are comparatively low, it may form a source of riverine DOC. The adsorption of DOC in the B horizon (podzolization) results in net retention of DOC. Subsequently, mature podzols appear to be a net source of riverine DOC in several instances (Aitkenhead and McDowell 2000 and references therein). The functioning of podzols as source or sink may largely depend on the specific hydrological features of the particular catchment. Fractionation of NOM components occur during transport: small (<3 kDa) and more hydrophilic components of NOM are more mobile than are larger (3–100 kDa), more hydrophobic components (Dunnivant et al. 1992; McCarthy et al. 1993).

During high-flow events, the soil is saturated with water, and the water flow in the soil column becomes more horizontal (Cronan and Aiken 1985). Shallow subsurface and surface runoff is common. The organic material which is transported in the subsurface water, has a relatively short retention time and is not affected by the same mechanisms as water percolating during base-flow periods. This altered flowpath results in higher DOC concentrations and in changes in DOC composition.

In addition to heavy rain events, the onset of the rainy season and snow-melt can also cause high-flow conditions. During the onset of high-flow events, the DOC concentrations rise rapidly with the discharge, as is described for the Ogeechee River (Georgia, USA) by Aiken and Cotsaris (1995). The DOC concentration increase from 4 mg/L during a base-flow periods to around 10 mg/L during high-flow events. However, each event flushes soluble organic matter from the soils, and eventually, the DOC concentrations no longer respond to the increased discharge.

Various studies show further how the composition of HS changes with increases in discharge: during onset of high-flow conditions, the organic material released has a high UV absorption. Aromaticity is the first variable to increase with rising discharges. Although the DOC concentrations increase to approximately twice the background levels, the aromaticity tripled. The mentioned processes are summarized schematically in Fig. 3.10. However in strongly structured soils, rainstorm events may lead to an almost conservative vertical transfer and may, thus overcome the separation of hydrophilic and hydrophobic compounds (Chap. 3.1.3.1).

The input of DOC can be a stochastic occurrence that depends to a large degree on the timing and intensity of snowmelt and rainfall. For lakes with large drainage area to volume ratio, snowmelt would be expected to replace much of the water in the lake, and therefore the dissolved HS characteristics for the lake will be reset from the previous year and from processes occurring during winter, to those of the meltwater input (Gibson et al. 2001). This can result in significant variations in dissolved HS concentrations from year-to-year. For larger, deeper lakes in which a smaller percentage of the volume is replaced, there may be a legacy of the processes and conditions occurring in the lake during the previous summer and winter (Gibson et al. 2001).

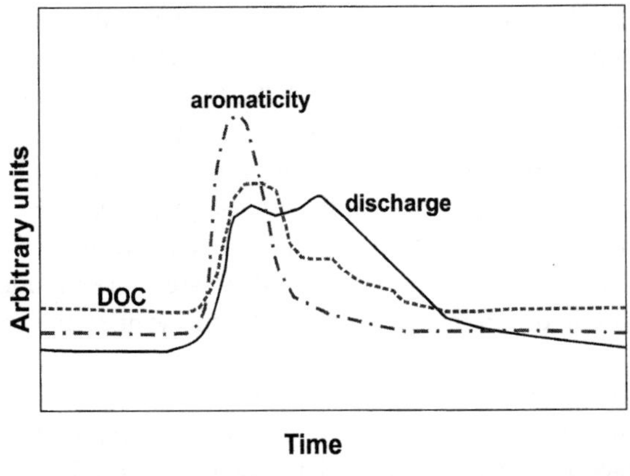

Fig. 3.10. Schematic display of temporal changes in DOC quality and quantity in a running-water during a flooding event (combined after several figures in Aiken and Cotsaris 1995, with kind permission of the American Water Works Association)

3.1.3 Influence on Humic Substance Quality

The influence of terrestrial vegetation on the chemical composition of HS is well documented. Terrestrial HS are characterized by a high lignin phenol content (Ertel et al. 1984; Rasyid et al. 1992; Wershaw et al. 1990). Even HS from the continental shelf possess a large quantity of lignin derivatives (Moran and Hodson 1994). In contrast, marine HS with no terrestrial influence are characterized by a lack of lignin derivatives, and are dominated by aliphatics rather than aromatics (Ertel and Hedges 1983; Harvey 1983; Harvey et al. 1983). Their aliphatic chemical nature confirms that they are derived from unsaturated lipids (Harvey et al. 1984).

The origin of HS from North Sea sediments is determined by Fooken and Liebezeit (2000) using a comparatively simple, elegant method, rather than expensive methods employing stable isotopes of C and nitrogen, or the use of lignin as a biogeochemical marker. The method studies the UV/VIS absorption at 270 and 407 nm. Marine samples are characterized by an absorption peak at 407 nm, terrestrial material is almost completely absent. The 270:407 nm absorption ratio or A2/A4 is used as an indicator of the origin of the material. Lignin has an A2/A4 of >6, whereas samples from purely marine locations have a value of <2. Studies on sediment core samples clearly show that early diagenesis has no influence on the A2/A4 ratio. Further studies must be carried out to determine if this easy method could be employed more generally.

The quality of DOC in freshwaters is also affected by catchment sizes. For instance, Engstrom (1987) states that the proportion of chromophoric dissolved organic carbon (cDOC) increases with an increased catchment area to lake area ratio for Labrador (Canada) lakes. The catchments of these lakes shared similar structures, one can infer that the larger the catchment, the higher the cDOC concentration in the associated lake. Further studies must determine if this statement is also valid for other regions or climates.

3.1.3.1 Major Processes in Soils and Forests Affecting DOC Quality

What happens to the DOC, in particular the quantity and quality of HS, in the various compartments of a woodland? Climate and soil drainage pathways play a decisive role for the DOC concentrations in the various soil horizons. However, there is no clear relationship between climate and the output of DOC from soil to water bodies.

In most studies on mobilization and transport of DOC in soils, only concentrations are considered. The DOC consists, however, of a highly heterogeneous mix of macro-molecules, polyeletrolytic materials, and a variety of small molecules. Precise statements on the dynamics, and ecological and ecochemical importance, of DOC in soils can only be made with knowledge of its structural composition. The functional composition, such as classes of different reactivity to surfaces, different acidity, and different bioavailability have to be elucidated. Frequently, the categorization of DOC is based on its charge and hydrophobicity in acids, bases and neutrals, according to Figs. 1.4, 1.6.

Hongve (1999, 2000) uses this relatively simple fractionation method (Fig. 1.6) to determine that in lithosol areas of a boreal climatic region, up to 50% of NOM in running waters may come from the canopy flow-through of the coniferous trees. The canopy flow-through of coniferous trees contains higher concentrations of hydrophilic and hydrophobic acids, than that from deciduous trees; (the latter has a higher proportion of neutral compounds). Furthermore, the runoff from coniferous forests is uniform, whereas deciduous forest soils deliver most DOC to water bodies in autumn. Therefore, it is very likely that the seasonality in stream water DOC concentrations is attributable to the deciduous litter (Hongve 1999), which imparts high DOC concentrations in the autumn, while coniferous litter and organic soil release DOC more evenly. Leaching from fresh deciduous litter may explain the seasonality in the concentration of DOC in discharge from forested catchments. Hongve (1999) shows that the impact of mires in the catchment of lakes on water color is an indirect one.

The above mentioned fractionation method (Fig. 1.6) is also applied by Guggenberger and Zech (1993a,b) who describe the DOC in rain water, in run off, and in the mineral soil solution of two spruce stands in the Fichtelgebirge (Bavaria, Germany). They obtained the following picture: hydrophilic neutrals (HiN) are mobilized in the tree canopy, and concentration decreases in the soil profile due to mineralization. For hydrophobic neutrals (HoN), the canopy is also a main source. Percolating through the soil, sorption of HoN on the organic material leads to a net immobilization in the runoff. Hydrophilic (HiA) and hydrophobic acids (HoA) are mainly released in the organic runoff. Both DOC fractions have a high affinity for mineral soils where the HoA is preferentially adsorbed. At low pH values in particular, the HiA dominates the mineral soil seepage.

The HoN fraction shows the closest relationship to plant litter, while the HiA and HiN fractions are strongly linked to microbial activity. The HiN fraction can even contain material which is produced by the microorganisms, thus both hydrophilic fractions exhibit a clear seasonal rhythm (Guggenberger and Zech 1993a). Even the HoA fraction is subject to changes due to microorganisms, but is comparatively refractory and strongly resembles soil FA.

Guggenberger et al. (1994) conclude that microbial activity is the most important factor controlling the mobilization of DOC in spruce forest soils. The HoA isolated from the forest soils are primary products of the decomposition of plant material or from complex microbial metabolites, and not the result of re-synthesis and polymerization of decomposition products.

Guggenberger et al. (1994) therefore bring into question whether or not the hydrophobic acids per se are already HS. They conclude that the hydrophobic acids are:

- **intermediates** in the decomposition of organic materials, which can be further degraded to hydrophilic acids (HiA) or mineralized to CO_2;
- nevertheless, still **precursors** of HS after precipitation on, or adsorption to, sesquioxides.

Guggenberger and Zech (1993a, 1994) conclude that in acidic forest soils, there are two distinct compartments which control the ecochemical fate of DOC. These are the biologically active forest soils from which the organic C is released into the soil water solution, and the storage B horizon in which the DOC is bound. The quantity and quality of DOC which is released in the top horizons is generally controlled by microbial activity, which is itself a function of environmental conditions such as temperature, nutrient supply and type of leaf litter.

The removal of DOC and changes in DOC structural relationships in the B horizon are controlled by abiotic processes which take place independently of seasonal variation in DOC inputs. The formation of insoluble DOC-metal complexes and the sorption of DOC by Fe and Al oxides, reduce the DOC concentration in the percolating water during podzolization. However, when matured, podzols may eventually become a source of groundwater and riverine DOC (Aitkenhead and McDowell 2000).

Soil development processes influence the DOC composition; for example, increasing saturation of sesquioxides by percolating DOC within the developing Bh horizon, leads to a reduced sorption capacity in this horizon and to sorption in deeper layers such as the Bs horizon. Guggenberger and Zech (1994) expect that the atmospheric deposition of sulfuric and nitric acids, which lead to acidification and podsolisation, enhance the passage of DOC from the soil to water bodies.

From his various studies, Guggenberger considers sorption on sesquioxides as the most important mechanism of DOC retention in soils rather than the changes in DOC solubility and reduction in microbial activity, and subsequent release of DOC from the solid phase. Guggenberger considers only relatively strong acidic conditions (pH values around 4). However, this statement is in contradiction to the limnological study of Lake Großer Arbersee below, carried out not far away from Guggenberger and Zech's study site. In contrast to the narrow pH range of Guggenberger's study, the paleolimnological pH reconstruction of the Holocene offers a comparison over a comparably wide pH range, from the circum-neutral and weakly acid conditions found in the late and early post-glacial periods, to the relatively strongly acid conditions due to anthropogenic acidification. It is likely that under strongly acidic conditions, the DOC release mechanisms will reverse in the mid-term as Guggenberger and Zech (1994) suggested, and that this process will be reflected in sediments of Lake Großer Arbersee via decreasing TOC concentrations (see Chap. 3.2.2.2, Fig. 3.30).

In a follow-up study, Kaiser et al. (2001a, 2002) report on the seasonal changes of DOM in soil solution under Scots pine and European beech. At the pine site, DOM released during winter and spring contains larger proportions of H associated with O-containing structures, and of low-molecular weight compounds. During summer and autumn, the contribution of O-containing structures and low-molecular weight compounds decline, and aromatic and aliphatic compounds of high-molecular weight dominate. Forest floor leachates under beech show a similar trend, but a larger number of low-molecular weight compounds is present. In spring, summer, and autumn, acetate and sometime succinate, are most prominent.

During the passage of DOM through the upper mineral soil at the pine site, the low-molecular weight compounds increase. These substances are thus more mobile than the macromolecules and/or are released due to microbial transformation of organic matter. At the beech site, subsoil porewater contains less low-molecular weight compounds than the forest floor solutions (Kaiser et al. 2002). Dissolved organic matter in winter and spring seems to be mainly controlled by leaching of fresh disrupted biomass debris with a large contribution of bacterial and fungal-derived carbohydrates and amino sugars. Dissolved organic matter leached from the forest floor in summer and autumn is controlled by decomposition processes in the forest floor resulting in the production of strongly oxidized, water-soluble aromatic and aliphatic compounds (Kaiser et al. 2001a). The chemical composition of DOM in forest floor seepage water in winter and spring indicates larger mobility, larger biodegradability, and less interaction with metals and organic pollutants than that released during summer and autumn. Thus, the impact of DOM on transport processes varies throughout the year due to changes in its composition (Kaiser et al. 2001a).

However, not all changes in the DOC are attributable to microbial activity. For instance, Kaiser et al. (2001b) show that isotopic fractionation of DOC percolating through the soil, that is often interpreted as due to microbial transformation, can be caused by selective sorption and desorption of organic C fractions to soil particles. That means that abiotic processes can mimic metabolic transformation and decomposition.

Organic forest floor layers are large potential sources for organic C and organically bound nutrients such as DON, dissolved organic phosphorus (DOP), and dissolved organic sulfur (DOS) (Kaiser et al. 2001c). The dissolved organic nutrients are mainly concentrated in the hydrophilic DOM fraction which is more mobile in mineral soil pore water than the hydrophobic fraction. Consequently, the concentrations and fluxes of dissolved organic nutrients decrease less with depth than those of DOC. Once released, DOC and organically bound nutrients are transferred into the underlying mineral soil with the percolating water. In strongly structured soils, rapid macropore flow induced by heavy rainstorms, leads to an almost conservative vertical transfer of hydrophilic and hydrophobic DOM due to a lack of interaction with mineral surfaces. Under these conditions, the distribution of organically bound nutrients between the hydrophilic and the hydrophobic fraction is no longer fundamental for the export from soil (Kaiser et al. 2001c).

3.1.3.2 Influence of Land-use on Humic Substance Quality

Land-use among other factors, changes the oxidative and hydrological regimes of soils. These changes must also be reflected in the qualitative composition of HS, an effect that is more conspicuous the higher the organic C content of the soil. Peatland soils are therefore the clearest indicators of DOC/HS changes resulting from landscape management. A few illustrative examples will follow.

Zsolnay and Görlitz (1994) report the effect of fertilization on the water extractable organic matter. Inorganic fertilization has no effect on this NOM fraction. Organic fertilization has a roughly linear effect. Its use results in increasing amount of water extractable organic matter in the system. This effect, however, is blurred presumably by transport and metabolism.

Kalbitz (2001) identifies differences in the composition of DOM in soils dependent on both land-use and soil depth. The DOC concentration is affected by land-use in topsoils only. Degraded peatlands have a high intensity of land-use and therefore a low soil C content and low DOC concentrations. The DOC concentrations decrease significantly with increasing soil depth. DOC retention through adsorption or decomposition is much higher at undisturbed peatlands than at degraded peatlands. In strongly degraded peatland soils, FA from the surface and sub-soils, as well as surface waters, possess lower C/N ratios, higher UV absorption, and higher absorption at 1620/cm in the FTIR spectrum, than FA from relatively undisturbed peatland soils. These former qualities indicate increased aromaticity and ongoing oxidative humification. Kalbitz et al. (2000a) quantify the land-use related humus degradation by means of the 'humification index' (Kalbitz et al. 1999; Kalbitz and Geyer 2001), which is the quotient of the relative fluorescence intensity at 470:360 nm.

It is commonly assumed for both soils and aquatic ecosystems, that DOC and DON are coupled to each other. In their review, Kalbitz et al. (2000b) conclude from initial evidence that the rate of release, and the fate of DOC, DON, and DOP in soils may differ to a far greater extent than previously assumed. Controls established for DOC, might thus not be valid for DON (and DOP) – the prevailing mechanism still being obscure. In a new study, Kalbitz et al. (2000c) question whether peatland degradation is reflected in different DOC and DON signatures, and whether differing land-use which effects humification, also effects the dynamics of DOC and DON. The results can be summarized as follows (Fig. 3.11):

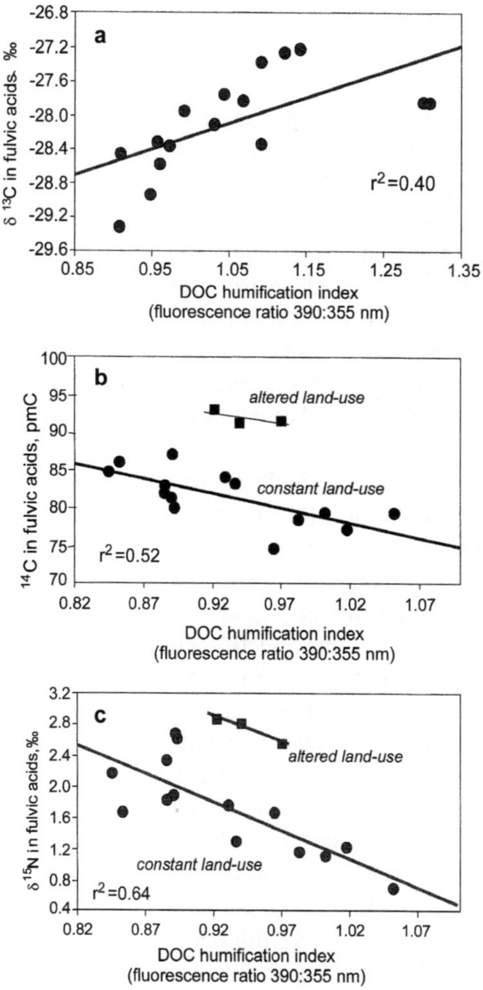

Fig. 3.11. Humification of soil fulvic acids as determinant of $\delta^{13}C$ (a), ^{14}C (b), and $\delta^{15}N$ (c) of soil organic matter (from Kalbitz et al. 2000c, with kind permission of the International Humic Substances Society)

1. enhanced humification causes an enrichment of stable heavy C-isotopes (Fig. 3.11a), because microorganisms prefer to use light isotopes (Kracht and Gleixner 2000);
2. peat decomposition in peatland soils following intensive long-term use leads to the release of old (^{14}C poor) DOC with a high humification index (Fig. 3.11b).

The sharp increase in ^{14}C concentration after land-use change shows a high C turnover and the release of young material;

3. changes in land-use also lead to enhanced nitrogen turnover, recognizable in the increase in heavy N isotopes in groundwater FA. It is interesting however, that FA with a high humification index exhibit **low** $\delta^{15}N$ values (Fig. 3.11c). This means that increasing humification does not automatically parallel an increased microbial modification of dissolved organic nitrogen components. In the humification processes – at least in this example, the C and N metabolism are uncoupled.

In a more recent study, Kalbitz and Geyer (2002) confirm that peat degradation in soils leads to lower DOC/DON ratios than the C/N ratio of the solid phase, indicating a preferential release of DON from soil organic matter. Furthermore, $\delta^{13}C$ ratios and the radiocarbon age of DOM increase with peat degradation and humification, indicating a high C turnover, an increased microbial modification and age of DOC. In contrast, $\delta^{15}N$ ratios decrease, probably as a result of N fertilization. Kalbitz and Geyer (2002) suggest a promoting effect of inorganic N on DON release, and a high humification of DOM at sites treated with inorganic N fertilizers. As a result, N fertilization promotes a release of amino acids depleted in ^{15}N and subsequent condensation with carbohydrates to HS (melanoidic pathway of the humification, Chap. 2.2).

FA which have percolated agricultural soils have a higher aromatic content than those from forest soil runoff (Geyer et al. 1996). The carbohydrate content is highly variable. In the FA from arable soils, the content of both aromatics and carbohydrates is reduced, compared to the above described peatland soils, indicating poorer aeration or a different degradation pathway.

Using a comparatively simple electrofocussing technique, Benedetti et al. (1994) show clear differences in HS from agricultural and forest soils. The agricultural soils have a small band at around pH 3.7, and a broad band between pH 4.3 and 4.8. Focusing profiles from forest soils are much more complex, possessing more bands, thus indicating different HS fractions than the agricultural soils. The low diversity of HS from agricultural soils can in part be explained by the specific land-use. These HS also have a low humification index. Application of fertilizers itself has a clear influence on the building blocks of the HS, for example, by increasing the proportion of carboxyl groups in the organic material (Ellerbrock et al. 1997), and impoverishing the aromatic structural units in the HS (Masciandaro et al. 1997). Plowing and conventional cultivation further increase the oxidation of the HS, which leads to fewer aromatic compounds, but more acidic functional groups (Arlauskas and Slepetiene 1997; Madari et al. 1997).

Considering organic-mineral interactions in soils, Schulten and Leinweber (2000) describe the effects of land-use on the distribution of HS in the soil profile. They consider the effect of soil texture on the proportions of functional groups in the HS. Based on the simplified concept that, depending on the specific hydrological conditions, the organic-mineral complex can act as a source or sink for dissolved organic C, one obtains an idea how land-use and soil texture can effect HS quality. Due to the clay mineral content in soils, a reciprocal relationship to

DOC concentrations in streams and rivers (Fig. 3.9) is to be predicted:

- the stronger the soil adsorption, the less DOC will reach the adjacent water. This results in a fractionation of mobile DOC;
- the higher the proportion of aromatics absorbed to the solid phase of the soil, the higher the relative aliphatic proportion in percolating DOC which finally reaches the surface water.

The organic C content of clay and silt in two soils, chernozems and vertisols is shown in Fig. 3.12. Relatively high organic C contents, 3–12% at the soil surface, are characteristic of chernozems, and these soils remain relatively rich in organic C to a depth of 1 m. The proportion of organic C in the clay fraction increases with depth (dark area under abscissa), indicating that the sorption sites increase with depth, while the organic material from root exudates or metabolites of microbial activity decrease (Schulten and Leinweber 2000).

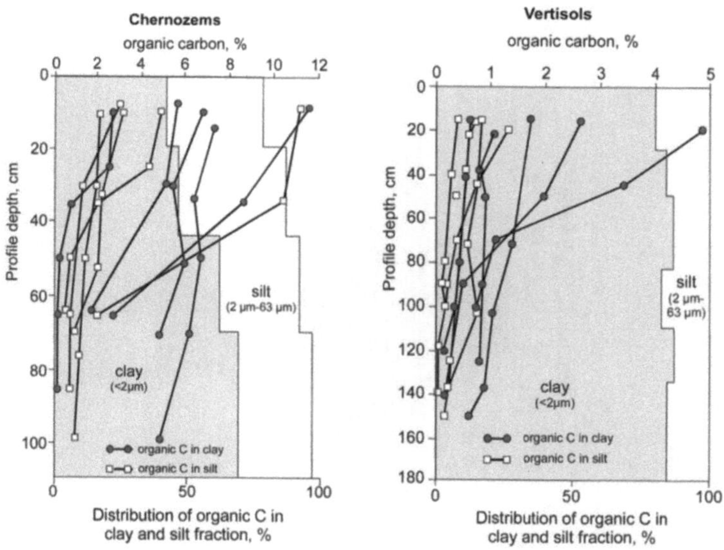

Fig. 3.12. Depth-profile of organic carbon in clay and silt particles in two differently organic-rich soils, a chernozem[2] and a vertisol[3] (after Schulten and Leinweber 2000, with kind permission of Springer Verlag)

In vertisols, the content of organic C in the clay and silt fractions is conspicuously low. These soils therefore contrast strongly with the chernozems. In vertisols, organic C also decreases with depth although the gradients are less steep than in other soils. The clay fraction contains the largest proportion (70–91%) of the

[2] Chernozems are deep, dark-colored soils rich in organics, calcareous in lower horizons, typical of grasslands and prairies.

[3] Vertisols are clay-rich soils that shrink and swell with changes in moisture content. They occur on every continent except Antarctica under climates that have a seasonal dry period.

organic C since this fraction itself comprises most of the soil.

Changes in land-use through, for instance, additional inputs of primary organic material or permanent plant cover, or by aeration, lead to measurable changes in the organic C content of organic-mineral fractions. Thus, significantly larger organic C contents are found in particle-size fractions from native soils than in the corresponding fractions of arable soils. Accordingly, fractions from manured soils had larger organic content than those from unfertilized soils and soils which receive only mineral fertilizers (Schulten and Leinweber 2000). Fig. 3.13 shows the distribution of the organic C in clay and silt fractions. The data pairs are arranged according to increasing clay contents from left to right. In general, the proportion of organic C in the clay increases and the proportion in silt decreases with increasing soil clay content, as can be seen for tilled and unmanured soils (upper half) as well as for native and manured soils.

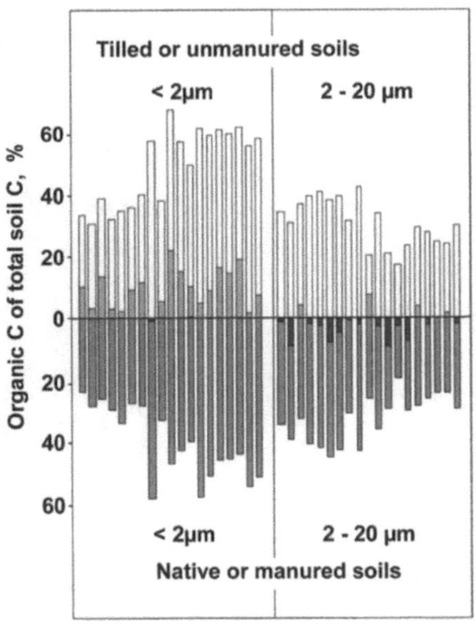

Fig. 3.13. Influence of different soil management on the distribution of whole soil organic carbon in organic-mineral particle-size fractions. The columns in the upper part of the diagram show the data for tilled or unmanured soils, and in the lower part for their native or manured counterparts. The management-induced differences are highlighted by darker signatures (from Schulten and Leinweber 2000, with kind permission of Springer Verlag)

This leads to the general rule that intensive land-use of soil results in an increase in the organic C content of the clay fraction, and at the same time a decrease in the organic C of the silt and sand fractions (the latter not shown in the Figure). This can, on the one hand, be explained by the binding of humified materials and microbial metabolites on the clay minerals and oxides, and the resistance of these complexes to mineralization and biodegradation. On the other hand, in conditions of increased inputs of organic materials from undecomposed or partly decomposed plant materials, an enrichment of organic C in the silt and sand fractions occurs (Schulten and Leinweber 2000).

Particle size has a definite influence on the composition of HS as shown in Fig. 3.14. Decreasing proportions of alkyl C, and increasing proportions of O-alkyl C and aromatic C occur with increasing particle size. Fractions >10 µm contain more O-alkyl C than does clay. The carboxyl groups show no dependence on particle size (Schulten and Leinweber 2000). Increasing retention of aromatic compounds has clear consequences for the reactivity of soils, and subsequently, riverine and lacustrine carbon input, since the aromatic fraction is responsible for the binding of lipophilic neutral chemicals (natural attenuation), for the absorption of electromagnetic radiation (UV and visible) at the soil surface and in waterbodies, and consequently the initiation of a sequence of photochemical and photomineralization processes (Box 1.1).

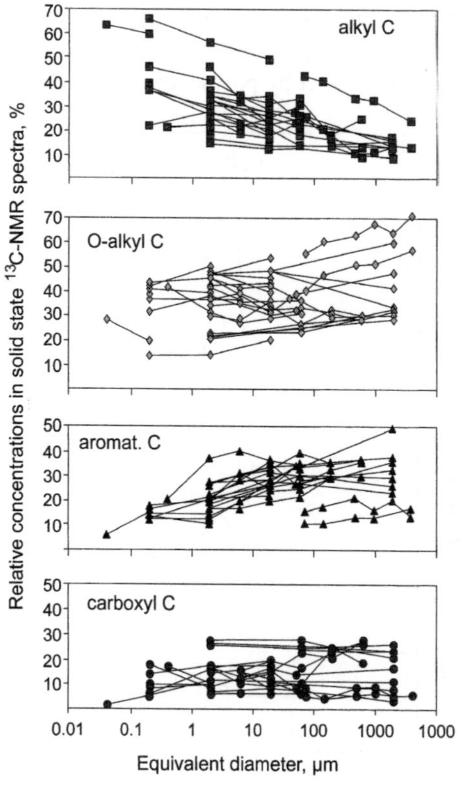

Fig. 3.14. Influence of soil particle size on the composition of HS in the organic-mineral phase (after Schulten and Leinweber 2000, with kind permission of Springer Verlag)

Fingerprint Studies

All dissolved substances in water can easily be separated by molecular size with size-exclusion (SEC)-chromatography and double detection (UV and DOC detection). Large molecules are eluted first from the column and small molecules last. In other words, the longer the retention time, the smaller the molecule. With a few DOC sources, some substances occur after the second exclusion boundary (for instance peak at 63 min retention time in the fish pond sample, Fig. 3.15). These substances interact with the matrix, and show that adsorption onto the matrix plays a role in separation, in addition to the molecular size of the DOC. With the SEC and double detection method, one obtains the following four organic component fractions: polysaccharides (no UV signal, detection only with DOC), HS (signals with both detectors), low-molecular weight fatty acids (signals with both detectors), and small molecules such as amino acids and oligopeptides (no UV signal, detection only with DOC). Characteristic separations of DOC from different sources are shown in Fig. 3.15. It is immediately apparent that the elution patterns vary quantitatively in the individual fractions. The various samples provide characteristic fingerprints: the marshland and drainage samples are clearly dominated by HS, while the sample from a sewage effluent is dominated by small fatty acids, and that from the lake outlet by polysaccharides.

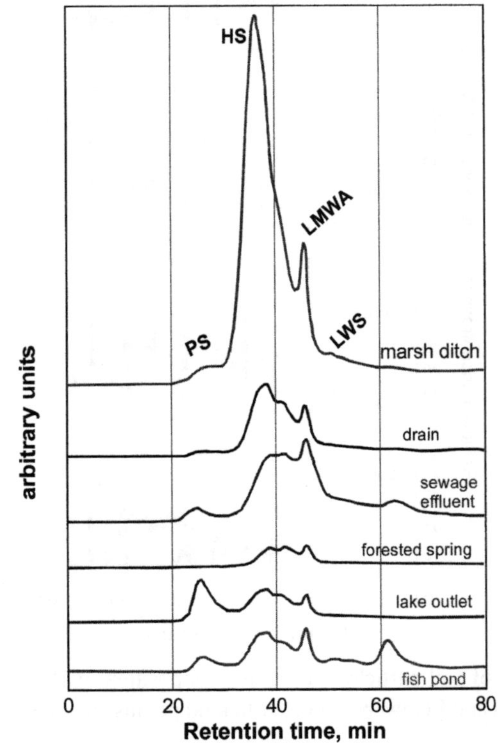

Fig. 3.15. DOC fingerprints of various origins with size-exclusion chromatography (SEC) and double detection (UV and DOC) (after Sachse et al. submitted) PS: polysaccharide; HS: humic substances; LMWA: low-molecular weight acids; LWS: low-molecular weight substances, such as amino acids as well as mono- and disaccharides

Provided that a representative database is available, multivariate statistical analysis of the DOC fingerprints can be carried out, for example by principle component analysis. The DOC fingerprints can then be further differentiated. In this way, the wetland samples, particularly those of acidic, oligotrophic peatland are separated from those from surface waters by higher molecular weight, high aromaticity, and a higher absolute DOC content (Fig. 3.16). In contrast, DOC samples from springs are characterized by low-molecular weight, low aromaticity, low absolute DOC concentrations, and low concentrations of small molecules.

Depending on the structure of the catchment, river samples (Demnitz Brook, small tributary to the River Spree) tend to be similar to the wetland samples, while other river samples (River Spree) are more similar to lake samples. Since the River Spree is flowing through a cascade of lakes, it is not surprising that its DOC resembles lake water DOC. It is conspicuous that the polysaccharide content is not clearly differentiated by the fingerprints. According to reports on humification from the composting process, young HS-like material is characterized by a high proportion of carbohydrates (Chefetz et al. 1998). It would be expected that in waters with significant autochthonous production, such as the lakes and slowly flowing river (Schlaube system in eastern Brandenburg, Germany), at least a seasonal dominance of polysaccharides would occur (as in the lake outlet sample in Fig. 3.15). For lake samples this is apparently more the exception than the rule. The reason for this could be a comparatively rapid microbial utilization of some DOC fractions.

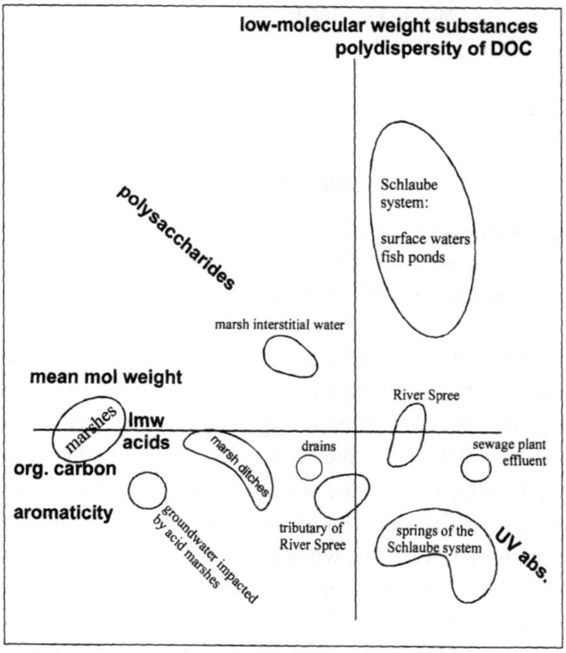

Fig. 3.16. Principal component analysis of DOC fingerprints of various origins (modified from Sachse et al. submitted)

Sachse et al. (2001b) report on the fingerprint characterization of dissolved HS in a small, artificially divided forest lake (Lake Große Fuchskuhle, Brandenburg State, Germany). Part of the catchment area is an acidic oligotrophic bog which drains into one section of the lake (Hehmann and Krienitz 1996; Šimek et al. 1998), the lake comprises regions with and without impact from this bog. The HS are differentiated according to molecular weight and absorption spectra. The molecular weight of samples from the acidic bog and its seepage water are in the range 4.0–7.5 kDa, with most between 5.5 and 6.0 kDa. The molecular weight of lake samples is in a lower and narrower range; 3.5–4.0 kDa. There is no clear difference between the lake part influenced by, or independent, of the acidic bog.

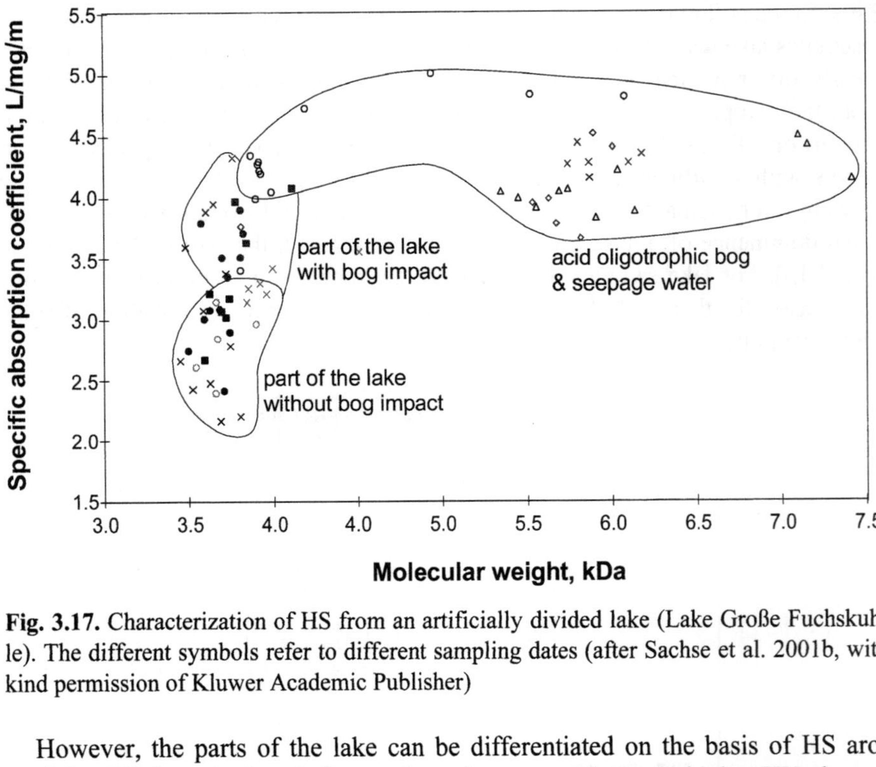

Fig. 3.17. Characterization of HS from an artificially divided lake (Lake Große Fuchskuhle). The different symbols refer to different sampling dates (after Sachse et al. 2001b, with kind permission of Kluwer Academic Publisher)

However, the parts of the lake can be differentiated on the basis of HS aromaticity. HS from the bog-influenced part have conspicuously higher UV-absorption than do HS from the control part (Fig. 3.17) (a measure of aromaticity of the DOC, Haitzer et al. 1999b). Furthermore, the HS from the bog-influenced lake part are not only smaller than in the wetland itself, but also possess clearly lower spectral absorption coefficients. One can explain these differences as follows: the high-molecular weight HS of the wetland are cleaved chemically or microbially en route to the lake or in the lake itself. Chemical cleavage by photolysis may occur if the HS are exposed to light. Photolysis can result in the loss of aromaticity (Bertilsson and Tranvik 2000; Donahue et al. 1998; Welker and Steinberg 2000). The HS in the lake parts not influenced by the bog have even lower aromaticity than those in the bog-impacted part.

If one looks again at the results of Kracht and Gleixner (2000) for Lake Hohlohsee (Black Forest, Germany) which contradict the accepted paradigms of humification (Chap. 2), then in bog-influenced waters, autochthonous humification pathways are of great importance. This metabolic route can also account for the low HS aromaticity in Lake Große Fuchskuhle.

In another paper, Sachse et al. (submitted) try to determine whether the above approach applied to Lake Große Fuchskuhle and its catchment is applicable even to whole landscapes, their land-use characteristics, and DOC export (Fig. 3.18). High-molecular weight and high aromaticity of the HA fraction are typical features of peat-influenced waters. Agricultural drainages are very low in labile DOC and are dominated by recalcitrant material as HS, due to biodegradation during soil passage. Interestingly, the HS characteristics (molecular weight and aromaticity) of the agricultural drainages are not in the range of peat-influenced HS, but are similar to aquagenic HS of lakes (Fig. 3.18), a result that agrees well with findings of Imai et al. (2001). Probably, that HS portion with higher aromaticity and molecular weights is specifically adsorbed to soil particles (Schulten and Leinweber 2000). In forest springs, a minor source of DOC, HS are high in aromaticity, which can be attributed to the outwash of pedogenic HS for the upper soil horizons, which are rich in HS originating from decomposed plant and leaf litter (Ertel et al. 1983). A special characteristic of sewage plant effluent appears to be that the molecular weight and the aromaticity of its HS (about 3.0 kDa; 2.5 L/mg/m)) are the lowest of all samples, as reported by Huber et al. (1994, 1996) and Hesse et al. (1997) for several anthropogenic point sources into German rivers (sewage plant effluents). Low aromaticity may be explained by associations of HS with partially degraded carbohydrates or proteins.

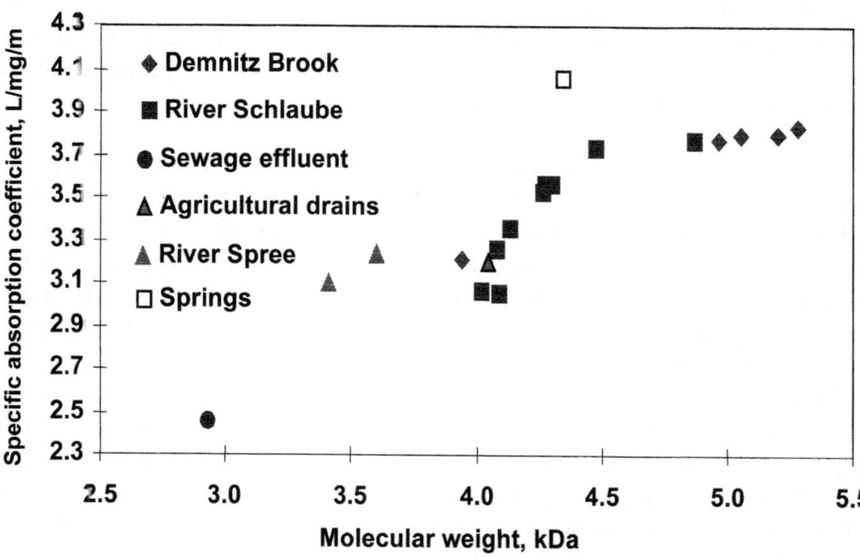

Fig. 3.18. Characterization of HS by peak molecular weights and specific absorption coefficients (after Sachse et al. submitted)

3.1.3.3 Humic Substances in Groundwater

Prior to the development of accelerated mass spectroscopy (AMS), studies on groundwater HS were expensive because water samples of more than 1000 L were usually required, and thus relatively few studies were conducted. By using less expensive AMS for determination of C isotopes, it is possible to get evidence of the origin, age, and evolution of HS in groundwater aquifers. Using AMS, Geyer et al. (1992) report that groundwater HS from around Gorleben district (Lower Saxony, Germany) probably derive from FA percolating through the soil and into the aquifer.

An overview of the most important components and common composition of groundwater with a low (<5 mg/L C) DOC content is shown in Fig. 3.19. In these waters, the hydrophilic organic acids (HiA) comprise the largest fraction in so-called 'black water' aquifers.

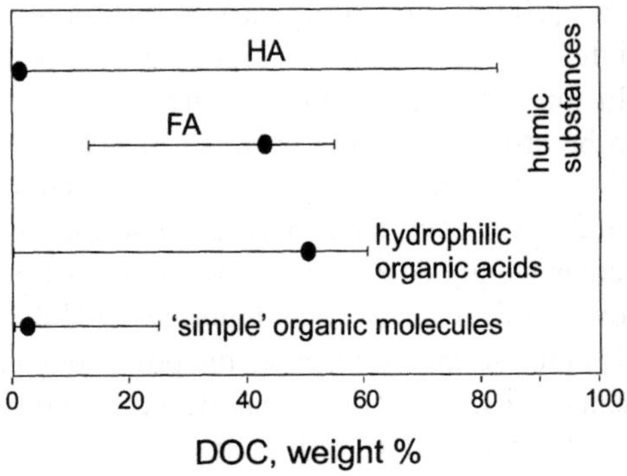

Fig. 3.19. Percentage of the most important DOC fractions in a groundwater with low DOC concentrations (black dots) and the possible variations in groundwater with high DOC concentrations (lines) (from various authors, after Geyer 1994, with kind permission of the GSF Research Center of Environment and Health)

A predominance of HiA is corroborated by Malcolm (1991) (Fig. 3.20c). The natural DOC content of groundwater is determined by the processes of formation and inputs, and decomposition and adsorption on aquifer sediments. The two most important groundwater DOC sources are percolating water, which passes from the soil surface through the unsaturated soil layers, and inputs from the aquifer itself. Inputs from flowing waters or surface water bodies are only of local importance according to Geyer (1994).

The DOC content of percolating water varies seasonally. The longer the dry period between two precipitation events, the higher the DOC concentrations in water passing through the A and B soil horizons, particularly at the beginning of a rain storm or with snow melt early in the year. The DOC can then be rapidly transported via the macropores in the soil profile to the C horizon. Studies on the degradability of DOC in the unsaturated zone of forest soils show that the strong decrease in percolating water DOC cannot be explained by biological decomposition alone; strong adsorption to Fe and Al oxides, as well as clay minerals in A and B horizon soil particles, must also be considered (Qualls and Haines 1992a; Tipping 1981). Qualls and Haines (1992b) suggest a two phase process in which most DOC is initially adsorbed and concentrated in the soil, and then subsequently hydrolyzed and biodegraded by exoenzymes from bacteria and fungi.

Another source for groundwater DOC is the organic material in the aquifer itself. The proportion of organic C generally increases with the size of sediment particles in the aquifer. The aquifer organic C enters the groundwater pool through dissolution and microbial activity. The dissolution of DOC is influenced by aquifer sediment surface conditions, retention time, and the chemical composition and temperature of the groundwater.

In a previous chapter, the supposedly recalcitrant, inert, nature of HS is mentioned as an indication of the age of this material in groundwater (Chap. 3.1.3.3). The HS can indeed be old. Geyer (1994) determines the age of various groundwaters from the FA isotope content. He describes an age of 20–30 years for the sub-surface groundwater in Fuhrberger Feld (Lüneburg Heath, the source of drinking water for Hannover, Germany). For a deeper aquifer, Ivanovich et al. (1996) determine an age of 5,000–25,000 years for FA. This colossal age apparently confirms the low reactivity of the HS, known from several soil studies (Chap. 1).

3.2 Dissolved Humic Substances and the Acid Status of Freshwaters with Low Acid Neutralizing Capacities

Operationally, DOC is composed of: hydrophobic acids (HoA), hydrophobic neutral substances (HoN), hydrophilic acids (HiA), hydrophilic bases (HiB), and hydrophilic neutral substances. The hydrophobic compounds are considered to be HS, which is only a part of the DOC. In non-colored freshwaters, FA and HA comprise about 40 or 45% of the DOC (Figs. 3.20a,b). The pie charts show that the DOC composition of samples of surface waters from the USA and Finland differ little from each other. In contrast, the composition of surface DOC is clearly different from that of groundwater; the latter is dominated by hydrophilic compounds (HiA, HiN, and HiB; Fig. 3.20c).

On the basis of fractionation of the various DOC components of waters from Finland and Maine (USA) which are very similar, regression equations can be constructed from which the proportion of hydrophobic and hydrophilic acids can be determined from the particular DOC concentrations (Mattsson et al. 1998). For the hydrophobic acids (Eq. 3.9):

$$\mathrm{HoA} = 0.371 + 0.0133\,[\mathrm{DOC}];\ r^2 = 0.53;\ P < 0.001;\ n = 46 \tag{3.9}$$

and for the hydrophilic acids (Eq. 3.10):

$$\mathrm{HiA} = 0.34 - 0.00303\,[\mathrm{DOC}];\ r^2 = 0.15;\ P < 0.01;\ n = 47. \tag{3.10}$$

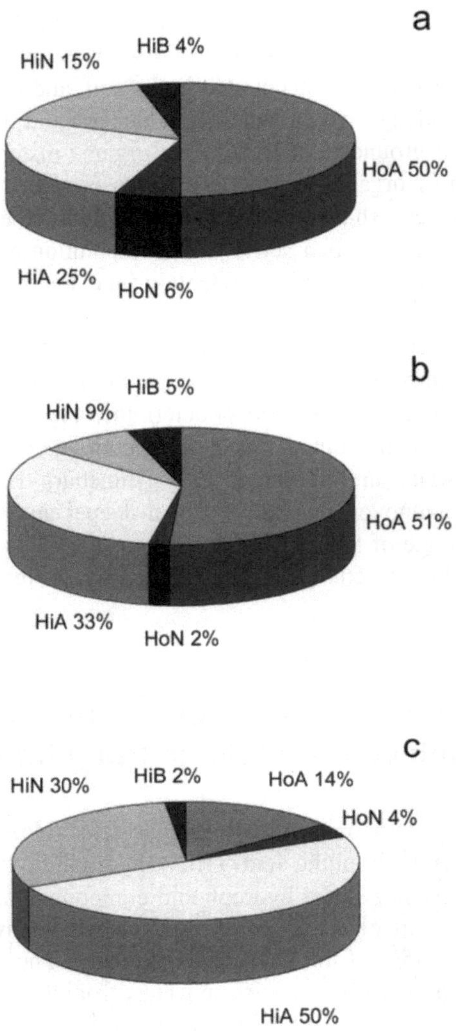

Fig. 3.20. DOC fractions in approximately 100 surface waters in the USA (**a**), in 16 Finnish lakes (**b**), and 25 groundwater samples from the USA (**c**). HiA: hydrophilic acids; HiN: hydrophilic neutrals; HiB: hydrophilic bases; HoA: hydrophobic acids; HoN: hydrophobic neutrals (from Peuravuori and Pihlaja 1999b, with kind permission of Backhuys Publishers)

The ratio of FA to HA is usually 9:1, but it may be as low as 4:1 (Peuravuori 1992). In humic-rich waters of boreal regions (Scandinavia, north Russia, Canada, Alaska), the proportion of HS (hydrophobic fraction) usually comprise a higher proportion of DOC (60–80%) than in humic-poor waters (Roila et al. 1994a). It may be assumed that the relatively small proportion of hydrophobic acids in humic-poor lakes is due to selective adsorption in soils and autochthonous production of hydrophilic compounds.

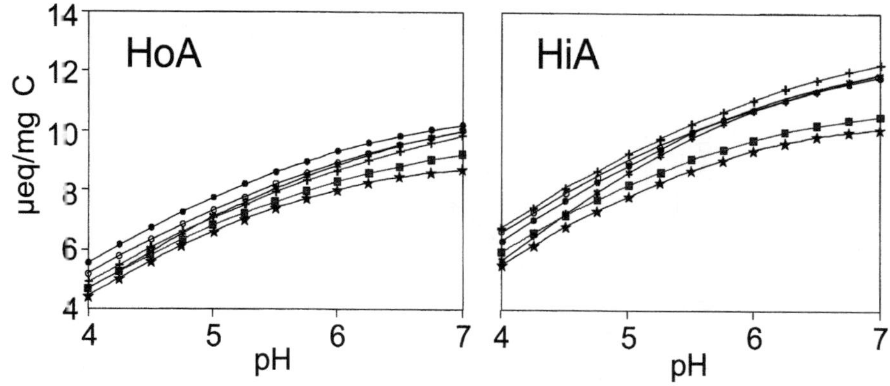

Fig. 3.21. Charge equivalents (per mg C) for hydrophobic acids (HoA) and hydrophilic acids (HiA) in six samples as a function of pH (after Roila et al. 1994a, with kind permission of Elsevier Science)

The acidity of hydrophobic and hydrophilic acids differ from each other. An example for six Finnish lakes is shown in Fig. 3.21. The titration curves of these two types of acids run close together, but in every lake, the hydrophilic acid fraction has a higher exchangeable acidity. For the hydrophobic acids, mean acidity is 10.4 µeq/mg C and for the hydrophilic acids, 12.1 µeq/mg C. Roila et al. (1994a) calculate an average charge density of 9.7 µeq/mg C. The dissociation of the organic acids increases with increasing pH from around 48% at pH 4 to around 92% at pH 7. From this data, a mean pK_S of 4.0 may be acquired (Kortelainen 1999b).

3.2.1 Acuto-Limnological Studies

Free acidity of freshwaters is the result of complex factors and many interactions, for which the process involved are not totally understood. Aquatic HS contribute to the acidity since they have a total acidity of approximately 10 mmol/g (Oliver et al. 1983), due to the presence of carboxyl and phenol hydroxyl groups. Other acidity components are carbon dioxide, hydrated Al, Fe, and silicic acid. However, Wilkinson et al. (1992) show that the model of Oliver et al. (1983) underestimates the contribution of organic acids to the acidity of Quebec lake waters. The organic matter in Quebec lakes with a higher charge density, differs from the reported (Roila et al. 1994a) and calculated values (Oliver et al. 1983). Wilkinson et al. (1992) also show that calibration on regionally discriminated data bases yield much improved predictions of the specific charge density of HS.

In a study of Finnish lakes (Kortelainen et al. 1986) (Fig. 3.22), the concentration of weak acids varies from 12 to 360 µeq/L. A clear dependence of weak acids on the TOC concentration is apparent, but the organic C cannot represent all weak acids. Weak inorganic acids such as silicic acid are also contributors (Kortelainen 1993; Kortelainen et al. 1986).

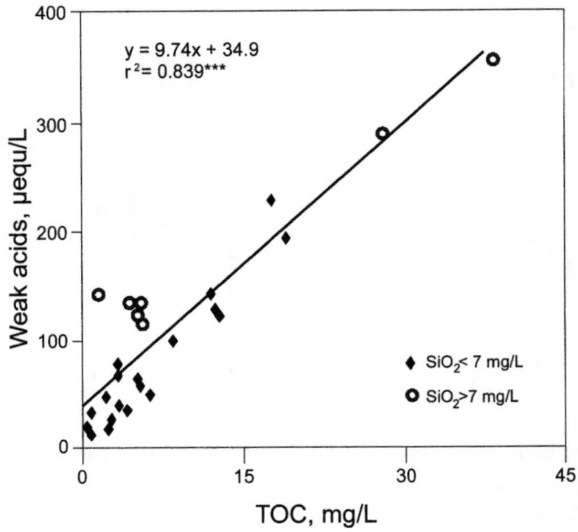

Fig. 3.22. Concentration of weak acids as a function of TOC (n=26) (after Kortelainen et al. 1986, with kind permission of the Water Association Finland and the Finnish Limnological Society)

Stepwise regression shows that in addition to TOC, SiO_2 and Al must be included in Eq. 3.11:

$$\text{Weak acids } (\mu eq/L) = 4.27 + 5.49\,[\text{TOC, mg/L}] + 9.72\,[SiO_2\text{, mg/L}] + 0.142\,[\text{Al}, \mu g/L];\ r^2 = 0.99 \quad (3.11)$$

Based on a survey of nearly 1000 Norwegian lakes, Lydersen (1998) confirms that the pH in lakes with low acid neutralizing capacities (between 0 and 50 μeq/L) is significantly correlated with the TOC contents (Fig. 3.23).

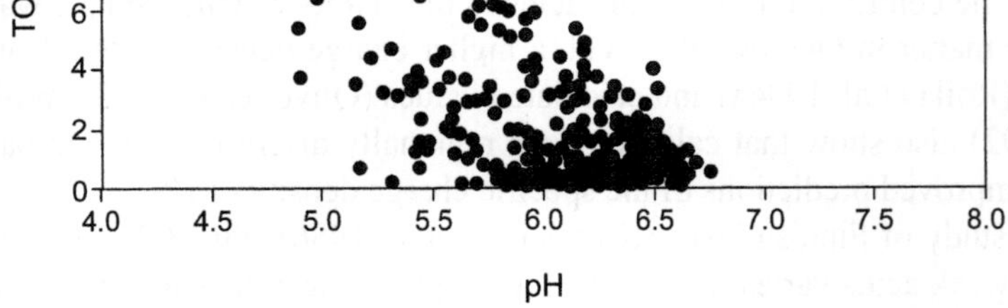

Fig. 3.23. Norwegian lake survey (1995) showing the relationship between pH and TOC for lakes 0–50 μeq/L ANC (from Lydersen 1998, with kind permission of Springer Verlag)

for lakes 0–50 μeq/L ANC (from Lydersen 1998, with kind permission of Springer Verlag)

3.2.1.1 Catchment Area and Acidity Status of Low ANC Lakes

In sensitive regions which are not affected by acid precipitation, acid inputs come from the catchment. From Finnish studies we already know that the TOC content is a function of the catchment and its characteristics. In turn, TOC determines acidity status. Accordingly, a relationship must exist where the catchment determines the acidity of a lake via TOC. In accordance with expectations, the lake area alone explains roughly 50% of the variance of the pH (Fig. 3.24). Lakes smaller than 0.1 km^2 have an average pH of 5.2; in lakes from 0.1–1.0 km^2 a pH of 5.7 prevails, and lakes >1.0 km^2 surface area have a pH of 6.4. Obviously a contributing factor is that the greater thickness and better development of soils in larger catchments provide a higher acid-buffering capacity. This means that the larger lakes possess a higher proportion of water that has percolated through less acidic soils and therefore have a greater loading of basic cations. Furthermore, according to expectations, the pH values of the lakes are negatively correlated with the proportion of peatlands in the catchment: the higher the proportion of peatland, the lower the pH in the lake water.

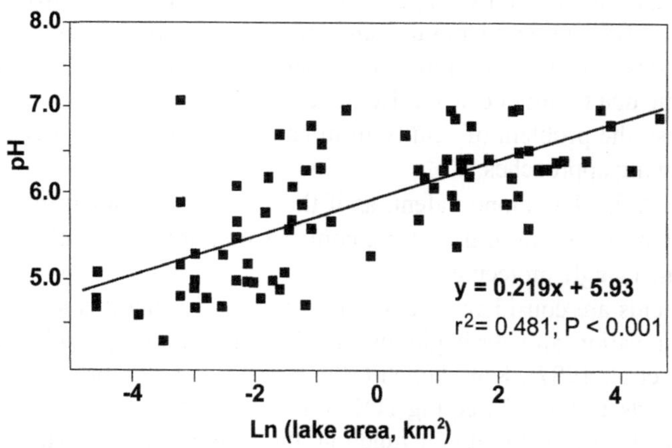

Fig. 3.24. Relationship between lake area (as natural logarithm, ln) and pH value of the water (n = 78) (after Kortelainen and Mannio 1988, with kind permission of Kluwer Academic Publishers)

In humic lakes, a multiple regression analysis (Eq. 3.12) shows that lake area, proportion of peatlands in the catchment, proportion of the catchment area that the lake surface represents, and the deposition potential of acidic substances (presented as latitude) are strong determinants of pH value, explaining 65% of the variation in pH values (Table 3.1).

$$\text{pH} = 5.4 + 0.16\,[\ln \text{lake surface area, km}^2] - 0.026\,[\text{peatland area, \%}] + 0.26\,[\ln \text{lake area, \%}] + 0.19\,[\ln \text{degrees latitude}]. \quad (3.12)$$

In comparison to other freshwaters in Scandinavia and North America, the contribution of organic acids to acidity status of the Finnish lakes is high.

Table 3.1. Stepwise multiple regression of pH value against lake area, peatlands percentage, lake percentage, and latitude (as substitute for acidifying potential, which applies to southern latitudes). The table gives the variable entered at each step and the cumulative percent of explainable variance (Kortelainen and Mannio 1988)

Variable	Step 1 lake area, ln	Step 2 peatland, %	Step 3 lake area %, ln	Step 4 degrees latitude, ln
Explainable variance in %	39	59	64	65

3.2.1.2 Buffer Against Mineral Acids

Until recently, organic acids have often been considered unimportant for the acidity of surface waters. Recent studies, however, clearly demonstrate the importance of organic acids in determining the acidity of surface waters. Norwegian lake surveys show that organic acids are able to decrease the pH of water from 0.5 to 2.5 pH units in the low alkalinity range of 0–50 μeq/L (Lydersen 1998). That means that dissolved HS can be a natural source of acidity in freshwaters. Another estimate calculates that, within temperate climate zones, up to 1/3 of acidic waters may be acidic due to organic acids (Hemond 1994).

In the past, the problem of acidity from natural organic acids was complicated by the following approaches:

- acids are defined as mono-valent, as if they are defined small molecular compounds. On the contrary, they are a complex mixture, with multiple acid components in a single molecule.
- organic acids are considered weak acids, despite the fact that some are strong, with dissociation already beginning at pH 3 and below (Brakke et al. 1987; Kullberg et al. 1987). Kortelainen (1999b) calculates the proportion of strong organic acids as 4.5–6.2 μeq/mg TOC in Finnish waters.

In contrast to the proton donor function, the acidic groups of HS are potential pH buffers in natural lakes in which the carbonate/bicarbonate buffer system is no longer active and in which strong mineral acids lead to anthropogenic acidification. The general ability of organic acids to act as buffers is shown in Fig. 3.25. In this figure, two freshwaters with pH 5.5 are shown. One contains no DOC, the other contains 0.5 mmol/L (6 mg/L). In order to lower the pH from 5.5 to 5.0, 8 μeq/L mineral acid for the DOC-free water is required, while 13 μeq/L (or 60% more) is required for the DOC containing water.

In comparison to the bicarbonate system, the organic acid buffer system is weak. In this regard, Kortelainen (1999b) determined an acid neutralization capacity of 1.6 μeq/g TOC. Other authors give 1.9 (Roila et al. 1994b), 2.0 (Hedin et al. 1990) or 2.4 to 2.7 μeq/g TOC (Mattson et al. 1995). The differences in the numbers given are not due to differences in materials, but to differing methodology. According to Kortelainen (1999b), the proportion of organic anions contributing to the buffering is about 16% of the average organic anion concentration which contributes to the acidity of Finnish lakes. This restricted capacity of the organic buffer system is the key to why humic lakes are more sensitive to the impact of mineral acids from acid precipitation than previously accepted.

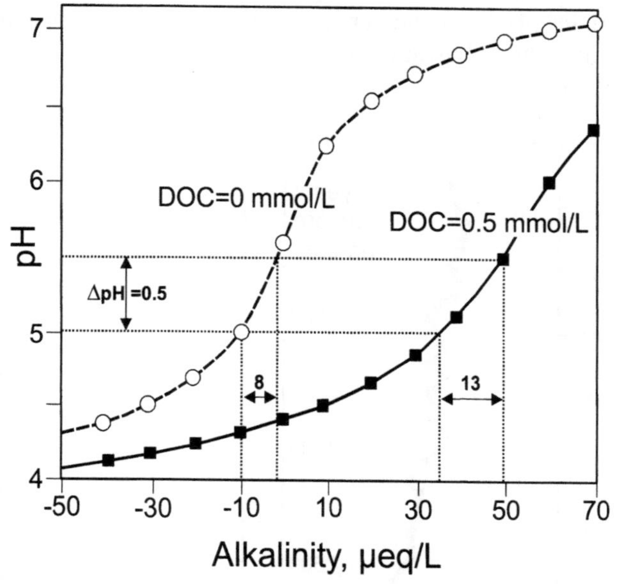

Fig. 3.25. Titration curves of two water samples differing in DOC content (after Hemond 1994, with kind permission of Wiley & Sons)

3.2.2 Paleolimnological Studies

3.2.2.1 Post-Glacial Acidification

Since HS contain a number of acid functional groups such as carboxyl and phenolic hydroxyl groups, these biogeopolymers make a considerable contribution to the acid-base status of freshwater bodies in non-calcareous areas. Thus, various authors consider the input of HS to be the main cause of documented acidification of sensitive waters in many regions (Krug and Frink 1983). However, the natural acidification of soils through vegetation development is one potential source of acidification for surface waters, but in most cases not the only one (Johnson et al. 1984).

In a paleolimnological study of Holocene sediments in Lake Großer Arbersee (Bavarian Forest, Germany), periods of natural and anthropogenic acidification can be differentiated by trends in TOC content. Late- and post-glacial diatom-inferred pH values and TOC data are shown in Fig. 3.26.

During the late-glacial phase, in sediments deeper than approximately 420 cm (roughly 10,000 years BP), the pH of the lake is in the range of natural soft waters (≥6.0). Melt water evidently delivers a comparatively large quantity of inorganic, acid neutralizing material to the lake since the organic C concentration is at a minimum (Fig. 3.26). In the post-glacial period (in sediments above approximately 420 cm depth), the pH drops more or less continuously from around 6 to around 5. At the same time, the sediment organic C content increases from around 40% to nearly 50% of the dry weight.

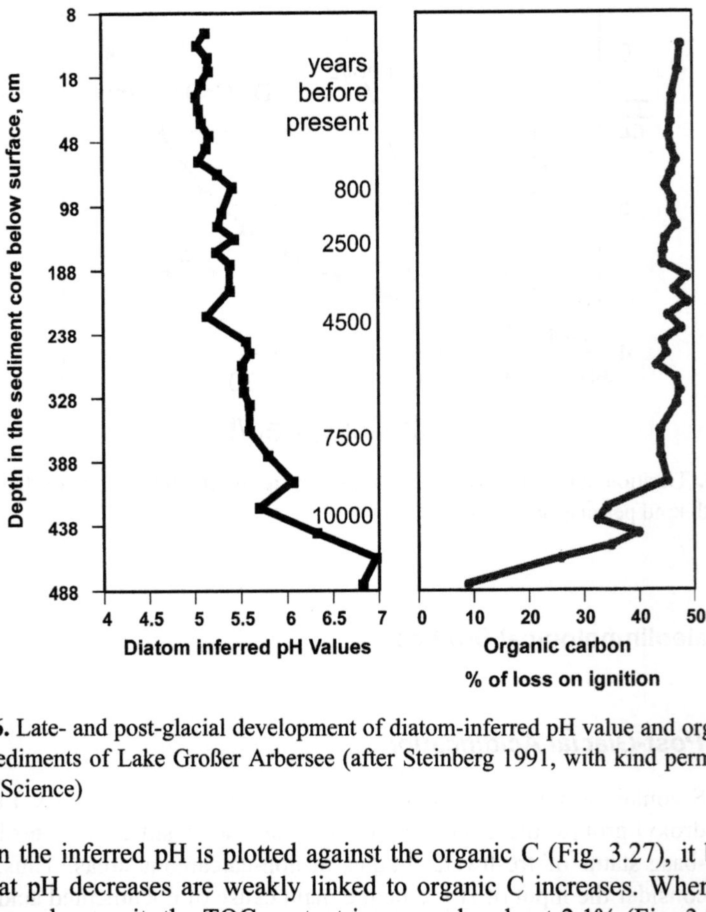

Fig. 3.26. Late- and post-glacial development of diatom-inferred pH value and organic carbon in sediments of Lake Großer Arbersee (after Steinberg 1991, with kind permission of Elsevier Science)

When the inferred pH is plotted against the organic C (Fig. 3.27), it becomes clear that pH decreases are weakly linked to organic C increases. When the pH value drops by a unit, the TOC content increases by about 2.1% (Fig. 3.27, right graph). A causal link is probable, and is found in the post-glacial development of soils and vegetation in the catchment. This development produces organic acids which percolate water to deeper soil layers (podzolization), and finally to groundwater or spring water. However, only 20% of the variance of the pH values can be explained by TOC variations. This comparably poor regression may be due to the fact that only bulk TOC is considered. However, there are some biological indications of acidification/dystrophication. In the subboreal period (approximately 230 cm depth), the change from a deciduous *Quercus* forest to a coniferous *Abies alba* and *Picea abies* forest leads to a distinct acidification and an increase in organic C. This development of non-deciduous woods, which is favored by cool climates, results in accumulation of coarse HS in the soil and subsequent accelerated leaching of organic acids. This acidification/dystrophication of Lake Großer Arbersee produces shifts in all planktonic and littoral biocoenoses studied; diatoms and chydorids, with an increase of typically dystrophic forms (Schirmer 1989), and mallomonadacean algae (Chrysophyceae), with a clear decline of *Mallomonas crassisquama*[4] (Hartmann and Steinberg 1986, 1989). The relative abundance of the latter species is reduced from approximately 60 to approximately 40%. Diatom-inferred pH is consistent with this finding.

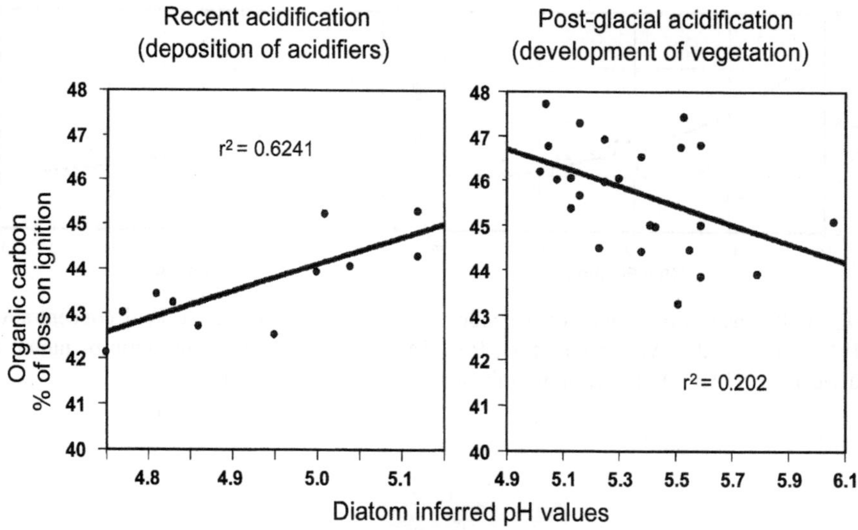

Fig. 3.27. Role of organic carbon in the acidification processes in Lake Großer Arbersee (Bavarian Forest, Germany) (after Steinberg 1991, with kind permission of Elsevier Science)

If HA and FA are considered separately as predictors of lake pH, a more precisely picture emerges than for the Lake Großer Arbersee sediments, where all fractions were considered together as TOC. The more precise approach is demonstrated with data from Lake Hirvaslompolo, Finnish Lapland (Fig. 3.28). Plotting diatom-inferred pH against HA and FA, it is evident that only HA appear to be responsible for the acids in the lake. However, this may be an preparation artifact, since a potential source of error is the inorganic impurities that are nearly always present, particularly in the FA fraction (Reinikainen and Hyvärinen 1997). Nevertheless, statistically more than 75% of the diatom inferred pH can be predicted by the HA concentrations.

---

[4] *M. crassisquama* is known to inhabit all kinds of localities, except acid ones with pH<5.0 (Hartmann and Steinberg 1986, 1989; Smol et al. 1984).

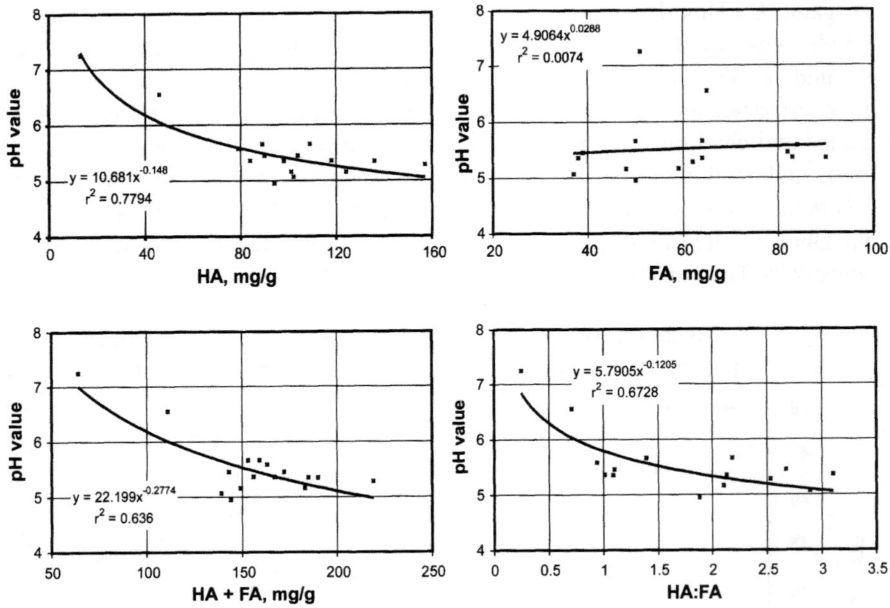

Fig. 3.28. Lake Hirvaslompolo: dependence of pH values on humic acid concentration (HA), fulvic acid (FA) concentration, HS (HA+FA) concentration, and ratio of humic to fulvic acids (HA:FA) (data from Reinikainen and Hyvärinen 1997)

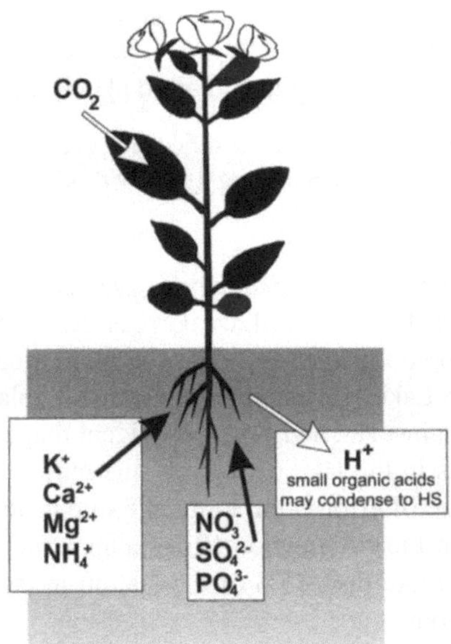

Fig. 3.29. Scheme of natural soil acidification through vegetation development

The mechanism of acidification by vegetation development can be seen from the ion balance of plants (Fig. 3.29), which take up more cations than anions from the soil solution since they take up an acid as anhydride (CO_2) from the atmosphere. To achieve electroneutrality, plants release small fatty acids into the soil water, which then percolate through the soil. With passage through the soil, HS-like materials can be condensed, and/or catalyzed by clay minerals (Larson and Hufnal 1980; Liu and Huang 2000) or manganese dioxide (Shindo and Huang 1984; Wang and Huang 1997). Such substance are probably a major cause for the increasing acidity in lakes mentioned above.

3.2.2.2 Anthropogenic Acidification

In the post-glacial period, natural acidification through vegetation development is linked to an increase in organic C in freshwater bodies (Figs. 3.26–3.28), in the recent acidification the increase in acidity is linked to a decrease in organic C (Figs. 3.27, 3.30). There are other than organic acids which reduce the pH in the most recent phase of acidification: natural organic acids were replaced by strong mineral acids, namely sulfuric acid and nitric acid which essentially derive from the burning of fossil fuels by man. The anthropogenic character of acidification from precipitation can be recognized through an analysis of residues from the burning of fossil fuels [for instance, Lake Großer Arbersee by Bruckmeier et al. (1997), and Black Forest lakes by Jüttner et al. (1997a,b)]. In Lake Großer Arbersee, approximately 60% of the organic C decrease can be statistically explained by the increase in acidity (Fig. 3.30).

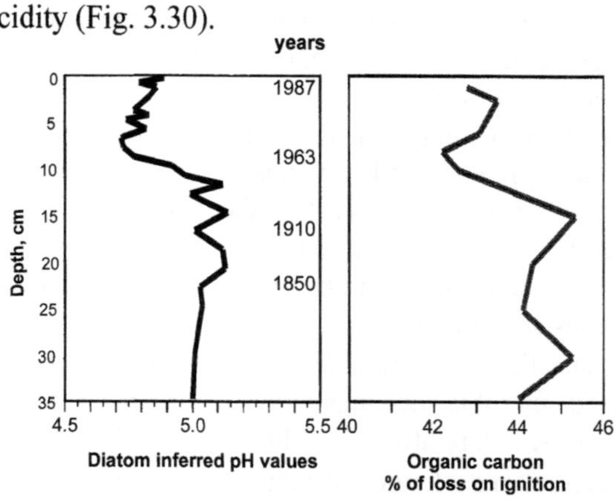

Fig. 3.30. Recent development of diatom-determined pH values and organic carbon in sediments from Lake Großer Arbersee (after Steinberg 1991, with kind permission of Elsevier Science)

In many lakes impacted by anthropogenic acidification, the intensity of the brown color of the water decreases. For instance at present, in all natural lakes in the Bavarian Forest, the water appears pale, almost glass-white with high transparencies, whereas historic reports state a dark-brown, 'weird' color (Gümbel 1868). The loss of HS in the water phase during the recent acidification period is obvious, a phenomenon that applies widely to acidified lakes (Marmorek et al.

1987).

There are two possible mechanisms for this phenomenon:

- coprecipitation of colored organic materials with Al or Fe hydroxides. Through the acidification of catchments, inorganic species of Fe and Al are dissolved from the soil and transported to the water body. The metal species partially hydrolyze and precipitate on entering lakes. Particulate and dissolved materials, including the chromophoric DOC (cDOC), then coprecipitate with the metal oxihydroxides. Thus the water looses much of its color (Dickson 1980; Persson and Broberg 1985; Schindler et al. 1997; Yan 1983);
- cleavage of high-molecular weight cDOC by Al ions. The resulting low-molecular weight dissolved HS products are less strongly colored (Steinberg and Kühnel 1987). Fe ions are able to cause similar effects as Al ions do. It is conceivable that, in extreme cases, lakes become paler through this cleavage mechanism without an absolute decrease in DOC concentration, because the low-molecular weight materials are more or less colorless.

For Lake Großer Arbersee, in the recent anthropogenic acidification (in contrast to natural acidification),

1. the increase in acidity is linked to a **quantitative** loss of organic C in the water. This agrees well with the hypothesis of Davis et al. (1985) (Chap. 3.2.2.2). **Qualitative** changes are still to be discussed (Chap. 3.2.2.3); and/or
2. the increase in the acid content of soils in the catchment area is conditional on a decrease in the inputs of organic C compounds into the lake.

For the latter processes, various potential mechanisms do exist (Marmorek et al. 1987; Schindler et al. 1997):

- increased deposition of mineral acids results in a reduced mobilization of organic compounds from soils and wetlands. For example, using high resolution X-ray microscopy, Myneni et al. (1999) show that in acidic sediments with a high C content, HS form dense structures with a low surface to volume ratio. These structures tighten the openings of macropores so that hydrological and washout processes are hampered;
- increased concentrations of mineral acids in soils, sediments and surface-waters result in reduced microbial decomposition of organic materials and thus in a reduced release of small, relatively soluble compounds;
- increased concentrations of mineral acids reduce the dissociation of organic acids and change the physical structure of dissolved HS. This leads finally to immobilization in the soil.

These hypotheses are by no means mutually exclusive, they overlap functionally and may reinforce one another.

3.2.2.3 Quantification of Organic Carbon Losses During Recent Acidification

There are reports of no change (Hedin et al. 1990) or even an increase in DOC concentrations (Wright et al. 1988) with increasing lake water acidity. Most studies, however, report decreased DOC concentrations in acidified waters (Marmorek et al. 1987). In cases where no decrease is found, the mechanism discussed by Guggenberger and Zech (1994) may apply. In a pioneering study of two Norwegian lakes, the development of both the pH and the DOC concentration are reconstructed by diatom inference (Davis et al. 1985). The authors find a DOC decrease from 6 to 3 mg/L.

In a study by Steinberg (1991), using a different approach to Davis et al. (1985), the diatom-inferred pH values are correlated to loss on ignition (a simple substitute of organic C). The loss per unit pH was determined from an exponential regression (Eq. 3.13):

$$\text{Loss on ignition} = a\, e^{b\,pH}, \tag{3.13}$$

where a and b are constants specific to the water body. Specific examples for lakes of various type are given in Fig. 3.31.

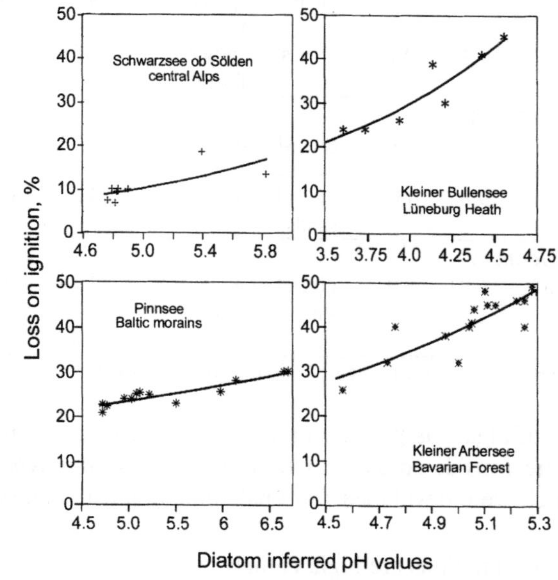

Fig. 3.31. Regression between diatom-inferred pH values and loss on ignition (marker of TOC) in central European lakes: Schwarzsse ob Sölden (central Alps), Kleiner Bullensee (Lüneburg Heath), Pinnsee (sands in the Baltic moraines) and Kleiner Arbersee (Bavarian Forest) (after Steinberg 1991, with kind permission of Elsevier Science)

All lakes in Fig. 3.31 exhibit a clear loss of organic substances during acidification. This process is more pronounced in lakes which are richer in organic compounds prior to acidification (Lake Kleiner Arbersee, Bavarian Forest; Lake Kleiner Bullensee, Lüneburg Heath). In addition, it is clear that there is no common trend in the loss of organic C. Per pH unit, the loss of organic C ranges from 10% (Lake Herrenwiesersee, Black Forest) to 50% (Lake Kleiner Bullensee) (Table 3.2). This is in contrast to the study of Davis et al. (1985), from which a unifying value can be extracted.

Table 3.2. Loss of organic carbon during the recent acidification in various central European lakes determined paleolimnologically (from Steinberg 1991)

Lake	Elevation, m a.s.l.	Area, ha	Max. depth, m	Geology	Vegetation	Organic carbon*	P level	pH-range
Central Alps, Austria								
Schwarzsee ob Sölden	2792	3.6	18.0	granite gneiss	little vegetation	45.5	0.021	6.0–5.0
Black Forest, Germany								
Herrenwiesersee	830	1.8	9.5	red sandstone	spruce/-fir forest	10.4	0.067	5.0–4.0
Bavarian Forest, Germany								
Großer Arbersee	935	7.7	15.9	gneiss	spruce/fir forest	14.8	<0.005	5.8–4.8
Kleiner Arbersee	918	9.4	9.7	gneiss	spruce/fir forest	50.5	0.0001	5.5–4.5
Lüneburg Heath, Germany								
Kleiner Bullensee	33	5.5	9.0	glacial sands	heath, pines	50.9	<0.005	5.0–4.0
Großer Bullensee	32	9.8	11.0	glacial sands	heath, pines	24.5	0.10	5.0–4.0
Baltic Moraines, Germany								
Pinnsee near Mölln	29.5	6.9	9.9	glacial sands	mixed forest	13.1	0.0001	6.8–5.8
						24.4		6.8–4.8

*Loss of organic carbon with decrease of 1 pH-unit, % from start value

The response of algae, as primary producers, to the changes in DOC contents and pH values are considered by Leavitt et al. (1999) in the hypotheses summarized below which were developed from paleolimnological studies and historical records:

- if the DOC concentrations are high, namely >5 mg/L DOC, a moderate acidification to a pH around 5.0 can reduce the DOC concentration, increase light penetration, and stimulate the production of benthic algae;
- if the starting DOC concentrations are low (around 2 mg/L DOC), or acidification is strong and sustained leading to a pH of <4.5, the main effect can be the increase in depth-penetration of damaging UV-radiation, thus inhibiting the potential benthic production increase.

The effect of the loss in dissolved HS on its role in UV radiation protection is considered in Chap. 4.

3.2.2.4 Compositional Changes During Acidification

During the present rapid acidification in sensitive geological regions, HS are affected by strong mineral acids and some of their properties are changed. These compositional changes are discussed in relation to results from the Humex project in Norway, artificial acidification experiments in Canada, and from paleolimnological studies in Lake Großer Arbersee, Germany. The results from experimental acidification field experiments in Norway and Canada confirm the laboratory results, and also the observations on the acid precipitation affected Lake Großer Arbersee described in the following section.

The Scandinavian Humex-Project

Water chemistry parameters such as pH value and ionic strength generally determine the properties of HS molecules. This means that the acidification of water bodies and catchment areas will also change the physico-chemical properties of HS. For example, under acid conditions HS form dense structures with a low surface area to volume ratio (Myneni et al. 1999), and the solubility of HS decreases. In addition, the solubility of HS is influenced by the heterogeneity of functional groups and differences in the molecular weight distribution (de Wit et al. 1993; Ephraim et al. 1996; Tipping et al. 1990).

While the above findings are based on laboratory studies, the Humex project (European **Hum**ic Lake Acidification **Ex**periment, Gjessing (1992)) sets out a field test in Lake Skjervatjern, Norway. The Humex project tries to determine the fate and role of HS during acidification of surface waters and changes in chemical and biological properties of HS. The project is carried out with the artificial acidification of the dystrophic, dark-colored Lake Skjervatjern and its catchment. The lake is divided with a plastic curtain into two parts. One half is acidified, the other acts as control (Gjessing 1994a,b). Following acidification, HS (TOC, color, and UV absorption) concentration increases in both parts. The HS increase in the control part is almost double that of the acid-treated part, and the TOC export from the control part exceeds that from the treated part of the lake. Significantly, the sulfuric acid treated catchment exports less HS than does the control area. This finding and the interpretation agree with paleolimnological studies of Lake Großer Arbersee. The reduced exports from the catchment area are also in good agreement with reports that in podsolized soils, acidity reduces the solubility of hydrophobic acids (Chap. 1.4) (David et al. 1989; Vance and David 1991).

Acidification leads to the protonation of HS functional groups, rendering them less water soluble and more lipophilic. The latter is measured as the *n*-octanol-water partition coefficient. In the Humex studies, the octanol soluble C in the acid treated part of the lake increases by 46–104 μg/L in the year following treatment, and is higher than in the lake prior to treatment (Kullberg et al. 1992). This is further evidence that the acidification of soils and water bodies changes the physico-chemical properties of HS.

Further details of changes of HS upon acidification in the Humex project are published by Malcolm and Hayes (1994). The authors find clear changes in HS composition. A year after acid application, the number of protons which are bound to the heteroaromatic nitrogen or oxygen decreased by about 13%, and the aromatic protons decreased by 27–31%. In the HA, the proton content is reduced by around 11%. In general, following acidification, 'aromatic protons' decrease, and 'aliphatic protons' increase in number. This statement is in good agreement with previous findings (Stabel and Steinberg 1976; Steinberg 1977), in which the coprecipitation with Al and Fe salts particularly affected the chromophoric DOC that is mainly comprised of aromatic compounds.

Artificial Acidification in the **Experimental Lakes Area,** *Ontario, Canada*

In the Canadian Experimental Lakes Area (ELA), two lakes are experimentally acidified. In these lakes and also in lakes around Sudbury, Ontario, which is acidified by waste gases from metal mining and smelting, the DOC decreases dramatically following acidification (Schindler et al. 1996; Yan et al. 1996). The DOC from the acidified and non-acidified areas of the ELA lakes is compared using fluorescence properties. DOC origins are deduced from the ratio of emissions at 450 nm (predominately allochthonous sources) to emissions at 500 nm (predominately autochthonous sources) (McKnight et al. 2001). L224 and L239 are circumneutral and serve as controls. During experimental acidification of L223 and L302S, the DOC quality changed strikingly (Fig. 3.32). Particularly in L302S, the fluorescence ratio shows a change in favor of autochthonous or autochthonous-like DOC.

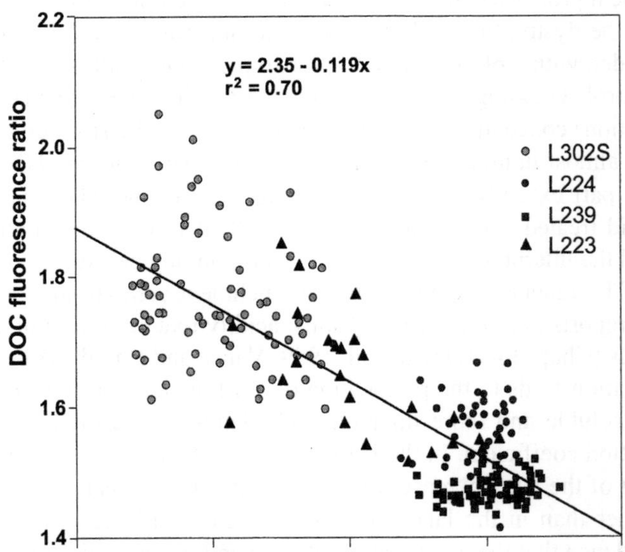

Fig. 3.32. General trends in DOC quality during the artificial acidification in Lakes 223 and 302S of the Experimental Lakes Area, Ontario, Canada. Data show the ratio of emissions at 450 nm (autochthonous sources) to emissions at 500 nm (allochthonous sources) (from Donahue et al. 1998, with kind permission of the American Chemical Society)

The acid-induced changes in DOC quality are probably caused by chemical oxidation or precipitation of the UV absorbing, aromatic fraction of the allochthonous DOC, thus leaving behind the more UV transparent aliphatic chains. As a consequence, the UV-B radiation can penetrate more deeply into the acidified lakes, which will lead to the extinction of UV sensitive organisms.

Paleolimnological Studies of Lake Großer Arbersee

A similar, complementary picture of the fate of HS during anthropogenic acidification comes from the paleolimnological study of Lake Großer Arbersee (Steinberg et al. 2001). The changes in HS fractions are investigated with ^{1}H-NMR along acidity gradients. The acidification history is reconstructed from subfossil diatoms (Bruckmeier 1994), and the anthropogenic character of the acidification is indicated by organic micropollutants, such as polychlorinated dioxins, furans, and biphenyls (Bruckmeier et al. 1997). The inferred pH values fall more or less continuously from 5.3 to 5.0–5.05 during last 150 years. Combined with diagenetic processes in the sediment, acidification alters the composition of HS as follows:

1. the contribution of HA to the total humic C increases from 12 to 17% (Fig. 3.33). Statistically, the reduced pH values account for 60% of this change;

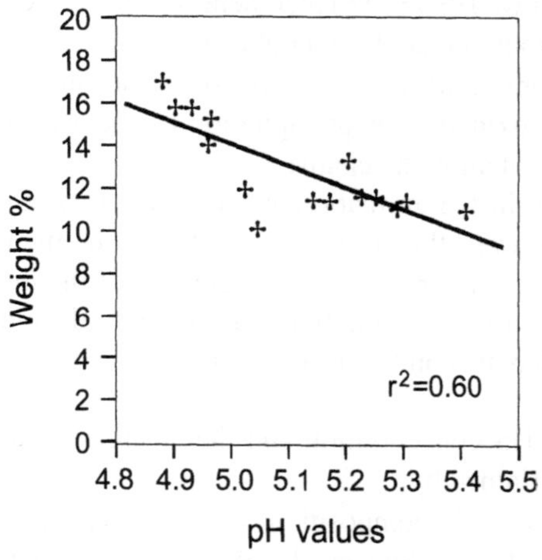

Fig. 3.33. Increase in proportion of humic acids with decreasing (diatom-inferred) pH value in sediments of Lake Großer Arbersee (after Steinberg et al. 2001, with kind permission of Wiley-VCH)

2. while the constituents changes in the FA fraction, they remain more or less unchanged in the HA fraction. Within the FA fraction (Fig. 3.34), percentages of carbohydrates increase by approximately 10%, whereas those of aliphatic and aromatic compounds decrease by approximately 10% and 3%, respectively. Acidification accounts for around 50% of the increase in carbohydrates and decrease in aromatics, and 30% of the decrease in aliphatics. Other processes af-

fecting HS composition may be diagenetic/pedogenetic changes in the sediment and deep soil layers within the catchment.

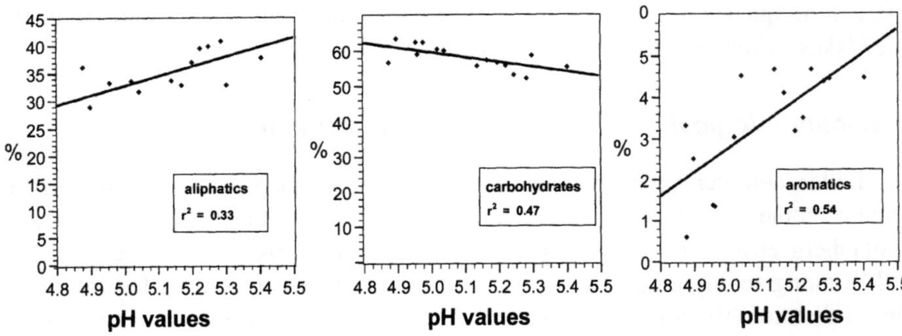

Fig. 3.34. Changes in the content of aliphatic, carbohydrate, and aromatic structures in the fulvic acid fraction as a function of (diatom-inferred) pH value in the sediments of Lake Großer Arbersee (from Steinberg et al. 2001, with kind permission of Wiley-VCH)

The described acidification-caused changes in HS constituents may be explained (**a**) by reductions in enzymatic activity, particularly of cellulases and proteases, in the raw HS in the catchment, (**b**) by protonation of acidic functional groups (carboxylic acids, phenolic compounds) which leads to an immobilization and retention of these compounds in the catchment, and/or (**c**) by increased chemical oxidation or precipitation of the aromatic fractions, leaving the more transparent aliphatic chains.

3. The 3% reduction in the aromatic content of the total C does not appear dramatic at first. However, the aromatics which provide the most important shield against UV radiation, are reduced by around 1/3 of the initial aromatic content, thus the penetration of UV radiation may increase drastically. The increase of UV penetration depends on the absolute DOC content of the water, as shown in Fig. 4.1.

The reduction in HS concentration and the changes in HS quality in acidified lakes can be best explained by:

- alterations of the HS composition due to deposition of mineral acids (Marmorek et al. 1987; Vogt et al. 1992, 1994). Since the HS store of the catchment area far exceeds the HS content of the lake and its sediment, catchment processes will characterize the features of the lake HS as well. However, closely related –but with respect to turnover, less important – chemical and microbial processes will also take place in lake;
- alterations in the processing of leaf litter mainly in the catchment and to a lesser degree also in the lake itself,
- selective retention of HS components in the catchment,
- selective precipitation of DOC constituents in the water column,
- increasing transparency, due to selective loss of DOC chromophores, will facilitate photodegradation of humic constituents (Schmitt-Kopplin et al. 1998)

even at increasing depths in the water column.

Chemical and physico-chemical, as well as litter processing mechanisms, are now considered in more detail.

Potential physico-chemical mechanisms in the catchment: several chemical and physico-chemical mechanisms may operate in the catchment soils and in the water column of the lake itself, leading to less colored, carbohydrate-rich HS in the lake water. With increasing proton concentration, the protonation of the HS increases, with subsequent decreases in solubility and increases in hydrophobicity of HA. The result is that with increasing acidity, an increasing proportion of the water soluble aromatic C is turned into more hydrophobic aromatic compounds and retained in the soils (Dunnivant et al. 1992; McCarthy et al. 1993). For HA and humins, input mechanisms distinct from those in the water phase may apply (for instance, particle erosion), so that only the FA fraction of HS show a acidity dependent decrease in aromatic content.

At least two additional mechanisms in the catchment may account for the mentioned changes in quality of HS. Stabel and Steinberg (1976) show that chromophoric DOC and aliphatic structures, such as small peptides, are cleaved upon acidification, by applying size exclusion chromatography (Sephadex). While the elution times of the peptides remain more or less unaffected, the chromophoric (mainly aromatic) DOC is eluted with a delay caused by interactions with the Sephadex matrix. Steinberg (1976) confirms the cleavage and the specific adsorption of chromophoric DOC on ferric Fe hydroxy-oxides when applying coprecipitation with Fe salts under slightly acidic conditions. Cleavage of aromatic and aliphatic compounds of the HS, and retardation of the aromatic moiety, may apply to processes in acidified soils. Interaction processes with the organic Sephadex matrix represents interactions with hydrophobic organic soil constituents, and coprecipitation with ferric Fe for adsorption on trivalent metal oxides.

Calculations by Schulten and Leinweber (2000) indicate that natural substrates, such as carbohydrates and oligopeptides, are probably bound only very loosely to the 'humic core', mainly by hydrogen bonds. These results support the empirical findings of the easy cleavage of HS and proteinaceous material.

In addition to the intrinsic reactivity and physicochemical binding mechanisms, seasonal variations will contribute to the changes in quality of HS. Clair et al. (1996), when studying the effects of seasonal variation on the composition of aquatic organic matter in some Nova Scotia humic-rich waters, show that (1) aromatic C increases slightly in the fall, (2) less structurally complex aliphatic C decreases from winter to spring, and remains at lower levels into later fall and (3) carbohydrates are at a maximum during the summer. One mechanism may be different hydrological processes which carry terrestrial and wetland DOM to streams and lakes (Fig. 3.10), allowing a selective adsorption process of DOM to occur in soils. Furthermore, the adsorption of aromatic C onto soil particles may be enhanced under acidic conditions.

Changes in leaf litter processing: increases in the concentrations of mineral acids in soils, sediments and surface waters will result in decreased microbial breakdown of organic matter (Kelly et al. 1984; Schindler 1994). In many instances, acidification changes the structures of microbial communities in favor of micromycetes, and against heterotrophic bacteria (Berg 1986). Furthermore, respiration and litter decomposition are reduced (Kok and van der Velde 1994; Lettl 1984; Prescott and Parkinson 1985; Skiba and Cresser 1986).

The acidification impacts enzymes which cleave polymeric substrates such as cellulose or proteins, rather than enzymes which metabolize comparatively small substrates, such as urea (Bewley and Parkinson 1986). This selective inhibition of enzymatic activity can explain the increase in carbohydrate content in the HS upon increase of acidity. A detailed study on leaf-maceration of aquatic macrophyte litter is presented by Kok et al. (1992). The authors find that all the fungi could develop at low pH (4.0), but maceration was only observed at pH 5.5 or even above. It is likely that inhibition of pectin degradation is an important factor causing suppression of leaf fragmentation at low pH. This inhibition may contribute to inhibited decomposition of macrophyte remains in acidic aquatic systems, and may lead to an increase in the carbohydrate content of organic matter in the sediments, as shown for Lake Großer Arbersee (Steinberg et al. 2001).

Kok and van der Velde (1994), when studying the decomposition of the aquatic macrophyte *Nymphaea alba* and the terrestrial tree *Betula pubescens*, emphasize that decomposition of both species is slower in acid than in circumneutral waters. Under acidic conditions, nitrogen and protein concentrations of *N. alba* litter remain high, whereas no effect of acid status is observed on the protein and nitrogen concentration in *B. pubescens* litter. Furthermore, the concentration of phenolic compounds is higher in *N. alba* litter from the acid than the circumneutral pond. No such difference is observed for *B. pubescens* litter. The high concentrations of phenolic compounds in *N. alba* litter from the acidic pond probably induces chemical immobilization (tanning reaction), resulting in high concentrations of nitrogen and protein in the litter.

This finding, however, is not in agreement with the results of Lake Großer Arbersee, where aromatic (including phenolic) compounds decrease with decreasing pH. This discrepancy may be due to (1) differences in litter quality: *N. alba* is not dominant in the floating-leaved macrophyte community of Lake Großer Abersee, (2) selective photochemical degradation of phenolic compounds in the water column (Schmitt-Kopplin et al. 1998) and, (3) other mechanisms.

3.3 Paleolimnological Reconstructions of Humic Substances Trends

By using a well characterized standard data set and transfer functions, diatoms can be used to reconstruct almost all their significant influencing environmental factors. Three paleolimnological studies using diatom remains, in Canadian tree line lakes, Finnish Lapland lakes, and northeast German lakes demonstrate the role of DOC and dissolved HS in freshwaters.

3.3.1 Lakes at the Canadian Tree Line

Natural long-term influences on lakes are strongly linked to climate and vegetation development. That means chemical (such as nutrient concentrations) and physical properties (such as mixing regime, water transparency) of lakes are tightly linked to processes in the catchment, as well as to climate changes, which alter the groundwater level, soil water content and outflow processes. These properties, in turn, will influence the structure of aquatic communities and their productivity. It follows that every climate-controlled change of the boreal forest boundary will influence the nutrient poor, non-calcareous lake of this ecotone boundary.

An interesting example of how the changes in the terrestrial catchment may impact lakes is presented by Pienitz et al. (1999) who describe sediment cores from lakes north-east of Yellowknife in the Northwest Territories of Canada, a region which today encompasses the northern forest boundary (Fig. 3.35).

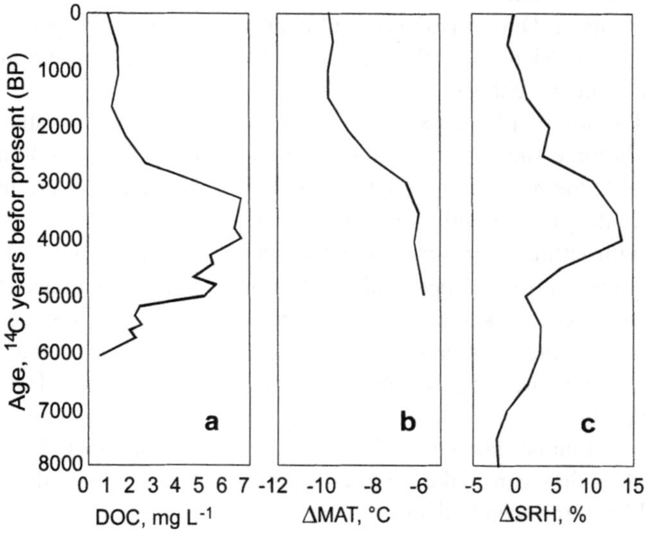

Fig. 3.35. Changes in DOC content in the water column of Queen's Lake (NWT, Canada) (determined by diatoms), mean annual temperature in central Canada (ΔMAT, reconstructed from $\delta^{18}O$ cellulose profiles), and relative summer humidity (ΔSRH, determined isotopically) in sediments from Queen's Lake, Toronto Lake and Whatever Lake during the last 8,000 years (after Pienitz et al. 1999, with kind permission of Regents of the University of Colorado)

The diatom data proves that in the mid Holocene (approximately 5,000 years BP), substantial limnological changes occur in the lakes at the tree line. These changes indicate maximal forest-tundra development 10,000–5,000 years BP, when temperatures and humidity are higher. The tree line moves northwards and the lakes have their highest DOC content and maximal productivity. The reconstructed, relative change in DOC is up to 5.8 mg/L between mid Holocene and present. The present vegetation in the region is, as in early tundra times, again dwarf bush tundra, and the tree line has moved southwards. The lakes decrease in productivity and become oligotrophic during this period. At the same time, peat bogs develop and a natural acidification process occurs. Increases in acidophilic and acid tolerant diatoms (*Achnanthes marginulata*, *Brachysira brebissonii*, *Cymbella* spp., *Eunotia* spp., *Frustulia* spp., *Navicula mediocris*, and *Pinnularia interrupta*) also take place.

Compared to other reconstructed environmental data, the boreal tree line lakes react extremely sensitively to climatic changes and the linked changes in vegetation. The loss of colored DOC caused wide reaching changes in the biocoenoses (Chap. 4).

3.3.2 Lakes in Finnish Lapland

Lakes in northern Fennoscandia are very similar to other north-boreal lakes, such as those in Canada. Due to underground bedrock, the lakes are poor in dissolved electrolytes and nutrients and, thus, poorly buffered. Hence, they are highly sensitive to environmental changes.

HS in lakes are, to a large extent, derived from terrestrial ecosystems. It is well known that latitude and the proportion of peatlands in the catchment, are the main determinants of the present organic C content in Finnish lakes (Kortelainen 1993). Similar to findings by Pienitz and Smol (1993), TOC concentrations are highly correlated with latitude and surrounding vegetation. Yet, in Finnish Lapland, despite the high proportion of peatlands, TOC concentrations in lakes are low compared to those in central and southern Finland, where there are less peatlands. The colder climate, longer frost period, and relatively low rates of primary production and decomposition, are the probable factors accounting for lower TOC concentrations in the north.

The remote Finnish lakes usually lack long-term monitoring, however paleolimnological studies can make long-term data available. Two paleolimnological studies will be introduced in detail.

3.3.2.1 Lake Hirvaslompolo

Different soil types and different vegetation exhibit characteristic HA:FA ratios, the ratio being generally higher in wet than in dry areas, and particularly high in mires (Kononova 1961). The stratigraphy of HS in lake sediments can provide useful information about previous environmental conditions, and the major determinants of lake water pH values. Reinikainen and Hyvärinen (1997) report HA–FA analyses and pH values inferred from subfossil diatoms in the Holocene sediment of a very small lake ('Hirvaslompolo') in eastern Finnish Lapland. A synopsis is given in Fig. 3.36.

In phase 1, initial terrestrial biomass growth leads to plant cover and development of organic soils after the deglaciation, and to acidification of the lake water.

During the early Holocene (zone 2), birch woodlands suggest dry conditions and an absence of peatland. The subsequent rapid rise in HA (phase 3) coincides with the rise in pine invasion. The following sharp decline in HA (phase 4) occurs simultaneously with a diatom inferred lake-level rise. The decline in HA at this point might reflect an increased input of poorly humified organic matter eroded and washed in from drier surfaces around the basin. The high HA:FA ratio in phase 5 is associated with the extension of *Sphagnum* peatlands in the catchment. HA values during this phase remain at a clearly lower level than during phase 3. Much of the HS is of peatland origin, but the rates of biomass production and decomposition are not particularly high. The clear drop in the HA:FA ratio near the top of the core (phase 6) might be explained by decreasing humidity.

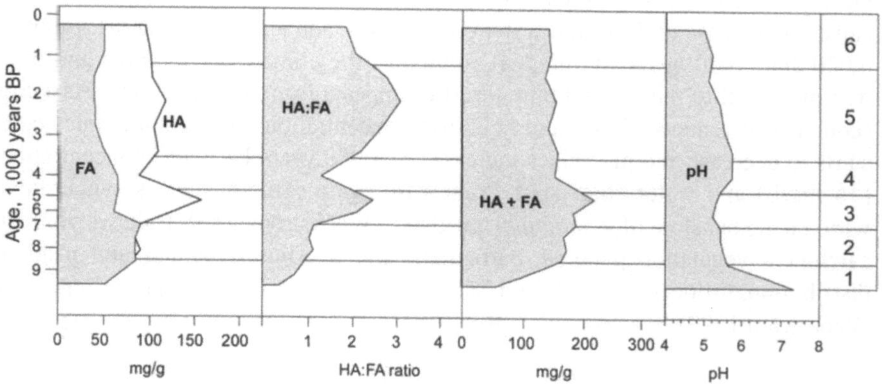

Fig. 3.36. Humic and fulvic acid[5] stratigraphy of Lake Hirvaslompolo and diatom inferred pH (Index B, Renberg and Hellberg 1982) (from Reinikainen and Hyvärinen 1997, with kind permission of Arnold Publishers). **1:** rising values for both HA and FA, **2:** equilibrium phase with maximal FA values and relatively low HA values, **3:** steep rise in HA and decline in FA, **4:** sharp decline in HA values, **5:** high HA:FA ratio caused by declining FA values, **6:** HA:FA ratio drops due to slight decline in HA and slight rise in FA. For more explanation of the phases, see also Fig. 3.37

From the acuto-limnological studies of Kortelainen (1993), we know that HA as well as FA determine the acid status of lakes situated on bedrock (if no carbonate buffer dominates). This also applies to previous conditions, as can be shown with Lake Hirvaslompolo sediments (Fig. 3.28).

The six phases observed in Lake Hirvaslompolo parallel well the phases found in sediments of Lake Tsuolbmajavri (Fig. 3.37) which lies in a more western region of Finnish Lapland. Thus, results from the investigation of Lake Hirvaslompolo may be more generally applicable.

[5] Humic and fulvic acids are separated by alkali extraction according to Schnitzer (1982). It is open to question whether diagenetic alterations of HA and FA in a Holocene sediment core can be excluded.

3.3.2.2 Lake Tsuolbmajavri

Weckström (2001) also uses subfossil diatoms as a tool for reconstructing the past key environmental variables, in order to document climate changes in north-western Fennoscandia (Fig. 3.37). Pollen data indicate that during the early Holocene, the catchment of Lake Tsuolbmajavri is characterized by birch forest. Gradual expansion of pine starts at 9,200 years BP. Pine reaches its maximum occurrence at 7,200–6,000 years BP, and the pine tree-line retreats from the area at 4,600 years BP. During the late Holocene, the cooler and moister climate leads to increasing peatland development in the catchment. The increasing TOC concentration (from 2.5 to >5 mg/L TOC), water color (from approximately 10 to >40 Pt units), and decreasing pH (from 7.7–7.5) from 8,500 to 6,000 years BP reflect the gradual development of forest and more humic soil. The disappearance of pine causes a decrease of TOC and water color values, and a slight rise in pH. Between 6,000 and 3,000 years BP, gradual cooling of the climate occurs, pine and birch tree-lines start to retreat, and pine forest disappears from the catchment. As a consequence of reduced lignin and cellulose concentrations in the soil, water color starts to decrease and pH values increase. At 3,000 years BP, TOC concentrations (>6 mg/L) and water color start to increase again. This change is synchronous with the expansion of *Sphagnum* peatland. Weckström (2001) shows that the catchment vegetation patterns, particularly the tree-line dynamics and peatland distribution, influence the pH- and TOC-status of Lake Tsuolbmajavri (Seppä and Weckström 1999).

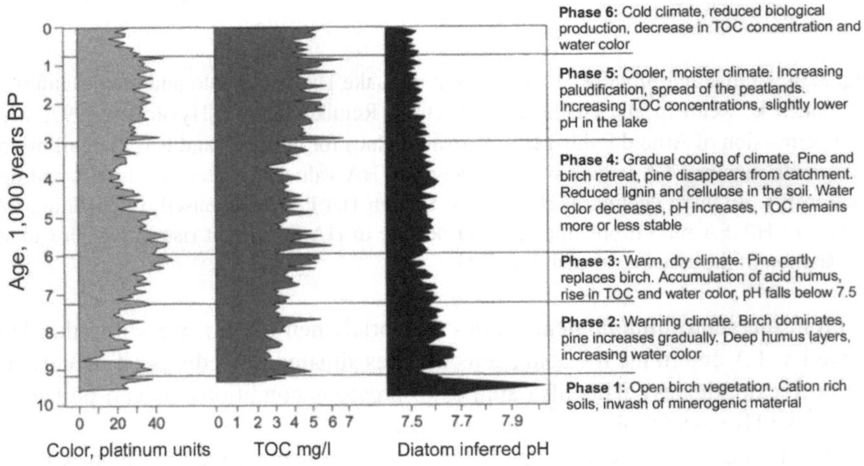

Fig. 3.37. Synthesis of the main Holocene limnological developments in Lake Tsuolbmajavri (after Seppä and Weckström 1999, with kind permission of the University Laval)

The TOC reaches its maximum in the mid Holocene in this Fennoscandian lake, when two maxima are observed (Fig. 3.37). In contrast, in Queen's Lake, Canada, a single mid-Holocene TOC maximum is observed (Fig. 3.35).

It is important to note that, during two periods (phase 3 and phase 4), the development of color and TOC concentration appears unrelated. During phase 3, when pines are invading, the color of the water markedly increases, whereas it markedly decreases in phase 4. Clearly rising or declining trends in TOC concentrations, however, are not obvious. Thus the quality of the TOC is changing more strongly than the quantity. In particular, in phase 3, TOC rich in chromophoric groups (deriving mainly from lignin, and causing darker colors in the lake water) and, in phase 4, TOC poor in chromophoric groups, enter the lake. Major changes in the catchment vegetation are due to the invasion and the retreat of pines. Thus, the detritus from the pines, as a source for the terrestrial HS is responsible the high moiety of chromophoric groups in the aquatic HA and FA.

3.3.3 Lakes in Northeast Germany

Based on regional transfer functions (Schönfelder et al. 2002), Steinberg and Schönfelder (2001) try to reconstruct climate and other environmental variables for north-east German lakes. However, Schönfelder (2002) does not succeed in establishing transfer functions for DOC or for the HS, and non-humic fraction. Neither DOC nor HS as a bulk parameter alone influence reproduction or physiology of the diatoms. This initially unexpected realization gives a decisive impulse to the search for further climate relevant limnological parameters. The absence of a unimodal relationship between DOC and the diatom communities seems to be a contradiction to the known fact that worldwide, wetland waters have a similar diatom flora dominated by *Pinnularia* spp. and *Eunotia* spp. (Schönfelder 2000), which are generally valid as indicators of so-called dystrophic water conditions.

By applying multivariate analysis, Schönfelder et al. (2002) now find that the specialization of many diatom taxa for peatland waters is accounted for, not by absolute DOC concentrations alone, but much more through the **surplus** of dissolved organic substances. When one considers the relative DOC concentration (ratio of DOC to phosphorus or nitrogen), rather than the absolute DOC concentrations, one gets very tight regressions to diatom species, typical of peatland waters. Hence, these ratios must be taken as the criterion for dystrophic conditions in northern German lowland freshwaters. For the reconstruction of climate related changes in freshwater ecosystems, this criterion is fundamental. An increase in absolute DOC concentrations can be due to increased phytoplankton growth as a consequence of climate warming, or a reduction in mineralization as a consequences of a climatic cooling.

Steinberg and Schönfelder (2001) consider that a climate reconstruction based on absolute DOC concentration alone, is not possible, either in general or for Brandenburg State, Germany, in particular. The DOC-P ratio is a much more relevant parameter as a measure of dystrophy. At low temperatures, there is in general a relative enrichment of DOC as a consequence of the difference between the high (allochthonous) input rates of DOC, and the low mineralization rates. This input mechanism also applies to the supply of inorganic nutrients, in particular plant-available phosphorus.

The transfer function is now applied to the reconstruction of climate events which left signals in the sediments of Lake Großer Treppelsee, Brandenburg State, Germany (Steinberg and Schönfelder 2001). This lake is part of a river-lake chain in the northeast German lowlands (near Eisenüttenstadt, River Odra) through which the River Schlaube flows. Climate-relevant parameters are shown in Fig. 3.38. There are two distinct phases in particular: the transition from the late Atlantic to the Subboreal period (5,800 years BP) when a clear jump in the DOC/TP ratios occurred, and around 1,400 years BP where a decrease in the DOC/TP ratios occurs before extensive clearance of woodlands by man which began around 750 years BP. In the first phase, the climate changes from very warm and moderately moist (late Atlantic), to cold and wet (Subboreal). This climate change leads to dystrophy in the River Schlaube waters. This is reversed around 1,400 years BP as the DOC/TP quotient falls to a minimum of 110:1 because the climate is warm and relatively dry.

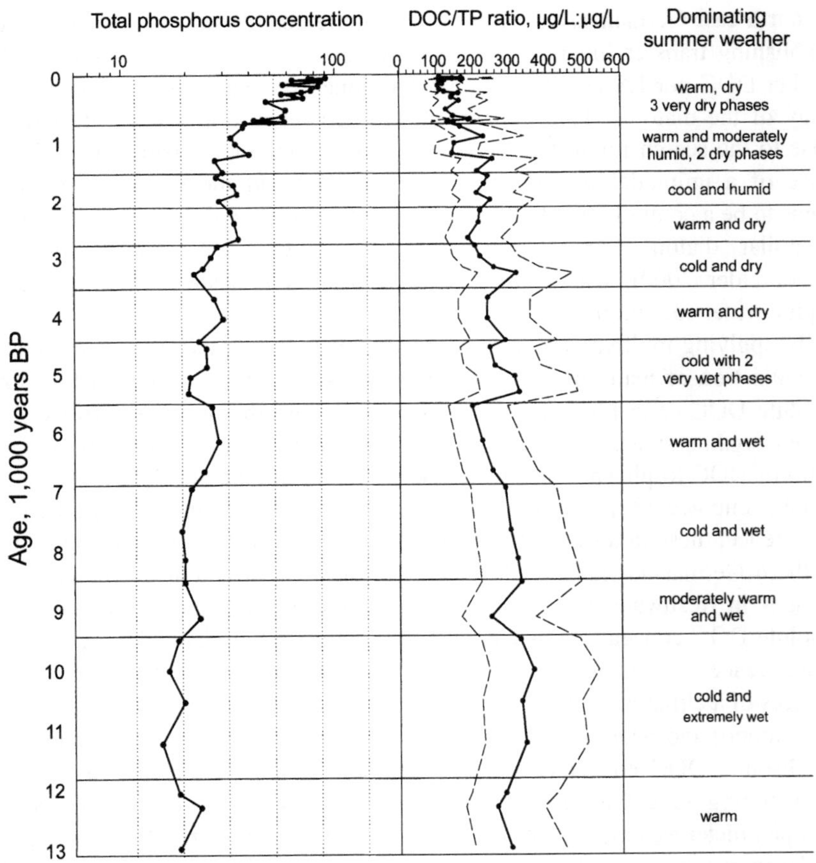

Fig. 3.38. Development of the climate, total phosphorus (TP), and the DOC/TP ratio (including 90 percentile) in Lake Großer Treppelsee (Brandenburg State, Germany), determined by subfossil sediment diatoms (after Steinberg and Schönfelder 2001, with kind permission of Blackwell Wissenschafts-Verlag)

It is evident that dissolved HS can have an important influence on the chemistry of freshwater bodies, in particular on their acid-base status. This is schematically shown in Fig. 3.39. The change is undoubtedly dependent on the climatic changes, and resulting vegetation changes. The graphically displayed regulatory role carried out by HS in freshwaters becomes more complete step by step and the pictures more complex.

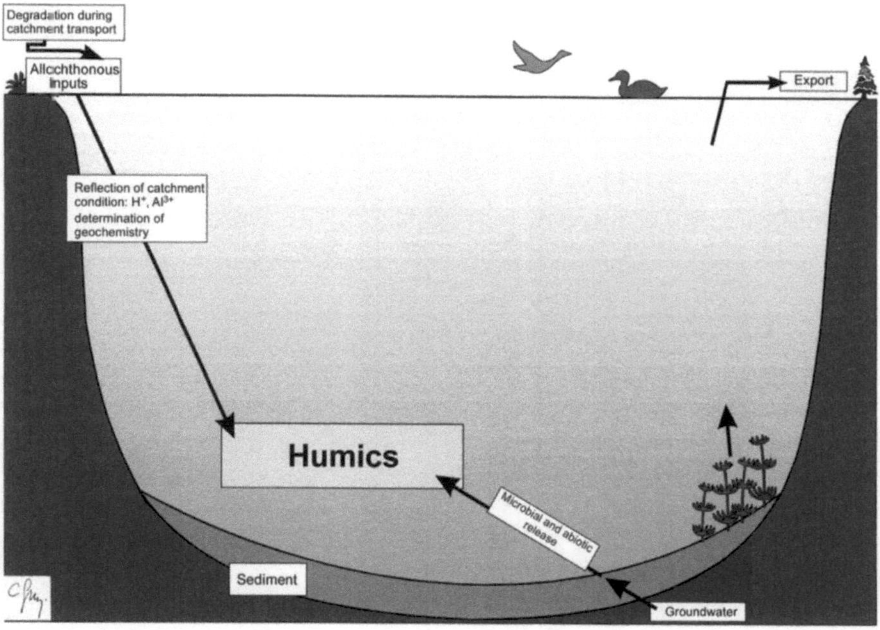

Fig. 3.39. Graphical summary of humic substances as a long-term determinant of water chemistry

4 Humic Substances and Global Climate Change

Global warming, precipitation of acidifying compounds, and increased UV radiation are clear symptoms of global change. The causes are not known in all cases and global climate change continues to be the subject of much discussion[1] (Corti et al. 1999). It is certain, however, that these climate changes inevitably leave traces in all ecosystems. It is to be expected that the most conspicuous changes take place in the most sensitive ecosystems, such as the boreal lakes, the Arctic and Antarctic lakes, and the pseudo-boreal lakes of high-mountain regions. For such lakes, a number of informative studies have been published.

Climatic variables affect vegetation, soils, hydrological conditions, and subsequently the organic carbon budget of landscapes and exports into water bodies. As a consequence, oscillations in quality and quantity of DOC occur in these water bodies. Furthermore, light penetration through the water can change. Overall, a self-accelerating DOC alteration is expected in boreal waters. Since in many areas, there are additional anthropogenic stresses such as eutrophication from increased nutrient loads, the consequences of global changes are not always clear at the local level. On a larger regional scale, various conflicting scenarios are possible, as demonstrated in studies from Canada and Europe.

Dissolved, colored (chromophoric) DOC (cDOC) provide a natural biogeochemical protection against UV radiation (Hessen and Faerovig 2001; Rasmussen et al. 1989; Scully and Lean 1994). This protection applies to a whole-lake scale as well as potentially to single cells: Campbell et al. (1997) discuss a rather speculative possibility of HS as a UV shield. The authors demonstrate that HS can precipitate and accumulate on the surfaces of aquatic organisms such as phytoplankton. These precipitates can protect from UV radiation and, under light expo-

[1] A fundamental question in the global warming debate concerns the extent to which recent climate change is caused by anthropogenic forcing, or is a manifestation of natural climate variability. It is commonly thought that the climate response to anthropogenic forcing should be distinct from the patterns of natural climate variability. But, on the basis of studies of nonlinear chaotic models with preferred states, it has been argued that the spatial patterns of the response to anthropogenic forcing may in fact project principally onto modes of natural climate variability. Conti et al. (1999) use atmospheric circulation data from the Northern Hemisphere to show that recent climate change can be interpreted in terms of changes in the frequency of occurrence of natural atmospheric circulation regimes. They conclude that recent Northern Hemisphere warming may be more directly related to the thermal structure of these circulation regimes than to any anthropogenic forcing pattern itself.

sure, can simultaneously release and supply organic and inorganic nutrients, providing a further advantage to the algae (Chap. 6). In contrast, aggressive oxygen radicals can also be released, leading to cell-surface and membrane damage (Chap. 5).

4.1 Susceptibility of Lakes to Climate Change

The photosynthetic underwater light climate (photosynthetically active radiation, PAR) is controlled by cDOC in two ways, by determining the quality of underwater light and the maximum depth at which photosynthesis can take place. The depth to which UV light penetrates (most meaningfully UV-B: 290–320 nm) is a function of cDOC components in water as Schindler et al. (1996) show for Canadian lakes. The data from 18 Canadian boreal and northern lakes give the following equation (Eq. 4.1):

$$1\%\ \text{UV-B} = 5.173\ [\text{DOC}]^{-0.706} - 1.029;\ r^2 = 0.98. \tag{4.1}$$

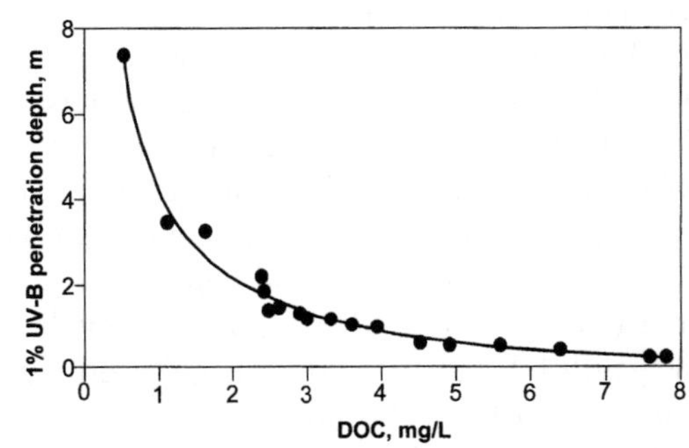

Fig. 4.1. Relationship between measured DOC concentrations and the UV-B penetration depth (1% of surface value) (after Schindler et al. 1996, with kind permission of Nature)

Detailed absorption relationships for the wavelengths 440 (PAR), 340 (UV-A) and 300 nm (UV-B) have recently been published by Bukaveckas and Robbins-Forbes (2000) for over 70 lakes in the Adirondack Mountains in New York State, USA. These waters have a wide range of DOC concentrations and are affected by acid precipitation to varying extents. The following absorption coefficients (a) are derived from this data:

440 nm	$a_{440} = 0.11\ [\text{DOC}]^{1.60}$	$r^2 = 0.85$
340 nm	$a_{340} = 0.62\ [\text{DOC}]^{1.59}$	$r^2 = 0.90$
300 nm	$a_{300} = 1.16\ [\text{DOC}]^{1.57}$	$r^2 = 0.91$

From these relationships, it is obvious that the absorption is more dependent on DOC concentration at shorter wavelengths than at longer wavelengths. According to Eq. (4.1), only a few milligrams of DOC per liter of water (for instance >4 mg/L) are sufficient to act as a biogeochemical shield against UV-B radiation, allowing the UV-B to penetrates only a few decimeters into the water column. From Fig. 4.1, one can also determine the consequences of DOC declines: while the DOC content decreases linearly, the UV-B penetration increases exponentially. This is especially striking at lower concentrations (<3 mg/L). Two factors are responsible for this:

1. autochthonously produced colorless to weakly colored DOC increases quantitatively, since the import of allochthonous strongly colored cDOC decreased due the climate changes; and
2. photodegradation and photobleaching of cDOC from both allochthonous and autochthonous sources increase as the retention time of the lake increases (Schindler et al. 1996) and also the exposure time of DOC to radiation increases.

Two restrictions apply:

1. this mechanism can take place only when the DOC is dominated by cDOC. For example, the saline prairie lakes of the semi-arid regions of western Canada have the highest known DOC for lakes (DOC concentrations in excess of 100 mg/L) (Arts et al. 2000). However, the waterbody is barely visibly colored and absorbs only a little UV radiation, although the inflowing stream water often is dark-brown. The cDOC of inflowing water is photolytically cleaved to components which have low aromaticity and low bioavailability. It follows that a high concentration of DOC can be maintained since no microbial utilization takes place;
2. 2hen the DOC in a water body is dominated by dissolved HS, the UV protection is not without side effects. Direct effects of UV-B can be replaced by the action of other noxious materials. For instance, Kaczmarska et al. (2000) studied enclosure in two lakes on Nova Scotia, Canada, differing in their cDOC content but otherwise with similar chemistry. In the clear-water lake with only 3.5 mg/L DOC, experimental exclusion of UV-B leads to no clear changes in the phytoplankton spectrum including sensitive cyanobacteria. In contrast, in the humic-rich lake with 13 mg/L DOC, the UV-B exclusion causes clear changes. Flagellated cryptomonads are among the sensitive phytoplankton. The H_2O_2 formation potential (Chap. 5) of the humic-rich lake is 10 times that of the clear-water lake. This photooxidation, and/or some other unidentified factor, may be responsible for observed changes in the phytoplankton spectrum.

4.1.1 Arctic and Antarctic lakes

At high latitudes, UV-light penetrates deeply into the water column, because the concentration and absorption capacity of the cDOC are low (Laurion et al. 1997). Measurements of underwater radiation along a latitudinal transect in northern North America showed a strong increase in UV transparency of lakes from the forested zones, north through the tundra to the Arctic polar desert (Pienitz et al. 1997a,b). These increases parallel a decrease in cDOC concentrations. Antarctic lakes are also characterized by large UV penetration depths and low cDOC contents.

The relationship of transparency (= attenuation length) to DOC concentration is shown in Table 4.1. Attenuation length is defined as: T(λ)=1/K, where K is the diffuse attenuation coefficient.

A plot of this data is given in Fig. 4.2. The linear relationship partly reflects the bio-optical uniformity of the sampled lakes: all the lakes are oligotrophic, and not turbid, which means that biotic and abiotic seston play no decisive role in light absorption. The role of cDOC in the light climate of sensitive lakes can be demonstrated with a few numbers. A decrease in cDOC from 2.0 to 1.6 mg/L increases the PAR transparency by only about 11%, whereas a decrease from 1.0 to 0.6 mg/L increases PAR transparency by as much as approximately 76%!

Table 4.1. Empirically determined relationships between transparency (= attenuation length, 1/K) and DOC concentration for Arctic, Subarctic, and Antarctic lakes (Vincent et al. 1998)

Wavelength	c	m	n	r^2
305 nm	0.12	−1.68	7	0.93
320 nm	0.42	−1.98	20	0.95
340 nm	0.58	−2.04	20	0.95
380 nm	0.75	−1.91	20	0.94
PAR	0.99	−1.11	20	0.90

The given coefficients apply for the equation:
log (1/K) = c + m log (DOC), n = number of lakes studied

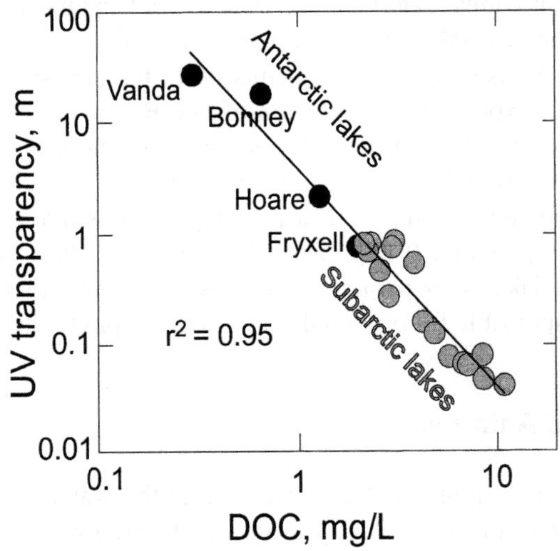

Fig. 4.2. Relationship between UV transparency (1/k, (λ), here as an example given for λ= 340 nm) and DOC concentrations for Antarctic (solid circles) and Subarctic (shaded circles) lakes (modified from Vincent et al. 1998, with kind permission of the International Glaciological Society). Note that both scales are logarithmic

Aquatic ecosystems suffer from several symptoms of global climate changes, such as ozone depletion and subsequent increase in UV-B radiation, and movement of the forest border. Yet, which of these symptoms has the most adverse effect on a lake or river? Vincent et al. (1998) develop a model for the risks of UV increases which can result from the decrease in thickness of the Earth's ozone layer, and the decreased cDOC inputs to lakes. This model explains that a 20% reduction in cDOC has a much greater effect on UV inhibition of phytoplankton, than a similar percentage depletion in stratospheric ozone.

4.1.2 Canadian Lakes

Quantification of the potential effects of UV fluxes on organisms in the past is precluded by meteorological and biological monitoring data which do not extend far enough back. Paleolimnological information is needed here. For instance, subfossil pigments, particularly protective pigments, from lake sediments can document the effects of increased UV radiation on biocoenoses: under UV exposure forms able to synthesize protective pigments, gain dominance. These protective pigments include scytonemin, a pigment which is produced by cyanobacteria following exposure to UV radiation (Garcia-Pichel and Castenholtz 1991).

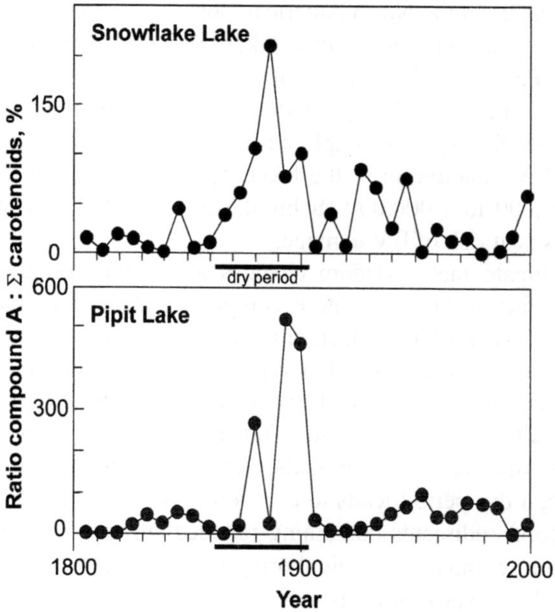

Fig. 4.3. Regional rise in UV penetration in the alpine lakes Snowflakes Lake and Pipit Lake in the Canadian Rocky Mountains during the dry period of approximately 1850 to approximately 1900 (shown by black bars). The rise in UV penetration is shown by the increase in UV protective pigment (compound A) (after Leavitt et al. 1997, with kind permission of Nature)

Leavitt et al. (1997) analyzed pigment profiles in two alpine lakes, Snowflake Lake and Pipit Lake, in the Canadian Rocky Mountains. The results show that the UV impact in these lakes has been significantly higher at many times in the past than in the present, even when the stratospheric ozone is strongly depleted. This agrees well with the above mentioned findings from lakes at the northern tree line. In both Rocky Mountains lakes, compound A which is thought be a UV protective pigment, increases abruptly between 1850 and 1900 (Fig. 4.3). During that time period, maximum concentrations of compound A are significantly higher than at present. The episode of increased UV penetration parallels dry periods at low altitudes and colder temperatures at the timber line of Alberta. It is assumed that this results in decreased inputs of allochthonous DOC. Evidence also comes from present-day studies in the soft-water lakes of the Precambrian Shield: the studies give evidence that even short dry periods can increase the UV penetration depth, since the cDOC concentrations are strongly reduced (Schindler et al. 1997; Yan et al. 1996). Also, it is certain that dry periods influence many lakes such as undisturbed, well buffered mountain lakes, and other lakes with low cDOC contents located north of the tree-line (taiga/tundra boundary) and in the Arctic as well as Antarctic ice deserts (Vincent et al. 1998).

The described mechanism of increased UV penetration after reduced DOC inputs from the catchment should be considered in relation to potential dramatic biozoenotic changes. In the North American boreal zone there are at least 100,000 to 200,000 lakes, and many uncounted lakes in the mountains, tundra, and high Arctic over the Earth with DOC concentrations of $\leq$2 mg/L (Leavitt et al. 1997). As the work of Vincent et al. (1998) shows, even comparably small changes in these low concentrations, for example during dry periods, already lead to significant increased UV penetration in the lakes. Other estimates of numbers of lakes range from 500,000 to 1,000,000 (Schindler et al. 1997), and clearly this huge number of lakes is at risk of UV damage.

Since one climate factor seldom changes alone, the scenario for a realistic combination of factors becomes more complex, and the feed-back mechanisms with reduction in lake DOC concentrations become more numerous. Yan et al. (1996) describe a dramatic case. They report on ten years of observations on the precipitation and water chemistry of Swan Lake, near the nickel-copper smelters and mines of Sudbury, Ontario, on the Canadian Precambrian Shield. A dramatic DOC decline is attributed to a combination of climate change and acidification. In this boreal lake, a dry period leads to a lowering of lake water level. As a consequence, the littoral sediments containing reduced sulfur (originating from acidic precipitation from chimneys of neighboring ore smelters) are exposed to atmospheric oxygen. This exposure leads to a re-oxidation of the sediment sulfur and a release of sulfuric acid into the lake water. The loss of DOC is sufficient to increase the UV penetration depth three-fold. In addition, the inorganic aluminum species which were released with the acidification, also reduced the cDOC content. The mechanism for this reduction has already been described in Chap. 3. A schematic presentation of the feed-back mechanisms in Swan Lake is shown in Fig. 4.4.

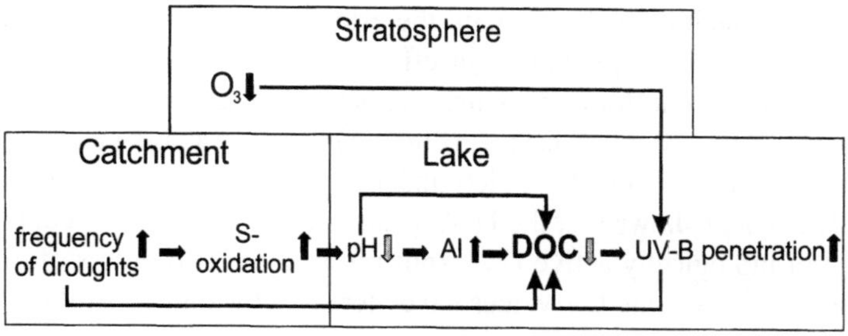

Fig. 4.4. Schematic representation of the link between dryness, acidification, DOC concentration, and UV penetration in lakes. Black linked processes, and grey arrows show changes in the frequency, intensity, or concentration of constituents or processes. The DOC concentrations are controlled by various mechanisms such as acidification, increase of aluminum concentrations, and/or through hydrological changes (dryness), and also through changes in DOC decomposition rates which can occur due to the increased UV-B penetration depth (after Yan et al. 1996, with kind permission of Nature)

Swan Lake may be an extreme example of the effects of multiple climate changes, however, the combination of both mechanisms acid precipitation and climatic changes may apply more generally to numerous lakes. In boreal regions, dry periods can lead to the penetration of oxygen to previously unexposed soil horizons. Through global warming, permafrost soils can partially thaw and thus be exposed to air. The dead organic biomass contains reduced organic sulfur, which on exposure to air, can be chemically or microbially oxidized. Thus the generation of sulfuric acids coincides with low cDOC import from the catchment area due to low precipitation.

From the Canadian studies (Clair and Ehrman 1996; Schindler et al. 1990, 1992), one can generally conclude that global climate change leads to increased evaporation and reduced precipitation. This, in turn, leads to reduced DOC imports and increased retention times. A study by Schindler et al. (1997) quantified the loss of DOC from boreal lakes: during the period 1970 to 1990, with a very warm climate, dryness, and an increased frequency of forest fires, the DOC concentration in lakes of the Experimental Lakes Area (ELA) is reduced by around 15–20%. Reduced DOC inputs from the catchment area are the main cause, although increased DOC consumption in the lakes themselves occurs, but is less important.

Of the various allochthonous sources, the reduced DOC input from rivers and streams is more important than the effects of forest fires. Similar to Yan et al. (1996), Schindler et al. (1997) also show that acidification – in this case experimental acidification – leads to an even greater DOC loss than would catchment mechanisms, since the internal lake elimination mechanisms are very effective (compare Chap. 3). In one ELA lake which was acidified to pH 4.5 in the 1980s, the DOC concentration drops to less than 10% of that before acidification. This DOC loss is higher than the values for mid-European lakes given in Table 3.2.

It has been mentioned several times that lakes at the tree-line are especially sensitive indicators of climatic change. Applying paleolimnology, Pienitz and

Vincent (2000) quantitatively compare the effects of increased UV radiation due to stratospheric ozone depletion, with effects caused by migrations of the tree-line. The mid Holocene forest maximum (Chap. 3) causes a high input of cDOC into the lakes. The UV–relationships (Fig. 4.5) can be reconstructed using a diatom transfer function for the paleo-underwater light relationships ('bio-optic model'). The model shows that the DOC peak in the mid Holocene coincides with a decrease of biologically active UV exposure by about two orders of magnitude. In the last 3,000 years, the UV values have increased by about 50-fold to reach the current values.

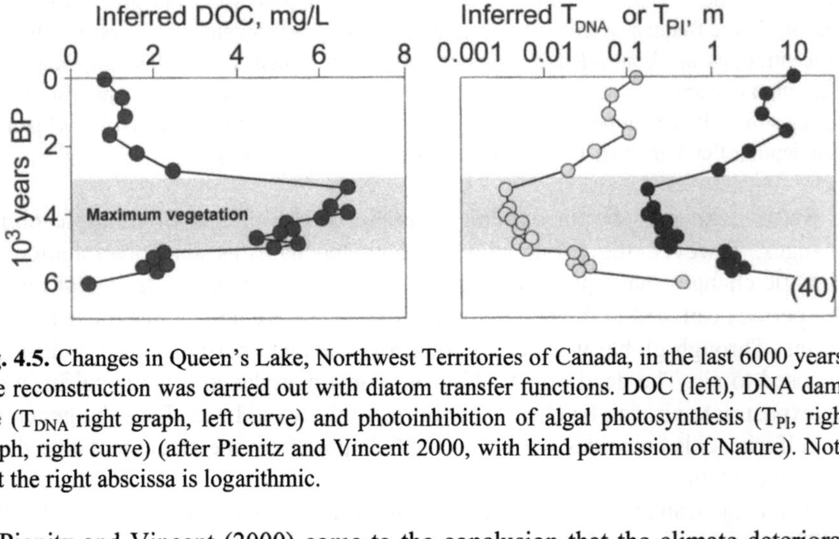

Fig. 4.5. Changes in Queen's Lake, Northwest Territories of Canada, in the last 6000 years. The reconstruction was carried out with diatom transfer functions. DOC (left), DNA damage (T_{DNA} right graph, left curve) and photoinhibition of algal photosynthesis (T_{PI}, right graph, right curve) (after Pienitz and Vincent 2000, with kind permission of Nature). Note that the right abscissa is logarithmic.

Pienitz and Vincent (2000) come to the conclusion that the climate deterioration and the withdrawal of the forests leads to a dramatic rise in exposure of the lakes to UV radiation, which is two orders of magnitude higher than the exposure which will result from a 30% depletion in stratospheric ozone.

4.1.3 European High Mountain Lakes

It is evident that the UV attenuation in the high mountains of Europe is controlled by cDOC. This clearly depends on the quality of DOC (Laurion et al. 2000) as shown in Fig. 4.6. The absorption of the cDOC is the best predictor for the light attenuation. The vegetation in the catchment of a particular lake has a great influence on the UV shield in the form of cDOC or HS, as we have already seen for the lakes of the tree-line in northern Canada (Chap. 3.3.1). Most sensitive are lakes with catchments dominated by bare rock. In contrast, the greatest UV protection is offered by HS which derive from trees, the lignin of which serves as precursors for aromatic-rich HS. The next most protected lakes are those surrounded by meadows. These lakes show medium DOC concentrations, medium absorption, and medium photobleaching coefficients, suggesting that these lakes are less vulnerable to UV radiation than lakes located on rocky terrain, but more vulnerable than lakes with trees in their catchment (Reche et al. 2001).

Chlorophyll exhibits a weak, but significant correlation with attenuation. This means that in deeper layers of the water column, phytoplankton cells contribute considerably to the UV attenuation. Photobleaching of cDOC (Chap. 5.3.4) is a serious problem in high alpine water bodies, particularly in shallow lakes in which the plankton cannot avoid UV radiation through vertical migration. Most sensitive are shallow high alpine lakes, particularly if there is no vegetation in the catchment area.

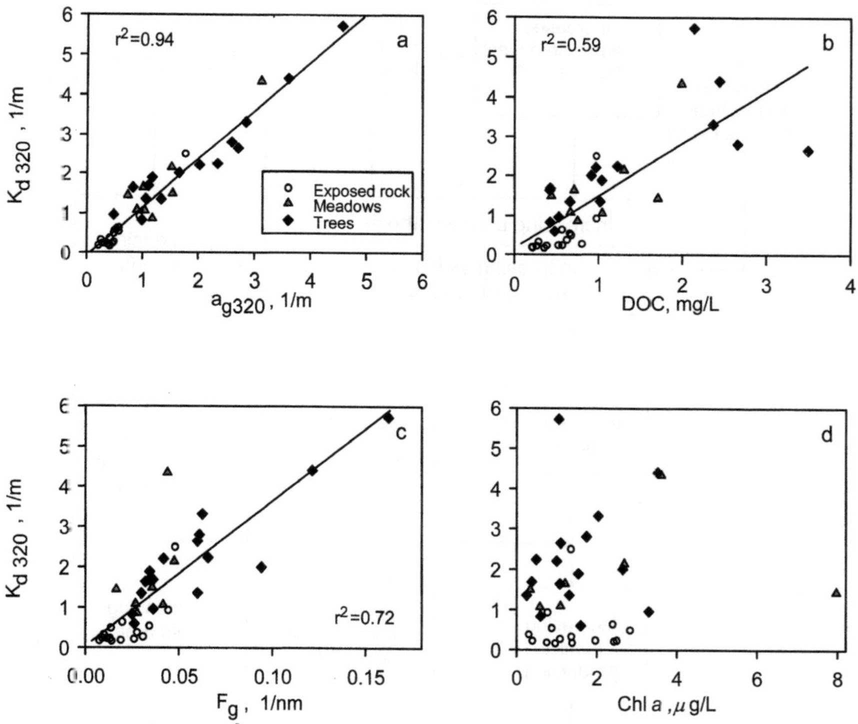

Fig. 4.6. Dependence of diffuse attenuation coefficient (measured at 320 nm, K_d320) of **a:** cDOM absorption at 320 nm (a_g320), **b:** DOC; **c:** cDOM fluorescence (F_g); **d:** chlorophyll *a*-concentrations (from Laurion et al. 2000, with kind permission of the American Society of Limnology and Oceanography)

4.2 Increase in DOC Concentration

The awareness of problems caused by DOC began in Scandinavia some 20 years ago, and is now spreading over Europe. This concern is directed to applied issues (DOC or NOM problems in drinking water processing) as well as to more pure limnological studies. We will refer to selected studies in Scandinavia, the United Kingdom, and Germany.

4.2.1 Lake and Rivers in Scandinavia

Our knowledge of possible regional consequences of global climate changes is still limited, and a general rule on effects cannot be derived from these limited number of studies. Contradictory statements regarding the fate of HS have to be anticipated.

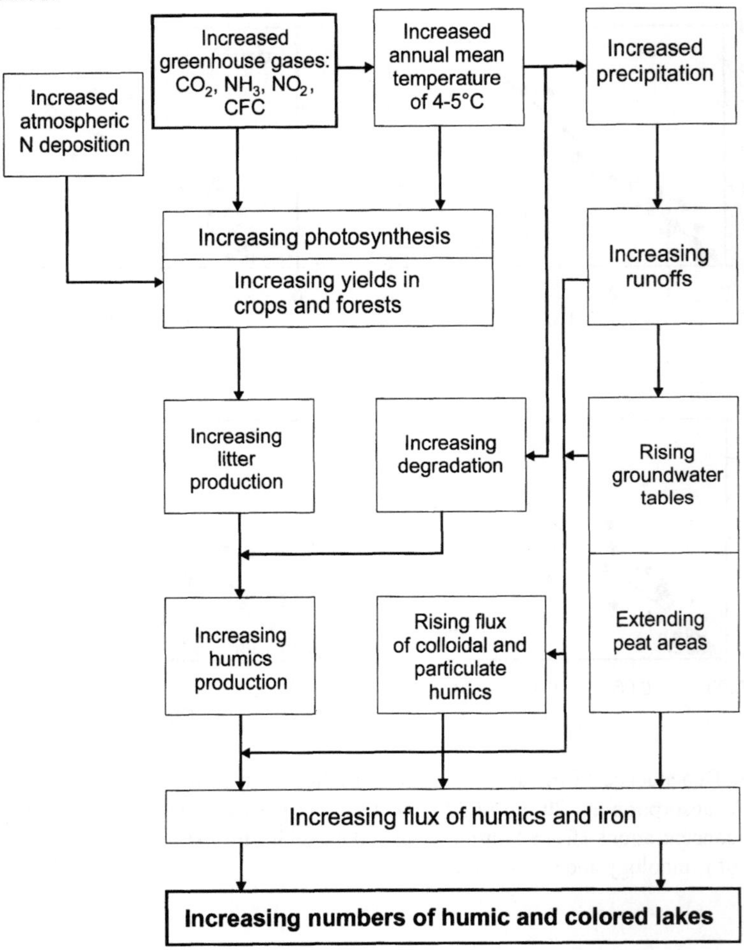

Fig. 4.7. Schematic representation of potential consequences of global warming and regional eutrophication on the water color of Scandinavian lakes (from Forsberg 1992, with kind permission of Kluwer Academic Publishers)

Pioneering studies on the fate of DOC or HS during global climate change are conducted by Forsberg and Petersen (1990) and Forsberg (1992) on Scandinavian lakes. These studies come to conclusions that clearly differ from the Canadian studies which predict decreases of DOC as the major effect during climate warming. The key factors in Forsberg's prognosis are the precipitation which regulates the flow of cDOC from soils to surface water bodies, and the eutrophication by airborne nutrients. Forsberg's conceptual model is shown in Fig. 4.7.

According to prevailing opinion, plants respond to increased atmospheric CO_2 content by increased rates of photosynthesis, as well as increased productivity and efficiency of water usage. The increased greenhouse effect extends the vegetation period and probably results in increased forest production. Forest growth which is usually nitrogen-limited forest growth (Abrahamsen 1980) is stimulated through increased nitrogen inputs from the air (Rodhe and Rood 1985). It is likely that such inputs will continue in the near future rather than will be reduced, since technical efforts to reduce nitrogen emissions are still comparatively poor. Increased temperatures will also increase the rate of decomposition and mineralization of the increased organic matter. Lastly, these eutrophication processes will lead to an increased production of HS which can be transported into surface waters.

As a result of increased temperatures, a greater proportion of precipitation will fall as rain rather than snow. This will certainly induce changes in the runoff regime and the lake water level, and may also enlarge wetland areas (Boer et al. 1990). Forsberg (1992) concludes that if the climate develops in this way, an increased production by vegetation and of HS is to be expected (Fig. 4.7). In Scandinavia, the greenhouse effect, coupled with regional eutrophication from atmospheric N deposition, will probably lead in the next decades to more humic and colored lakes in which many physical, chemical, and biological processes will change.

The model assessment has been corroborated by some Norwegian field surveys. Recent data from monitoring programs indicate temporal rises in color and DOC/TOC for several surface waters. Data from southern Norway on TOC show a significant increase for the last 10 years (Fig. 4.8).

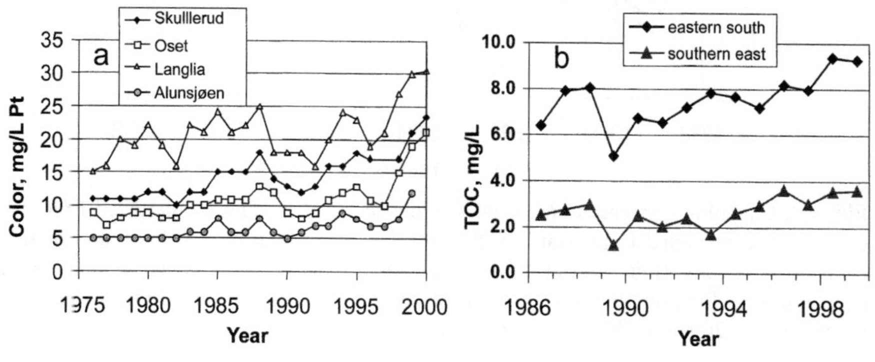

Fig. 4.8. Left: time series for color in the four drinking water sources for the city of Oslo, Norway. Right: time trends for TOC concentrations in lakes from southern parts of Norway (yearly average) (Becher pers. comm. 2001, with kind permission of the author)

For instance, in four drinking water sources for the city of Oslo, the color starts to increase between 1980 and 1985, with a peak in 1987/88, followed by a drop in 1990/91. Later in the 1990s, the color increases to the highest values in 1999/2000. The increase is not governed by environmental damage such as anthropogenic acidification, as areas both affected and unaffected by acid rain, as well as waters with low and high pH values, show the same trend. Also, the relative increase is higher for color than for TOC, which indicates a change in NOM quality. Furthermore, there is a covariance between NOM trends and precipitation and temperature. That means that climate change may be one main reason for the observed trend (Becher et al. 2001). As discussed above, increased temperature and moisture will accelerate the degradation of organic material in soil to humic matter, which again will be washed out to a larger extent by increased precipitation (Liltveld et al. 2001). Since the regions which show the highest increase in TOC (southern Norway) are also those most impacted by airborne nutrients, the complex mechanism described by Forsberg may also apply.

4.2.2 Lakes and Streams in the United Kingdom

Recently, Freeman et al. (2001b) describe a 65% rise in the DOC concentrations in the freshwater draining from upland catchments in the United Kingdom over the past 12 years (Fig. 4.9).

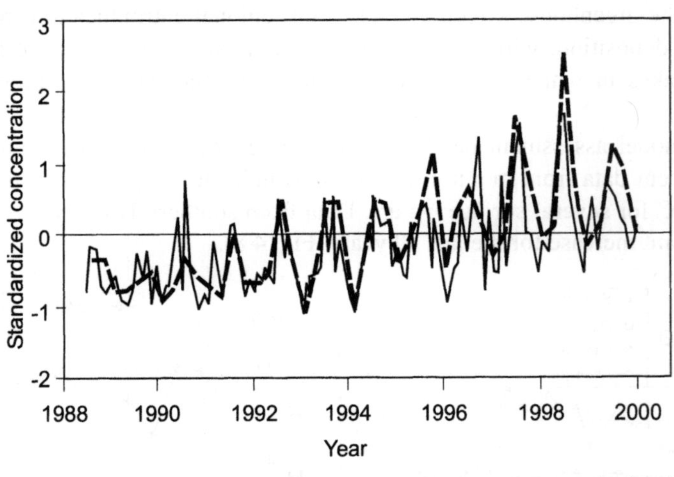

Fig. 4.9. Changing concentrations of DOC: time series of median standardized DOC concentrations determined from quarterly data for 11 lakes (thick broken line), and monthly data for 11 streams (thin line) in the UK Acid Waters Monitoring Network (standardized concentrations for each site have a mean of zero and a standard deviation of one) (after Freeman et al. 2001b, with kind permission of Nature)

Similar to the Norwegian findings, the increase applies to remote, unacidified sites, as well as to those recovering from anthropogenic acidification. Changes in land use or river discharge do not account for the observed increases. The authors assume that rising temperature may drive this process by stimulating the export of DOC from peatland. Laboratory studies indicate that temperature increase may play a role, since an increase of 10 °C leads to a 36% (Q_{10} = 1.36) increase in activity of a phenoloxidase. This is accompanied by an equivalent increase in DOC (Q_{10} = 1.33), and an even greater increase in release of phenolic compounds (Q_{10} = 1.72) from the soil matrix. However, the observed temperature rise in the upland catchments in the United Kingdom is only 0.66 °C, thus temperature increase can only be one among multiple mechanisms contributing to a rise in DOC concentrations, and at present it is doubtful that it is the major one.

4.2.3 German Reservoirs

Since the beginning of the 1990's, German drinking water reservoirs of the middle range mountains also show a clear increase in the specific absorption (SAC at 254 nm) (Fig. 4.10). This indicates an elevated import of DOC from the surroundings. The highest increase occurs in Saxonian reservoirs, for instance in the Rauschenbach reservoir. The forest in the catchment of this reservoir exhibited the most severe symptoms in Germany of the central European forest die back. Die back and re-establishment of forests may be one major reason for the SAC increase.

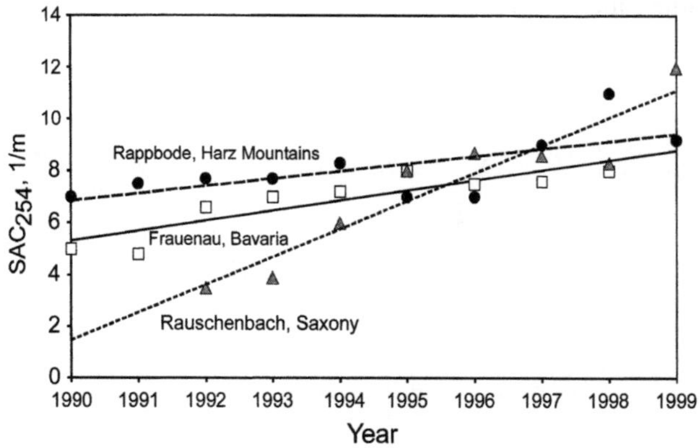

Fig. 4.10. Increase of specific absorption coefficient (SAC) at 254 nm in three German drinking water reservoirs located in middle range mountains (after Grunewald et al. 2002 in press, with kind permission of Blackwell Wissenschafts-Verlag)

From Canadian paleolimnological studies at the boreal tree line, we know that the development of forests is connected with an increase of DOC in the water of lakes and streams (Chap. 3). This mechanism also applies to smaller scales of space and time. For instance, Micka et al. (1985) report that the 70% to 35% decrease of forest cover in the catchment of the Fláje reservoir (Czech Republic), adjacent to the Rauschenbach reservoir, leads within a few years to a reduction in concentrations of humic matter from 8 to 4 mg/L. Hence *vice versa*, the forest re-establishment must increase the DOC concentrations in the water again. However, because the SAC rises also in other German middle range mountain regions, such as the Bavarian Forest and Harz Mountain (Fig. 4.10), which are less impacted by forest die back, additional mechanisms must apply. Probably effects from global climate change also contribute to the SAC increase, but the mechanisms behind the phenomenon may be even more complex and still obscure. At present, the causes and mechanisms of the DOC increase are subject to a extensive ongoing study (Grunewald et al. 2002 in press).

A simultaneous increase in DOC (or as surrogates: color or SAC) in the various regions of Europe which could be attributed to a common trend in climate change cannot currently be found, mainly due to inconsistencies in the monitoring programs and the data sets. Whether the climate changes will lead to a reduction in UV protection by dissolved HS in a specific region, depends particularly on the specific hydrological changes and the nutrient status. Increases in precipitation will probably lead to greater humus content, and consequently more highly colored lakes, while reductions will cause the opposite effect. Therefore, evapotranspiration can play a decisive role (Clair and Ehrman 1996). However, regionally based prognoses remain difficult to develop since the climate, for example in mountain regions, is determined by local conditions and consequently can be asynchronous with regional or global climate trends (Williams et al. 1996).

Fig. 4.11 summarizes the overwhelming role of HS as biogeochemical shield against UV radiation.

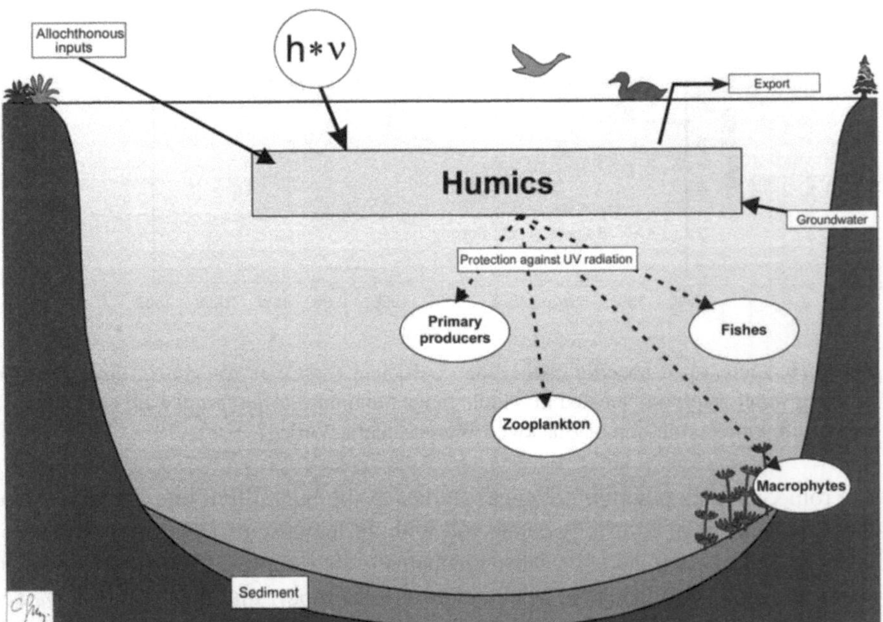

Fig. 4.11. Grafical summary of humic substances as biogeochemical shield against adverse radation

5 Source of Inorganic and Organic Nutrients and Interaction with Photons

Dissolved HS comprise incompletely oxidized carbon. Therefore, HS have the potential to act as substrates for oxidations. Thess processes can be carried out by light or by microorganisms under oxygenated conditions. Usually photooxidation precedes microbial usage. When light is absorbed by the HS, a series of physical and above all, physico-chemical processes such as photolysis occurs. These processes have direct and indirect effects on the biota of lakes and rivers, especially on the heterotrophs. The complexity of these photochemical processes are described and their ecological role understood incrementally. Our present knowledge still gives an incomplete picture, which will gradually be completed.

In contrast, the influence of HS on the underwater light climate and as a determinant of several ecological key processes is already almost classical in the established limnological knowledge, and is covered at the beginning of this chapter.

5.1 Underwater Light Climate

HS change the underwater light climate through absorption of light, and consequently change the living conditions for autotrophic organisms. Jones (1998) summarizes this process succinctly. In clean water, such as in non-productive ocean zones, red light is more strongly attenuated than green or blue light, since it is strongly absorbed by the water itself (Fig. 5.1, left graph).

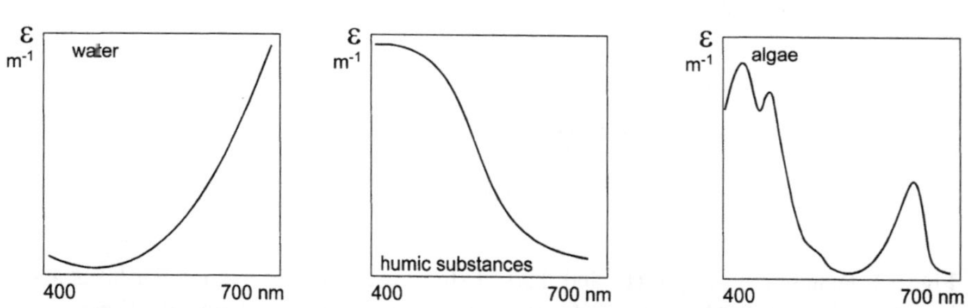

Fig. 5.1. Absorption curves of distilled water, humic substances, and algae (after Eloranta 1999a, with kind permission of Backhuys Publisher)

The presence of low concentrations of HS changes this situation: these substances principally absorb the short wavelengths, namely the blue light of the PAR spectrum (Fig. 5.1, central graph), so that in many lakes, green light penetrates deepest. This is shown schematically in Figs. 5.1 and 5.2. If the HS concentration is high, the red light penetrates more deeply than the green (Fig. 5.2, right).

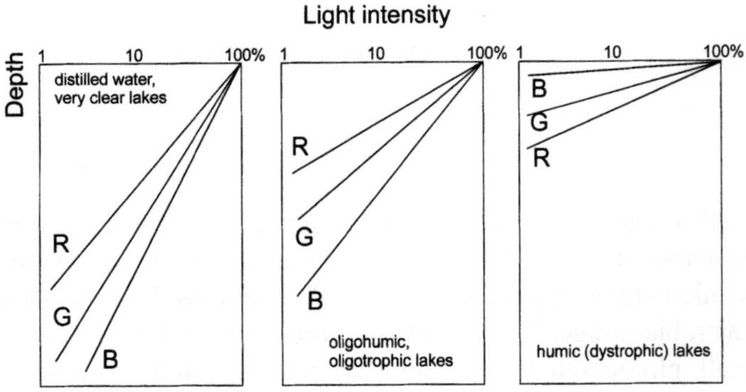

Fig. 5.2. Penetration of red (R), green (G), and blue (B) light radiation into water of different type of lakes (after Eloranta 1999b, with kind permission of Backhuys Publisher)

The underwater light available to phytoplankton can therefore be dominated by blue/green light in humic-poor oligotrophic waters, and by red light in (polyhumic) lakes with high HS concentrations. Phytoplankton, with their accessory pigments, make use of the absorption window of chlorophyll *a* (in the blue and red zones) (Fig. 5.2, right graph). This usage should apply in particular to the phycocyanin-containing cyanobacteria. This pigment absorbs strongly in the 600–650 nm region and is well suited to the red/orange light of humic lakes. There is some relatively weak evidence for a dominance of cyanobacteria in humic lakes from Canadian field studies on picocyanobacteria (Jones 1998).

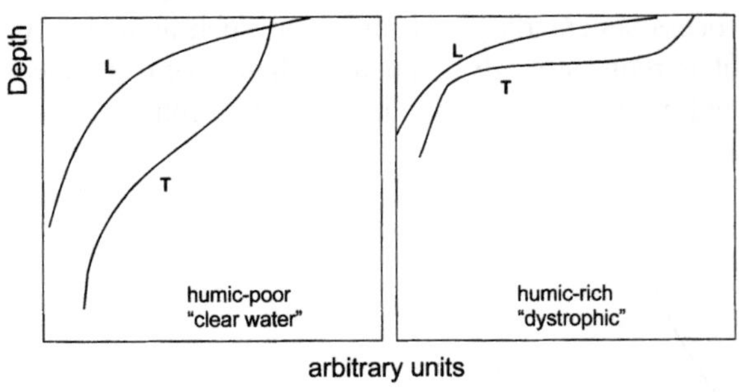

Fig. 5.3. Penetration of light (L) and thermal stratification (T) in humic-poor lakes and in humic-rich lakes (after Eloranta 1999c, with kind permission of Backhuys Publisher)

HS influence light quantity even more than light quality. This means that in humic lakes, the euphotic zone is reduced in depth. In such lakes, the major part of the absorbed energy leads to a warming of the surface water layers, and the minor part leads to photochemical processes. Humic-rich lakes warm up much more quickly in early spring than do humic-poor lakes, and have a sharper, vertical, thermal gradient (Fig. 5.3). This results in a sharp physical barrier between the epilimnion and hypolimnion which inhibits the vertical migration of phytoplankton and zooplankton (Eloranta 1999c). These organisms have developed special strategies to cope with the underwater light and thermal climates of humic lakes, as will be shown in Chap. 9.

More specifically, Jones (1998) shows how this effect is apparent in forest lakes in the Evo region of southern Finland (Fig. 5.4). In these small, sheltered lakes, increasing water color reduces the euphotic zone (Z_{eu}), but also leads to a decrease in the mixing depth (Z_m), so that the effective light climate experienced by the phytoplankton in these lakes varies relatively little with the water color. However, in larger humic lakes in the same region, although the relation between Z_{eu} and water color is consistent (Fig. 5.4 upper panel), the greater susceptibility to wind mixing destroys any correlation between Z_m and water color (Fig. 5.4 lower panel). A similar relationship is reported for 21 Canadian Shield lakes by Fee et al. (1996).

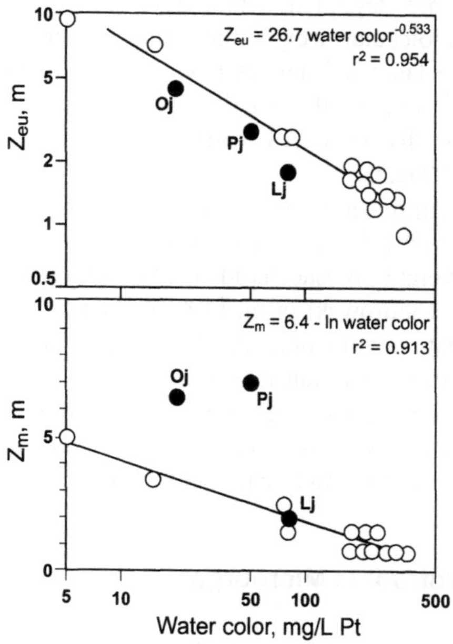

Fig. 5.4. Dependence of a summer euphotic zone depth (upper panel) and a summer mixing depth (lower panel) on water color for 14 small forest lakes in southern Finland. Also indicated are data from three larger lakes in the same vicinity: Pääjärvi (**Pj**), Ormajärvi (**Oj**), and Lovojärvi (**Lj**) (after Jones 1998, with kind permission of Springer Verlag)

In addition to the warming, the penetrating energy brings about a series of photochemical processes, and renders a large proportion of the dissolved HS bioavailable. However, more and more findings indicate that an, as yet hard to quantify, fraction of dissolved HS, is directly bioavailable to heterotrophic microorganisms.

5.2 Source of Carbon and Energy

Dissolved organic matter (DOM) is an important potential energy source for microorganisms. Because of their refractory nature, the biological role of DOM is the subject of contradictory academic opinions and perceptions (Jørgensen 1976). For example, one of these opinions sees humic-rich lakes as unproductive and starved (dystrophic). However, this opinion has been changed by current knowledge, as will be shown (see also Chap. 9).

One of the most important factors which regulates the microbial bioavailability is molecular size, often less precisely expressed as molecular weight. The simplest organic molecules are utilized first, within hours; this material is referred to as 'light fuel' (Münster et al. 1999a,b). The high-molecular weight DOC, deriving from phytoplankton, are degraded within days (Stabel 1977; Steinberg 1977), whereas other high-molecular weight molecules require months to be cleaved (Saunders 1976). The latter substances have highly diverse chemical structures and composition. This implies that small quantities of defined, repeatable structural properties in the dissolved HS matrices make them resistant to microbial definition (Münster 1999a,b).

Recent studies certainly show that a change in the paradigms on this theme is called for. Already Sherr (1988) shows that marine heterotrophic flagellates utilize high-molecular weight polysaccharide (>500 kDa) as C source better than do bacteria. In addition, Ciborowski et al. (1997) describe that simuliid larvae can grow in a medium that contains only allochthonous DOC, meaning that they can use DOC as a direct or indirect nutrient source. Small DOC components are also taken up by other animals, for instance nematodes, through their cuticles, (Lopez et al. 1979). For the bacterivorous nematode *Adoncholaimus thalassophygas,* this source can be even more important than microorganisms in some circumstances.

5.2.1 Use by Heterotrophic Microorganisms

The above described claim that humic lakes have low productivity is commonly true for primary production. However, it does not generally apply for heterotrophic microorganisms. New measurements of community respiration and bacterial production confirm that heterotrophy is at least as high in humic lakes as in eutrophic lakes. In humic lakes, it exceeds primary production. Preliminary findings show that this also applies for other waters, such as many humic-poor lakes (Cotner et al. 2000), and also for extremely acidic mining lakes (Nixdorf et al. in press). Chapter 9.1 summarizes the general applicability of 'net heterotrophy'. One can assume that bacteria are the most important component of community respiration. How can one explain this mechanism if HS are the principle fuel (Münster et al. 1999b)?

5.2.1.1 Molecular Size and Degradability

For lakes, there is increasing evidence that autochthonously produced C alone is insufficient to support bacterial production. Therefore, the possibility that allochthonous DOC is responsible for the higher bacterial biomass is discussed (Münster 1999a,b, with further references). Tranvik (1988, 1989, 1990, 1992a,b, 1993, 1998) is one of the first authors to test this hypothesis. He tests the capacity of two DOC molecular weight fractions (>10 kDa and >1 kDa) to support bacterial growth. High-molecular weight DOC (>10 kDa) is generally more available than low-molecular weight DOC (<10 kDa), and supports growth of more bacteria per unit organic C (bacterial carrying capacity). However, when the HS content rises, the availability of this high-molecular weight DOC to bacteria decreases. A general result of Tranvik's studies is that humic-rich lakes in southern Sweden possess a higher carrying capacity than humic-poor lakes in this region. According to the work of Tulonen et al. (1992), this also applies to (small) Finnish humic lakes.

Later, Amon and Benner (1994) establish that the findings from Scandinavian lakes also apply to marine systems. They study the bioavailability of high- and low-molecular weight DOC in water samples from the northern Gulf of Mexico during a diatom bloom. Bacterial growth and respiration is three to six times higher in the presence of high-molecular weight DOC than with low-molecular weight DOC. Although both DOC pools contain poorly defined mixtures of compounds with differing reactivities and turnover times, the results of Amon and Benner (1994) show that in oceanic DOC, there are small molecules present which turnover very slowly and are relatively unavailable to microorganisms.

In another study, Amon and Benner (1996a) expand these studies to waters with various DOC sources and compositions. These include waters with DOC from rivers, estuaries, and unpolluted, non-eutrophicated marine systems. They describe that in the Amazon, more than 80% of the DOC belong to the high-molecular weight fraction, while up to 70% of marine DOC is in the low-molecular weight fraction. In all experiments, they find that bacterial growth and respiration rates are significantly higher with high-molecular weight DOC than with low-molecular weight DOC. The high-molecular weight DOC is utilized at a rate of 0.7–22.5% per day and low-molecular weight DOC at only 0.5–6.6% per day. From this, it follows that the traditional model of DOC decomposition, in which low-molecular weight compounds are the most bioavailable, is not generally applicable for many natural DOC mixtures.

From the above described independent studies, a new conceptual model is called for in which the bioreactivity of organic material follows a continuum from large to small molecules. High-molecular weight fractions are biodegraded faster than smaller fractions, leaving behind, or even producing, the smaller and more refractory molecules. The bioreactivity also depends on the diagenetic situation.

In other words, fresh material is degraded more rapidly than older material. The size-reactivity continuum model is shown in Fig. 5.5.

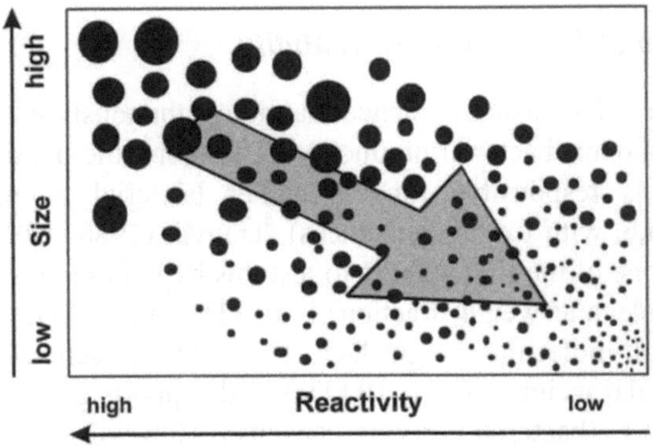

Fig. 5.5. Schematic diagram of the size-reactivity continuum model for organic matter decomposition in aquatic environments. Arrow indicates the major pathway of degradation from bioreactive organic particles and macromolecules to refractory low-molecular weight compounds. The size of the dots is representative of the size of organic matter, with a gradient from large dots for particulate organic matter, and medium-sized dots for high-molecular weight DOM, to small dots for low-molecular weight DOM. The distribution of dots indicates that most larger sized organic matter is more reactive than most smaller sized organic matter (after Amon and Benner 1996a, with kind permission of the American Society of Limnology and Oceanography). See, however, comment on the high turnover of low-molecular weight compounds such as free amino acids and mono- and disaccharides in the text.

This diagram clearly explains the existence of more or less refractory high-molecular weight and particularly low-molecular weight fractions of organic C. However, it does obviously not take into account that low-molecular weight organic nutrients, like freely dissolved amino acids and carbohydrates as well as oligopeptides and disaccharides, have very high turnover times (Bunte and Simon 1999; Grossart and Plough 2001; Grossart and Simon 1998a,b; Fischer 2002; Münster 1984, 1985; Münster et al. 1999a,b). That means that the residual concentrations do not represent the reactivity of these low-molecular weight organic compounds.

Hatcher and Spiker (1988) present a general scheme for the humification process (Chap. 2) which corresponds well to the size-reactivity continuum model. In their scheme, the labile molecules are rapidly used up during the early diagenesis phase. Refractory compounds are selectively conserved and eventually converted to humin. Further degradation leads to oxidation of humins and the formation of HA and FA. Humification leads, overall, to strongly oxidized substrates of lower molecular weight and increased solubility. These are reasons for the assumption that the refractory low-molecular weight components in the water have hardly any similarity to biomolecules, and therefore 'escape' microbial utilization (Amon and Benner 1996a).

In contrast to the refractory low-molecular weight compounds, the high-molecular weight fractions have a higher proportion of identifiable and probably reactive biomolecules, as shown for amino acids and carbohydrates in FA and HA in Fig. B.2.1.2. The results from molecular modeling of the binding type between 'light fuel' such as peptides and carbohydrates on the one hand, and HS on the other (Fig. B.2.1.7) show that only weak bonds exist between the HS and the microbial substrates. Hence it is not only likely, but certain, that the latter have a high bioavailability. What Amon and Benner discuss hypothetically in their model shows therewith to also have a mechanistic basis – except for the high turnover low-molecular weight compounds.

A series of studies deals with the direct and indirect utilization of dissolved HS by aquatic microorganisms (de Haan 1974; Geller 1985a,b, Geller 1986; Moran and Hodson 1990; Stabel et al. 1979; Tranvik and Höfle 1987; Tranvik 1988, 1993, 1998). Stabel et al. (1979) describe, for example, that in decomposition studies with bacterial cultures, up to 30% of dissolved HS is utilized. Obviously, non-HS support bacterial growth better than HS (Moran and Hodson 1990; Steinberg 1977). The total decomposition of dissolved HS is slow, but continuous, as shown by Geller (1986). Ryhänen (1968) attributes the low rates of biological decomposition of HS in Finnish lakes to a shortage of inorganic nutrients. As we shall see, this is only one of several possible explanations.

5.2.1.2 Enzymatic Decomposition

In many humic-rich freshwaters, there are three principal ways in which dissolved HS decomposition by microbial extracellular enzymes can be defined (Münster and de Haan 1998; Pflugmacher et al. 1999a):

1. **hydrolytic cleavage** by enzymes, such as acid and/or alkaline phosphatase, aminopeptidase, lipase/esterase and glucosidase, which can cleave phosphoester-, peptide-, lipid-ester and glycoside bonds respectively. The hydrolytic cleavage takes place in the side chains, which are bound to the humic backbone by ester or amino bonds,
2. **oxidative cleavage** of the humic backbone itself. This can, for example, be carried out by phenoloxidases or phenolperoxidases, which can cleave particular aromatic ring structures. These oxidative enzyme systems can also carry out the reverse reactions; and
3. **metabolism within the transformation system**. The enzymes of this system transform hydrophobic compounds into more hydrophilic ones which are excreted (animals) or stored in the vacuoles (plants).

The enzymatic cleavage of side chains by hydrolases, proteases, phosphatases[1] and aminopeptidases is comparatively well studied (Burns and Ryder 2001; Freeman et al. 1990; Hendel and Marxsen 1997, 2000; Jahnel et al. 1994; Jahnel and

[1] Observations of Siuda and Chrost (2001) suggest that bacterial phosphatases participate substantially in processes of DOC decomposition in lake water, whereas the specific 5'-nucleotidase is mainly responsible for bacterial P demand.

Frimmel 1995; Kang et al. 2001; Marxsen and Fiebig 1993; Marxsen and Schmidt 1993; Münster 1994, 1999a; Münster et al. 1992a,b; Scholz and Marxsen 1996; Sinsabaugh and Foreman 2001). Clear evidence exists that the hydrolytic enzymes may be inhibited in the presence of HS, particularly the phenolic compounds of the HS (Freeman et al. 1990, 2001a).

In contrast, studies on enzymatic attack on the C skeleton itself are rare, such as on phenoloxidases, ligninolytic enzymes, or phenolases (de Haan 1976; Freeman et al. 2001a,b; Hendel and Marxsen 2000; Münster et al. 1998), probably related to methodological difficulties. It is probable that these enzymes play a greater role in the turnover of the DOC reserve than hitherto suggested. It is, for example, plausible that wood decomposing enzymes (for degradation of cellulose and lignin) can also degrade HS. In this regard, the best-studied model organism is the white rot fungus (*Phanerochaete chrysosporium*). However, other white rot fungi (such as *Nematoloma frowardii*) and litter decomposing fungi, like *Stropharia rugosa-annulata* or *Agrocybe praecox,* can also degrade lignin (Fritsche 1998).

Wood decomposition via the cellobiose oxidase system leads, through various stages, to the Fenton reaction (reduction of ferric Fe with hydrogen superoxide), with the hydroxyl radicals, which have a high oxidation potential and oxidize non-specifically. Lignin is broken down by lignin peroxidase (syn. ligninase) and other enzymes[2] (Fritsche 1998): the extracellular peroxidase requires so-called mediators, as alone this enzyme cannot make direct contact with the lignin network. These mediators, such as veratryl alcohol (3,4-dimethosybenzalcohol), are oxidized, in this case to veratryl aldehyde and the oxidation products react with the lignin. The intervention of mediators explains the evident non-specificity of lignin peroxidases, with their only substrate being H_2O_2. This H_2O_2 is delivered by various oxidases including the above-mentioned cellobiose oxidase. During daytime, H_2O_2 in lakes is produced via photolytic release from radiated HS (Chap. 5.3).

The situation with H_2O_2 as a mediator indicates that in addition to lignin, a diversity of xenobiotics can be oxidized; these include chlorinated hydrocarbons, polycyclic aromatic hydrocarbons, polychlorinated biphenyls, polychlorinated dioxins and furans, as well as explosives (TNT), and pesticides (Achtnich et al. 1999; Haemmerli et al. 1986; Hofrichter et al. 1998; May 1996; Mougin et al. 1994; Valli and Gold 1991). HS can certainly also be oxidized, as shown by Haider and Martin (1988) and Dehorter and Blondeau (1992) for soils. This pathway also applies for decomposition in freshwaters, particularly the sediment surfaces, which one can consider as water-saturated soils. Hence, there are no plausible reasons why the coarsely sketched processes of cellulose and lignin degradation, cannot proceed in sediments as it does in soils. Fritsche (1998) emphasizes that lignin decomposition is a co-metabolic process, in which sugars from the degradation of cellulose and hemicellulose serve as growth substrates. In this instance, there may be a difference between soils and sediments, as the provi-

[2] Recent findings suggest that the diversity of the genes encoding a single step in lignin degradation is high, as exemplified with a salt marsh bacterial community (Buchan et al. 2001).

sion of growth substrates for sediment bacteria is not totally guaranteed, since usable substrates from the seston are utilized during sedimentation.

The white rot fungus *P. chrysosporium* has the 'weakness' that its lignin peroxidase activity requires low N conditions (Li et al. 1994; May 1996 with further references), which is generally the situation near the surface of sediments due to denitrification. Because of this, there may be a thin layer within sediments where there are sufficient organic nutrients for the development of *P. chrysosporium*, and a medium so poor in N that the fungal lignin peroxidase can operate. Experimental evidence to support this suggestion does not yet exist.

Münster et al. (1998) are the first limnological group to test an unspecific peroxidase in a humic lake. This particular peroxidase uses the previously mentioned veratryl alcohol as substrate. The peroxidase activity in the epilimnion of the polyhumic lake Mekkojärvi varies seasonally from spring (74 nmol/L/h) to autumn (273 nmol/L/h), and in the water column near the surface (800 nmol/L/h) to the low light zone (110 nmol/L/h) (Fig. 5.6). The necessary H_2O_2 can be delivered through the illumination of HS (Chap. 5.3), which means it is a light-driven process, since the dark activity of the peroxidase is very low in this lake. Enzymes which directly attack the C skeleton of HS, rather than the side chains, evidently play a fundamental role also in aquatic systems.

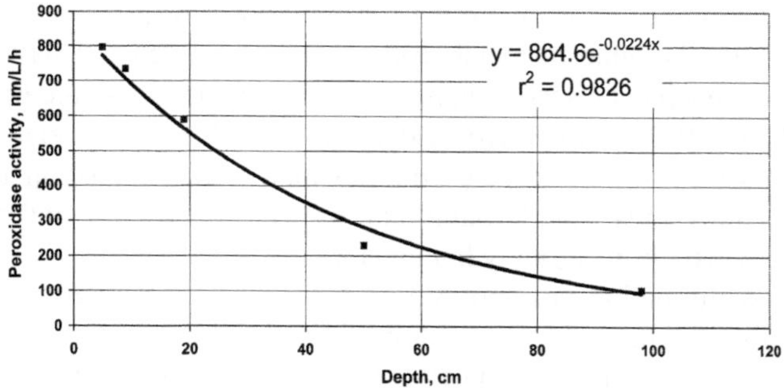

Fig. 5.6. Decreasing peroxidase activity in a depth profile of the polyhumic Lake Mekkojärvi, Finland, for the 2.0-µm pre-filtered size fraction in summer 1994 (after Münster et al. 1998, with kind permission of Wiley-VCH)

An even less specific peroxidase is under study in lake Fuchskuhle (Fig. 5.7) (Burkert in prep.). It is evident that the peroxidase activity is – with one exception – elevated in that part of the lake which is influenced by an oligotrophic bog in the catchment. Interestingly, the highest activities are found during fall and winter months, which have the highest inflow of surface and seepage waters. Because the search for aquatic microorganisms responsible for the peroxidase activity has failed thus far, Burkert (pers. comm.) concludes that the peroxidase is probably washed in from the catchment during the high flow periods. If this assumption can be corroborated, is would mean that non-eutrophicated lakes and their metabolism depend on their catchments in two ways: 1) C and energy is washed in as HS, and 2) the means to handle these substances are imported.

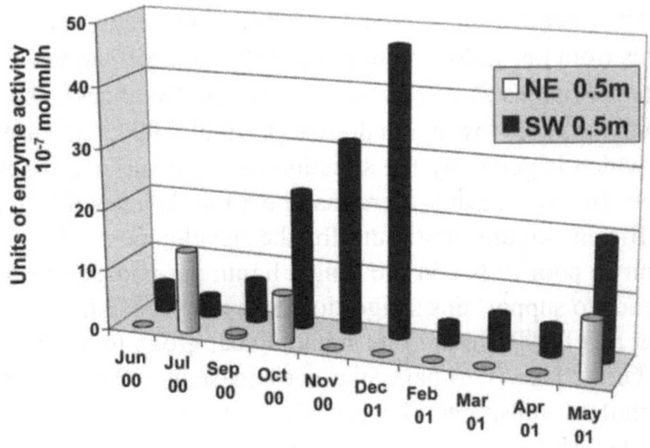

Fig. 5.7. Seasonal changes in peroxidase activity in the artificially divided forest Lake Fuchskuhle (Brandenburg State, Germany). Samples were taken 0.5 m below the surface from the northeast (NE) and southwest (SW) compartment. The SW basin is influence by an oligotrophic bog in its catchment (after Burkert in prep, after Steinberg and Burkert 2002, with kind permission of Blackwell Wissenschafts-Verlag)

Since HS contain aromatic and aliphatic structures which resemble several classes of xenobiotics, it is suggested that microorganisms which are able metabolize HS are also able to metabolize several xenobiotics. This suggestion is supported by the almost classic finding that a *Pseudomonas* isolate grows on benzoate (de Haan 1977). This microorganism is also able to co-metabolize HS. In addition, Larsson et al. (1988) describe an increased degradation of particular chlorinated compounds (trichloroguaiacol, as well as di-, tri-, and pentachlorophenol) if consortia from strongly humic lakes are added.

The biotransformation enzyme system represents another group of enzymes potentially able to attack the C skeleton of HS. Studies on the role of these enzymes in the C cycle of lakes and rivers are very rare. It has been known for a long time that these enzyme systems can degrade xenobiotic chemicals such as polycyclic aromatic hydrocarbons, dioxins, furans, and biphenyls. Transformation enzymes are naturally non-specific in their activity. In reality, these enzymes act in two phases (Pflugmacher et al. 1999a):

- **Phase I** – oxidizing or reducing enzymes which provide the hydrophobic substrate with a reactive group, and are thereby making the substrate slightly less hydrophobic, and
- **Phase II** – conjugating enzymes which conjugate a water soluble compound to the already altered substrate, making it even more hydrophilic.

First results from lake sediments show that these type of enzymes act not only within organisms, but also outside them (Chap. 8, Fig. 8.40). It is therefore expected that these enzymes can degrade 'heavy fuels' senso Münster, that is to say HS. It is also very plausible that (micro)-organisms which are pre-exposed to xenobiotic chemicals and, hence, have their transformation systems already activated, possess a higher capacity to degrade refractory dissolved HS than non-pre-exposed microorganisms. Even highly refractory dissolved HS contain a few functional groups that are exposed to the action of the transformation enzymes, especially those of phase II. In addition, it is very likely that a known fraction of functional groups of dissolved HS are newly built photochemically, if exposed to sunlight. Future studies on the transformation enzymes, such as the conjugating enzymes of phase II, which include the well known glutathione-*S* transferases (Pflugmacher and Steinberg 1997; Pflugmacher et al. 1997, 1999a, 2001), hold much promise for understanding the HS metabolism in addition to that of hydrolytic and oxidative enzymes (Jackson et al. 1995; Münster et al. 1998). A detailed description of the functions of phase II biotransformation enzymes in HS-exposed organisms, with particular reference to glutathione-*S* transferases, will be given in Chap. 8.

5.2.2 Predictability of Microbial Utilization

If one considers the variation of HS in terms of structure or composition, as already described, the question automatically arises as to whether one can derive a consensus of the degree to which HS can be broken down chemically, and the degree to which they can be broken down by aquatic microorganisms. In other words, are there particular structural elements or chemical compositions accountable for the chemical or microbial attack?

Sun et al. (1997) answer this question by studying the degradability of DOC from various locations in the catchment of the Ogeechee River in Georgia, USA. The authors determine how much microbial biomass grows per unit DOC. Elemental composition is used to describe the DOC composition. Surprisingly, a multiple regression analysis yields a simple general relationship between microbial growth and elemental composition (Eq. 5.1), a_0–a_3 being coefficients:

$$\text{microbial growth} = a_0 + a_1(\text{H/C}) + a_2(\text{O/C}) + a_3(\text{N/C}) \tag{5.1}$$

For the Ogeechee River samples, $a_0 = 38.4$; $a_1 = 10.6$; $a_2 = -70.9$ and $a_3 = 183.2$ with an $r^2 = 0.933$.

This regression shows that bacterial growth is positively correlated with the H/C ratio, indicating the relative proportion of aliphatic and aromatic structures in the DOC. The negative correlation with the O/C ratio shows that the more weathered DOC, which generally has a higher content of carboxyl groups, is less bioavailable than fresh DOC. The strongest influence on microbial growth is the N/C ratio, which indicates the proportion of protein derivatives in the DOC. Amino acids, peptides, and proteins in the DOC support microbial growth better than do carbohydrates. This finding is supported by studies of soil organic matter (Marschner and Kalbitz in press; Ohno and Doolan 2001). Intrinsic DOM characteristics that generally enhance its biodegradability are high contents of carbohydrates, organic acids, and proteins for which the hydrophilic neutral fraction seems to be a good estimate. In contrast, aromatic and hydrophobic structures that can also be assessed by UV absorbance decrease DOM biodegradability, either due to their recalcitrance or due to inhibiting effects on enzyme activity.

Fresh, algal-derived DOM from an Arctic ice floe revealed – not surprisingly – slightly different results. Approximately 30% of the DOC was used by bacteria, indicating the highly reactive nature of this fresh material. Over half of the DOC consumption is accounted for by losses of combined neutral sugars and amino acids. The initial composition of the DOC was characterized by high neutral sugar (14% DOC) and amino acid (7.4% DOC) yields, and the dominance of glucose and glutamic acid. During microbial degradation, the neutral sugar and amino acid yields decreased, and the molecular composition of the DOC became more uniform. The relatively constant abundance of D-amino acids, and the dramatic changes in the neutral sugar and amino acid compositions, indicated that bacteria are important in shaping the chemical composition of marine DOC by selectively removing bioreactive components and by leaving behind biorefractory components (Amon et al. 2001).

Very recently, Kisand et al. (2002) provide information on the phylogenetic affiliation of culturable bacteria capable of catabolizing riverine DOM. Additions of riverine DOM consistently promote the growth of estuarine bacteria in C-limited dilution cultures. At least 42 different taxa are culturable. Five species of the *Cytophaga-Flexibacter-Bacteriodes* group and one in the γ-proteobacteria phylogenetic group (*Marinomonas* sp.) are numerically dominant. All dominating isolates are determined to be new species. The same group of species dominate, independent of the season investigated, suggesting a low diversity of bacteria catabolizing riverine DOM in the estuary. It also suggests a broad tolerance of the dominating species to seasonal variation in hydrography, chemistry, and competition with other species.

5.2.3 Hotspots of Carbon Turnover in a Lake

Following Forbes' (1887) fascinating concept of a 'lake as a microcosm', the scientific attention of classical freshwater ecology has been focussed on the pelagic zone of lakes, irrespective of the actual contribution of this zone in the total carbon turnover. However, freshwater wetlands, such as littoral zone of lakes dominated by emergent macrophytes (mainly *Phragmites australis*) are among the most productive ecosystems worldwide with annual above-ground production of up to 10 kg dry mass per m^2 (Hocking 1989; Mitsch and Gosselink 2001; Kvet and Westlake 1998; Wetzel 1990; Wetzel and Howe 1999). Furthermore, most lakes of the world are relatively small in area and shallow (Wetzel 1990). In such lakes, a substantial portion of the carbon produced in the whole lake may derive from above-ground macrophyte production in littoral areas.

Very recently, Buesing (2002) presented the first estimates of bacterial and fungal biomass and production at the aquatic ecosystem level in a non-pelagic environment, demonstrating that the detrital layer of the littoral zone is inhabited by an extremely active microbial assemblage with a respiration of 875 $gC/m^2/a$ and a production of 800 gC/m^2/a. Bacteria are the most important organisms recycling organic matter within the littoral reed stand, in spite of their low biomass. Fungi contributed only a minor fraction to the secondary microbial production, regardless of their approximately 20-fold greater biomass than that of bacteria. To sustain the observed carbon demand for microbial production, more than the total annual above-ground primary production (of *P. australis*) is needed. That means that additional allochthonous sources of carbon are neccessary to sustain the observed heterotrophic production, especially in view of the potential export of particulate and dissolved organic matter from the littoral into the pelagic zone (Wetzel 1990).

Littoral zones are hotspots of particulate and dissolved carbon metabolism from autochthonous and allochthonous sources. Although Buesing (2002) does not characterize DOC qualities, for instance by fingerprint techniques (see Chaps. 3, 5.2.6), it is easy to presume that only the most recalcitrant fractions, HS and their building blocks, resist the microbial attack. Moreover, due to the high metabolic activity, the production of refractory dissolved organic matter from rapidly utilizable labile compounds, such as glucose or glutamate, by bacteria is feasible (Amon and Benner 1996a; Ogawa et al. 2001). Ogawa et al. (2001) describe the potential underlying mechanism: exoenzymes play a critical role in both the microbial utilization and the formation of refractory DOC. It is possible that nonspecific or promiscuous activities of enzymes, that occur with much lower efficiency than primary activities, occasionally produce fragments from macromolecules that escape recognition by bacterial enzymes. Given this scenario, the rate of formation of refractory DOC is dependent on the rate of microbial activity: the higher the microbial activity the higher the production of refractory, HS-like substances. That means that the littoral zone with its vegetation may act as a net sink for autochthonous and allochthonous DOC, and simultaneously as a source of autochthonously produced or transformed allochthonous HS-like substances, which may be released into the pelagic zone of the lake. This fact may serve as a striking example of the latent ambiguity of the ecological role of HS in freshwater systems (see also Chaps. 5.2.4, 8, and 9).

5.2.4 Indirect Utilization

In addition to direct utilization of dissolved HS, secondary (indirect) mechanisms are probably also developed. Within the microbial community, some organisms can produce so called co-factors such as growth stimulating substrates, vitamins, which cannot be produced by other organisms, but are necessary for the primary users of the 'heavy fuels'. In this way, the necessary enzymes and transport systems can be generated, or the required energy status of the cells provided (Münster 1999a,b with reference to original papers).

An example in which algae probably benefit from bacteria produced co-factors is described by Steinberg and Bach (1996). The dissolved HS effects are compared in an algae-bacterial system, with those in an axenic algal system. FA from the north German waterworks in Fuhrberg (Fuhrberg, Lüneburg Heath) are the tested dissolved HS material. Axenic alga *Scenedesmus subspicatus* (Chlorococcales) are generally unable to use these FA: increasing dissolved HS concentrations strikingly reduce the algal yield. This is not surprising as it is often found in field studies that primary production in polyhumic waters is reduced through light absorption (Bledsoe and Phlips 2000; Carpenter et al. 1998; del Giorgio and Peters 1994; Guildford et al. 1987; Jackson and Hecky 1980; Klug 2002; Klug and Cottingham 2001; Phlips et al. 2000). HS clearly compete with the pigments of the algae for photons penetrating the water. In a comparison of the influence of various factors limiting primary production, Carpenter et al. (1998) find that the natural variability in cDOC concentration has the greatest influence. However, there are other possible mechanisms for the reduction in photosynthesis, namely the release of potentially toxic photooxidants (Chap. 5.3.2), or a direct adverse effect on the algae (Chap. 8.3). However, this impact is rarely studied or even merely discussed, since as often quoted, many limnologists consider HS both geochemically and ecophysiologically inert.

The overall inhibitory effect of increasing dissolved HS concentration on axenic cultures of *S. subspicatus* is shown in Fig. 5.8.

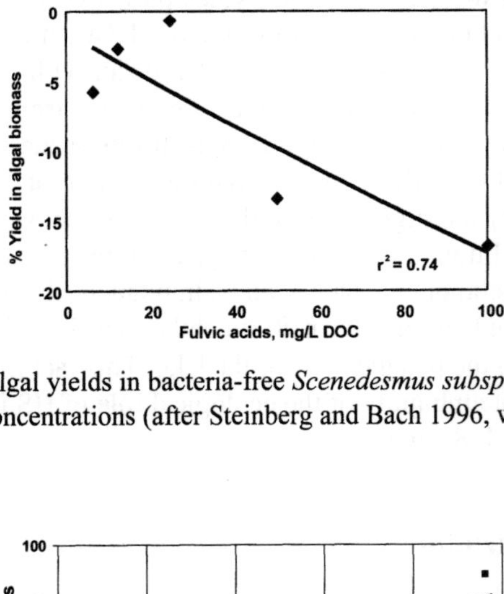

Fig. 5.8. Decreasing algal yields in bacteria-free *Scenedesmus subspicatus* cultures with increasing fulvic acid concentrations (after Steinberg and Bach 1996, with kind permission of Wiley-VCH)

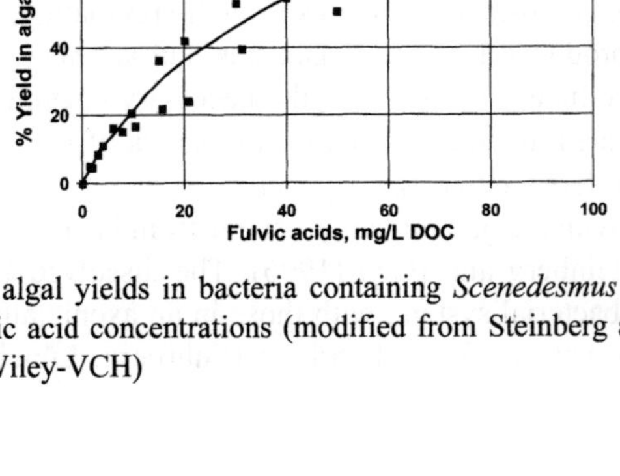

Fig. 5.9. Increasing algal yields in bacteria containing *Scenedesmus subspicatus* cultures with increasing fulvic acid concentrations (modified from Steinberg and Bach 1996, with kind permission of Wiley-VCH)

However, if the algal cultures contain bacteria, the cultures exhibit an unexpected increase in growth with increasing FA concentrations, as clearly shown in Fig. 5.9. For example, 20 mg/L DOC supports approximately 30% more algal growth than the DOC-free control. Apparently the dissolved HS-supported algal growth follows a saturation curve. This means that increase in algal growth is limited by the lack of light via FA light absorption and self-shading.

The results with the Fuhrberg FA show that support of algal growth by HS can generally occur. Such a stimulation of phytoplankton growth in culture and field studies is well known. Most authors attribute this to improved bioavailability of inorganic nutrients, in particular trace metals (Giesy 1976; Gjessing et al. 1998; Horth et al. 1988; Jones et al. 1998; Steinberg and Herrmann 1981; Steinberg and Münster 1985; Wojciechowski and Górniak 1990). However, this is only one of several possible explanations. In the account of Steinberg and Bach (1996), nutrients are well in excess of algal requirements, even at the end of the experiment. The growth supporting effect for *Scenedesmus* is only found in the bacteria-containing cultures. From this, it must be concluded that the stimulating effect of FA is attributable to microbial activity. This occurs through as yet unidentified growth factors.

In addition to this mechanism, co-factors can also be mediated through non-chromosomal DNA such as plasmids. In recent studies, many plasmids are identified which code specifically for the degradation of refractory organic pollutants and heavy metal tolerance (Münster et al. 1999a,b with further references). The overall effects, both stimulating and inhibitory, of HS on aquatic organisms, in particular on primary producers, are described in more detail in Chap. 8.

5.2.5 Changes during Transport and Input into Water Bodies

Studies on the retention times of dissolved HS and of water color measurements in Canadian freshwaters indicate that the dissolved HS looses its chromophores during passage through the catchment, the rivers, and connected lakes. In receiving water bodies, dissolved HS accumulates as less differentiated, less structured material (Rasmussen et al. 1989). This means that the biogeochemical changes in dissolved HS in freshwaters lead to a relative enrichment of refractory components ('heavy fuels'). At the same time as structural and functional groups are lost, the interactions and affinities of the microbial degradation and transport system for refractory dissolved HS compounds must become weaker, less specific. This finally leads to reduced decomposition and utilization rates. This conclusion is plausible and explains the great persistence and the generally great age of refractory dissolved HS in humic waters (Münster 1999a,b).

Processes during transport through a catchment have further effects on the organic C compounds as described by Egeberg et al. (2002). In the Norwegian 'NOM-typing project', carried out to obtain a better physico-chemical understanding of natural organic substances in raw water and in the water during drinking water processing (Gjessing et al. 1999), interesting basic knowledge is obtained.

For example, Egeberg et al. (2002) infer the effect of catchment size on molecular size of NOM-isolates. The mean molecular size of NOM ranged from 2.2 kDa for small DOC-rich wetlands, to 1.0 kDa in large, humic-poor lakes. Since the above-mentioned results obtained by diffusimetry accord very well with those obtained by effective diffusivities (Fettig 1999), one can estimate the mean molecular mass of DOM from the correlation with catchment size. Fig. 3.7 shows this relationship for the 9 water bodies which are studied in the NOM-typing project. There is a highly significant logarithmic relationship between the catchment size and the mean NOM molecular mass. Statistically, the size of the catchment explains 90% of the variability of the NOM molecular mass. The remaining 10% of the variability is determined by other characteristics of the catchment such as vegetation, geological background, and rainfall or other climatic factors.

The structural attributes of DOC from the Ogeechee River (a black water river in Georgia, USA) and its catchment (Sun et al. 1997) are shown in Fig. 5.10. The DOC from algae and macrophytes is rich in aliphatic compounds and therefore has a high bioavailability for microbes. In contrast, DOC from pines, cypresses and hardwoods is much more aromatic. This signifies that DOC derived from terrestrial sources is less bioavailable than algal or macrophyte derived DOC. Along the flow of the Ogeechee River, the proportion of aromatics steadily increases. The DOC, thus, loses its property of being a good source for microbial growth.

The work of Sun et al. (1997) considers the availability of DOC only in the absence of light. The photolytic cleavage of the aromatic, more chromophoric fraction of the DOC, and the connected bioavailability of the products are not subjects of the study. It is evident in the Ogeechee River study, that the proportion of aromatics increases rather than decreases along the river course. This is consistent with the work of Opsahl and Benner (1998) who describe a clear photolytic degradation of lignin phenols in Mississippi water.

To date, we only consider the changes in DOM in the catchment and running waters as a substrate for microbial processes. However, along the flow path, not only does the substrate change, but the microflora changes, too. What is the bioavailability of the same DOM with changes in microbial biocoenoses? This question is addressed by Stepanauskas et al. (1999) for dissolved organic nitrogen (DON) which is transported from the wetlands of southern Sweden to the Baltic Sea. It is concluded that the susceptibility of DON to microbial mineralization increases during the passage from freshwater bodies to the marine environment. This implies that DOM from terrestrial wetlands can be an important N source for coastal ecosystems during the summer if the concentrations of inorganic N are low. This study also shows that marine bacterioplankton can utilize a considerably larger proportion of DON than can lake bacterioplankton.

A recent preliminary seasonal budget of bioavailable N in catchments of central and northern New Jersey, USA, suggest that 20–60% of total dissolved N is bioavailable from forests and pastures in coastal marine ecosystems, and that the response of the plankton community to bioavailable DON is more complex than to DIN. There is a linear increase in bacterial production when either DIN or bioavailable DON is added. However, the increase in bacterial production per micromole N added is apparently five times greater when N is added as bioavailable DON compared to when it is added as DIN (Seitzinger et al. 2002).

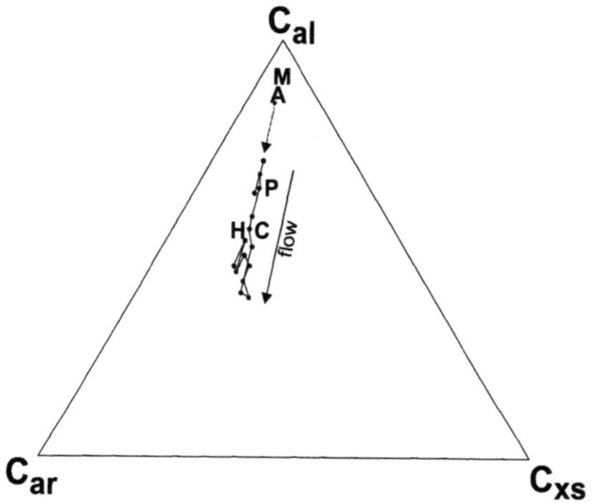

Fig. 5.10. Most probable structural distribution of organic carbon in biomass leachates (**A**: algae; **C**: cypress, **H**: hardwood; **M**: macrophytes; **P**: pine) and in Ogeechee River DOM samples. The apices in this ternary plot correspond to 100% aliphatic C (C_{al}), 100% aromatic C (C_{ar}), and 100% 'excess' C (C_{xs}). The opposite edge from an apex represents 0% of that form of C (after Sun et al. 1997, with kind permission of the American Society of Limnology and Oceanography)

5.2.6 Retention in Benthic Biofilms

In streams and rivers, the surface-bound bacterial activity greatly exceeds the activity of free-living bacteria. The share of the metabolism that can be attributed to biofilm-coated surfaces is calculated for small streams and larger rivers and lakes, with more than 99% occurring in the hyporheïc zone of small streams, slightly less for the hyporheïc zone in larger rivers and lakes; less than 1% of the community respiration in the water column (Fischer 2002). Hence, it is worthwhile to consider the biofilm activity in more detail. The River Spree near Berlin, Germany, is a 6th order lowland river with sandy sediments and a mean depth of 1m. The sediments present an ideal colonization substrate for heterotrophic bacteria (biofilm), due to their high porosity and consequent permeability. Through the intimate contact between the colonized surfaces and the through-flowing water, an extensive exchange between water and sediment is possible. This can affect the retention of HS components, which can be separated by a fingerprint technique (Chap. 3).

Along the river, phytoplankton development increases (Köhler 1994). Hence, two main sources for the microbial C metabolism in the sediments occur; autochthonous production of DOC, and allochthonous input from the catchment. Samples are taken during algal blooms (May 12), and shortly afterwards (May 26) and during the subsequent clear-water phase (June 17). The DOC content of the river water decreases from 7.4 to 5.9 mg/L during the study period (Fig. 5.11).

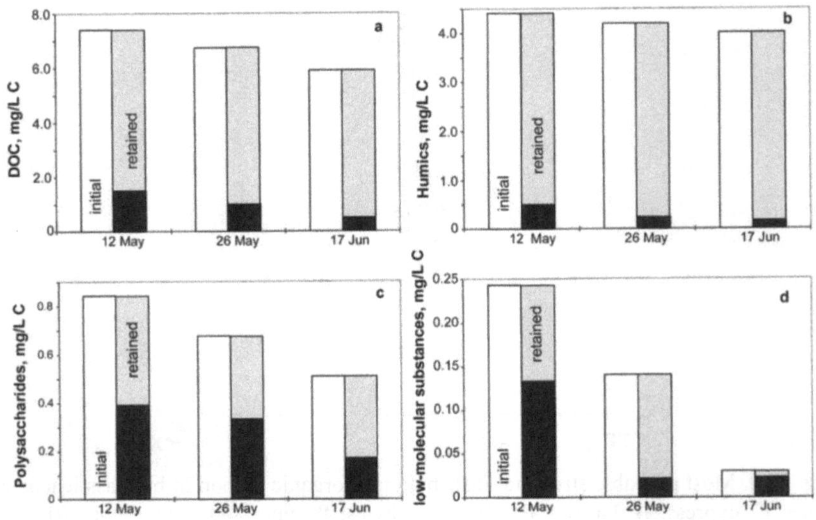

Fig. 5.11. Changes in DOC concentrations and composition in the sediment of River Spree during (12 May), and shortly after the spring phytoplankton bloom (26 May), as well as during the clear-water phase (17 June). Changes are characterized by the fingerprint technique (Chap. 3). **a**: unfractionated DOC; **b**: HS; **c**: polysaccharides; **d**: low-molecular weight substances (from Fischer et al. 2002, with kind permission of the American Society of Limnology and Oceanography)

During the algal bloom, DOC contains higher concentrations of polysaccharides and low-molecular weight substances, than during the clear-water phase. Passage through the sediment, which takes around half an hour, clearly changes the composition of the DOC. During the algal bloom (May 12), up to 20% of DOC is retained by the sediment or degraded. During the clear-water phase, this proportion is reduced to about 9% (Fig. 5.11a). The retained portion of HS is smaller: 11% during the algal bloom, and 4% in the clear-water phase. Markedly higher proportions of polysaccharides (40–50%), and low-molecular weight substances (up to 77%) are retained; the latter include the amino acids and peptides of the HS. Similar fingerprint patterns of the DOC have been reported for water columns (many studies including Münster 1982, 1984, 1985; Stabel 1977; Steinberg 1977; Volk et al. 1997), but the present report appears to be the first one for a sediment.

Changes in the amounts and composition of DOC from the River Spree after an algal bloom, are strikingly similar to elution diagrams obtained in laboratory incubation before and after sediment core perfusion. During the algal bloom, a considerable amount of labile DOC, especially polysaccharides, is released into the river water. After the breakdown of the algal bloom, the amount of labile substances concomitantly strongly decreases in the river water. It is concluded that this change in riverine DOC composition is mainly caused by biofilm bacteria, analogous to the bacterial decomposition of DOC in the laboratory incubation (Fischer 2002).

Besides bacterial utilization, which further retention mechanisms may be responsible for the changes in the DOC fingerprints? Only aromatics and non-polar aliphatics can be absorbed on clay particles. Such moieties are found in HS, occasionally in amino acids and peptides, but never in polysaccharides. In other words, neither carbohydrates nor amino acids can be absorbed on clay particles unless they are bound to HS.

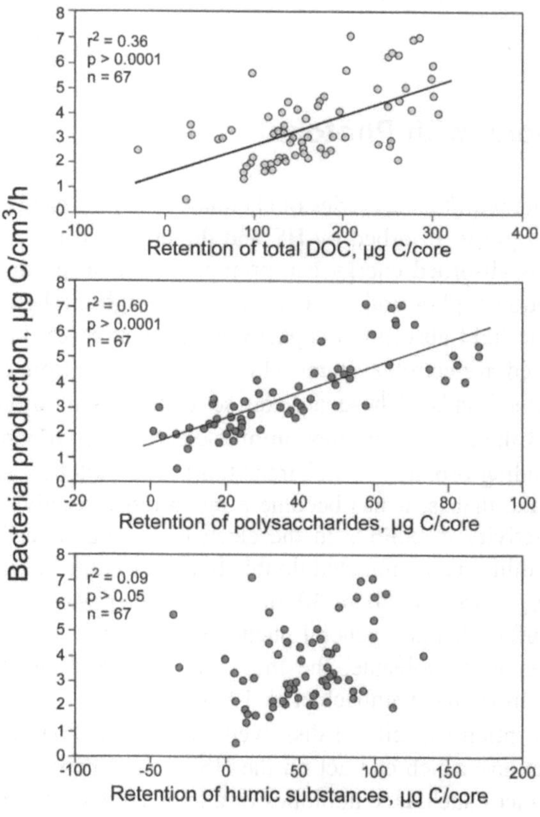

Fig. 5.12. Relationship between DOC retention and bacterial production in sediments of the River Spree, Germany, February–November 1998 at temperatures ranging from 7 to 22 °C (after Fischer 2002, with kind permission of Academic Press)

A similar situation applies for the interactions with the organic matrix in sediment. Here the aromatic compounds are particularly important. For the binding of saccharides and amino acids/peptides, hydrogen bonds play a predominant role. These should exhibit no barrier for exoenzymes as long as they are physically able to reach their potential substrates. This means that a clearly higher retention rate for saccharides and peptides than for HS, is an important factor for microbial activity. Fig. 5.12 presents regression analyses of DOC and its main fractions, polysaccharides and HS, as predictors for microbial growth for a 10 month period. It appears that the findings during and after the algal bloom in the River Spree may be generalized: the microbially mediated retention is generally higher for low-molecular weight organic matter, most notably amino acids and mono- and disaccharides (not shown), and for the polysaccharide fraction of high-molecular weight organic matter. The retention is relatively low for the HS. However, the HS form the largest portion of the DOC, and can therefore make up a substantial proportion of the total organic matter retained in the cores (Fischer et al. 2002; Fischer 2002).

5.3 Interactions with Photons

Where HS interact with light, a series of chemical reactions can occur. Most of the energy from sunlight is absorbed by HS and directly used for warming. A small proportion of this absorbed energy can prompt the HS to promote a chain reaction. The absorption applies both for UV and visible light in the wavelength range 290–600 nm. The light absorption capacity is, in most cases, linked to the presence of delocalised π-electron systems which are available from aromatic rings or conjugated double bonds[3]. These are the so called **chromophores**. The most likely stimulus is the electron promotion of π-electrons from bonding orbitals, to so-called antibonding orbitals ($\pi \rightarrow \pi^*$-transition). The molecule is then said to be in an excited state, that is, it has become a much more reactive species as compared to the reactivity it exhibits in the electron ground state. In straight-chain polyenes, each additional conjugated double bond shifts the absorption maximum of lowest energy, $\pi \rightarrow \pi^*$, by about 30 nm to the higher wavelengths (a so-called bathochromic shift). This is a general phenomenon, and it may be stated that the more conjugation in a molecule, the more the absorption is displaced toward higher wavelengths (Schwarzenbach et al. 1993).

- The light absorption capacity of dissolved HS itself leads to a series of photochemical reactions which can act on the HS itself (Chaps. 5.3.3, 5.3.4), or on foreign substances such as xenobiotics or allelochemicals (Chap. 5.3.5). Miller (1998) summarizes these reactions as follows:
- decrease in average molecular weight of DOC,
- changes in the optical properties of the water as a result of the changes in DOC (Morris and Hargreaves 1997),
- release of a complex mixture of reactive oxygen species (ROS), the photooxidants (Blough and Zepp 1995),
- release of oxidized C products such as CO, CO_2 or short-chain fatty acids (Ber-

[3] Conjugated double bonds are very widespread in natural compounds, for example in photosynthetic and accessory pigments such as carotenoids and porphyrins.

tilsson and Tranvik 1998, 2000; Miller 1994; Valentine and Zepp 1993; Wetzel et al. 1995; Zepp et al. 1995); the underlying mechanism is decarboxylation, and

- release of N- and P-rich products, and finally ammonia, phosphate ions, amino acids, and dissolved primary amines (Bronk et al. 2001; Moran and Zepp 1997; Morell and Corredor 2001; Tarr et al. 2001; Wang et al. 2000).

Understanding of the photochemistry of HS is restricted due to the indefinable nature of the materials. One must therefore use integrated data or generalized parameters to describe the HS, derivatives, and the chromophores. Only the reactions of small, well-defined species and their products can reliably be determined, if they are sufficiently long-lived. The principal routes of photoreactions with HS involve the transfer of their energy on reactive species (RS). Very reactive species (RRS, such as radicals) are formed which can also be directly formed from the reactive HS*. The multifunctional character of HS in their ground or activated state leads to reactions with RRS which leads to further formation of HS and partial degradation to CO_2 (Fig. 5.13). The altered HS (HS', H'', HS*) can also participate in the primary photoreactions (according to Frimmel 1994, 1998).

As mentioned, the chromophores of DOC (FA and HA) are the most important light absorbers. These groups induce indirect phototransformation, and are activated many times a day according to Schwarzenbach et al. (1993), based on calculations for Lake Greifensee (Switzerland). On a sunny day, each chromophore in the epilimnion is activated 270 times, that is ten times or more per hour. The reactive oxidative species include singlet oxygen ($^1O_2$) (Frimmel et al. 1987; Wolff et al. 1981; Zepp et al. 1977), free peroxide radical ($ROO^\bullet$) (Faust and Hoigné 1987; Mill et al. 1980,), hydrogen peroxide (H_2O_2) (Cooper and Zika 1983; Scully et al. 1996), hydrated electron (e_{aq}^-) (Zepp et al. 1987a,b), hydroxyl radical ($HO^\bullet$), organic peroxides (Larson et al. 1981; Vaughan and Blough 1998), and superoxide anion (O_2^-) (Baxter and Carey 1983; Petasne and Zika 1987). In addition to dissolved HS, nitrate can also absorb sunlight, which also leads to the formation of $HO^\bullet$ and a further chain reaction is facilitated (Münster et al. 1999b; Zepp et al. 1987a,b).

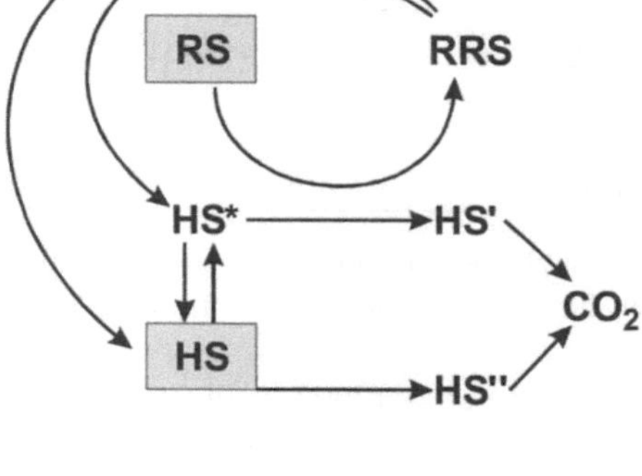

Fig. 5.13. Pathway of photoreactions with HS (after Frimmel 1994, with kind permission of Elsevier Science)

$HO^{\bullet}$ can also be formed from cDOC by direct photolysis (Mopper and Zhou 1990; Vaughan and Blough 1998). This mechanism is dominant in oligotrophic humic waters which have very little nitrate in the euphotic zone (Münster et al. 1999b).

A schematic overview of the reactive oxygen species, ROS, is provided in Fig. 5.14. An important reaction is the classical Fenton-reaction (Paciolla et al. 1999) in which hydrogen superoxide is reduced and hydroxyl radicals produced:

$Fe(II) + H_2O_2 \rightarrow Fe(III) + OH^- + HO^{\bullet}$.

When this reaction runs with light energy, where H_2O_2 comes from the photolysis of cDOC, it is known as a photo-Fenton reaction. For photolytic processes, H_2O_2 acts not only as an oxidizing agent, but also as a reducing agent, as in the photoreduction of metals. Formally, the oxidation state of oxygen in H_2O_2 is only –1 and therefore can release further electrons for reduction.

Vaughan and Blough (1998) discuss the possibility of $OH^{\bullet}$-radicals being directly produced by the photolysis of cDOC. Here the induced triplet state of quinones, which are also structural components of HS (Box 2.1), remove a hydrogen atom from the water molecule. The individual chemical species have very different half-lives, from only a few microseconds, as for singlet oxygen (Haag and Hoigné 1986), to well over an hour as for hydrogen superoxide (Scully et al. 1998).

Educts		Products
HS	$\overset{h\nu}{\Rightarrow}$	HS*
HS*	$\Rightarrow$	$HS + {}^1O_2$
HS	$\overset{h\nu}{\Rightarrow}$	$HS^+ + e^-_{(aq)}$
$e^-_{(aq)} + O_2$	$\overset{h\nu}{\Rightarrow}$	O_2^-
$2O_2^- + 2H^+$	$\Rightarrow$	$H_2O_2 + O_2$
H_2O_2	$\overset{h\nu}{\Rightarrow}$	$OH\bullet + OH\bullet$

Fig. 5.14. Scheme of the photochemically active chemical species (reactive oxygen species, ROS), which may be produced from HS upon radiation ($h\nu$)

Very recently, Goldstone et al. (2002) evaluate the role of an individual ROS, the $HO^{\bullet}$ radical, as a mechanism for the photodegradation of cDOC in sunlit surface waters, and present evidence that $HO^{\bullet}$ reactions with HS do not result in measurable formation of bioavailable C substrates, other than low-molecular weight acids. Also, bleaching of humic chromophores by HO• is relatively slow. That indicates that HO• reactions with HS are unlikely to contribute significantly to observed rates of dissolved HS photomineralization and low-molecular weight acid production in sunlit waters.

By far the most important acceptor of the activated chromophores in natural waters is oxygen in its ground state (triplet oxygen, $^3O_2$). Since the activation of $^3O_2$ to its first activated state (singlet oxygen, $^1O_2$) requires only 94 kJ/mol, almost all chromophores that can absorb UV or visible wavelength light generally transfer their absorbed energy to an oxygen molecule (Schwarzenbach et al. 1993). The $^1O_2$-molecules can then oxidize a series of substrates including xenobiotic chemicals and allelochemicals (Chap. 5.3.5).

In a classic study, Haag and Hoigné (1986) established that the various types of aquatic DOC give very different quantum yield for $^1O_2$-production (Fig. 5.15). If one considers Türlersee to be an outrider in Fig. 5.15, then there is a clear linear dependence of $^1O_2$-production on the DOC concentration (Eq. 5.2). It is evident that this relationship would be even clearer if the DOC were related to the chromophore groups (cDOC).

$$[^1O_2] = 1.52 + 1.91[DOC], r^2=0.91 \qquad (5.2)$$

Draper and Crosby (1983) are among the first to discover that hydrogen peroxide (H_2O_2) is common, and at relatively high concentrations acts as a photooxidant. After only short exposure to sunlight, H_2O_2 appears at concentrations of up to 6.8 µmol/L. In highly eutrophic waters, concentrations in excess of 30 µmol/L can be produced. Being exposed to even lower H_2O_2 concentrations than approximately 7 µmol/L, different effects may occur in algae, zooplankton, or bacteria (Hessen and Faerovig 2001; Xenopoulos and Bird 1997) (Chaps. 5.3.5, 9).

The mechanism of H_2O_2 formation is described by Scully et al. (1996): when UV radiation is absorbed in aquatic ecosystems by dissolved HS, superoxide is formed which reacts with itself to produce H_2O_2. Scully et al. (1996) also describe H_2O_2 formation in relation to DOC over a broad spectrum of aquatic ecosystems, from strongly colored dystrophic lakes, to humic-poor lakes.

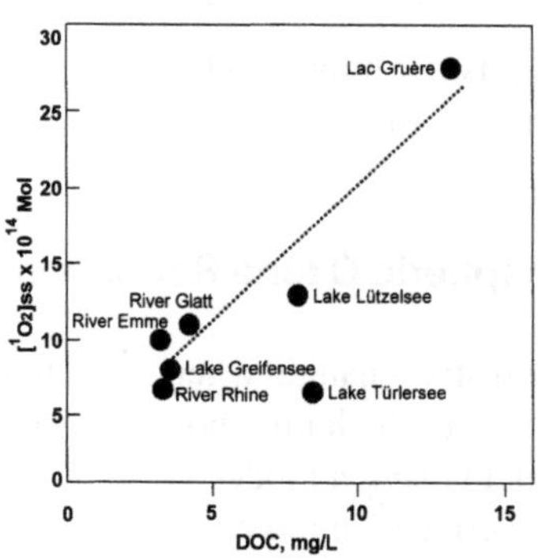

Fig. 5.15. Singlet oxygen production in various Swiss rivers and lakes as a function of DOC concentrations at noon on a clear day (graph with data from Haag and Hoigné 1986)

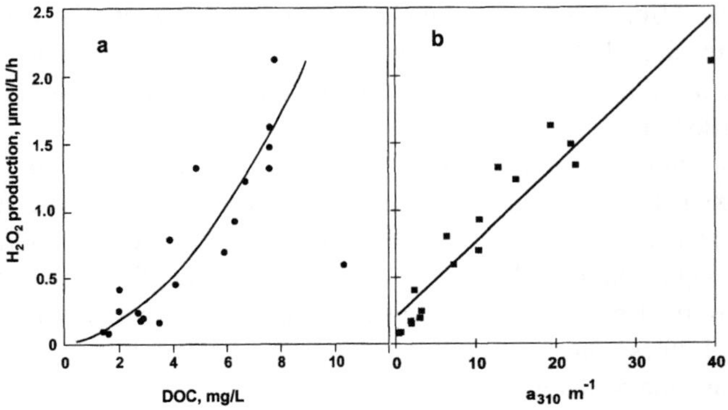

Fig. 5.16. Relationship between H_2O_2 production rates and DOC quantity (**a**) and DOC quality (**b**) given as absorption at 310 nm (after Scully et al. 1996, with kind permission of the American Society of Limnology and Oceanography)

Hydrogen peroxide shows a clear relationship to DOC concentration (Fig. 5.16a, Eq. 5.3).

$$[H_2O_2] = 49.65\ [DOC]^{1.71},\ r^2 = 0.94 \qquad (5.3)$$

In a recent marine study, the oxygen requirement for H_2O_2 production is described. Approximately 45% of oxygen is transformed to H_2O_2, and less than 1% to singlet oxygen (Andrews et al. 2000), which again shows the predominant quantitative role of H_2O_2. This role probably also applies to freshwaters.

The absorption of light by chromophore groups is mediated mainly through their aromatic structures. It is evident that the more aromatic structural units are present, the greater the release of ROS. Accordingly, a clear relationship is expected between the content of aromatic structural elements in HS (given as absorption at 310 nm), and the release of ROS (as shown for H_2O_2 by the study of Scully et al. 1996). This provides a further example for the important role of the aromatic components in the reactivity of HS in aquatic ecosystems (Eq. 5.4):

$$[H_2O_2] = 185 + 55.9\ a_{310},\ r^2 = 0.91. \qquad (5.4)$$

That means that the cDOC is not only quantitatively (Fig. 5.16a), but also qualitatively (Fig. 5.16b), responsible for the release of the reactive chemical species.

5.3.1 Effects of Stratospheric Ozone Reduction on Photochemistry

How does the photochemistry change when the stratospheric ozone declines? Which lakes are more effected, the humic-poor lakes or humic lakes? Scully et al. (1997) consider this problem using a model. It is known that through light, HS are expected to have negative effects on biota, via both the free radicals and the ROS produced. Both can be harmful to biological membranes, proteins, and DNA. The production of ROS can take place within the cells of aquatic organisms, as well as in the surrounding water. Intracellular ROS production can cause many toxic effects (Karentz et al. 1994). Little is known about the extracellular production.

Scully et al. (1997) model the possible effects of increasing UV-B radiation on the formation of H_2O_2 and singlet oxygen ($^1O_2$) for lakes with a wide range of DOC concentrations. The model results indicate that the **relative** increase in H_2O_2- and $^1O_2$-production in humic-poor lakes such as Lake Erie and Lake Ontario is greater than in colored lakes. While up to 70% of the relative increase at the surface will be in humic-poor lakes, only 25% of the relative increase will occur at the surface of humic-rich waters. This is attributable to the fact that the quantum yield is higher in humic-poor lakes than in colored waters. This does not mean, however, that the photochemical impacts will be reduced in colored waters. On the contrary, the absolute increase in photochemical reaction products is at least as great at the surface of these colored waters as in humic-poor lakes. This increase is restricted to the water layers near the surface. It remains the case from these results that humic-poor waters and their organisms would be harmed more by the expected stratospheric ozone increase than would humic-rich waters.

5.3.2 Toxic Effects after Radiation

Apart from the above described model study on the effects of ROS on biological systems, only a few, and in part contradictory, empirical findings are available. The short-lived ROS are not generally accumulated, and act only in the top-most few centimeters of the water column. With the longer-lived species, such as H_2O_2, there can be a partial mixing throughout the epilimnion. One of the few papers on this issue studies the growth inhibition of the green alga *Ankistrodesmus bibraianus* (Gjessing and Källqvist 1991). The inhibitory effect generally lasts a few days or weeks. Lund and Hongve (1994) observe a similar effect on heterotrophic bacteria which are exposed to UV-impacted HS, this effect also lasts a few days to a week.

In an enclosure study in two lakes in Nova Scotia (Canada) with contrasting cDOC content, Kaczmarska et al. (2000) describe distinct effects of UV-B in the colored lake. Photosensitive phytoplankton, in particular Cryptomonads, disappear with penetration of UV-B. As described in Chap. 4, these authors attribute this finding to the formation of H_2O_2 or other ROS.

A complex picture of the toxic action of irradiated HS is presented by Hessen and van Donk (1994). These authors describe *A. bibraianus* bioassays with UV-radiated surface water from two humic waters:

- with moderate UV doses (1.1–5.4 J/cm^2 at 312 nm), phytoplankton production is stimulated, while
- higher UV doses (>10 J/cm^2) lead to minor to serious growth impairment.

Since the phytoplankton is added after light exposure, the results support the hypothesis of the long-lived algicidal effect induced by UV radiation.

In contrast to the findings of Lund and Hongve (1994), Hessen and van Donk (1994) do not find any effects of ROS on dark respiration. Since the dosage used by Hessen and van Donk (1994) is representative of the sun intensity in high summer, chemical (ROS release), ecotoxicological (growth inhibition), and ecological effects (growth promotion) are expected to occur concurrently. This means that the exposure of HS to UV radiation does not follow a dose-effect relationship – the mechanisms run concurrently.

Upon UV radiation, nutrients are also released which are complexed by HS (Francko 1990; Steinberg and Herrmann 1981). This can explain the growth promotion with low UV doses and short-term radiation in the experiments of Hessen and van Donk. Yet, potentially toxic metals, for instance Al, may also be released. These metals can also lead to toxic long-term effects. It is also conceivable that with short-term exposure, the enhanced bioavailability of metals and/or nutrients compensate for the negative effect of long-lived oxidants such as H_2O_2. With longer exposure times, the effects could be reversed.

The most important conclusion is that natural UV doses can have an inhibitory effect on primary production. This supports the supposition that significant toxic effects can accumulate with a few days of strong sunshine. The key factor for the real situation in the ecosystem is the half-life of individual effects. However, these data are not yet available. The findings of Hessen and van Donk (1994) underline the potential danger of increasing UV radiation related to global climate change.

The situation is still more complex, as Xenopoulos and Bird (1997) show by adding environmentally realistic concentrations of H_2O_2 (<1 µmol/L) to plankton communities in Lac Cromwell, a small humic lake in the Laurentian Hills (Quebec, Canada). The results indicate that even small amounts of added H_2O_2 inhibit bacterial production in this lake. A 0.1 µmol/L addition inhibits bacteria by as much as 40%. In contrast, low concentrations of added of H_2O_2 usually stimulate photosynthesis. These results indicate different sensitivities of phytoplankton and bacteria to ROS, and provide one potential explanation for net-autotrophy in some non-eutrophicated lakes, as reported for Canadian Shield lakes (Carignan et al. 2000) (Chap. 9).

5.3.3 Cleavage and Bioavailability

Lönnerblad (1931) already notes that oxygen is undersaturated in the epilimnion of humic lakes. At that time, this situation was attributed to high heterotrophic activity. Subsequently, as methods are developed for the determination of bacterial numbers and heterotrophic activity, it becomes apparent that such activity is probably not the cause for the oxygen under-saturation (Hutchinson 1957). In addition, Miles and Brezonik (1981) show that the abiotic oxygen deficit increases with HS content following light exposure. Meanwhile, the photodegradation of HS and DOM in general is intensively studied (Anderson et al. 1985; Backlund 1992; Bauer and Frimmel 1987; Gilbert 1980; Moran et al. 2000; Norwood et al. 1987; Sörensen et al. 1995). Strome and Miller (1978) are among the first authors to suggest that an increased degradation of dissolved organic substances is possible when the HS are split by UV radiation. The photodegradation leads to smaller and more easily bioavailable organic compounds. One of the pioneering studies on photodegradation and improved bioavailability of aquatic HS (refractory organic substances) is published by Geller (1985a). She studies the abiotic and microbial persistence of refractory high-molecular weight DOC from various lakes, and finds that the exposure of the macromolecules, both to daylight in the laboratory as well as to sunlight, clearly increases the bioavailability compared to the dark control in which slow decomposition also takes place. During a 6 week exposure and under weakly photolytic conditions (daylight >300 nm), 15% of winter DOC and 25% of summer DOC from the high-molecular weight fraction is degraded. The concentration of the low-molecular weight fraction rises accordingly. The smaller macromolecules also play an important role in the biodegradation and are responsible for a further 10% of the DOC decrease (Geller 1986).

Since the work of Geller, the influence of radiation, in particular UV radiation, on the composition and function of DOC in surface waters has received much attention. This applies in particular to the transformation and degradation of organic materials. For example, the formation of CO_2 is proved by Kulovaara and Backlund (1993) and Salonen and Vähätalo (1994) following exposure of high-molecular weight DOC to UV radiation. In addition, the release of various intermediates such as CO (Conrad and Seiler 1980) and short-chain fatty acids (Allard et al. 1994; Backlund 1992; Bertilsson and Allard 1996; Wetzel et al. 1995) is described. The photolytic products have a very high bioavailability (Lindell et al. 1995, 1996). The increased bioavailability is associated with significant increased bacterial numbers and biovolume of the bacteria.

The bioavailability of photolytic products is already known from pioneering studies of Kieber et al. (1989, 1990) who show that the biological uptake of pyruvate, a valuable C_3-fatty acid, is strongly correlated with its photochemical production (in seawater). In addition, they find that the photochemical precursors of pyruvate come from the DOC fraction with a nominal molecular weight of 0.5 kDa.

In humic waters with high DOC concentrations, the photochemical release of low-molecular weight substances plays an important role in the nutrition of bacteria (de Haan 1992; Salonen and Vähätalo 1994; Tranvik and Höfle 1987; Vähätalo and Salonen 1997). The direct coupling of photochemical production to biological use of organic C accounts for the comparatively constant activity of bacteria that is independent of phytoplankton primary production (Salonen et al. 1992).

In a detailed study, Wetzel et al. (1995) determine that bacterial production increases when bacteria grow on NOM (HS fractions of decaying rush *Juncus effusus*) are subject to mild UV-B. Exposure to natural UV shows an immediate and sustained stimulation of bacterial growth, which is attributed to the release of many short-chain fatty acids, including acetic, citric, formic, levulinic, and propionic acids. The dominant fatty acids are acetic and levulinic acids. The short-chain fatty acids are rapidly metabolized by bacteria. In their review, Moran and Zepp (1997) list the following organic compounds released from sunlit DOC: acetaldehyde, acetate, acetone, citrate, formaldehyde, formiate, glyoxal, glyoxalate, levulinate, malonate, methylglyoxal, oxalate, propanal, and pyruvate.

Dahlén et al. (1996) determine how the quality and quantity of DOM in humic surface waters is influenced by prolonged (89 h) low-intensity UV-A radiation. The UV-A dosage is realistic, being only about a third of that of the natural sunlight on a cloudless day in southern Sweden. The short-chain fatty acids, namely oxalic, malonic, formic, and acetic acids, are identified as photolytic products (Fig. 5.17). The photochemically produced fatty acids comprise in total about 0.18 mg/L C. Such quantities certainly affect the chemistry of the studied waters: metal speciation can be changed by some of these fatty acids, in particular the divalent fatty acids, which have metal complexation properties similar to FA. In addition, microorganisms can also be affected since these acids are good substrates for growth. Furthermore, the DOC concentration is decreased through the transformation of organic molecules to CO and CO_2, through photo-decarboxylation of DOC following the UV radiation. This decrease is 0.83 mg/L C, which is four times as high as the sum of the low-molecular weight fatty acids.

Similar short-chain fatty acid spectra are found by Bertilsson and Tranvik (1998) for other Swedish humic lakes. Here the smallest acid, formic acid, is dominant. In total, the four short-chain fatty acids are released at a rate of 19 µg/L/h C. These acids are the most important substrates for bacterial production. Interestingly, the bacterial production is less than the rate of photochemical release. We will return to this phenomenon later.

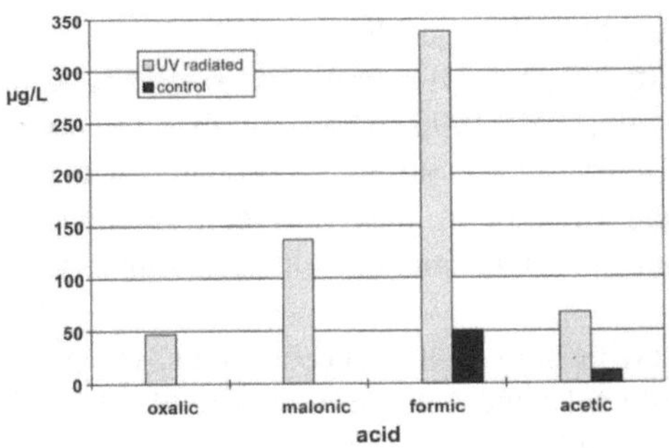

Fig. 5.17. Production of oxalic, malonic, formic, and acetic acids in samples radiated with UV for 89 h compared to non-radiated controls (modified from Dahlén et al. 1996, with kind permission of Elsevier Science)

The above-described studies show the photolytic role of the dissolved HS, and explain the Geller phenomena by showing mechanisms of the stimulation of heterotrophic production. This is an important aspect of the ecological understanding of dissolved HS as a main component of freshwater ecosystems. However, the contributions of Wetzel et al. (1995), Dahlén et al. (1996), and Bertilsson & Tranvik (1998) remain still phenomenological, and are little oriented towards ecological chemistry.

The following questions remain open:

- which structures are responsible for the observed release of fatty acids? Are the chromophoric structures or the aliphatic side chains split?

- is there a dose-effect relationship between UV exposure, the composition of HS, and the release of short-chain fatty acids?

The answers to these questions require the use of chemically well characterized HS, the subsequent application of quantitative structure property relationship (QSPR) methodology to link the various described effects with the various structures, or the application of molecular modeling.

The low-molecular weight C photoproducts are mostly, but not always, singularly responsible for the stimulation of bacterial growth. Miller and Moran (1997) do not find bioavailable materials in sufficient quantities to explain the increase in bacterial production. These authors therefore conclude that the chemically uncharacterized high-molecular weight DOM fractions are also modified to more biologically available forms upon radiation with natural sunlight.

Following the pioneering work of Lönnerblad (1931) which while not correctly describing the underlying mechanisms, does indicate a very interesting limnological phenomenon, little work is published which calculates the gross effect of photolysis on ecosystems. Such studies include those of Vähätalo and Salonen (1997) on the weakly acid lake Valkea-Kotinen, Finland. These workers calculate the proportion of photochemical dissolved HS degradation in relation to bacterial production. They find that:

- 2.5 mmo/m^2/d C of the cDOC (dissolved HS) is photochemically degraded to products which have less absorption than the initial dissolved HS,
- about 0.51 mmol/m^2/d C is photochemically transformed to DIC, and
- approximately 2 mmol/m^2/d C of non-chromophoric organic degradation products are formed

Overall, 5 mmol/m^2/d C is degraded or transformed. The mean rate of bacterial respiration of 7.4 mmol /m^2/d C in the epilimnion of this lake is of a similar magnitude. These numbers underline the importance of dissolved HS photolysis for pelagic C cycling.

Reitner et al. (1997) also provide data emphasizing the importance of photolysis through studies on the microbial and photolytic degradation of dissolved HS in very humic water of the reed belts of Lake Neusiedl (Austria/Hungary). They estimate that photooxidation in the top 5 cm of the water column leads to a dissolved HS turnover time of approximately 45 days. The non-HS remains twice as long in this layer.

Amon and Benner (1996b) carry out studies in the Amazon basin, similar to those conducted by Vähätalo and Salonen (1997) in Finland. The photoreactive and biologically refractory DOC in surface waters of the Rio Negro exhibit the high photochemical turnover rate of 4.0 μmol/h C. The oxygen requirement is similar at 3.6 μmol/h O_2. In contrast to the boreal lake Valkea-Kotinen, in this tropical water, the photochemical degradation of DOC (release of fatty acids plus photomineralization) is seven times greater than bacterial production. It is apparent that photomineralization of biologically refractory DOM in running waters is much more important than previously accepted. This pathway can be one of the most important mechanisms for the transfer of land-derived DOM to the coastal zone.

Also from marine microbiology, comes the first clear evidence that illumination of DOC does not always lead to an increase in bioavailable organic substrates. Benner and Biddanda (1998) and Obernosterer et al. (1999) find that microbial production under the action of light is usually markedly less than the production in the dark. UV inhibition can only explain part of the inhibition. The authors suggest the following scenarios. The increase in fresh, labile DOC in near-surface water is tightly linked to phytoplankton production, either via direct excretion or through grazing of zooplankton. Exposure to sunlight affects DOC as well as bacteria. In part, the bacteria are only reversibly inhibited by UV radiation (Herndl et al. 1993; Karentz et al. 1994). The effect is that originally labile DOC can be rendered more refractory upon radiation (Fig. 5.18).

The low bioavailability of marine DOM and HS in the euphotic zone is explicable, considering their formation (Harvey et al. 1983; Harvey and Boran 1985). Marine HS – in contrast to terrestrial and most freshwater HS – originate from lipids or lipid-like structures, and are linked together via reactions with ROS through various oxygen bridges. Two to four of such fatty acid chains are linked to each other. This process makes scarcely bioavailable substances even less bioavailable.

If the bacteria from the surface layers are mixed into deeper layers in the sea, which are not impacted by UV radiation, they rapidly recover (Kaiser and Herndl 1997). At these depths, the activity of bacteria depends on the bioavailability of the products. In contrast, the mixing process in the late afternoon can bring refractory DOC from the UV-free depths, and recovered bacteria to the surface, and photolysis of DOC allows activity of the bacteria to increase.

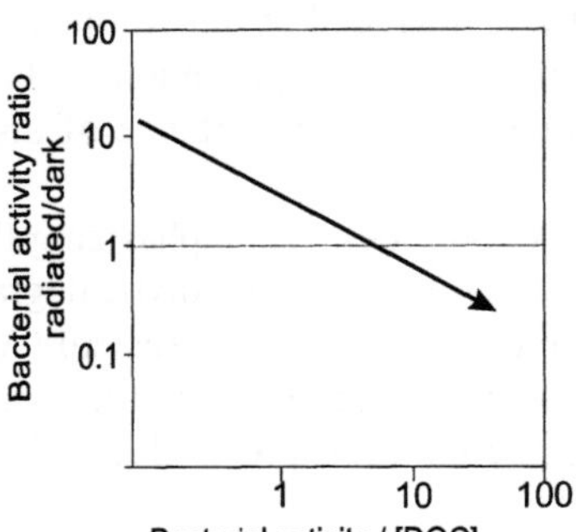

Fig. 5.18. Relationship between bacterial activity in the dark normalized to DOC, and the ratio of bacterial activity in the irradiated treatment to bacterial activity in the dark treatment. A bacterial activity ratio of <1 indicates lower bacterial activity in the treatment exposed to solar radiation than in the dark treatment (modified from Obernosterer et al. 1999, with kind permission of the American Society of Limnology and Oceanography)

The exposure of DOC to solar radiation causes a reduced availability of exposed DOC for bacteria, if the ratio of bacterial activity to DOC concentration is high. This implies an originally labile DOC which cannot be used by bacteria. If the ratio of bacterial activity to DOC concentration is low (indicative of more refractory DOC), bacterial activity can increase following solar radiation. In other words, although freshly released, labile DOM might be photochemically rendered more refractory upon exposure to radiation; more refractory deep-water DOM, mixed from deeper layers to the surface layers, might become susceptible to bacterioplankton uptake after photoinduced alteration.

First indications are that the explanation of Obernosterer et al. (1999), described above, is probably also applicable for freshwaters (Bertilsson and Tranvik 1998; Tranvik and Bertilsson 2001). Bertilsson and Tranvik (1998) find that the photochemical release of short-chain fatty acids at the surface of humic lakes can exceed use by microorganisms, and can thus lead to an accumulation of these materials in this layer. One explanation is inactivation of bacteria by natural UV radiation. Another is that some of the short-chain fatty acids, such as oxalic and formic acids, are not good microbial substrates due to relatively low energy contents (high oxidation states). The incorporation of these acids into the potentially bioavailable pool can lead to a clear overestimate of the microbial production potential.

Tranvik and Bertilsson (2001) show that the effect of radiation on the ability of DOM to promote bacterial growth is a positive function of the terrestrial humic matter, and a negative function of indigenous algal production. They suggest that the net effect of radiation is a result of counteracting, but concurrent processes rendering DOM either labile or refractory. Humic DOM is predominantly transformed into forms of increased lability, whereas in algal derived DOM photochemical transformation into compounds of decreased bacterial substrate quality dominate. One potential mode of action, whereby labile DOC is rendered more refractory, has been outlined above.

5.3.4 Photobleaching and Photomineralization

A recent survey in northern Michigan, USA, reveals that temporal dynamics in DOC and color are related to ice break date, as well as spring and summer precipitation. Years of late ice break and high spring rain are associated with high DOC and color in spring. Summer droughts appear to lead to declines in color and DOC (see the paleolimnological studies referred to in Chap. 4). The common temporal dynamics of DOC and color are probably the result of climatic conditions that affected loading of allochthonous C, as well as losses due to photodegradation (Pace and Cole 2002).

Photochemical reactions in water lead to a change in the cDOC, which is seen as a decrease in water color (photobleaching), which is visible to the naked eye, particularly during the summer months (Morris and Hargreaves 1997). According to the mechanism described by Vähätalo and Salonen (1997), a large proportion of no longer chromophoric substances, but aliphatic C, and to a lesser degree carbon oxides are produced. The aromatic ring structures of the dissolved HS are opened. In general, the water color is reduced during the summer months. It is estimated that about 20% of the reduction in lake color is caused by natural radiation (Gjessing and Gjerdahl 1970). In addition, due to the degradation of the global ozone layer in the troposphere, and the consequent increase in UV-B radiation, the proportion of cDOC in lakes is probably decreasing worldwide. Therefore the natural UV shield declines, and the UV penetration depth consequently increases exponentially (Chap. 4). In general, energy transmission changes. The light penetration increases, and the absorption of heat extends to a greater water depth. The strong temperature gradient during summer thermal stratification decreases.

In their study of the shallow Subarctic Lake Kachishayoot in northern Québec, Canada, Gibson et al. (2001) report a number of important factors in the cDOC dynamics of lakes which may generally be applicable. Photodegradation of cDOC proceeds at an approximately constant rate during ice-free periods. In the absence of significant cDOC inputs, this results in a steady decrease in cDOC concentration. The input of cDOC depends to a large degree on the timing and intensity of hydrological processes (Chap. 3).

Schmitt-Kopplin et al. (1998) describe in detail, how the structure of dissolved HS changes by photobleaching. This process is oxygen dependent. The pH in solution falls, due to the formation of carboxyl groups. The average molecular weight declines. The reaction starts with a disaggregation, probably due to the oxidative cleavage of hydrogen bonds and easily oxidizable phenolic bonds. This is followed by a slower photodegradation. Structural analysis indicates a selective degradation. Structures which derive from lignins and lipids are the most photolabile, while N components, alkylaromatics, and carbohydrates are the most stable, and consequently accumulated in the system. The results are shown in Fig. 5.19.

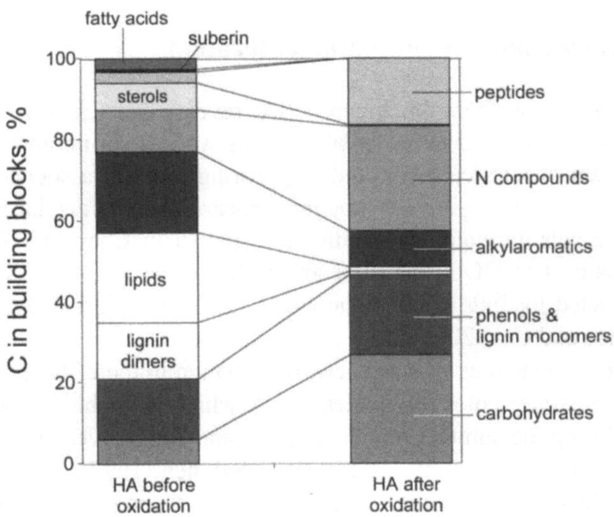

Fig. 5.19. Quantitative distribution of structural subunits of humic acids after oxidative degradation (after Schmitt-Kopplin et al. 1998, with kind permission of the American Chemical Society)

5.3.4.1 Role of Lignins

Terrestrial dissolved HS is predominantly derived from lignin, the major building blocks of which are substituted phenols. This fraction is mainly, but not exclusively, responsible for the aromaticity of HS. Hence, determining the environmental fate of the lignin fraction it is fundamental, in particular tracking the photochemical reactivity and the photobleaching behavior after it arrives in surface waters. Opsahl and Benner (1998) show that Mississippi water loses about 75% of its total dissolved lignin fraction within 28 days of exposure to sunlight (Fig. 5.20). Photooxidation is the most important mechanism, photodegradation splits the macromolecules (>1.0 kDa) and oxidizes some lignin-phenols, an example is the formation of vanillic acid from vanillin. The photochemical stability of the lignin phenols and their derivatives can be ranked as follows: acetovanillone>ferulic acid>syringic acid>vanillin>acetosyringone>syringaldehyde>vanillic acid>p-coumaric acid. The residuals of dissolved lignin are much less subject to photooxidation than the start material, and does not change much more en route to the open ocean. The qualitative composition of the photochemically altered DOC is very similar to the DOC of the open ocean.

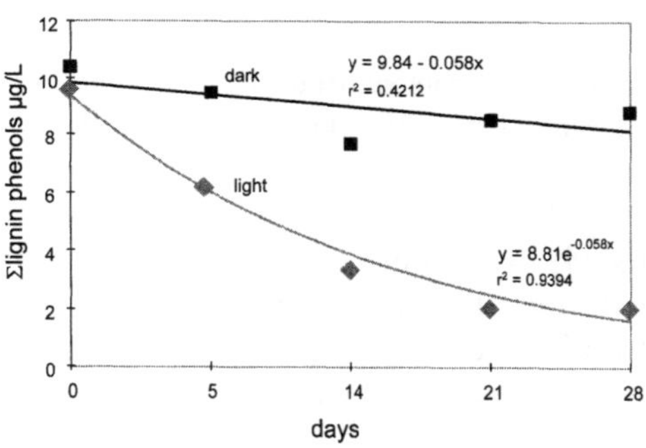

Fig. 5.20. Temporal changes in lignin phenol concentrations (measured as the sum of the six vanillyl and syringyl phenols) in light and dark treatments (after Opsahl and Benner 1998, with kind permission of the American Society of Limnology and Oceanography)

5.3.4.2 Rates and Balances

The most important end products of the photochemical reactions of dissolved HS are CO_2 (Chen et al. 1978; Miller and Zepp 1995) and CO (Moran and Zepp 1997). The illumination of dissolved HS can therefore have a short-term positive effect on organisms in surface waters, because the oxidation of organic components in acid lakes can stimulate primary production through the increased CO_2 concentration (de Haan 1992, 1993; Granéli et al. 1996; Lindell and Rai 1994; Salonen and Tulonen 1990).

The quantum yield (moles of produced CO_2 per mole of absorbed photons) increases strongly with decreasing UV wavelength (Gao and Zepp 1998). Since the energy rich radiation only reaches the top-most layer of the water-body, photomineralization is limited to this zone (Fig. 5.21). Granéli et al. (1998) assume that UV-B, UV-A and photosynthetically active radiation (PAR) contribute 17, 39 and 44% respectively to the photomineralization of dissolved HS. Vähätalo and Salonen (quoted in Münster et al. 1999b) describe the following proportion for lake Valkea-Kotinen, Finland: 10% UV-B, 67% UV-A and 23% PAR. Münster et al. attribute the large difference from the results of Granéli et al. (1998) to the use of filters by the latter, which could overestimate the proportion of UV-B and PAR. However, in both studies, the importance of UV-A for photomineralization is clear.

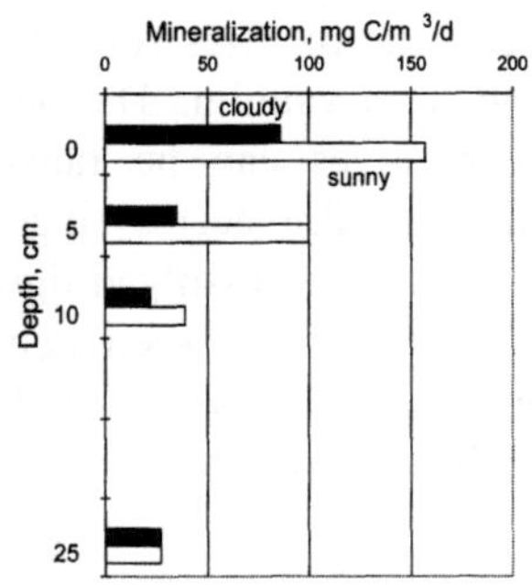

Fig. 5.21. Vertical distribution of photomineralization rates in Lake Skjervatjern, Norway, during a cloudy and a sunny day (after Münster et al. 1999b, with kind permission of Backhuys Publisher)

In a more recent study, Vähätalo et al. (2000) find a maximum photomineralization rate of approximately 1 mmol/m^2/d C. Half of the mineralization is attributed to radiation of wavelengths >360 nm. The authors develop the following relationship of photomineralization (in mol/m^2/d C) to global radiation (in MJ/m^2/d) (Eq. 5.5):

$$\text{photomineralization} = 0.0647 \text{ global radiation}; \quad r^2 = 0.77. \tag{5.5}$$

This relations indicates, for instance, that while the top-most 37 m in the Sargasso Sea supports 10% of the mineralization, a depth of only 2.4 cm supports 10% of the mineralization in the most dystrophic lakes in Finland.

In addition to the work of Vähätalo and Salonen (1997), de Haan (1993) gives a balance for whole water bodies which yields somewhat different results. He finds that the degradation of aquatic HS occurs down to a depth to which UV with wavelengths <320 nm can penetrate. De Haan calculates that the UV-related degradation of aquatic HS on a yearly basis, in relatively shallow mid-sized lakes, is low compared to primary production of phytoplankton (only 8%) and macrophytes (only 3%) if there is no nutrient limitation to the production. Differences in the calculations of Vähätalo and Salonen (1997) and de Haan (1993) are not surprising since it is extremely difficult to determine the true quantity of light penetrating the water column.

Swedish lake are comparable to Finnish lakes. In five southern Swedish lakes with differing HS concentrations (3.9–19 mg/L DOC), the DIC production is of the same order of magnitude as primary production or respiration of the plankton community (Granéli et al. 1996). Photodegradation is estimated as 86–410 mg C/m^3/d or 44–171 mg C/m^2/d, while the respiration rate for the entire epilimnion is 393–860 C/m^2/d. These numbers show that photoproduction of DIC from DOC is an important process in humic lakes of the boreal zone. Two further results of this study are worth noting. DIC production and photomineralization is clearly linked to the loss of fluorescent properties of the HS (Fig. 5.22). DIC production through photomineralization can thus account for the regularly observed CO_2 supersaturation in lakes. Granéli et al. (1996) conclude that not only UV-B accounts for the photomineralization process, but also longer wavelength light. At least the total energy content of visible light is about 10 times that of the UV radiation (Kirk 1994). The assumption of Granéli et al. (1996) is in contrast to that of de Haan (1993).

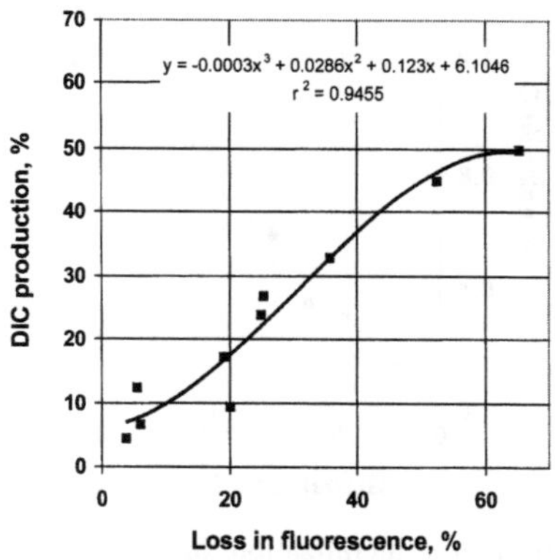

Fig. 5.22. Dissolved inorganic carbon (DIC) production in relation to loss in fluorescence in five lakes (modified from Granéli et al. 1996, with kind permission of the American Society of Limnology and Oceanography)

This putative contradiction is resolved by the experimentation, and above all in the mathematical modeling, of Reche et al. (2000). The authors show that UV-B, UV-A and also PAR can cause photobleaching in the water-column; of these, UV-A is the most important light component. If the DOC concentration is above 3 mg/L C, the share of the three wavebands is independent of the organic C content. Extrapolations show that summer photobleaching of cDOC requires 18–55 days to account for the 50% reduction of absorption in the entire water column. In UV-rich conditions such as the southern hemisphere, photobleaching is 3–5 times more rapid than in temperate climates.

Moran et al. (2000) present new data on the quantitative aspects of the photobleaching of DOC for the coastal zone impacted by input of terrestrial DOC. Photobleaching reduces the total DOC by only about 31%, while the cDOC is reduced by about 50%, and the fluorescent DOC (fDOC) by about 56%. From a kinetic analysis, the authors conclude that following prolonged photobleaching, a pool of unchanged DOC resists this process. It is very likely that this is related to aliphatic structures. Further quantification of the photobleaching/photomineralization is carried out by Vähätalo et al. (2000). These authors find a significant linear dependence of photomineralization in the Finnish lake Valkea-Kotinen upon the global radiation (Fig. 5.23).

The qualitative cDOC action spectra for the release of photomineralization products and organic products are modeled. Miller (1998) describes action spectra for CO and formaldehyde (Fig. 5.24). It is obvious that CO is released by UV-A radiation, while the formaldehyde release requires more energy and occurs only with UV-B radiation. The action spectra are evidently product-specific. Furthermore, the particular action spectra are dependent on the quality of the cDOC, and the examples shown are only two of many potential photoproducts.

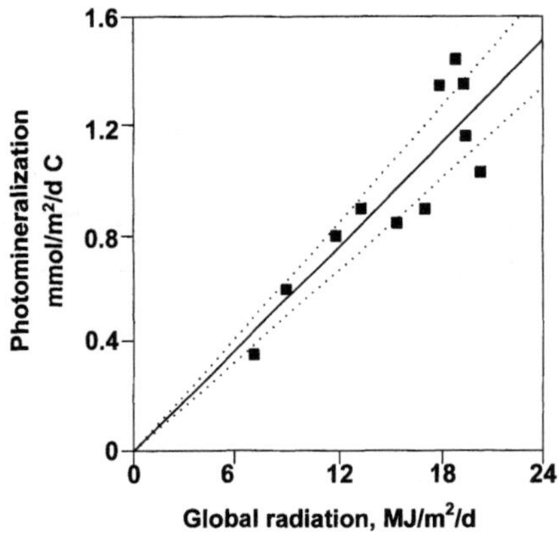

Fig. 5.23. Dependence of photomineralization of cDOC on global radiation in Lake Valkea-Kotinen, Finland. Dotted lines are 95% confidence intervals (after Vähätalo et al. 2000, with kind permission of the American Society of Limnology and Oceanography)

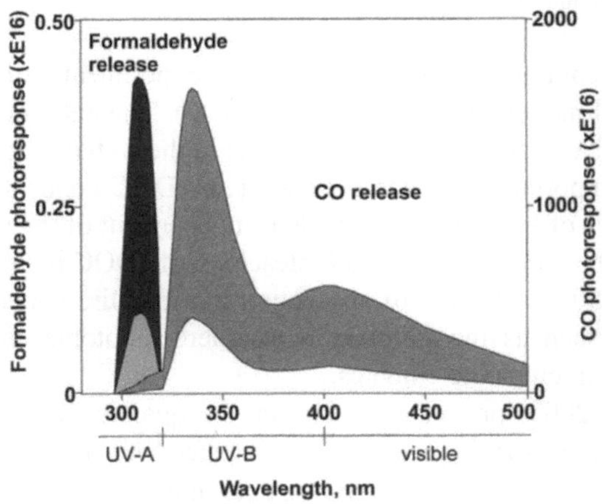

Fig. 5.24. Modeled photoresponse of carbon monoxide and formaldehyde release as a function of wavelength and DOC concentration (after Miller 1998, with kind permission of Springer Verlag). Dark areas: 20 mg/L DOC; light areas: 5 mg/L DOC

Formaldehyde and CO are however, not substrates for microbial growth. Rather, such substrates include short-chain fatty acids for which action spectra have become available very recently (Miller et al. 2002). Calculations of DOC photoproduct formation in southeastern U.S. coastal surface waters indicate a formation ratio for biologically labile photoproducts: CO of 13:1 (Fig. 5.25). That means that the quantum yields for biologically labile photoproducts formation show a greater contribution from UV-A and visible wavelengths, than does CO.

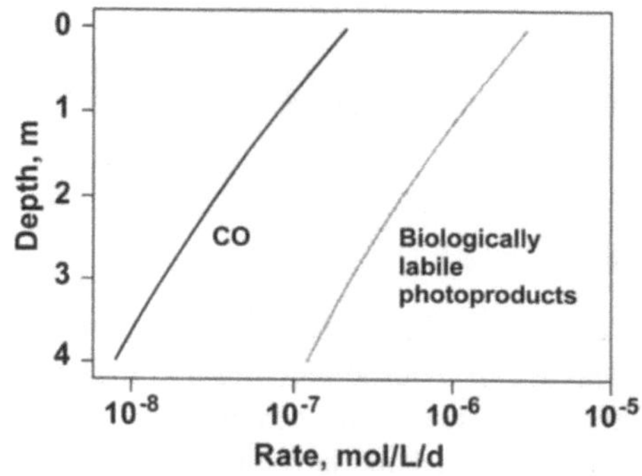

Fig. 5.25. Rates of photochemical formation of biologically labile photoproducts as a function of depth on the southeastern U.S. coastal shelf. Rates of photochemical formation of CO are shown for comparison (modified from Miller et al. 2002, with kind permission of the American Society of Limnology and Oceanography)

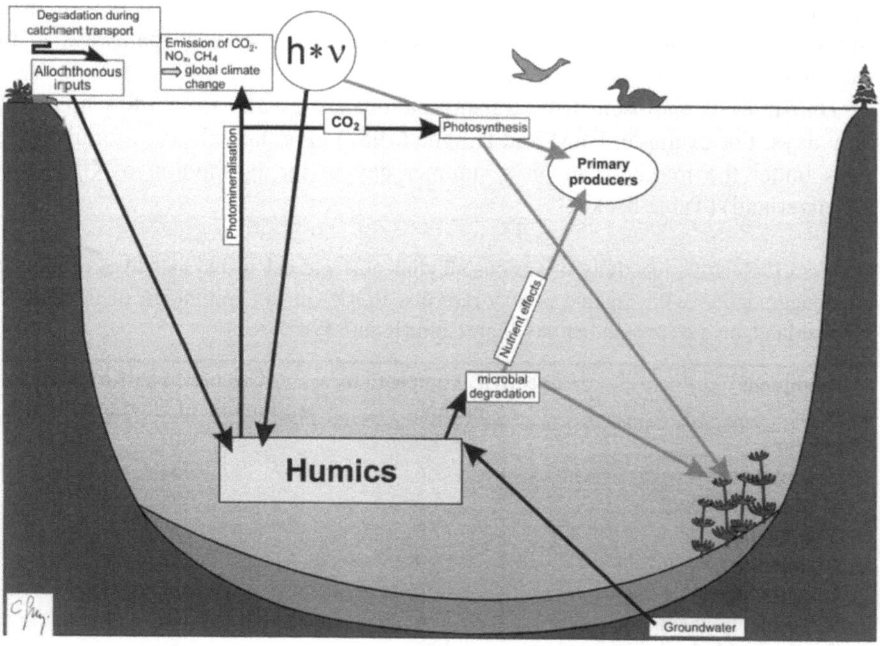

Fig. 5.26. Synopsis of HS as an energy and CO_2 source. h*ν is solar energy

However, there is also recent evidence that apparent quantum spectra for DOC photoproducts vary spatially, and that DOC source or light exposure history can influence formation rates (Miller et al. 2002). Nevertheless, application of the apparent quantum yield to coastal regions worldwide predicts an annual formation of biologically labile photoproducts in coastal waters of $206 \cdot 10^{12}$ g C.

Fig. 5.26 summarizes these results in relation to HS as an energy and CO_2 source in lakes, and is thus another piece of puzzle in the whole picture of HS as important regulators in lake systems.

5.3.5 Indirect Photolysis of Xenobiotics and Allelochemicals

5.3.5.1 Xenobiotics

Thus far, the effects of light on dissolved HS are discussed in relation to improved bioavailability, by cleavage of macromolecules and the provision of organic nutrients (short-chain fatty acids and amino acids), and CO_2. However, the oxidative potential of the above mentioned oxidative chemical species (ROS) can now be extended to xenobiotic and allelochemical compounds. The energy required to excite these organic compounds is much higher than for triplet oxygen. This process is known as indirect or sensitized photolysis (Schwarzenbach et al. 1993). In addition to the adsorption to inorganic and organic seston, photolysis is a very important elimination pathway for many persistent xenobiotics in water, which cannot directly absorb radiation with a wavelength >290 nm. Only a few of many studies which provide information on the intervention of such pathways in the ecological framework are cited here.

Half-lives of photosensitized transformations are usually from a few hours to a few days. For example, Faust and Hoigné (1987) give the half-lives for alkylphenols under the midday sun on a summer day in the epilimnion of Greifensee (Switzerland) (Table 5.1).

Table 5.1. Relative reactivity of various alkylphenols and calculated half-lives of indirect photodegradation with organic peroxy-radicals, ROO•) in the epilimnion of Greifensee, Switzerland, on a clear summer day (Faust and Hoigné 1987)

Compound	Reactivity compared to phenol	Calculated half-lives in the epilimnion of Greifensee[a] (days)
phenol	1.0	200
4-nonylphenol	2.6	78
4-isopropylphenol	5.3	38
2-methylphenol	7.3	27
4-ethylphenol	8.9	22
4-methylphenol	12.4	16
2,6-dimethylphenol	23.8	8
2,4,6-trimethylphenol	40.6	5

[a] mean 24-h light intensity on a clear June day at a latitude of 47°N.

It is clear that with an increasing number of alkyl substitutions, the reactivity of a particular phenol increases. Another interesting fact is that if one compares the various 4-alkylphenols with each other, the reactivity decreases with increasing numbers of C atoms, in other words, with increasing hydrophobicity of the molecule. An explanation offered by Faust and Hoigné (1987) for these findings is that because of hydrophobic interactions of the alkyl group with parts of the DOM, the probability of encounters of the phenol with the more polar peroxy radical decreases.

In many cases, indirect photolysis is a function of dissolved HS concentration. For example, Mills and Schwind (1990) find that for two organo-borates, the dissolved HS mediated photolysis is directly proportional to the HA concentrations up to 10 mg/L. At higher concentrations, there is a negative deviation from linearity, presumably due to trapping of ROS by HS (Faust and Hoigné 1987; Mill et al. 1980). Mills and Schwind (1990) also report that the removal of dissolved oxygen doubles the effectiveness of the reaction, which indicates that singlet oxygen does not take part in the reaction.

Such xenobiotic chemicals which themselves have the potential for direct photolysis, are subjected to indirect photolysis in natural freshwaters with a sufficient dissolved HS content. This can for example be illustrated by the studies of Kulovaara et al. (1995) who compare the degradation of DDT by UV radiation with intense simulated sunlight radiation. The degradation rate is clearly faster with the UV radiation (minutes) than with the simulated sunlight radiation (hours). The presence of natural dissolved HS has a retarding effect on the photolytic degradation of DDT in the UV radiation, but an accelerating effect with artificial sunlight. This shows that two different reaction pathways occur in the experiments:

1. fast, probably direct photolysis of DDT, caused by the UV radiation;
2. relatively slow, sunlight-mediated direct degradation of DDT, superimposed by the indirect, dissolved HS sensitized processes.

Corin et al. (2000) report on another two-pathway degradation for the resin acid dehydroabietic acid. Again, in the UV experiments, the degradation is substantially slower in humic water than in humus-free control water, whereas in the simulated sunlight experiments the degradation rate is accelerated by the presence of DOM. These differences are obviously due to different reaction pathways in the experiments. Radiation of the aqueous resin acid (dehydroabietic acid, DHAA) solutions gives rise to a great number of degradation products of which, for example, 7-oxodehydroabietic acid and 7-oxodehydroabietin are formed in high amounts. During photolysis of DHAA in humic water, decarboxylation of DHAA to dehydroabietin appears to be one of the main reactions. The bacterial toxicity of the aqueous DHAA solutions decrease with increasing radiation time. Consequently, the photolysis of DHAA does not generate any notable amounts of toxic intermediates, and/or the intermediates formed are rapidly further degraded into compounds of lower toxicity than the parent compounds or are incorporated into the HS.

A clearly accelerated photolytic degradation of herbicides, if HS are present, is found by Chiron et al. (1995). With 4 mg/L HS, a half-life of 84 minutes for alochlor and 150 minutes for bentazon, are found with radiation of 550 W/m^2 at 300–800 nm. The degradation of alachlor follows a pseudo-second order reaction, and for bentazon a first order reaction.

Another example of indirect humic-mediated degradation is the photolysis of the common fungicide vinclozolin (Fig. 5.27) (Moza and Hustert 1997). With visible light (UV filtered), in the absence of HS, there is only a very slow degradation: approximately10% in 8 hours. Sunlight in the absence of HS approximately doubles the degradation rate. With visible light (UV filtered), and the presence of HS, there is a 50% degradation in 8 hours, while sunlight and HS together support another 10% degradation in this time period. Certainly, the initial speed of degradation is higher in the presence of UV. In both cases with HS in the system, a two phase degradation is observed. Following a rapid degradation, in which with sunlight radiation, 50% of the fungicide is broken down in 3 hours, a second, slower degradation phase occurs in which only a further 10% is broken down within the next 5 hours.

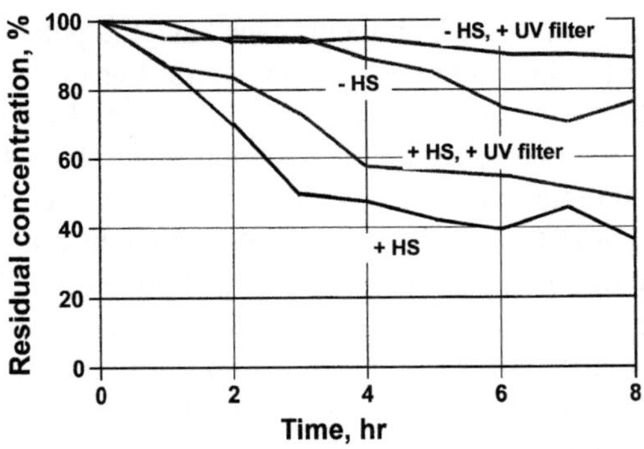

Fig. 5.27. Direct and indirect (+HS) photolysis of the fungicide vinclozolin with UV radiation and visible light, and visible light only (+UV filter) (after Moza and Hustert 1997, with kind permission of the GSF Research Center of Environment and Health, Munich, Germany). Vinclozolin: 1 mg/L; dissolved HS: 5 mg/L DOC

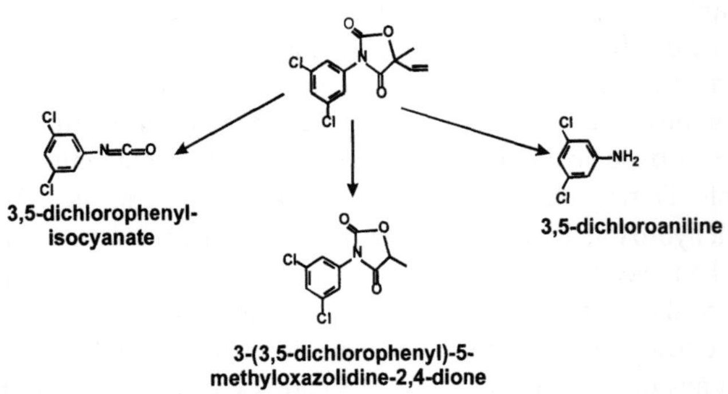

Fig. 5.28. Major photolysis products of vinclozolin (after Moza and Hustert 1997, with kind permission of the GSF Research Center of Environment and Health, Munich, Germany)

Major photolysis products are shown in Fig. 5.28. No complete mineralization takes place, but rather a decomposition. The toxic potential of the products is interesting: Moza and Hustert (1997) state that all three identified products are toxicologically more hazardous than the parent material. It can be postulated that a comparable effect also underlies the findings that in the presence of HS, an increased irritation of the test fish occurred (Chap. 8.5). This means that incomplete mineralization of xenobiotics and probably also of allelochemicals certainly can lead to an increase in the toxic potential. There are, however, very few reports in the literature on this subject.

The opposite phenomenon may also occur in the presence of HS, namely the inhibition of toxic compound production. Fukushima and Tatsumi (2001a) describe this interesting indirect degradation pathway of chlorinated compounds, when comparing the Fenton reaction (H_2O_2/Fe(III)) with the photo-Fenton reaction (H_2O_2/Fe(III)/HA). They study pentachlorophenol (PCP). Although 40% of the PCP is degraded after 5 h of radiation in the H_2O_2/Fe(III) system, more than 90% is degraded after 5 h of radiation in the H_2O_2/Fe(III)/HA system. This shows that the degradation of PCP is clearly enhanced by the presence of HA in the photo-Fenton system. In the system without HA, the production of octachlorodibenzo-*p*-dioxin (OCDD) is detected, and 2-hydroxy nonachlorodiphenyl ether is also identified as a precursor of OCDD. However, no OCDD production is observed in the H_2O_2/Fe(III)/HA system. This indicates that the presence of HA suppresses the production of OCDD during the degradation of PCP by the photo-Fenton system. Such an effect by HA can be attributed to a reaction sequence wherein reaction intermediates derived from PCP, such as PCP•, are incorporated into HA by covalent bonds. Since illuminated HS release ROS, such as H_2O_2, a similar result can be obtained in systems without added H_2O_2 (Fukushima et al. 2000 b).

Table 5.2. Degradation of various pesticides in the presence or absence of HS (from Moza and Hustert 1997) n.d. = not determined

Pesticide	Control of	Degradation, %		Half-life, h	
		– HS	+ HS	– HS	+ HS
Metalaxyl	Oomycetes	<10	65	n.d.	43.3
Vinclozolin	*Botrytis cinerea*	<1	50	5.8	3.8
Carboxin	Black rust, fireblight	80	90	0.65	0.5
Triadimefon	Powdery mildew, rust	90	90	1.36	1.25
Thiabendazol	dermatophytes, *Microsporum*	90	90	1.01	1.16

The degradation of other pesticides, in the presence or absence of HS, is documented in Table 5.2. If one accepts that photolysis follows a first order rate of reaction in dilute aqueous solutions, one can determine the half-life which provides a clear measure of the photochemical degradability of the particular pesticide in water. Table 5.2 indicates:

1. different pesticides are degraded at different rates. For the degradation with or without HS, different structures within the xenobiotic molecules are responsible. For direct photolysis, the structures are chromophoric groups, and for indirect photolysis, the structures are those that are sensitive to the photooxidants. With some radical reactions, such as with OH• radicals, the aromatic structures are preferentially attacked. These are essentially the chromophores. A nucleophilic attack occurs at locations of low electron density, such as the C-atom in sterically accessible carbonyl groups, whereas electrophilic attacks are at locations of higher electron density. In the carbonyl example, the location of higher electron density is the oxygen atom;
2. the presence of HS usually, but not always accelerates the photolytic process (Moza and Hustert 1997 vs Corin et al. 2000; Kulovaara et al. 1995). Different HS isolates have different effects, thus HS-mediated degradation rates are not comparable, unless a quantitative structure property relationship can be established relating specific structural features to the photolytic efficiency.
3. Westerhoff et al. (1999) pursue the question of sensitive structures from the viewpoint of water treatment. The presence of HS, the effectiveness of oxidation and disinfection measures is markedly reduced, and leads to disinfectant by-products, which are a health concern. It is highlighted that the requirement of oxidants depends on the HS structure. Of all studied structural parameters (aliphatic fraction, aromatic fraction, elemental composition, UV absorption at 254 and 280 nm), the specific UV absorption (254 and 280 nm) is best correlated with the amount of oxidants needed for disinfection. This means that in turn, it is the organic π-electrons (in the aromatic and conjugated double bonds) of the HS molecule that determine the reactivity to oxidants.

The first clear indication of which structural units in HS may affect indirect photodegradation of the model xenobiotic atrazine, is worked out by Hapeman et al. (1998). In the absence of nitrate, most of the studied mimics (oxalic acid, pyrogallol, quinone, and salicylic acid) cause an increase in the atrazine degradation rate relative to the control (no mimic present) (Fig. 5.29). Minimal effects, however, are observed in the presence of 2-octanol and coniferyl alcohol. To further elucidate the pathways, atrazine photodegradation experiments are conducted in the presence of the DOC mimics, and with or without *t*-butanol, a known OH• scavenger. The rate of atrazine degradation for each of the mimics with *t*-butanol present, is the same in all cases as the rate of atrazine degradation in the control without mimics or *t*-butanol. This suggests that the observed enhancement in the atrazine degradation rate in the absence of *t*-butanol, is due to the generation of OH• or other radical species, produced from the degradation of the DOC mimics, that may be quenched when *t*-butanol is present.

The results indicate that the DOC mimics affect not only the overall degradation rate, but the degradation pathways as well. Each mimic behaves differently and often can take on several concurrent roles. The chemical behavior cannot simply be predicted by the light absorption (extinction coefficient) (Fig. 5.29). Coniferyl alcohol has no effect on the photodegradation rate of atrazine. Pyrogallol and salicylic acid cause a slight increase in atrazine degradation rate if sufficient UV radiation is available. Oxalic acid and quinone increase the atrazine degradation strongly, whereby quinone may serve as a sensitizer in addition to an OH • generator. Oxalic acid is a know OH• generator. It exists as oxalate at environmentally relevant pH and may form a complex with excited atrazine, which can undergo electron transfer to yield oxalate radical anion, which in turn can react with oxygen to form oxygen anion $O^{\bullet}{}_2^-$. This then is readily protonated (twice) to form hydrogen peroxide, which upon radiation gives $OH^{\bullet}$. Furthermore, oxidation of oxalic acid can yield $CO^{\bullet}{}_3^-$ which can also react with atrazine. Thus, a rate increase is observed in the atrazine degradation in the absence of more efficient $OH^{\bullet}$ generators, such as nitrate.

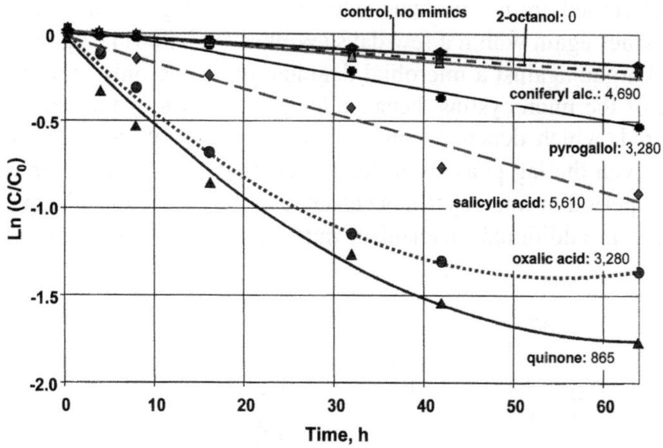

Fig. 5.29. Influence of dissolved organic carbon mimics (5 mg/L) on the atrazine (30 µM) degradation rate. The figures behind the compound names are extinction coefficients at 290 nm (from Hapeman et al. 1998, with kind permission of SETAC)

Generally, DOC can act as an OH• generator, OH• scavenger, photosensitizer, or even as a radical generator and scavenger. Not only the rate but also the processes involved, and thus the product profiles, are a function of the DOC structural properties. Knowledge of DOC concentration and the type of functional groups on the DOC are thus required to accurately predict the photolytic fate of xenobiotics in surface waters.

The structural characterization of the structural building blocks of DOC appears to be a prerequisite to understanding its environmental behavior. Otherwise, one will still obtain ambiguous results with DOC enhancing photodegradation – as with the fungicide propiconazole (Vialaton et al. 2001), and DOC inhibiting photodegradation – as with another fungicide dichlofluanid (Sakkas et al. 2001). The DOC from one source is by no means identical to the DOC from another. Furthermore, paramagnetic metals [Cr(III), Co(II), Mn(II), Cu(II)] have an strong impact on HS as a fluorescence quencher as well as radical scavenger and, thus, inhibit aquatic photodegradation as recently shown with organo-P (Kamiya and Kameyama 2001). This effect is apparently not adequately addressed.

5.3.5.2 Cyanotoxins as Examples of Allelochemicals

Indirect photolysis through dissolved HS is not restricted to xenobiotic chemicals. Biopolymers such as exoenzymes or DNA, kairomones (for example, fish kairomones which trigger the daily vertical migration, or induce the development of long helmets in *Daphnia*), and allelochemicals such as cyanotoxins, can be equally good objects for photolytic attacks, which are mediated by dissolved HS or, more rarely, by plant pigments. Some examples of selected microcystins (toxins from cyanobacteria) are given for elucidation.

In many instances, peak microcystin concentrations are short-lived. For instance, in an artificical bay of lake Müggelsee, the concentration of dissolved mircrocystins increased from non-detectable to over 70 µg/L and dropped to non-detectable values again within a few days (Welker et al. 2001) (Fig. 5.30). Welker et al. (2001) argue against a microbial degradation as the only mechanism of the fast decline of the microcystins, because long lag phases of the microbial degradation occured, which depended on the initial concentration of the microcystin (Fig. 5.31). Even the lag phase with the lowest microcystin concentration clearly exceeded the period of microcystin occurrence in the bay of Lake Müggelsee (Fig. 5.30). Hence, an additional mechanism must apply, this being the indirect photolysis.

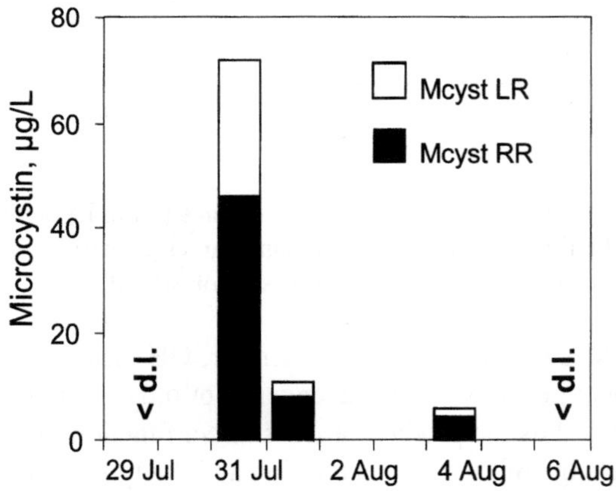

Fig. 5.30. Temporal changes in dissolved microcystins in a small, unflushed bay of Lake Müggelsee, Berlin, during an algal bloom in summer 1997. < d.l. = below detection limit of 0.5 µg/L (from Welker et al. 2001, with kind permission of Springer Verlag)

Indirect photolysis may be mediated by dissolved photosynthetic pigments and HS. With one microcystin variant, Tsuji et al. (1994) show that photosynthetic pigments can induce photolysis. However, the pigment concentrations employed are so high that the results are valid for algal scums, but not for conditions in the water column. Whether or not HS-sensitized photodegradation of microcystins may be responsible for their short-term excistence has been tested by Welker and Steinberg 2001).

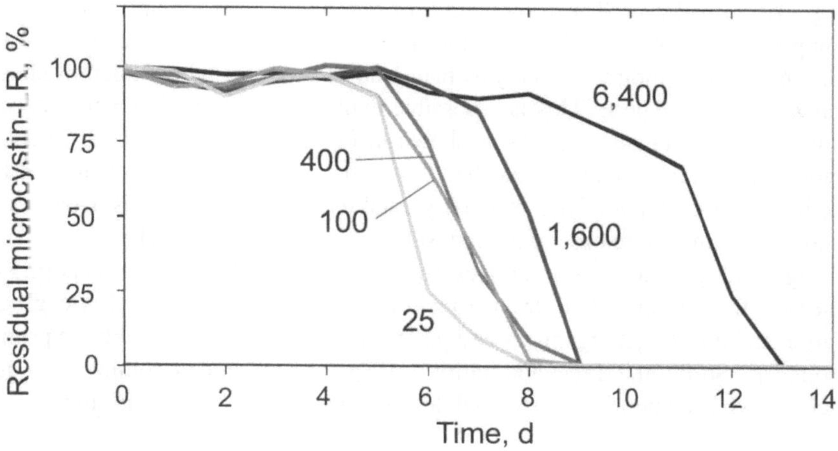

Fig. 5.31. Temporal changes in degradation of microcystin-LR by a microbial consortium from an oxidation pond near Griffith, Australia. Note that with increasing microcystin-LR concentration (in µg/L denoted by the numbers in the graph), the lag time increases (from Welker et al. 2001, with kind permission of Springer Verlag)

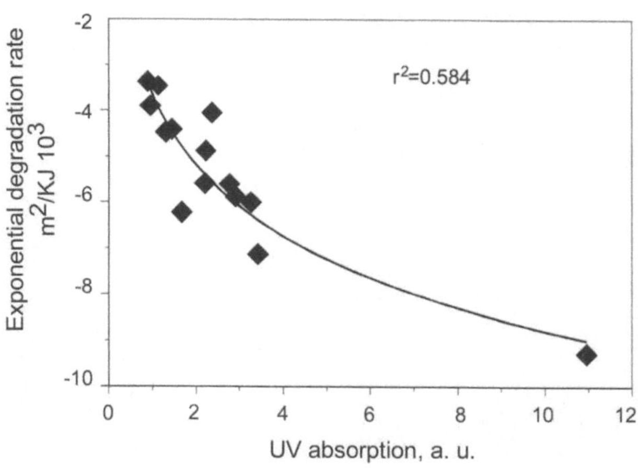

Fig. 5.32. Dependence of first-order rates of photosensitized degradation of microcystin-LR on UV absorption in natural waters from Berlin and Brandenburg. The fitted curve is a logistic growth function. a.u. = arbitrary units (from Welker and Steinberg 2000, with kind permission of the American Chemical Society)

Indirect photolysis, mediated by dissolved HS, can be responsible for an accelerated degradation of microcystins. Under optimal conditions and in the presence of 7 mg/L DOC Fuhrberg groundwater FA, microcystin-LR has a half-life of about 10 hours (Fig. 5.32). Such rapid rates are not found in the field, because of poor penetration of UV radiation into the water column. Such indirect photolysis is possible, but under the conditions of a eutrophic lake, it is not a very effective degradation pathway. The combination of photolytic cleavage and microbial decomposition is possibly the way that microcystins disappear from open water. That means that indirect photolysis may lead to products which microorganisms can rapidly decompose. This is a possible explanation for the low persistence of microcystins in aquatic systems as shown in Fig. 5.30. Although the FA concentration in most lakes may be less than that in the experiment, the potential of HS-mediated indirect photolysis is clearly established. This mechanism occurs directly following the release of microcystin from cyanobacterial cells.

Fig. 5.33 graphically summarizes potential direct and indirect effects of ROS released from illuminated HS, particularly those rich in chromophoric groups. Target and non-target organisms and guilds are exemplified by planktonic primary producers. So far, the direct and indirect effects of ROS in determining ecological niches have been poorly studied, and deserve much more attention in future.

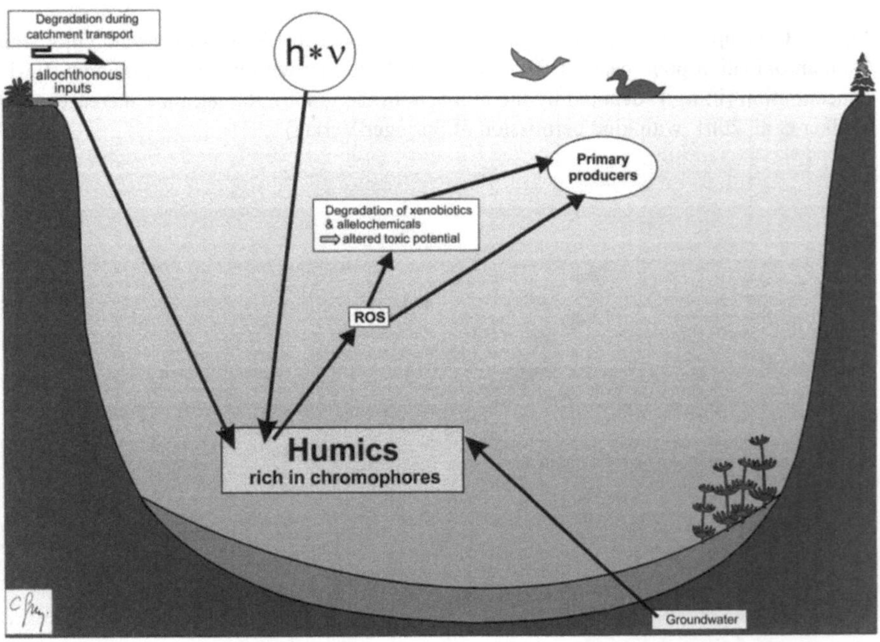

Fig. 5.33. Graphical summary of potential direct and indirect effects of reactive oxygen species (ROS) released from irradiated HS. h∗ν is solar energy

6 Interactions with Nutrients, Metals, Halogens, Biopolymers, Pheromones, and Electrons

By their functional groups and molecular configurations, HS can change the chemical speciation of nutrients, metals, and many organic molecules. HS change not only the bioavailability, but also the toxicity, of metals and xenobiotics. A well known example is mercury, which forms organic species, the toxicity of which far exceeds that of the inorganic species. Hence, HS clearly participate in biogeochemical cycles as will be shown in the chosen examples of nutrients, metals, biopolymers, and pheromones. Furthermore, HS may act as a redox catalyst.

6.1 Nutrients

HS interact with key nutrients, such as P and N, as well as with the essential metals, exemplified by Fe and Mn.

6.1.1 Phosphorus

6.1.1.1 Water Column

In many freshwaters, P is the most important growth-limiting nutrient. It is therefore no surprise that much of the work on HS-nutrient interactions is carried out with P. The concentration of HS can influence the rate at which ionic ortho-P is taken up by plankton. For instance, Brassard and Auclair (1984) report that the 1–10 kDa DOC fraction modifies ortho-P uptake in Canadian Shield lakes. This modification can be stimulation or repression, as Francko (1986) describes for plankton in the presence of HS from *Typha* (cattail). The stimulation or repression of P-uptake is generally lake-specific and depends on particle size.

While there are some reports from soil science on the chemical species of P which can be incorporated into soil organic matter[1], comparable information for

---

[1] From Russian soils, Makarov et al. (1997a,b) describe the following phosphorus species in humic substances: phosphonates, phosphate monoesters, glucose-1-phosphates, phosphate diesters, pyrophosphates, polyphosphates and unknown species. HA contain more

aquatic ecosystems is sparse. Although, in principle, the same chemical species must apply for soils, sediments, and water, there is however, no direct information for the latter. Probably the best studied interaction between HS and P is the complexation of ortho-P, in particular in the presence of Fe. Many studies show that at relatively low pH values and low redox potentials, aggregates of HS, Fe(III), and ortho-P form colloidal aggregates. P bound in this manner will inhibit direct assimilation by organisms (Francko and Heath 1982, 1983; Jones et al. 1988, 1993; Shaw 1994; Steinberg and Münster 1985). HS alone probably do not react with P in significant quantities; the major complexation is due to the presence of Fe.

In humic-rich, Fe-rich natural freshwaters, a large proportion of ortho-P is bound to high-molecular weight aggregates. The physico-chemical nature of the HS-metal-P association is not yet fully understood. Some aspects are described by Jones et al. (1993) and Shaw (1994), who find that Fe and P form colloids only in the presence of HS. Their studies also show that the transfer of Fe and P to larger sized molecules decreases with lowering of the pH value (Shaw et al. 1992). In contrast, the formation of particulate Fe is suppressed at higher pH values, in the presence of dissolved HS. This observation provides a key to a possible mechanism: it is well known that the pH value influences the protonation of HS functional groups, as well as the conformation of the HS aggregate (Box 2.1, Fig. B.2.1.3). It follows that the lower protonation and smaller conformation of HS at higher pH values, favors truly dissolved forms of Fe and P. This preference could be caused by the formation of stable Fe(III) oxide colloids. At lower pH values, the higher protonation leads to globular and ring-like HS aggregates, with an increased quantity of Fe.

Direct evidence for the HS-Fe-P complexation in humic-rich waters is provided by de Haan et al. (1990), who show that in the presence of HS, both Fe and P are simultaneously incorporated in a fraction with a nominal molecular mass of 10–20 kDa. Without HS, this type of complexation does not occur. The transfer of Fe and P to larger molecular weight fractions appears to be due to the complexation reactions, and involves the formation of HS–Fe–P complexes.

In a series of experiments, de Haan et al. (1990) and Wetzel (1993) observe that labeled PO_4^{3-} is released from high-molecular weight complexes, probably through displacement reactions in which low-molecular weight P esters participate, thus altering the balance between HS and isotope concentrations. This finding has become a generally accepted model. Other possible mechanisms which alter the speciation of HS-Fe-P compounds are photoreactions. For instance, Steinberg and Baltes (1984) examine the effect of UV radiation on a complex of FA and SRP. In water from a peat bog, they find a bimodal separation pattern of SRP, into high-molecular weight and low-molecular weight components The high-molecular weight, rather than the low-molecular weight SRP fraction, is sensitive to UV radiation. The addition of Fe^{2+} and low concentrations of Mn^{2+} reduces the concentration of low-molecular weight SRP in favor of high-molecular weight

phosphonates, phosphates diesters and polyphosphates, while FA contain more glucose-1-phosphate.

SRP. Higher concentrations of Mn^{2+} result in the opposite effect. The underlying mechanism can be either a cleavage of the high-molecular weight SRP, or displacement reactions in the aggregates.

In a study with water from a wetland lake, Francko and Heath (1982, 1983) report that Fe(III) is also part of a UV sensitive complex, and that UV radiation releases P through the reduction of Fe(III) to Fe(II). Both the photoreduction and the P release are reversible, and in darkness the UV sensitive P complex and Fe(III) are regenerated. The turnover time for the P release from the UV sensitive complexes and the photoreduction of Fe(III) is about an hour. As an additional P release mechanism, it is suggested that the cleavage of phenolic compounds in the HS can lead to a reduced affinity of Fe for P (Steinberg and Baltes 1984).

Corroborating the results of Francko and Heath (1983), Jones et al. (1988) outline that P is absorbed by two high-molecular weight fractions (>100 kDa and 10–20 kDa). Radiation with sunlight leads to a slow release of P. This release can be attributed to the HS sensitized photoreduction of Fe(III). Also, complexes between extracellular phosphatases and HS can be destroyed photolytically, whereby the phosphatases are reactivated and may cleave P esters (Boavida and Wetzel 1998; Wetzel 1991) (Chap. 6.4.1).

To date, most of the studies on HS-P associations are conducted in the laboratory. In one of only a few studies which translate these mechanisms to the ecosystem scale, Cotner and Heath (1990) study the factors which control the release of P from the HS-Fe-complexes. Addition of Fe^{3+}, and radiation with UV, increases P release. In a lake, the oxidation of Fe^{2+} is slower than the photoreduction, leading to the conclusion that with UV radiation, Fe accumulates as Fe^{2+}. During summertime, Fe^{2+} concentrations decrease with phytoplankton development. This means that the biological uptake of Fe^{2+} in the water markedly decreases both Fe^{2+} and Fe^{3+} concentrations, such that the role of photosensitive P regeneration becomes less important.

One can conclude that models of the potential influences of HS on epilimnetic P cycling are, as yet, still rather rudimentary. There appears to be more than one P pathway in the epilimnion. Hence, Francko (1986) proposes a model with three complexed P sources (phosphomonoesters, high-molecular weight colloidal-P, and Fe-HS P-complexes), and with three main mechanisms of P regeneration (alkaline phosphatase, displacement from colloids, UV radiation and photoreduction). The quality and quantity of HS and Fe concentrations are the most important chemical parameters which influence P cycling, as displayed in Fig. 6.1. It is obvious that HS control many aspects of P cycling.

Shaw et al. (2000) report that the interaction of HS with Fe and P in humus-rich lake waters is rather more complex than hitherto considered, and cannot be attributed to a single and specific mechanism of interaction. Nevertheless, the authors suggest that HS promote the association of Fe^{3+} and P with high-molecular weight fractions, primarily by acting as peptizing agents for inorganic colloids containing Fe and P. Differences between samples in terms of their ionic composition, for example, clearly lead to rather different relationships between the speciation of Fe and P, and the molecular weight distribution of HS. Moreover, the influence of HS on the speciation of Fe and P is more marked at low pH in higher ionic strength regimes, and at higher pH in the lower ionic strength regimes. Shaw et al. (2000) conclude that much remains to be elucidated regarding the molecular weight distributions of Fe and P in humic lake waters and their interactions with HS. Moreover, it is frequently suggested that the P associated with larger molecular weight fractions may provide a longer-term source of nutrition for primary producers in humus-rich lakes (Jones 1998). Yet, detailed study of the mechanisms of both association and dissociation is still needed to ascertain both the potential of high and medium molecular weight fractions of P as a source of P available to phytoplankton, and the implications of interactions between Fe, P and HS for the ecology of primary producers in humus-rich lakes.

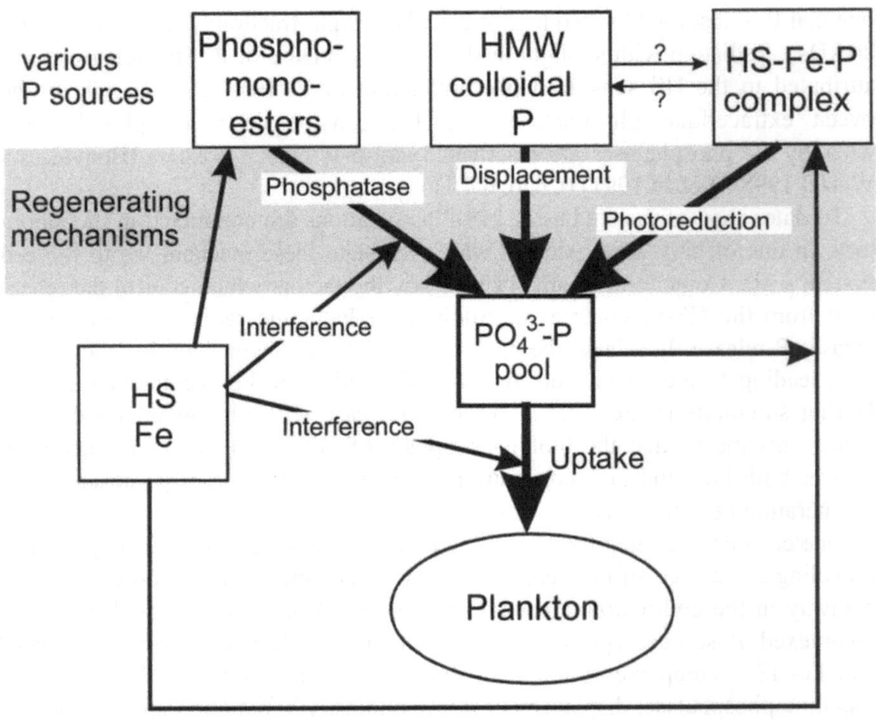

Fig. 6.1. A diagram of the continuum model of epilimnetic P cycling. In this model, three groups of complexed P sources and P-regenerative mechanisms may co-occur in a given lake system. The relative importance of each regenerative mechanism is not fixed in time, but is rather dependent on dynamic changes in physicochemical and biotic parameters. Central to this diagram is the concept that HS not only sequester P into a complexed form, but also interact with alkaline phosphatase (Chap. 6.4.1.1) and ortho-P uptake, as well as with the pool size of phosphomonoesters. In this manner, and perhaps by other interactions not shown, HS control multiple aspects of the P, determining the position of a lake in the hypothetical continuum of P and HS content (after Francko 1990, with kind permission of Wiley & Sons)

6.1.1.2 Redox Interfaces

Humic substances can interact with P at redox interfaces. Contrasting effects, probably dependent on the quality (origin) of HS, will result in peatlands serving as sources or as sinks for P for adjacent surface waters. This service is of major concern in landscapes where degraded fens are recharged by groundwater, such as the peatlands in the glacial landscapes of northeast Germany, Poland, and the Baltic countries (Zak et al. in press).

At the redox interface, precipitation of Fe(III) oxihydroxides is the main process for P retention. If the source of HS is dominated by angiosperm plants such as *Carex* spp., *Phragmites australis*, or *Typha* spp., large amounts of DOC is coprecipitated with precipitating Fe. If the source of HS, however, is dominated by mosses such as *Sphagnum* (as occurs in oligotrophic and mesotrophic fens), the precipitation of Fe(III) oxihydroxides can be inhibited by the formation of stable Fe–humic complexes, with diameters smaller than 0.45 μm. The mechanism behind the inhibition is still obscure, and future studies should determine, if a mechanism such as the formation of high-molecular weight fractions with increased aromaticity as described by Haiber et al. (2001b) (see Box 2.1) also applies.

6.1.2 Nitrogen

The environmental fate of N bound to HS in aquatic ecosystems, and its sensitivity to photochemical and microbial decomposition, is as yet much less studied than that of HS-P interactions. Previously, it has been considered unlikely that HS can be a N source for the microbial web, since the bioavailability of HS-bound N is thought to be very low. In addition, in HS there is generally a very high C:N ratio, of around 50:1 (Thurman 1985). Therefore it has been considered almost impossible that bacteria could gain significant quantities of N from HS. It was seen as unlikely that bacteria can extract HS ammonia. However, in a more recent study, Bushaw et al. (1996) show that these views must be revised. These authors report that exposure of DOC to sunlight forces the release of nitrogen-rich compounds of high bioavailability. These compounds include ammonia, which is the most effectively released.

According to the findings of Wang et al. (2000), the release of ammonia is a two or multi-step process; a photoprocess, followed by one or more dark reactions. One mechanism proceeds through an hydroxyl radical intermediate, and continues in the dark, through decomposition of photochemically produced H_2O_2 to form hydroxyl radicals. In addition to the hydroxyl radical pathway, there is also a non-hydroxyl radical dependent pathway.

The photoproduction of ammonia does not depend on DOC, but on dissolved organic N (DON). The low-molecular weight fractions (<1 kDa) are more photolabile than are the high-molecular weight fractions (>1 kDa). Up to 38% of the DON is released photochemically in the euphotic zone per day. With the dark reactions, there is both a continuous release and a continuous uptake of ammonia by

DOC. Tarr et al. (2001) give more details on the photolytic products and the mode of action of ammonia and amino compounds release. At least 20 amines are observed upon radiation of water samples containing Suwannee River FA. Among the amino acids identified are alanine, asparagine, citrulline, glutamic acid, histidine, norvaline, and serine. Although NOM photosensitized degradation of amino acids produces ammonia, amino acids do not appear to be a quantitatively important intermediate in the photochemical formation of ammonia from NOM. This is concluded from experiments where hydroxyl radical scavengers are added to irradiated NOM samples. Since the addition of these scavengers does not increase the net photoproduction of amino acids, it is believed that photochemical amino acid release is not a significant pathway in the photochemical formation of ammonia from NOM. Other functional groups appear to be responsible for the majority of ammonia photoproduction from NOM – the nature of these functional groups remains unknown (Tarr et al. 2001).

The fluorescent DOC fraction is the major source of ammonia release. Studying the Orinoco River plume, Morell and Corredor (2001) present experimental evidence identifying photomineralization of riverine fluorescent DOC as a potential fertilization mechanism, capable of sustaining phytoplankton biomass up to several hundred kilometers from the river delta. Waters with a high content of fluorescent DOC are observed to decrease their fluorescence yield, while releasing ammonia upon exposure to sunlight. This mechanism, along with N retention through recycling, explains enhanced surface phytoplankton biomass up to 1000 km from the river delta.

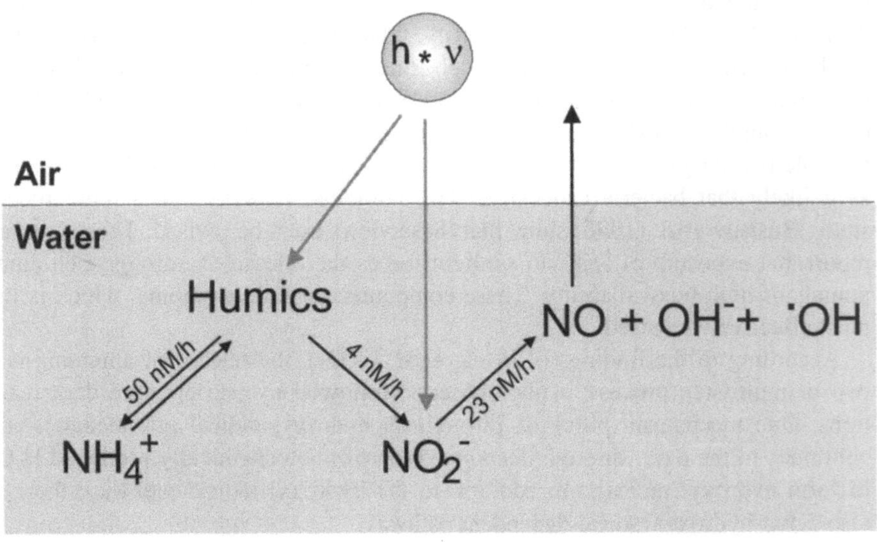

Fig. 6.2. Model of light-driven release of inorganic nitrogen from HS (from Kieber et al. 1999, with kind permission of the American Chemical Society). h*v is solar enegy, affecting humic substances and nitrite molecules

Nitrite is released from HS through illumination. The quantity of nitrite released depends on the concentration of HS, and the duration of light exposure. The quantity in solution is less than loss of nitrite directly through photolysis, and much less than the release of ammonia from HS. The quantitative relationships are shown in Fig. 6.2.

Evidently amino acids and primary amines are also released from illuminated HS. Wang et al. (2000) propose a mechanism whereby free amino acids are photochemically released. These amino acids are then subject to further breakdown to ammonia due to strong UV radiation and long-term exposure. Since dissolved amino acids are found only at low concentrations (Münster et al. 1999a; Steinberg 1977), their conspicuously short life is probably due to photochemical as well as microbial breakdown.

6.1.3 Metals

Mobility and bioavailability of metals in aquatic environments are strongly dependent on the nature and the behavior of their carrier compounds. In Box 1.2, evidence is presented that particularly high-molecular weight or colloidal fractions with their carbohydrate moieties, associate most efficiently with metals (Quigley et al. 2002). Macromolecular DOC and colloids in humic-rich surface waters consist of irregular networks of organic and inorganic entities, ranging in size from some nanometers up to the micron level (Buffle and Leppard 1995a,b; Wilkinson et al. 1997).

In a recent study, Burba et al. (2001) characterize colloidal DOC and metal species by multistage ultrafiltration and by exchange reactions on-site, immediately after sampling. The assessed metals (Al, Cu, Fe, Mn, Pb, and Zn) are predominantly enriched in the macromolecular and sub-particulate range. Strongly competing ligands (EDTA) and metal ions (Cu(II)), added on-site to the water, are considered to be suitable discriminators between available and inert metal species in humic-rich DOC and colloids. The conceptual model is presented in Fig. 6.3. The kinetics and the equilibria of the exchange reactions are used for operational characterization of the kinetic and thermodynamic stability of the metal species of interest. Burba et al. (2001) show that the conditional exchange constants, K_{ex}, of the metal species are in the order Al<Fe<<Zn<Mn≈Mg≈Ca, indicative of particularly strong competition between Cu(II) ions and Mn, Mg, and Ca, but only weak competition between Cu(II) and Al and Fe. In contrast to EDTA exchange, the Cu(II) exchange equilibria are established rather quickly, usually within 5–10 min. This leads to the assumption that distinctly different mechanisms govern the exchange reactions. It is highly probable that the relatively rapid Cu(II) exchange occurs preferentially with macromolecular metal complexes, whereas the chelator EDTA might also react with inorganic substructures, for example the Al and Fe oxide hydrates often contained in natural high-molecular weight and colloidal organic carbon (Buffle and Leppard 1995a,b; Wilkinson et al. 1997). The exchange

kinetics and equilibria greatly influence the bioavailability, since only freely dissolved metals are bioavailable, and may act either as nutrients (this chapter) or as toxicants (Chap. 7), depending on the metal.

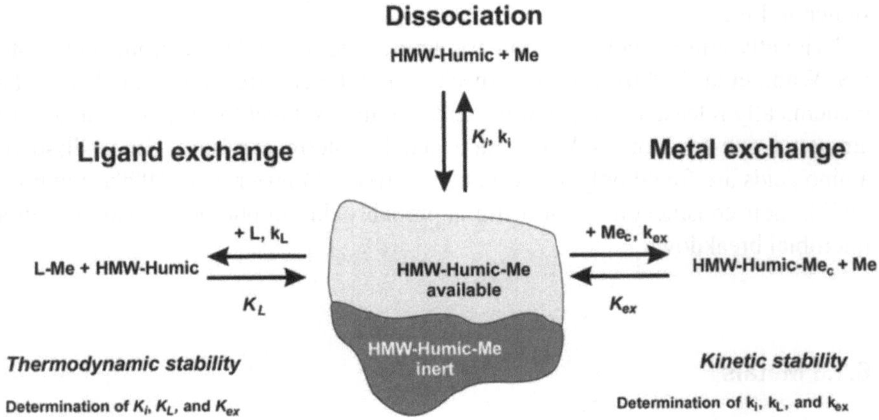

Fig. 6.3. Conditional stability of high-molecular weight and colloidal metal species (high-molecular weight-HS-Me) characterized by competitive ligand and metal exchange (from Burba et al. 2001, with kind permission of Springer Verlag). Me = metal, Me_c = competing metal, L = organic ligand. K, k = conditional rate constants

6.1.3.1 Iron and Manganese

In natural waters, the bioavailability of metals, such as Fe and Mn, which do not form organometallic compounds, is largely affected by the amount and composition of humic material and dissolved inorganics, under positive redox potential and circum-neutral pH. One piece of evidence for the role of HS is the finding of Vörös et al. (2000) who describe that with UV exposure, breakdown of HS can result in an inhibition of algal growth because the bioavailable Fe is precipitated.

Ionic structures such as carboxyl-, alcohol- and phenolic hydroxyl groups form complexes with metals (Linnik 1996, 1998; Linnik and Iskra 1996; Linnik and Nabivanets 1984). However, this mechanism may function only in high metal concentrations, which seldom or never occur in natural freshwaters (Christman and Gjessing 1983). At natural HS concentrations, the metal binding partner is not oxygen, but N atoms and reduced S atoms (quantitatively relativley unimportant humus constituents).

Iron

In addition to the previously described effects of Fe and HS on the P, light and HS both have direct effects on many aspects of the Fe itself. These include the oxidation kinetics of Fe(II), the reduction of Fe(III), and the reductive dissolution of oxides.

The controversy concerning whether Fe bound to HS is more or less bioavailable (Francko and Heath 1982 *vs.* Jackson and Hecky 1980) is at least partially solved, if one considers the sunlight-dependent, reversible bonding of Fe to HS (Münster et al. 1999b). This process may be considered an adverse effect on HS, since the photoreduction of Fe oxidizes organic carbon in lakes and even streams (McKnight et al. 1988). Important aspects of the interactions of Fe, HS, and sunlight are studied by Voelker (1994), Voelker et al. (1997), and Emmenegger et al. (2001). The authors outline that:

1. at pH 5, the Fe(II) oxidation rate increases with increasing FA concentrations. The main oxidants of dissolved Fe(II) are $HO_2/\dot{O}_2^-$, produced via reduction of O_2 by photo-excited FA, and H_2O_2, the product of Fe(II) reaction with HO_2/O_2^-. Following illumination, the FA provide reactive oxygen species (ROS).
2. Fe(III) is reduced by FA in the dark[2]. The Fe(III) is in particulate or dissolved forms.
3. Fe(III), particulate or dissolved, is reduced in the light through FA via photochemical ligand-to-metal charge-transfer reactions.

Both reactions (2 and 3) play almost equally significant roles in the reduction of dissolved Fe(III) (Voelker et al. 1997). It is evident that $HO_2/\dot{O}_2^-$ can be both an oxidant for Fe(II), and a reductant for Fe(III). In water irradiated by sunlight, the reaction of inorganic Fe(III) species with photochemically derived $HO_2/\dot{O}_2^-$ can be an important source of Fe(II). This finding accounts for the fact that in aquatic systems, the predominant species of Fe during day-time is Fe(II).

There few field studies on Fe cycling in freshwaters. One is by Emmenegger et al. (2001) in two circumneutral Swiss lakes: Lake Greifensee, a eutrophic, natural water body, and Lake Melchsee, an oligotrophic, artificial mountain lake. Radiation by simulated sunlight leads to pH dependent (pH 6.9–9.1), steady-state Fe(II) concentrations which are similar in both lakes. Superoxide appears to be a key parameter for light-induced Fe redox cycling. It originates from radiated cDOC:

$$cDOC + O_2 + h*v \rightarrow cDOC^+ + O\bullet_2^-.$$

Field measurements of Fe(II) concentrations show a pronounced day/night cycle, with Fe(II) concentrations of approximately 0.1–0.2 nmol/L at night, and up to 0.9 nmol/L near the surface during the day.

A further freshwater study is by Herzsprung et al. (1998) with amendments by Friese et al. (2002) in three extremely acidic mining lakes in Lusatia, Germany, with a pH between 2.35 and 3.0. The results are summarized in Fig. 6.4. In these extremely acidic lakes, a suboxic or anoxic hypolimnion forms over the sediment, and Fe(III) in infiltrating groundwater can be reduced by bacteria using Fe(III) as an electron sink or terminal electron acceptor. The reduction process also requires molecular oxygen or organic carbon compounds. In other words, microbial processes lead to an increase in Fe(II) concentration in the hypolimnion. In the

[2] Humic substances can also reduce metallic iron, as shown by Steinberg (1980), and Lovley et al. (1996)

epilimnion, Fe reduction is achieved by photoreduction through HS.

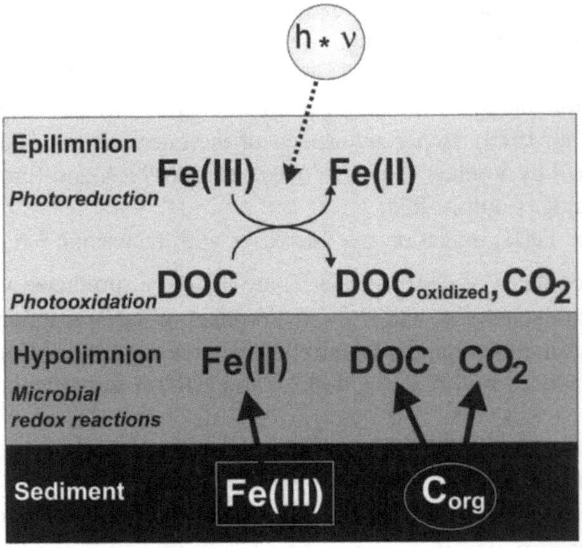

Fig. 6.4. Potential sources of Fe(II) are microbial redox reactions in the hypolimnion as well as photoreduction, which requires DOC and simultaneously oxidizes the organic carbon. h*ν is solar energy. This photooxidation of DOC is the main reason for its very low concentrations in the extremely acidic Lausatian post-mining lakes (modified from Herzsprung et al. 1998, and Friese et al. 2002, with kind permission of Wiley-VCH)

For the extremely acidic open-cast mining lakes, the described interaction between Fe and organic carbon has far-reaching consequences. With pH values below 4.0, inorganic carbon exists as CO_2. At these pH values, a pool of HCO_3^- cannot exist. Primary production of phytoplankton in such acidic conditions will be carbon limited. As seen, DOC concentrations are also low, and in summer are often below analytical detection limits (Herzsprung et al. 1998). This situation drives the phytoplankton to seek zones with a relatively high CO_2 concentration, namely the littoral zone and the layers near the sediment, where CO_2 enters via seepage, or where CO_2 concentrations are increased through heterotrophic activity (Steinberg et al. 1999). Another strategy to accommodate carbon demand is mixotrophy, enabling phytoplankton to feed on bacteria. Mixotrophic phytoplankters are indeed highly abundant in the carbon-poor, acid post-mining lakes (Nixdorf et al. 1998, in press).

Complexes of Fe and HS can also act as oxidants. Luther et al. (1996) determine Fe(II) and Fe(III) concentrations in salt marsh porewater by gel filtration. Fe(II) is mostly found in the <0.1 kDa fraction, but also occurs in the 0.1–5 kDa fraction. Fe(III), the most important oxidant of sulfide, is typically found in fractions with a molecular mass of 0.1–5 kDa. This fraction is formed by HS and other natural organic complexes. When the porewater pH falls below 3, both Fe species coprecipitate with HS, via oxidation. Since the authors find no free O_2 at depths of more than 2 mm, the Fe(III)-HS complexes act as oxidants.

Pracht et al. (2001) show which structural features of the organic substances may be responsible for the Fe(III) reduction. The authors describe the redox process between Fe(III), in dissolved form and as mineral phase ferrihydrite ($5Fe_2O_3$ $9H_2O$), and phenolic substances, using phenolic model compounds, namely the dihydrobenzene reductants catechol, hydroquinone, resorcine, and 2-methoxyphenol guaiacol. Pracht et al. (2001) show that catechol and guaiacol are effectively oxidized to CO_2 by reducing Fe(III). Hydroquinone shows a reduction of Fe(III), but no accompanying mineralization is detected. In contrast, resorcine shows no reaction with Fe(III). The observed order of reactivity for the investigated phenolic compounds is catechol>guaiacol>hydroquinone>>resorcine. Whether or not mineralization occurs is determined by the position of the hydroxy groups: phenolic substances with two hydroxy groups in the ortho-position, or at least one hydroxy group and a methoxy group, can be oxidized to CO_2, while Fe(III) is reduced.

Manganese

Under oxidizing conditions in freshwaters, MnO_2 is the thermodynamically stable form of Mn. However, it has been known for a long time that Mn also exists in dissolved form in the euphotic zone. Mn(II) is highly bioavailable. As for Fe, HS play a fundamental role in the photoreduction of Mn (Matsunaga et al. 1995; Spokes and Liss 1995). A possible mechanism is the reduction of MnO_2 by H_2O_2, which is present due to the photoreduction of O_2 in the presence of HS (Sunda and Huntsman 1994). If, however, the enzyme catalase is added, which removes the H_2O_2, the photoreduction of MnO_2 ceases. Other authors suggest there is a competition for absorption between the organic material and the photochemically produced reductant H_2O_2 (Spokes and Liss 1995).

In a study with the mineral birnessite and HS, Banerjee and Nesbitt (2001) show that Mn is reduced to Mn(III), while HS are oxidized to CO_2. The authors do not find evidence for reduction of Mn(III) to Mn(II) in the presence of HS. The carboxyl group of HS apparently inhibits this step, probably through formation of strong Mn(III)-carboxyl surface complexes. In addition, the proton promoted dissolution of the soluble Mn(II) component of birnessite in distilled water, would appear to be impeded by the addition of HS. This can be attributed to the formation of strong, multinuclear surface complexes between Mn(II, III) and the adsorbed carboxyl groups. That means that without light, HS reduce Mn(IV), but the reduction does not necessarily lead to Mn(II) compounds.

In addition to HS, hydroxycarbonic acids can also play a role in the photoreduction of MnO_2 (Matsunaga et al. 1995). The maximum percentage of reduced Mn (Mn(II)) is 65% when glucaric acid (dicarbonic acid of glucose) is present, 30% when gluconic acid (monocarbonic acid of glucose) is present, and approximately 26% when tartraric acid (2 C atom shorter than glucaric acid) is present (Fig. 6.5). These acids are derived from the decomposition of phytoplankton. No reduction of Mn(IV) occurred without light. Thus, Mn(IV) reduction is a photo-process, the mechanism however remains as yet obscure.

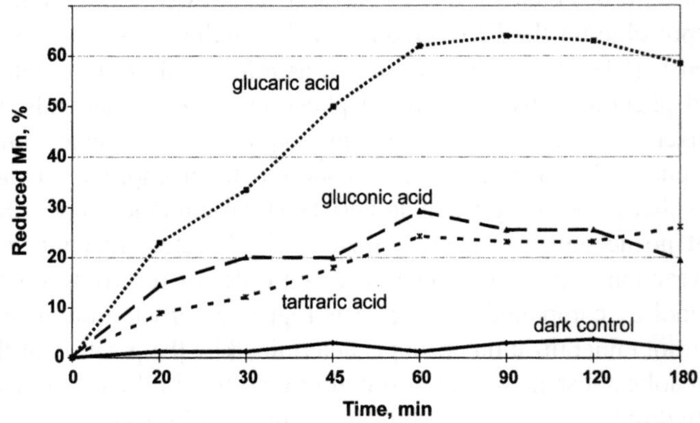

Fig. 6.5. Photoreduction of MnO_2 in seawater in the presence of hydroxycarbonic acid and UV radiation (from Matsunaga et al. 1995, with kind permission of Elsevier Science)

Studies by Stone and Morgan (1984) and Waite et al. (1988) indicate that MnO_2 reduction in freshwaters particularly depends on aromatic compounds and FA. Furthermore, HS mediated Mn reduction has a special role in freshwaters impacted by acid mine drainage from metal mines (Scott et al. 2001). In many instances, particularly the headwater regions are impacted.

6.2 Mercury

Under natural conditions, several metals such a mercury (Hg) and lead (Pb), and metalloids such as arsenic (As) and antimony (Sb), may form organic compounds which can greatly alter the metal availability and toxic properties. Hg is used here as an example to describe these transformations in detail.

Hg is mainly released into the environment as a result of human activities such as the burning of fossil fuels and industrial use of Hg. Håkanson (1996) provides informative data on environmental Hg in Sweden: of around 83,000 lakes, approximately 40,000 harbor fish (>1 kg pike) with Hg concentration in excess of the guideline of 0.5 mg/kg (Hg per fresh weight); in about 10,000 of these lakes, the critical value of 1.0 mg/kg is exceeded. There are also alarming reports from North America and Asia of increased Hg concentrations in edible fish (Driscoll et al. 1994; Grieb et al. 1990; Haines et al. 1994, 1995; Lindquist et al. 1991). In the state of Michigan alone, there are more than 10,000 lakes for which fish consumption is not recommended because of Hg concentrations of greater than 0.5 mg/kg. Most contaminated sites are soft-water lakes.

It is striking that there are hardly any point sources of pollution for Hg contamination. Consequently, Hg has become an environmental contaminant of concern particularly for the boreal zone, in which soft-water lakes with relatively high DOC concentrations dominate. DOC is fundamental to the Hg contamination issue. According to Nilsson and Håkanson (1992), there is a positive relationship between the Hg content of fish (pike, perch) and the HS content (as water color) in Swedish lakes of more than 5 m depth. Such lakes can develop an anoxic or anaerobic hypolimnion, thus the potential bioavailability of Hg from HS depends on the absence of molecular or chemically bound oxygen in the water.

In a more recent study, Sonesten (2001) emphasizes that the Hg content in fish is heavily influenced by the land use in the surroundings. The highest Hg levels in roach (*Rutilus rutilus*) are recorded in fish from boreal forest lakes, whereas lower levels occur in fish from lakes surrounded by arable land. However, the Hg levels in fish from lakes influenced by extensive wetlands are less well explained by these environmental variables, suggesting that the Hg burden in fish from this kind of lake is governed by other factors.

Since Hg contamination is seen as an important environmental problem, interest has recently increased, in particular in relation to Hg release into the environment, its transport and environmental fate. As a result of new sampling and analytical methods, studies at the pmol/L concentrations are now possible (Bloom and Fitzgerald 1988). Consequently, it reported that Hg is present in extremely low concentrations (1–10 ng/L) in all natural waters (Gill and Bruland 1990), and is significantly biomagnified in the aquatic food web (Driscoll et al. 1994; Mason and Sullivan 1997; Watras and Bloom 1992).

Since Hg is persistent, significantly bioaccumulated and biomagnified, and very toxic to humans and animals (Clarkson 1994; Wolfe et al. 1998, respectively), Hg is designated by the American Environmental Protection Agency (EPA 1997) as a first priority target, and at present intensive efforts are being made to increase and improve knowledge of Hg, particularly in the aquatic environment (Balogh et al. 1998; Benoit et al. 1999a; Bloom et al. 1999; Gill et al. 1999; Hurley et al. 1998; Mason and Sullivan 1998; Meyer 1998; Quémerais et al. 1999; Ravichandran 1999; Ravichandran et al. 1998, 1999).

A schematic overview of Hg cycling in lakes is presented (Fig. 6.6): atmospheric deposition is predominately inorganic Hg, with some methyl Hg. In oxic water, Hg(II) is complexed with inorganic ligands (such as Cl^- or OH^-) linked to DOC or particles. Hg ions can be reduced to elemental Hg (Hg(0)) by microorganisms. Most freshwaters are oversaturated for elemental Hg, so Hg(0) is volatilized. In anoxic waters, Hg forms strong aqueous complexes with sulfide, and in high sulfide concentrations, HgS precipitates. At low sulfide concentrations ($\leq 10^{-5}$ mol/L), uncharged complexes such as HgS^0_{aq} or $Hg(SH)(OH)^0$ form, which readily pass through biological membranes[3] (Benoit et al. 1999a,b, 2001a) and are thus available for bacterial methylation. For the latter process, sulfate reducing bacteria play an important role (Benoit et al. 2001a). Methyl Hg can bind to DOC,

[3] As a measure of the bioavailability of inorganic Hg species, the well known *n*-octanol-water-partitioning coefficient is used (Benoit et al. 1999b). With inorganic Hg-species this coefficient decreases with increasing sulfide concentrations.

or can be demethylated by microbial processes.

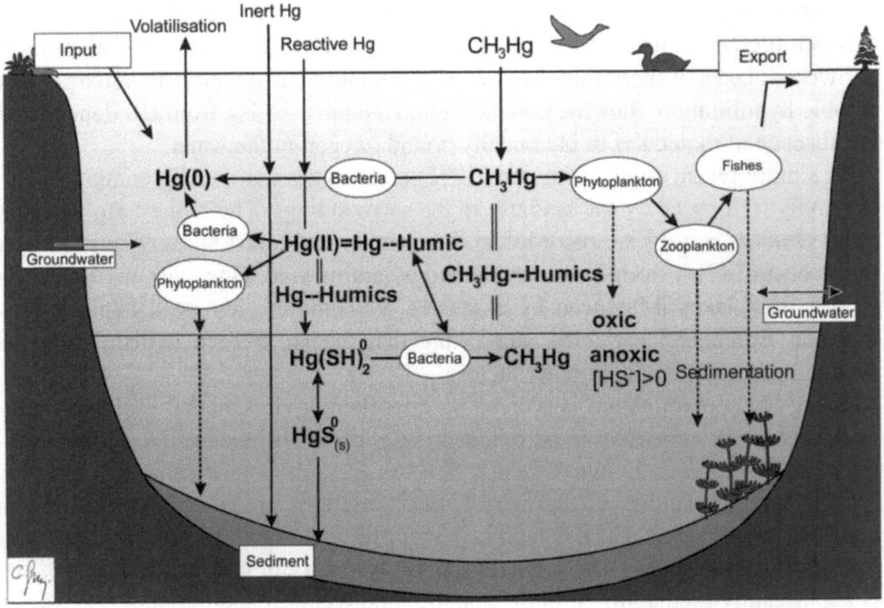

Fig. 6.6. Hg cycling in a lake (modified from Driscoll et al. 1994, with kind permission of the American Chemical Society). Hg-Humics stands for all organically bound Hg. Solid lines are direct and dashed line are indirect mercury pathways

6.2.1 Fish Mercury Content and Water Chemistry

For an understanding of the processes which control Hg accumulation in aquatic biota, many studies relate Hg concentrations in fish and zooplankton to chemical parameters which themselves show a relationship to the biogeochemistry of the catchment area (Back and Watras 1995; Cope et al. 1990; Grieb et al. 1990; Haines et al. 1995; Hurley et al. 1995; Louchouarn et al. 1993; McMurty et al. 1989; Meili 1991; Wren et al. 1991; Watras and Bloom 1992; Westcott and Kalff 1996). The two factors which evidently control bioaccumulation of Hg are pH value and dissolved HS concentrations (water color). However, correlations between DOC concentrations and fish tissue Hg are inconsistent. Concentrations of DOC explain a significant amount of the variation of Hg in lake trout in Ontario (McMurty et al. 1989), and there is a positive correlation between lake color and concentrations of Hg in perch in the former Soviet Union (Haines et al. 1994, 1995). In contrast, increasing DOC concentration is associated with decreasing Hg content of yellow perch in seepage lakes of the Michigan peninsula (Grieb et al. 1990).

6.2.2 Mercury Speciation in Freshwaters

In freshwaters, Hg exists mainly in two forms, inorganic Hg(II) and monomethyl Hg (MeHg(I)). The latter, MeHg(I), being a neurotoxin, is the more hazardous form (Clarkson 1994); furthermore, it is biomagnified in the aquatic food webs (Lawson and Mason 1998; Mason and Sullivan 1997; Watras and Bloom 1992). In natural aquatic systems, the concentrations of the free ions (Hg^{2+} and CH_3Hg^+) are extremely low because both forms are almost completely complexed with various inorganic (mainly HS^-, OH^- and Cl^-) and organic ligands (such as HS) (Hudson et al. 1994).

It is accepted that both Hg(II) and MeHg(I) form strong complexes with organic macromolecules, and this complexation has marked effects on the transport and bioavailability of Hg in aquatic ecosystems. Many field studies have found a clear correlation between DOC (or water color) and concentrations of dissolved Hg (Driscoll et al. 1994, 1995; Hurley et al. 1995; Lee and Hultberg 1990; Mierle and Ingram 1991; Meili 1991). Further information comes from model calculations which are calibrated with DOC and Hg field data (Hudson et al. 1994). Loux (1998) assesses the potential binding of Hg(II), Hg^0, MeHg(I), and Me_2Hg with NOM by hydrophobic and ionic mechanisms. His findings suggest: (1) although potentially hydrophobic binding with NOM will probably occur in the following sequence $Me_2Hg > Hg^0 > MeHgOH$, MeHgCl, none of these species appear to be strongly hydrophobic, and hence hydrophobic processes cannot explain the observed NOM-Hg associations; (2) hydrophobic binding mechanisms of the various Hg species are insufficient to explain the observed environmental bioconcentration factors; (3) ionic Hg binding with carboxyl-type sites on natural organic carbon is not sufficiently energetic to explain the observed Hg-NOM associations in the environment; and (4) environmental Hg binding with organic matter probably occurs through ionic reactions with naturally occurring sulfhydryl binding sites.

Studies using photochemical methods (Frimmel et al. 1980), or X-ray absorption spectroscopy (Morra et al. 1997; Xia et al. 1998) show that aquatic HS contain significant quantities of S in reduced forms (thio, thiol, or sulfide). These groups play an important role in the complexation of Hg(II), and are functionally dominant over the oxygen ligands (Xia et al. 1999). Since the reduced S ligands occur at relatively low densities, the more numerous oxygen ligands can also play a role in the complexation of Hg(II). As soon as Hg(II) is bound to a reduced S ligand, it is more likely that an oxygen ligand will bind next on the Hg atom to complete the complex. In a recent report, Skyllberg et al. (2000) corroborate the bidentate bonding in HA, by which Hg^{2+} ions bind in two-fold coordination involving one reduced S and one O or N.

As a result of methodological difficulties in determining Hg speciation, it is not surprising that quantitative data such as conditional stability constants for Hg-DOC compounds are calculated in only a few studies (Cheam and Gamble 1974; Mantoura et al. 1978; Lövgren and Sjöberg 1989; Strohal and Huljev 1971; Yin et al. 1997). The published constants differ markedly. For instance, for the Hg(II)-

DOC complex the constants vary by 25 orders of magnitude. In some studies, the stability constants are surprisingly small (Table 6.1).

If the low values of the stability constants are valid, DOC could not control the speciation of Hg, and the inorganic ligands such as Cl^- and OH^- will bind to Hg(II) more effectively. This is however, unlikely as indicated by the powerful indirect evidence for Hg(II)/DOC interactions in the above outlined field studies. Further experimental evidence comes from the dissertation of Ravichandran (Ravichandran 1999; Ravichandran et al. 1998, 1999) who describes the conspicuous increase in the solubility of cinnabar (red HgS), and the inhibition of precipitation and aggregation of meta cinnabar (black HgS), two extremely insoluble Hg minerals, if hydrophobic acids (HA and FA) are present. This means that several stability constants must actually be much higher than those reported in Table 6.1. This discrepancy probably exists because many old studies, and some recent studies, do not use environmentally relevant Hg concentrations; the applied concentrations are much too high. Hence, complexation of Hg depends not only on reduced S of HS, but predominately, on the oxygen containing functional groups (Bloom et al. 1999). Consequently, excessively small stability constants are determined.

Table 6.1. Conditional stability constants of Hg(II) with HS

HS source	pH	log K	Reference
Sediment HS	5	5.2	Strohal and Huljev (1971)
Seawater HS	8	18.1	Mantoura et al. (1978)
River HS	8	19.3–19.7	Mantoura et al. (1978)
Lake HS	8	18.4–20.1	Mantoura et al. (1978)
Peat HS	8	18.3	Mantoura et al. (1978)
FA from the River Moscwa	6.5	11	Zhilin (1998)
Soil HS	4–6	4.7	Zhilin (1998)
Diverse sources	8	13.7–14.9	Zhilin (1998)
Soil organic matter	3.0–3.4	31.6–32.2*	Skyllberg et al. (2000)
Hydrophilic DOC, oligotrophic, low-sulfidic Everglade site	not given	10.6	Benoit et al. (2001b)
Hydrophobic DOC, eutrophic, high-sulfidic Everglade site	not given	11.8	Benoit et al. (2001b)

* surface complex formation constants

Recently, Benoit et al. (2001b) using a competitive ligand approach, show that the conditional stability constants for the hydrophilic and hydrophobic DOC fractions from contrasting Everglade sites are 10.6 and 11.8, respectively (Table 6.1), values which are similar to previously published stability constants for Hg binding to low-molecular-weight thiols. Furthermore, the hydrophobic DOC from the eutrophic, sulfidic site, shows a pH-dependent decline in the octanol-water partitioning coefficients, that is consistent with the model of Hg complexation with thiol groups as the dominant Hg binding sites in DOC. The experiments demonstrate that the DOC isolates are stronger ligands for Hg than are the chloride ion or ethylenediamine-tetraacetic acid (EDTA). Hence, in oxic surface waters containing a low sulfide ion concentration, Hg speciation will be dominated by FA. If, however, the sulfide ion concentrations in the water exceed about 10^{-7} mol, dissolved inorganic mercury solution speciation will be dominated by Hg-S complexes, as speciation calculations with a modified PHREEQC[4] indicate (Reddy and Aiken 2001).

There are only a few studies on the complexation of MeHg(I) with DOC. Miskimmin (1991) gives evidence that the MeHg(I) concentration in water rises when DOC is added to a sediment/water system. Using an equilibrium dialysis method, Hintelmann et al. (1995, 1997) determine the conditional stability constants for MeHg(I) with one HA and two FA with log K = 12.2–14.5. In addition, Amirbahman et al. (2002) show that estimated binding constants for complexes of MeHg(I) with HA are similar in magnitude to those of MeHg(I) with thiol-containing compounds, suggesting that binding of MeHg(I) involves the thiol group of HA. The results also show that only a small fraction of the reduced S species in HS may take part in binding MeHg(I), but in most natural systems, this subfraction is considerably higher in concentration than is the ambient MeHg(I).

Summarizing, one can say that the database for a quantitative description of Hg complexation with natural DOC is surprisingly small. At present, there are no reliable conditional stability constants for the complexation of Hg(II) and MeHg(I) with DOC. Determination of conditional distribution coefficients by, for instance, an equilibrium dialysis ligand exchange method is the subject of current studies (Haitzer and Aiken, pers. comm.). In addition, Reddy and Aiken (2001) use a chemical speciation model to describe and quantify the interactions between Hg and FA. Negatively charged functional groups of FA compete with inorganic sulfide ions for Hg ion binding. This competition is evaluated by using a discrete site-electrostatic model to calculate Hg solution speciation in the presence of FA. Model calculated species distributions are used to estimate a Hg-FA apparent binding constant, in order to quantify FA and sulfide ion competition for dissolved inorganic Hg (Hg(II)) ion binding. Speciation calculations done with PHREEQC suggest that for very low sulfide ion concentrations (about 10^{-11} mol/L) in Everglades' surface water, concentrations of Hg-FA and Hg-S complex are similar. Where total sulfide concentration is measurable (about 10^{-7} mol/L or greater) in Everglades' surface water, Hg-S complexes should dominate dissolved inorganic Hg solution speciation. In the absence of sulfide ions (for example, in the Everglades' oxic surface water), FA binding should dominate dissolved inorganic Hg speciation.

---

[4] PHREEQC: A computer program for (geo)-chemical speciation, batch-reaction, one-dimensional transport, and inverse geochemical calculations provided by the US Geological Survey (wwwbrr.cr.usgs.gov/projects/GWC_coupled/phreeqc/)

6.2.3.1 Reduction of Mercury(II)

An old paradigm is that the reduction of Hg(II) in soils and sediments, to elementary Hg, is carried out only by microorganisms. The first indication that HS are also involved in this process comes from the work of Alberts et al. (1974) and Miller (1975), who show that an abiotic reduction of Hg(II) is feasible. Meanwhile views on the roles of biotic and abiotic formation pathways clearly change. Allard and Arsenie (1991) and Wallschläger et al. (1998) prove that reduction of Hg(II) to Hg(0) by HS is thermodynamically possible. The reaction is hindered by competing ions such as Cl^- which can build complexes with Hg(II). The number of available complexation sites is reduced following methylation of HS, subsequently Hg(0) formation is inhibited by the intramolecular process. Air hinders the reduction, whereas light augments the process. In addition, Wallschläger et al. (1998) show that Cd may act as catalyst in Hg reduction.

6.2.3.2 Methyl Mercury Formation

Formation of MeHg(I) is reported from many aquatic habitats. The methylation can be carried out both biotically by microorganisms, and abiotically. However, the role of HS in the latter is not clear. In many studies, increased DOC concentrations reduce the rate of methylation. For example, Matilainen and Verta (1995) find that in aerobic humic waters, MeHg(I) formation is an incidental process carried out by bacterial exo-enzymes or other dissolved compounds such as dissolved HS. In the study of Porvari and Verta (1995), HS concentration is negatively correlated with methylation, but positively with MeHg(I) concentration. The percentage of free MeHg(I) generally decreases with lowering pH (Hintelmann et al. 1995), probably due to the protonation of thiol ligands. Anoxic conditions increase methylation in waters, but not in soils or sediments (Porvari and Verta 1995). This contradiction can in part be explained mechanistically; according to Benoit et al. (1999a,b, 2001a), methylation depends on Hg speciation with S: HgS^0 is bioavailable, whereas HgS_2H^- is not. Since the Hg species present in an individual habitat is dependent on sulfide concentration, and sulfide concentration is probably higher in the soils and sediments than in water, MeHg(I) can be formed only in the water column, whereas in soils and sediments, the formation of Hg-S prevails.

In his review, Weber (1993) emphasizes the potential of abiotic methylation of Hg(II) through methyl zinc components and HS. Methyl zinc compounds, such as mono-, di- and trimethyl zinc, occur in all freshwater compartments, including water organisms and sediments. In contrast to the findings of Porvari and Verta, Weber shows that HS are the most important methylation agents.

Driscoll et al. (1994) also emphasize that in the Adirondack[5] lakes, there is a strong relationship between the total methyl Hg concentration and DOC concen-

[5] This area contains many lakes acidified through acid deposition and receives much scientific attention since the 1980s (Charles and Whitehead 1986).

tration, as well as between the Hg and the methylated Hg in solution. The relationship between total methyl Hg and DOC in 16 Adirondack lakes is shown in Fig. 6.7 and has an r^2 of 0.90 (Eq. 6.1):

$$[\text{total methyl Hg, ng/L}] = 0.01 + 0.02\,[\text{DOC, mg/L C}]. \tag{6.1}$$

Statistics give further information on the role of wetlands in the catchments on both the total Hg and total methyl Hg concentrations (Eqs. 6.2, 6.3):

$$[\text{total Hg, ng/L}] = 0.82 + 0.28\,[\%\ \text{wetland area}]\ ;\ r^2=0.65; \tag{6.2}$$

$$[\text{total methyl Hg, ng/L}] = 0.03 + 0.03\,[\%\ \text{wetland area}];\ r^2=0.83. \tag{6.3}$$

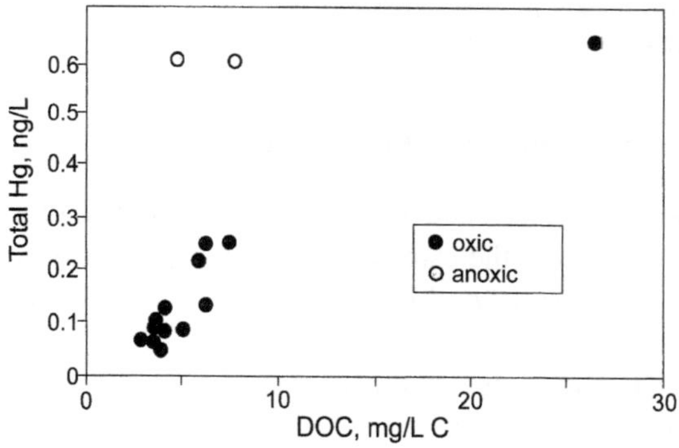

Fig. 6.7. Concentrations of total Hg as a function of dissolved organic carbon (DOC) concentration in Adirondack lakes, New York, USA (after Driscoll et al. 1994, with kind permission of the American Chemical Society)

These regressions appear to contradict the finding of Sonesten (2001) who does not find a clear relationship between the Hg content in fish and water of 78 Swedish lake, and the amounts of wetland in the catchment areas. Unfortunately, Sonesten (2001) does not discuss the contrasting findings, thus, this contradiction cannot be explained.

Even in elemental Hg spill sites, soil organic matter controls speciation of the Hg. For instance, Renneberg and Dudas (2001) show that decades after the original spill, the elemental Hg is transformed and is dominantly (up to 85% in an individual site) associated with soil organic matter, and to a lesser extent the soil mineral fraction. Further, binding to HS generally transfers Hg between various environmental compartments. For example, Wallschläger et al. (1996) describe that Hg is transported from industrial sites via HA complexes; floodplain soil particles and FA appear not to participate in the transport process. This accords well with the studies of Mierle and Ingram (1991) on diffuse contamination sources. The former authors come to the conclusion that the seasonal pattern and the estimated export of Hg from catchments is tightly linked with HS quality and quantity, which means that HS generally control both the solubility and the export of Hg.

6.2.4 Fate of Mercury in Aquatic Ecosystems

6.2.4.1 Mercury(II) vs. Methyl Mercury(I)

The chemical speciation of Hg is one of the most important determinants of biomagnification in aquatic food webs. Zooplankton fed with MeHg(I) contaminated algae, accumulate methyl Hg which is mainly associated with the algal cytoplasm. In contrast, inorganic Hg, associated with the algal membranes, is hardly taken up by zooplankton (Mason et al. 1996). The transfer of Hg within the food web from zooplankton to planktivorous fishes, and finally to piscivorous fishes, leads to a further increase in total Hg concentration, of which methyl Hg (the neurotoxic form) comprises almost 100% in the top predators (Lawson and Mason 1998; Mason and Sullivan 1997; Watras and Bloom 1992).

6.2.4.2 Complexation by Organic Ligands

The complexation of Hg(II) and MeHg(I) by organic ligands clearly influences

- the transport of Hg from the catchments into aquatic ecosystems (Hurley et al. 1995; Lee and Hultberg 1990; Meili 1991; Mierle and Ingram 1991; Pettersson et al. 1995),
- the release of Hg from sediments (Miskimmin 1991; Ravichandran et al. 1998; Wallschläger et al. 1996),
- the adsorption of Hg to soil particles: the adsorption of MeHg(I) is enhanced in the presence of FA, whereas that of Hg(II) is clearly diminished (de Diego et al. 2001) and
- the bioavailability of Hg.

In general, the accumulation of metals in aquatic ecosystems appears to occur via two contrasting mechanisms, linked to the complexation with organic ligands (Driscoll et al. 1994, 1995):

1. natural organic material accelerates the transport of Hg out of the catchment and into a lake, and probably strongly increases the abiotic methylation of Hg within the lake (Nagase et al. 1982; Wallschläger et al. 1998); consequently there are **increased** concentrations of Hg and methyl Hg in lakes;
2. in contrast, the complexation of metals to natural organic macromolecules leads to large hydrophilic aggregates which are not thought to be bioavailable (Brezonik et al. 1991; McCarthy 1989; Sjöblom et al. 2000); this results in a **reduction** of the bioavailable fraction of metals in aquatic ecosystems.

For Hg, the second mechanism is qualitatively demonstrated for Hg(II) and MeHg(I) in laboratory experiments, particularly for bacteria (Barkay et al. 1997), oligochaetes (Nuutinen and Kukkonen 1998), water fleas (Monson and Brezonik 1999), insect larvae (Rouleau et al. 1998; Sjöblom et al. 2000), mussels (Gagnon and Fisher 1997), and fish (Choi et al. 1998). Sjöblom et al. (2000) show that a strong negative influence of dissolved HS on the bioavailability of Hg is due to both inorganic and MeHg(I) in freshwaters (measured as uptake by *Chaoborus* larvae). This negative influence occurs with Hg and MeHg(I) with DOC concentrations >0.1 mg/L and 1 mg/L, and suggests that the high Hg levels often found in fish from humic lakes in the boreal forest zone cannot be explained solely by direct uptake of MeHg(I) from the water phase into biota at low trophic levels.

The total Hg concentration in most freshwaters is taken as 1 ng/L. Of this, 20 to 30% should be MeHg(I) (O'Driscoll and Evans 2000). According to this assumption, MeHg(I) must be totally bound to HS. If one accepts a MeHg(I) concentration of 1 ng/L in freshwaters, 7.3 µg/L DOC is sufficient for this bonding. This DOC concentration is exceeded by far in all freshwaters. However, according to the findings of O'Driscoll and Evans, only part of the MeHg(I) is bound to DOC. Hence, it remains an open question, why these findings and simple calculation are thus incompatible. The given numbers do demonstrate, however, that various processes in the Hg are not yet understood and therefore can not be precisely modeled.

The Hg-HS item will remain a subject of great scientific interest in the future. Two important questions are: The binding capacity of HA and FA and the interrelated speciation of S, as well as the effect of competing cations and pH.

6.2.5 Mercury in Sediments and Floodplain Soils

In surface waters, there is competition between sulfide ions and hydrophobic acids (FA and HA) for binding with mercury. If sulfide ion concentrations are well below 10^{-7} mol (Reddy and Aiken 2001), as is the case in oxic water, FA are the most successful competitors (Chap. 6.2.3.2). In sediments and floodplain soils, however, the concentrations of sulfides, HS, and Hg are considerably higher than in the water column, and hence the question about Hg speciation arises. From studies of the River Elbe, Germany, Wallschläger et al. (1998) present evidence that an intimate coupling exists between the geochemical cycles of Hg, and organic carbon between sediments and floodplain soils of the Hg-contaminated stretch of this river. HS exert a dominant influence on several important parallel geochemical pathways of Hg, including binding, transformation, and transport processes.

Despite the elevated concentrations of Hg in the Hg-contaminated stretch of the River Elbe, Wallschläger et al. (1998) assume that Hg is bound to HA via S-containing ligands. Only Hg in the smallest molecular weight fraction is retained on such a cation exchanger, presumably because it consists either of free Hg^{2+}, or of low-molecular weight Hg–HS complexes, which do not have sufficient S-containing ligands to compete with the functional groups on the exchanger. These findings agree well with the facts that Ha does not get displaced from the Hg-HA complexes at extremely acidic pH, by changes in redox potential, or in competition with a hundredfold excess of competing metals such as Cd.

Significant differences exist between the Hg-HS associations in sediments and floodplains. Both HA and FA contribute to Hg binding in the sediments. In contrast, Hg in the floodplains soils is almost exclusively bound to very large HA, with a nominal molecular weight >300 kDa. This is unusually large for a soil HA. Wallschläger et al. (1998) suggest that this observation is due to the fact that the Elbe system is heavily polluted with (heavy) metals. The presence of the heavy metal ions could lead to a shift in the molecular weight distribution of the soil HA, by binding individual smaller HA molecules together. This would mean that not only have HS a very significant effect on the geochemistry of Hg, but in turn, Hg and other (heavy) metals exhibit a reverse effect on HS molecular weight distribution, thereby indirectly partially regulating their own geochemical cycles. Wallschläger et al. (1998) also observe that the large HS molecules are apparently electroneutral, which is of fundamental importance for the mobility and availability of the Hg bound in the River Elbe floodplains: under normal environmental conditions, the mobility should be very low. The bonding appears to be chemically inert and irreversible under environmental conditions.

Furthermore, mobilization of Hg from the solid phase is chemically coupled to HA molecules, thereby affecting availability, controlling transport processes, and turning the River Elbe into a dynamic system with respect to Hg speciation. The formation of volatile Hg compounds is probably due to abiotic reduction of Hg^{2+} to Hg^0 by HS (Alberts et al. 1974). Although Cd^{2+} is unable to displace Hg from Hg-HS, it appears to have a catalytic effect on the Hg reduction. In addition, Wallschläger et al. (1998) describe that sediment and floodplain HS have the potential to enhance the water-solubility of HgS (cinnabar) by many orders of magnitude. The agrees well with the HS study from the Florida Everglades (Ravichandran 1999) and shows that even cinnabar may not be resistant to transformation, and can easily be mobilized by simple matrix (HS) induced reactions. Field observations also show that Hg association with water-soluble HA and FA continuously increase downstream in the River Elbe, indicating that HS play a key role in the longitudinal Hg transport in the Elbe system (Wallschläger et al. 1998).

6.3 Other Trace Elements

6.3.1 Trace Metals

Dissolved HS very strongly influence the speciation of trace metals, and their environmental fate in freshwaters. In particular, HS greatly influence the retention and mobility of many metal ions in soils and natural waters. Many metals are strongly bound to HS, and in order to understand metal transport and bioavailability in natural environments, the interactions with HS must be considered. Humin and HA in soils participate in the retention of metals, while FA, and to a lesser extent HA, and still unknown compounds are involved in their transport in soils and natural waters (Clapp et al. 2001).

Determination of the capacity of water samples to form complexes with a model heavy metal, such as Cu, is long established (Alberts and Giesy 1983), and generally the principle objective is to distinguish between free and complexed forms of a metal cation. Such a distinction is relevant because of the common strong correlation between the free concentration of a metal cation and its biological effects (Erickson et al. 1996; Morel and Hering 1993; Perdue 2001; Sunda and Guillard 1976).

Several studies show that strong ligands mostly consist of compounds in the <10 kDa range, but that colloidal ligands with similar properties also exist (Sigg et al. 2000). A comparison of the complexation standard heavy metals, such as Cu and Cd, by standard HA or FA (Suwannee River), with those of surface water samples indicate that stronger ligands than FA and HA are present at low concentrations in surface waters. Specific strong ligands occur in particular in eutrophic lake waters, whereas in lakes with low biological productivity, the ligands more closely match the FA characteristics (Xue and Sigg 1999). These former ligands, probably of biological origin, exhibit a selectivity for Cu over, for instance, Zn (Xue et al. 1995). Also for nickel, strong organic complexes with conditional stability constants of log K = 12–15 (pH 7.2–8.2) are found to play an important role for speciation (Xue et al. 2001). Applying ^{113}Cd NMR, Otto et al. (2001) show that Cd predominately binds to the oxygen containing functional groups of HS. Furthermore, a fast exchange between free and complexed Cd species is observed.

Two details complicate the quantification of metal-HS complexation. One is that the metal-DOC stability constants are strongly conditional, and can vary both with the properties of the dissolved matter or chemical matrix, as well as with time. The other factor is the low concentrations of the complexing ligands and metals, which render analysis difficult. There are only a few analytical techniques available which are sufficiently sensitive and selective to reliably measure free metal ions at extremely low concentrations. Some recent analytical developments are described by Rozan et al. (1999).

6.3.1.1 Modeling Metal-Ion binding

Metal ion complexation studies often culminate in an effort to summarize knowledge of metal binding by HS through the development of chemical speciation models. While for proton binding, reasonably successful models exist, for metal-ion binding the situation is more complex. A useful model for metal-ion binding should, in principle, be able to describe and predict ion binding over a wide range of conditions as a function of pH, the concentrations of the HS, and the ionic strength. However, the perceived distribution of binding sites is directly related to experimental conditions, and to the type of conceptual model that is forced on the experimental data (Milne 2000; Perdue 2001).

There are three key areas of importance in understanding ion binding to natural organic matter (Milne 2000):

1. variable charge and potential. As a result of ion binding, the variable electric charge on the humic molecules creates an electric field around the particle,

which in turn influences further ion binding;
2. chemical heterogeneity. Ion binding to HS can only be described satisfactorily using a distribution of affinity constants;
3. competition. In natural systems, several ions are in competition for the same reactive sites, and for metal ions there is always competition with protons.

According to Perdue (2001), there are three models which very successfully describe the interaction of HS and metal cations: (1) competitive Gaussian distribution model (Perdue 2001), (2) the NICA (non-ideal competitive adsorption) (Koopal et al. 1994; Milne 2000; Milne et al. 2001), and (3) Humic Ion-Binding Model VI (Tipping 1998). The models all assume that HS contain two classes of binding sites, presumably attributable mainly to the carboxyl and the phenolic hydroxyl groups (Milne 2000; Perdue 1998, 2001; Tipping 1998). At extremely low metal cation concentrations, the nature of binding sites is non-uniform, and nitrogen-containing functional groups, dissimilar to those of amino acids, are likely to be engaged in the complexation (Frenkel and Korshin 1999; Frenkel et al. 2000).

In the models, each class of sites contains multiple sites, the relative concentrations of which, are distributed symmetrically around a central log K value, and the width of the distribution is controlled by a width parameter. To varying degrees, these models consider the intrinsic heterogeneity of binding sites, electrostatic effects on the effective reactivities of binding sites, and Donnan effects[6] that alter concentrations of small ions near a large polyion (Perdue 2001).

In the **competitive Gaussian model**, only monodentate (1:1) reactions can occur, that is, 1 mol of either a proton or a metal ion can react with 1 mol of a binding site. The relative concentration of a binding site is related to its log K value for proton or metal binding, by a Gaussian distribution function. Because proton binding studies strongly indicate that HS contain two classes of acidic functional groups, two Gaussian distributions of binding sites are used in the model (Perdue 2001).

The **NICA model** resembles the competitive Gaussian distribution model, both in its use of only 1:1 reaction stoichiometries, and in the use of continuous symmetrical distributions of binding sites (Milne 2000; Perdue 2001). In a Donnan type model, to describe the electrostatic behavior, the HA is assumed to behave as a gel, rather than a particle, with the electrostatic potential distributed throughout the gel volume, rather than being concentrated at the particle surface. Milne (2000) finds that the NICA-Donnan model[7] convincingly describes metal-ion

---

[6] Originally, the Donnan effect is described as the extra colloid osmotic pressure of proteins caused by the uneven distribution of small, diffusible cations and anions in the plasma. It arises because semi-permeable membranes in the body separate biological fluids containing mixtures of small, diffusible ions (electrolytes) and larger non-diffusible proteins (usually negatively charged). With HS, the term Donnan effect is used in a more generic mode.

[7] The Consistent NICA-Donnan model (NICCA-Donnan model) is the latest and most rigorous version. It combines thermodynamic consistency in its treatment of competitive adsorption, and a bimodal form of the NICA isotherm with a Donnan description of the

binding over a wide range of pH and metal-ion concentrations. This model is also able to adequately predict the binding stoichiometries using an empirical, rather than mechanistic, approach.

As a next step, Milne (2000) fits 124 data sets to the NICA-Donnan model in order to obtain generic parameter values. Agreements between metal-binding data for different materials and techniques are predominantly good, lending confidence to the strategy of deriving generic descriptions. The fitted generic parameters represent the behavior of a 'typical' FA or HA, and can describe most observed metal-binding over up to 16 orders of magnitude of metal concentrations, pH (<2–10), and ionic strength. Milne (2000) successfully obtains generic log K values for the two binding sites.

Humic Ion-Binding Model VI is a multidentate site model, and pictures HS as rigid spheres of uniform size, with ion-binding groups positioned on the surface. Proton binding is described with a site density, two median intrinsic equilibrium constants, two parameters defining the spread of equilibrium constants around the medians, and an electrostatic constant. Intrinsic equilibrium constants for metal binding are defined by two median constants, log K_{MA} and log K_{MB}, which refer to carboxyl and weaker-acid sites respectively, together with a parameter defining the spreads of values around the medians. A further parameter takes account of small numbers of strong binding sites. Model VI belongs to the most developed models of metal ion-HS interaction and it will therefore now be considered in more detail.

The distribution of the elemental composition, and the distribution of structural features both follow a Gaussian distribution (Perdue 1998; Rice and MacCarthy 1991; Box 1.1); it is therefore not surprising that Tipping (1998) is able to show that strong and weak complexes are interrelated. Hence, if the complexation by carboxyl groups is known, the total metal complexation capacity of a given HS can adequately be determined by (Eq. 6.4):

$$\log K_{MB} = 3.39 \log K_{MA} - 1.15; r^2 = 0.80. \quad (6.4)$$

For 22 metals and metal species, Tipping (1998) calculates log K_{MA} values (Table 6.2). The labile protons involved in the complexation of metals are only available in limited numbers in the FA molecules. Since in the aquatic medium, there is more than one metal species present, there is competition for binding sites. The new equilibrium is determined by the binding constants of the particular metal species as well as their concentrations. Model VI also considers these effects, as shown when applied to the data of Alberts et al. (1992), who describe the Cu binding capacity of FA and HA from groundwater and soils in the presence and absence of competing metals (Mg, Ca, Al and Fe(III)) (Table 6.3). It is evident that the model predicts the Cu concentrations very well.

electrostatic behavior of the humic particle. The consistency affects only the description of competitive or multicomponent binding.

Table 6.2. Intrinsic equilibrium constants of fulvic and humic acids for metal bonding on carboxyl groups (log K_{MA}) (from Tipping 1998)

Metal	Fulvic acids log K_{MA}	Humic acids log K_{MA}
Mg	1.1	0.7
Al	2.5	2.6
Ca	1.3	0.7
V(IV)O	2.4	2.5
Cr(III)	2.2	2.2
Mn	1.7	0.6
Fe(II)	1.6	1.3
Fe(III)	2.4	2.5
Co	1.4	1.1
Ni	1.4	1.1
Cu	2.1	2.0
Zn	1.6	1.5
Sr	1.3	1.11
Cd	1.6	1.3
Ba	0.6	–0.2
Eu	2.4	2.1
Dy	2.5	2.9
Pb	2.2	2.0
Th	2.7	2.8
U(VI)O_2	2.1	2.2
Am	2.6	2.5
Cm	2.0	2.2

The results further show that under the experimental conditions, the divalent metal ions (Ca^{2+} and Mg^{2+}) do not significantly displace the Cu ions at the concentrations used. This situation can differ with low ionic strengths, low dissolved HS concentrations, and high concentrations of competing metals as we shall discuss for the case of Al. For example, weak displacement of Cu by Ca (Buffle et al. 1980; Hering and Morel 1988; McKnight and Wershaw 1989), Mg by Al (Cabaniss and Shuman 1988), or Pb by Al (Mota et al. 1996) are very well reproduced by Model VI. The pH dependence of such competition processes is determined by Town and Powell (1993) for the bonding of Cu in the presence or absence of Mg, and subsequently by Tipping using Model VI. Here also, a very good agreement occurs between measured and calculated values.

With increasing pH, the differences between the negative logarithm of the free Cu concentration and all bound metals (p(Cu) – pν) becomes smaller (Fig. 6.8). However, the decrease is smaller in the presence of Mg ions than in their absence. This means that either the concentration of the Cu ions increases, or the quantity of bound metals decreases, or both occur simultaneously. This is indicative of a net release of Cu from complexes with FA at neutral pH values, as well as after the addition of basic cations (Mg^{2+}, Ca^{2+}).

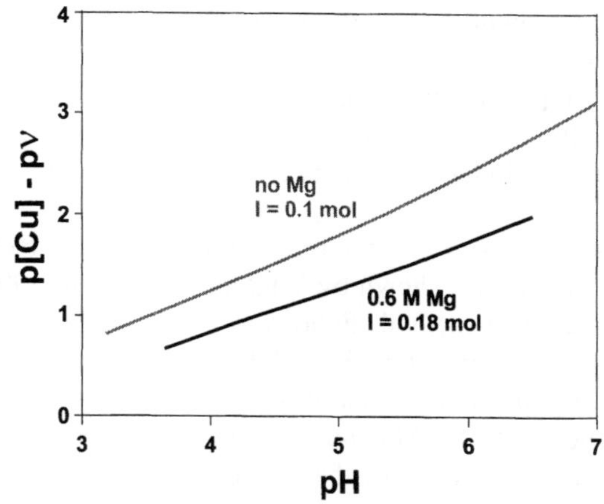

Fig. 6.8. Binding of Cu to FA in the presence and absence of Mg. Humic Ion-Binding Model VI fits the data from Town and Powell (1993) so well that the data points are omitted. Log K_{MA} for Cu is 2.01 and for Mg 1.13. (Cu) is the concentration of freely dissolved Cu ions, ν means moles of metal (Cu and Mg) bound per g FA or HA, and p signifies $-\log_{10}$ (after Tipping 1998, with kind permission of Kluwer Academic Publishers)

Table 6.3. Application of Humic Ion-Binding Model VI to the competition data of Alberts et al. (1992). Cu binding to FA and HA is measured, in the absence and presence of competing metals. HS: 75–95 mg/L DOC, Cu: 31 µmol/L. Ionic strength: 0.1 mol/L, pH 5.1 (Tipping 1998)

Sample	log K_{MA} Cu	Competitor, µmol/L	$[Cu^{2+}]_{measured}$	$[Cu^{2+}]_{calculated}$
Groundwater FA	2.13	—	3.3	3.3
		29.3 Mg	3.3	3.3
		29.3 Ca	3.3	3.3
		19.6 Al	5.9	5.3
		19.5 Fe	7.0	5.8
Groundwater HA	1.94	—	7.4	7.4
		22.6 Mg	6.8	7.5
		22.7 Ca	7.4	7.6
		15.2 Al	11.1	11.5
		15.1 Fe	12.1	11.6
Soil FA	2.13	—	5.9	5.9
		19.1 Mg	6.4	5.9
		19.1 Ca	6.5	5.9
		12.8 Al	9.0	7.7
		12.8 Fe	9.8	8.2
Soil HA	1.81	—	11.1	11.1
		15.9 Mg	9.3	11.2
		15.9 Ca	10.3	11.3
		10.7 Al	14.1	13.4
		10.6 Fe	16.5	13.8

When studying the binding of metals to HS under field conditions, it is fundamental to know the concentrations of strong binding metals, such as Fe(III). The study of Peters et al. (2001) highlights this critical feature of strong binding metals, which potentially affects binding of other metals to the high-affinity sites of HS. From the current Model VI database, Cu and Fe(III) are most important in this respect.

As stated above, at low concentrations, the various trace metals do not bind uniformly to HS. This is also shown by Vogl and Heumann (1997) using isotope dilution, a new analytical method. Following a size exclusion chromatography (SEC), the cDOC is determined by UV, and the total DOC determined by ^{12}C. Coelution of metals with DOC is indicative of a physical and chemical binding of metals to DOC. The results for Mo and Cu from a raised peat bog lake are shown in Fig. 6.9. The results clearly show that Cu forms complexes with all UV-active HS and FA fractions, as shown by the similar shape of the Cu and UV curves. In contrast to Cu, Mo shows two distinctly separated peaks, the second of which cannot be identified by UV absorption, but by ^{12}C detection. Because preferably aromatic compounds with π electron systems absorb at 254 nm, the UV-inactive Mo-DOC complex is probably formed by aliphatic compounds.

The ecological relevance of the chemical speciation of metals is discussed by Marx and Heumann (1999) with reference to Cr and Cu. The authors conclude that Cr forms stable complexes, whereas Cu forms kinetically unstable complexes. This means that a complete exchange of Cu ions and HS complexed Cu ions is possible, but this does not apply to Cr.

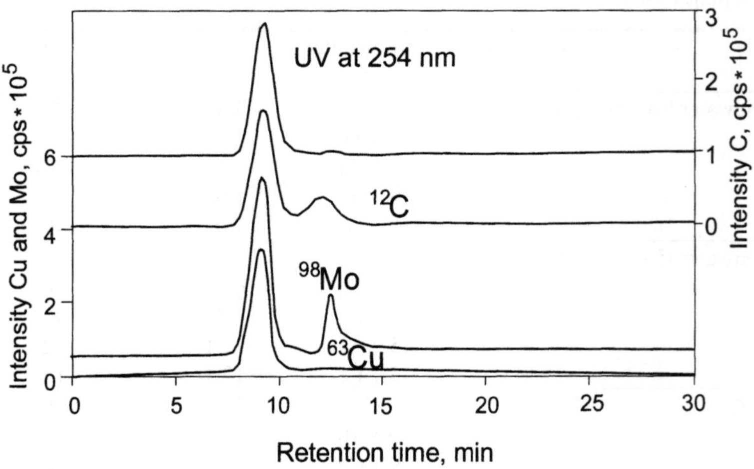

Fig. 6.9. Size exclusion chromatogram of cDOC (UV detection at 254 nm), total C (^{12}C), Mo, and Cu of a raised peat bog sample (Lake Hohlohsee, northern Black Forest, Germany) (after Vogl and Heumann 1997, with permission of Springer Verlag)

In general, the environmental behavior of metals is determined by the complex stability constants, which are controlled by thermodynamic equilibria under particular conditions. For example, the concentrations of free metal species not bound to HS determine the uptake by organisms (free ion activity model = FIAM, Morel and Hering 1993). Under thermodynamically unstable conditions, the kinetic behavior of complexes gains importance in the environmental behavior of metals. This is indicated, for instance, by the kinetic long-term stability of Cr(III)-HS complexes which are very mobile after they have been formed (Marx and Heumann 1999).

In addition to trace metals, the salts of main group metals are also important in water bodies. The various metal cations present in HS containing waters compete for binding sites in or on the HS. A well known example is the binding of Cu to FA, which is different in hard and soft waters. In hard-waters, there is competition between Cu and Ca for binding sites on FA (Breault et al. 1996). This competition has significant effects on the potential toxicity of heavy metals (Chap. 7.3).

6.3.2 Aluminum

Between the 1970's and the 1990's, the interactions between dissolved HS and Al received much attention in relation to the acid rain problem. With decreasing pH in the soil solution, increasing quantities of inorganic Al are dissolved from clay minerals. Many studies (Driscoll 1984; Driscoll et al. 1980; Hall et al. 1985, 1987; Rosseland et al. 1990; Schecher and Driscoll 1987; Wood and McDonald 1987) show that the toxicity of acid water to freshwater organisms is determined by three parameters: H^+ (pH) and Al, both as toxicants, and Ca as the antidote. The toxicity of Al is in turn, determined by its speciation, with only the inorganic species being toxic, these include free Al^{3+} ions as well as fluoro-, chloro- and sulfato complexes. As soon as Al is complexed to dissolved HS, it looses its toxicity (Driscoll et al. 1980). This complexation is a function of pH, and of the available inorganic and organic ligands. Since the bonding of Al to HS is important not only for detoxification, but also for its transport in the environment (Choudry 1984) the molecular mechanisms of this complexation are now comprehensively considered.

Preliminary studies with SEC show that Al is co-chromatographed with both high-molecular weight and low-molecular weight DOC, preferentially in the nominal ≥5 kDa fraction (Steinberg 1980). This is a clear indication, but not a real proof, for binding between the organic molecules and the Al. Confirmation of the bonding comes from improved analytical methods (Alomary et al. 2000; Driscoll 1984) and speciation modeling (Tipping et al. 1988). The metal complexation occurs via hydrogen atoms at particular binding sites. Labile hydrogen atoms are exchanged with metals including Al. These labile hydrogen atoms are mainly found in OH-, NH- and COOH- groups (Alomary et al. 2000). Through the reaction, the number of exchangeable hydrogen atoms on HS molecules is reduced. One type of proton group participates in metal complexation, while others do not. Earlier studies (Backes and Tipping 1987; Lövgren et al. 1988; Tipping et al. 1988) ascribe the metal-complexing and non-complexing protons to carboxyl and phenyl groups, respectively. In addition to these findings, new mathematical modeling by Tipping (1998) and Perdue (1998, 2001) indicate two intrinsic equilibrium constants for metal binding. Alomary et al. (2000) determine the number of active hydrogen atoms in various HS isolates using a combined mass spectrometry technique, and surprisingly, they find a fairly constant proportion: only approximately 7–9 active hydrogen atoms exist per one FA molecule.

The distribution of organic and inorganic Al species is exemplified in a German stream draining a damaged forest on underground bedrock. Haag et al. (2001) analyze factors controlling the total concentration and aqueous speciation of Al in the River Große Ohe (Bavarian Forest), using a thermodynamic equilibrium model and a mixing approach. A model compound for HS is derived on the basis of the relationship between anion deficit and the organic carbon content in the river, as well as literature data. An equilibrium speciation model for Al is established, considering this model compound and relevant inorganic solutes. Applying the model to measured stream water samples, highlights that aqueous speciation of Al is mainly controlled by the pH value and discharge, and that free Al concentrations reach toxic levels during acidic episodes. The calculated concentrations of Al_{org} and Al_{inorg} as a function of the pH value are given in Fig. 6.10.

Al_{org} only appears in appreciable amounts at pH values greater than 5, at lower pH the role of organic complexes is negligible, so that toxicity is barely mitigated by complexation of organic ligands. In contrast, high concentrations of Al_{inorg} (the toxic species) reaching approximatley 25 µmol/L (Haag et al. 2001), are associtated with low pH. Since both H^+ and total Al concentrations are positively correlated with discharge, high discharge events lead to aqueous Al concentrations which are considered to be toxic for aquatic biota.

Furthermore, comparing measured concentrations of sulfate and H^+, and calculated concentrations of Al^{3+}, with solubility curves of gibbsite-like minerals and jurbanite, clearly shows that total Al concentrations are not controlled by equilibria with these mineral phases alone. The observed relationship can be better explained from a mixture of two distinct waters, representing base-flow and high-flow chemistry, and the resulting equilibrium concentrations. This indicates that total Al concentrations, in particular during high discharge events, is mainly controlled by the mixture of water with differing chemistry and flow paths (Haag et al. 2001).

If the pH of an acidic body of water rises, as for instance when acidic Al-rich water mixes with limed or neutral waters, low-molecular inorganic forms of Al will be transformed to high-molecular weight forms, and hence precipitate. The water of these mixing zones may have strong adverse effects on the biota (Henriksen et al. 1984), via rapid Al precipitation onto susceptible organs, such as fish gills, leading to osmoregulation failure, inhibition of enzyme activities, and subsequent gill lesions (Poléo et al. 1994; Rosseland and Staurnes 1994; Rosseland et al. 1992). The water in the mixing zones is more toxic than the original water. An additional mechanism may be that large excess of Ca and Mg ions may result in a displacement of Al from the FA complexes, as shown in Fig. 6.8. As a result, the concentration of toxic inorganic Al species in the water increases.

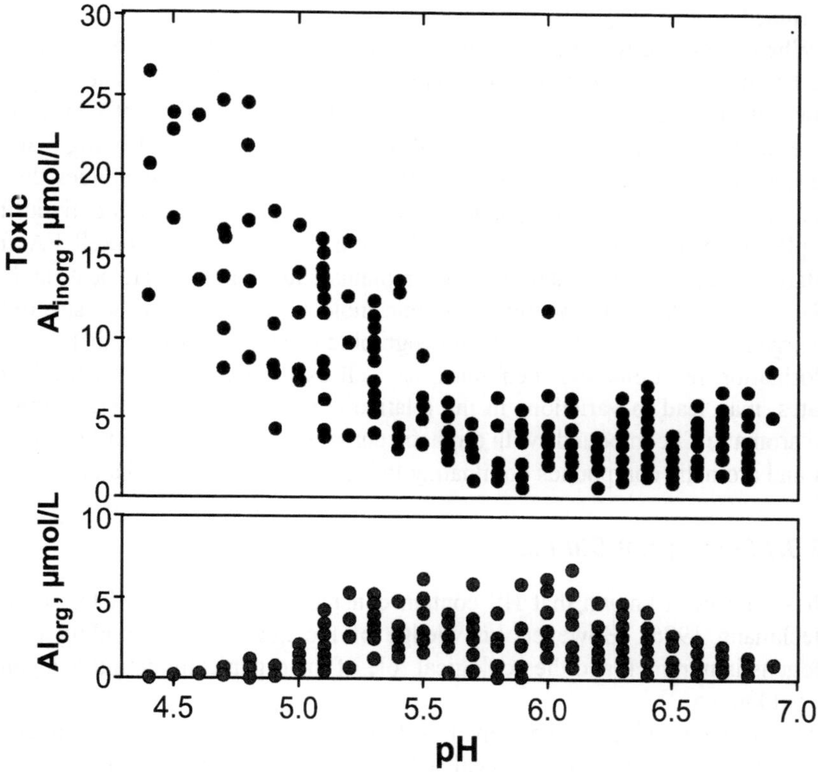

Fig. 6.10. Concentrations of organically complexed and inorganic aluminum species from stream water samples as a function of pH (after Haag et al. 2001, with kind permission of Wiley-VCH)

6.3.3 Halogens

The paradigm that halogens are inert under natural conditions is obsolete, particularly since recognition of global ozone holes. Halogens in fact play a very active role in global transformation processes, and it becomes ever clearer that HS play an important role in their speciation.

Myneni (2002) presents evidence that organo-Cl compounds are the dominant forms of Cl in the organic fraction of soils, sediments, aquatic systems and humified organics of all examined plant samples. Chlorinated phenols and mono- and dichlorinated aliphatic compounds are the most likely compounds in the organo-Cl fraction in humics and weathered plant materials. Polychlorinated acetic acid, alkanes and phenols, and mono-chlorinated cyclic compounds are not very likely, but may not be ruled out. The percentage of aromatic organo-Cl compounds increase with humification. The concentration of aliphatic Cl remains constant or starts to decline relative to the aromatic Cl as the humification of plant material continued.

When compared with the Cl forms in humified plant material, the live plant tissue has hydrated and H-bound Cl^- without any detectable chlorinated organics. However, their occurrence at increasing concentrations in the recently fallen plant leaves and in humified leaves in the leaf litter suggests that Cl^- of living plant tissue is rapidly converted to organo-Cl (Myneni 2002). Myneni continues that the occurrence of organo-Cl compounds as the major Cl fraction in HS derived from simple organisms, such as phytoplankton and bacteria (Lake Fryxell FA from Antarctica) and lignin-containing higher plants (Suwannee River, peat, and soil HS), suggests that chlorination of organic molecules is universal irrespective of the organic substrate sources and the geographic location. However, differences in chlorination reactions and their rates, as well as the chemistry of organic substrates, may lead to variations in the relative concentration of different aliphatic and aromatic Cl compounds with aliphatic compounds dominating in Lake Fryxell FA and aromatic compounds dominating in lignin-rich terrestrial plants.

6.3.3.1 **Fingerprint** ***Studies***

It has long been known that HS contain considerable quantities of O, N, and S (Ziechmann 1996). However, knowledge of the quantity and binding type of halogens is minimal, and the ecological role of the organically bound halogens is not yet known.

Using the previously mentioned isotope dilution technique, Rädlinger and Heumann (1997) study the chemical speciation of Cl, Br, and I in the presence of HS. The three halogens give different partitioning into the UV-active fractions of DOC. In a municpal wastewater sample, Cl and Br favor the high-molecular weight, weakly UV-active fraction, while I exhibits the strongest interaction with the low-molecular weight, most UV-active fraction. For a lignite wastewater, the partitioning is different: none of the three halogens bind to the high-molecular weight fraction, but all favor the medium-weight fraction. With a forest seepage water, Br and I are associated with the high-molecular weight fraction and its declining flank, while Cl is found only in the declining flank of the high-molecular weight fraction.

Over a period of one to eight weeks, some HS/halogen species change with age. Although there are clearly structural changes in the HS with increasing age, as shown in the UV absorption curves, Cl remains bound in its starting fraction, and little change is observed with Br after eight weeks. In contrast, a clear change occurs with the HS-I species (Fig. 6.11), which is explained by a low carbon-halogen binding strength (Heumann et al. 2000). HS utilizing microorganisms are decisively involved in the aging and changing of HS-halogen fractions, as Heumann et al. (2000) and Rädlinger and Heumann (2000) show for I. With the microbial transformation of HS, I favors the newly formed high-molecular weight, UV-active fraction. Overall, no I is lost from the HS fractions (Fig. 6.11).

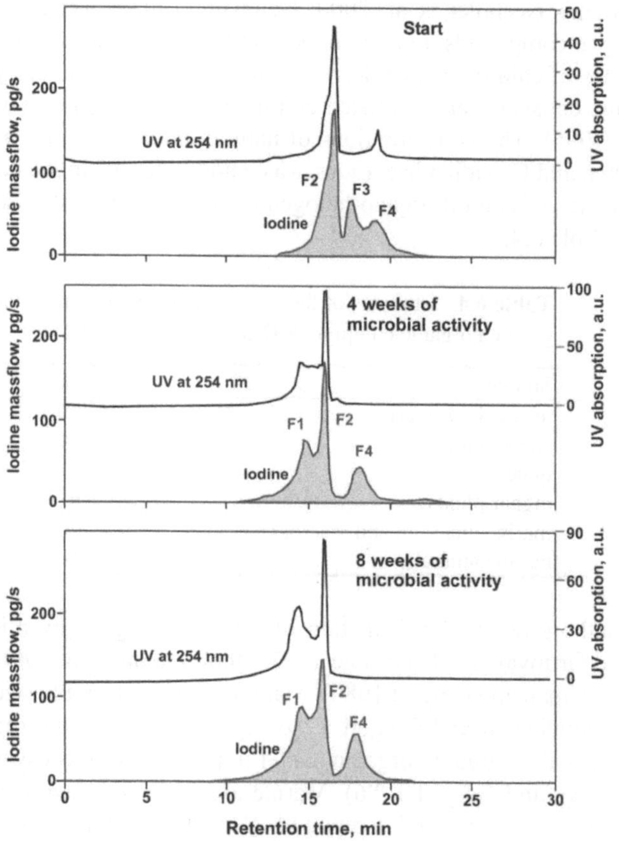

Fig. 6.11. Distribution of ^{129}I in different HS fractions of a wastewater sample from a sewage plant in Mainz (Germany) during 8 weeks of microbiological activity. UV absorption in arbitrary units, a.u. (after Rädlinger and Heumann 2000, with kind permission of the American Chemical Society)

6.3.3.2 Formation of Halogen Compounds

Halogens are known to participate in global cycling in considerable quantities; not only due to anthropogenic emissions. It has been known for a long time that organic halogen compounds can be formed under natural conditions; they are not only products of the chemical industry. Evidence comes from pre-industrial sediments in which polychlorinated-dibenzo-*p*-dioxins and furans occur (Gaus et al. 2001; Isosaari et al. 2002; Jüttner et al. 1997a; Schramm et al. 1994, 1997). In this non-industrial process, enzymatic transformation of inorganic halogen ions plays an important role (Geckeler and Eberhardt 1995; Rappe 2000). The oceans are implicated as one of the main natural sources of organo-halogen compounds, where organisms such microalgae and macroalgae can release large quantities of these compounds (Keppler et al. 2000; Naumann 1993)[8]. Terrestrial sources of organo-halogen compounds, too, are important (Keppler et al. 2000). Globally, 3–6 million tones of chloromethane derive from natural sources every year, whereas anthropogenic emissions are estimated at a negligible 26 thousand tones annually (Field et al. 1995). The concentrations of natural chloroaromatic metabolites exceed the Dutch and Canadian hazardous-waste norms for analogous chlorophenols in soil. Estimates of natural organo-halogen compounds from the various sources are listed in Table 6.4.

Table 6.4. Estimates of the various sources of natural halogenated organic compounds (Naumann 1993, 1994)

Source	%
terrestrial bacteria	29
terrestrial fungi	19
lichens	8
higher plants	9
marine plankton and macroalgae	20
marine animals	16

Several authors report the formation of organic halogens by microorganisms (Asplund and Grimvall 1991; Gribble 1992, 1994; Hjelm et al. 1996; Hoekstra et al. 1999; Neidleman and Geigert 1986). Many bacteria can synthesize halogenated metabolites (Neidleman and Geigert 1986). Fungi may produce organohalogens concomitantly with degrading organic matter in the terrestrial ecosystem (Asplund 1992; Neidleman and Geigert 1986). Marine algae are reported to produce halometabolites, mainly brominated compounds (Neidleman and Geigert 1986). Plant organic matter can contain 10 to 100 µg/g Cl per dry weight (Asplund 1992). It is uncertain, whether higher plants are able to produce halometabolites.

The spectra of halogenated compounds produced by bacteria, fungi, and algae is large and includes aromatic and aliphatic compounds (Gribble 1994; Hjelm et al. 1996, 1999; Naumann 1993, 1994; Neidleman and Geigert 1986). The number of naturally produced halometabolites is estimated to exceed 3,000 (Eurochlor). Field et al. (1995) report that higher fungi synthesize three families of organohalogen metabolites: halomethanes, halogenated aromatics, and haloaliphatic compounds, such as chloroform. It is suggested that fungi are important sources of elevated concentrations of chloroform in soil air (Hoekstra et al. 1998). These organohalogen metabolites have demonstrable physiological roles as antibiotics, as methyl donors and as substrates for H_2O_2-generating oxidases. That means that these compounds, such as 2-chloro-1,4-dimethoxybenzene, can replace veratryl

[8] Naumann (1993) writes that it was not until 1934, that diploicin isolated from lichens by W. Zopf in 1904 is identified as the first chlorine-containing natural substance. The paradigm: '*Nature does not have a chlorine chemistry*' has probably been considered valid for too long. Even Thurman (1985) writes: '*Halogens such as chlorine, bromine, and fluorine do not occur naturally in aquatic humic substances*'.

alcohol as a redox mediator in lignin peroxidase catalyzed oxidations (Teunissen and Field 1998).

Only recently, Putschew et al. (2001) show that organic bromine compounds can be found in surface freshwaters. In the waterways of Berlin, Germany, up to 35 µg/L of an organic bromide (not yet identified) are present in summer, and around 10 µg/L are present in spring, autumn, and winter. From indirect evidence, Putschew et al. (2001) conclude that the organically bound bromine is of biotic origin and formed by haloperoxidase activity.

Two enzymes catalyzing biohalogenation are known: heme-containing haloperoxidase, and non-heme haloperoxidase. Peroxidases catalyze the oxidation of a wide spectrum of substrates with hydrogen peroxide or other hydroperoxides as the oxidant (Neidleman and Geigert 1986). From the reaction below (Neidleman and Geigert 1986) it can be seen that a hydrogen ion (acidity) is needed for haloperoxidases to work.

$$\text{substrate} + H_2O_2 + X^- + H^+ \xrightarrow{\text{haloperoxidase}} \text{halogenated product} + 2\,H_2O.$$

In addition to the anthropogenic and biotic pathways, an abiotic pathway of organic Cl formation, via exchange of Cl with organic substances in soils and surface waters, is discovered for organochlorine compounds (Öberg 1998). Öberg realizes that the organic bound Cl comprises the main pool of this element in soils and sediments, a fact which has been overlooked until recently. Organic materials contain Cl in similar concentrations as P, that is in the lower percent range. For example, the deeper layers of peat in a *Sphagnum* bog contain on average 0.6 mg/g organically-bound Cl, and in a *Carex* bog, 0.2 mg/g. These figures are supported by Grøn and Raben-Lange (1992) and Krog and Grøn (1995), who applied an improved procedure and find up to 0.24% halogens in HA from the B horizon and from groundwater.

There are two potential pathways of organo-Cl formation: the incorporation of microbially formed organochlorines in the organic material, and the microbially-induced formation of reactive Cl, which is subsequently incorporated in the organic material (Fig. 6.12).

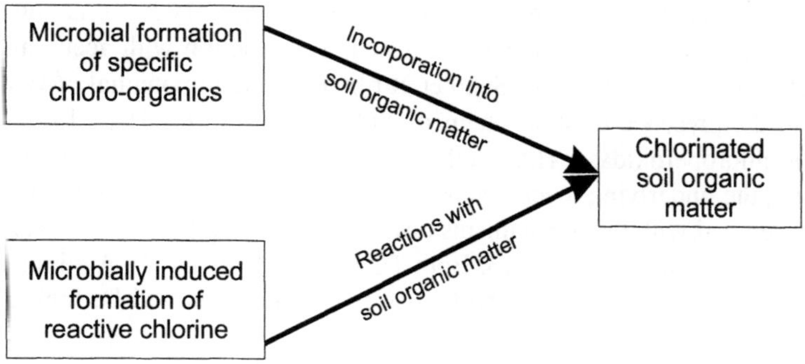

Fig. 6.12. Two potential pathways of organic chlorine formation in soils and sediments (after Öberg 1998, with kind permission of Wiley-VCH)

Although the formation rates of organo-Cl compounds in sediments are unknown, one can make estimates from the flow rates from the soils, and therefore the loading to water bodies. Taking an average output of organic material of about 50 kg/ha/a, and content of 2 mg Cl_{org} per gram C, one can estimate a flow of 100 g Cl_{org}/ha/a from the catchment into receiving surface waters. This numbers is verified from field studies.

Obviously, net formation of organohalogens in forest soils is closely related to degradation of organic matter. Recent findings are that the amount of organically bound halogens in soil increases with decreasing pH, and seems to be related to lignin degradation (Öberg 2001).

Trichloroacetic Acid

Asplund and Grimvall (1991) report that concentrations of organohalogens in lakes are related to the concentration of HS. Since most of the organic matter in humic forest lakes probably originates from the drainage area (Meili 1992), the terrestrial ecosystem in the drainage area is a likely source of halogenated organic material in lakes. Recently, Niedan et al. (2000) describe the potential of soil FA to be chlorinated by chloroperoxidase activity, whereby chlorinated aromatic compounds are produced. This potential is demonstrated with trichloroacetic acid (TCA). In addition to industrial production and abiotic (photooxidative) formation from chlorinated ethenes and ethanes, there are some indications that TCA[9] can also be produced by biota (Schöler 1998). Hoekstra et al. (1995) find dichloroacetic acid (DCA), TCA, and trichloromethane after the chloroperoxidase catalyzed reaction of chloride and H_2O_2 with HS. Also, chlorobenzoic acid may be formed biologically and may comprise up to 0.5 µg/L in humus-rich waters (Niedan and Schöler 1997).

Schöler (1998) finds high TCA concentrations in bog water, for instance up to 1 µg/l in Lake Hohlohsee (Black Forest). He assumes that the measured TCA contents represents a steady state concentration resulting from decomposition and formation processes. Schöler also proposes a connection between the pH value of the water sample and the TCA concentration, because several fungal chloroperoxidases have activity maxima in the acidic range.

To assess natural production, potential precursors of TCA formation, namely short chain fatty acids and a HA, are tested. The reaction with acetic and HA is most effective (Schöler 1998). Schöler attributes the high potential of HA to form TCA, to the presence of short chained aliphatic acids in the HA. However, even without chloroperoxidase, HA itself possesses a comparatively high chlorination potential, the underlying mechanisms of which are not understood, but may possibly be identical with the mechanisms described by Öberg (1998). An example of a TCA flux is given in in Fig. 6.13. Both TCA production and TCA elimination depend on the organic matter content of soils and the rates will be higher in soils

[9] TCA displays a comparatively low toxicity to aquatic animals as demonstrated with the zebrafish *Danio rerio* (Wiegand et al. 1999a).

rich in organic matter.

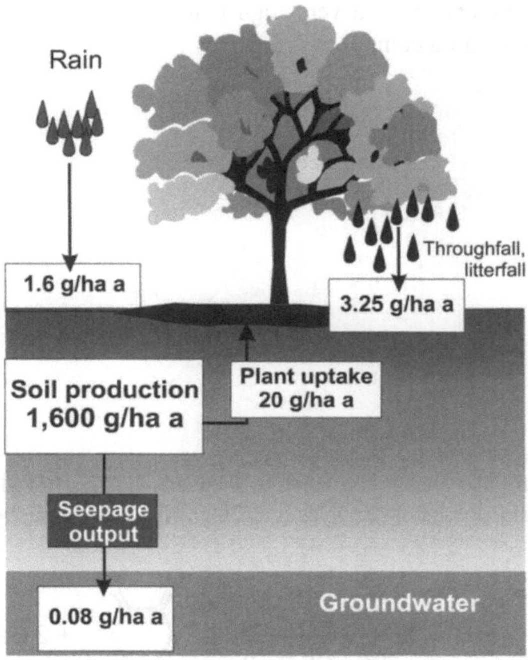

Fig. 6.13. Simplified TCA flux demonstrating the significance of soil organic matter in TCA production (modified from Schöler 1998, with kind permission of Eurochlor Brussels)

Volatile Halogenated Organic Compounds

In a recent study, Keppler et al. (2000) explain the formation of climate-relevant, volatile halogenated organic compounds from HS. In soils and sediments, halide ions can be alkylated during the oxidation of organic substances by electron acceptors such as Fe(III). When the available halide ion is chloride, the reaction products are CH_2Cl, C_2H_5Cl, C_3H_7Cl and C_4H_9Cl. The corresponding alkyl bromides or alkyl iodides are produced when bromide or iodide are present. This series is formed in the ratio 17:4.5:0.5:1 for I in a wetland water. In this case, the formation of the halo-carbon compounds is a non-linear function of the Fe(II) concentration. The authors assume that a fraction of the soil halogen content is methylated by natural oxidation processes, and that this fraction increases in the sequence Cl<Br<I. The suggested reaction scheme is shown in Fig. 6.14. Interestingly, neither sunlight nor microbial mediation is required for these reactions. Keppler et al. (2000) discuss the alkyloxyphenols as the initial structure in the HS, as shown in Fig. 6.15.

The oxidation of guaiacol by Fe(III), and the nucleophilic substitution of the methyl groups by halides, occur almost synchronously. The authors estimate that a maximum of 1 of 30,000 carbon atoms in the HS is released from the soil as volatile halogenated organic compounds within one hour. Therefore, they conclude that soils and sediments have a very high potential for formation of volatile organic halogen containing compounds.

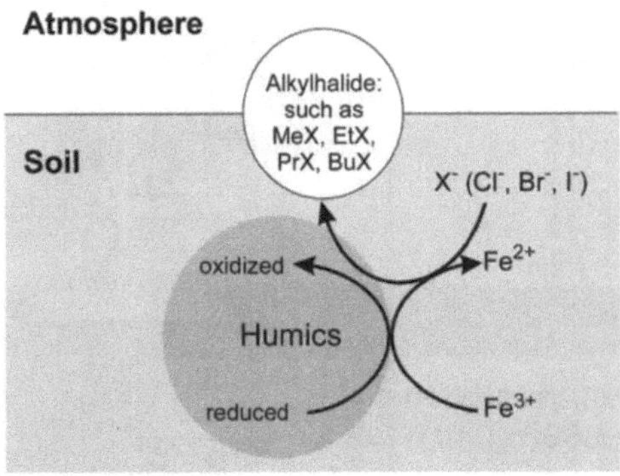

Fig. 6.14. Model for alkyl halide formation by the reaction of Fe(III) and organic matter in the presence of halide ions (from Keppler et al. 2000, with kind permission of Nature)

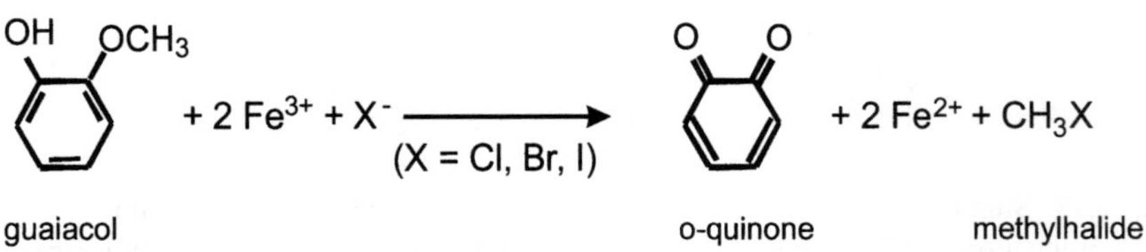

Fig. 6.15. Formation of methylhalides from guaiacol, ferrihydrite, and halide ions (after Keppler et al. 2000, with kind permission of Nature)

Polychlorinated Dibenzo-p-Dioxins

It is well known that in sediment–water systems, the rates of alkyl halide reduction increase with organic matter content (Peijnenburg et al. 1992). With quinonic and phenolic hydroxyl groups ranging from 13 to 56% (molar concentration) of all oxygen-containing functional groups in natural organic matter (Schnitzer and Khan 1972), organic mediated dechlorination reactions may potentially contribute significantly to the fate of chlorinated contaminants. Hydroquinone/quinone-type pairs are shown to dominate redox properties (Lovley et al. 1996; Scott et al. 1998; Chap. 6.5).

Even polychlorinated dibenzo-*p*-dioxins (PCDDs) are subject to humic-mediated dechlorination, with the appearance a new pattern and the increase of tetra- (including the most toxic 2,3,7,8-tetra-CDD isomer) to mono-CDD congener concentrations among the products of dechlorination. The 2,3,7,8-tetra-CDD preferentially accumulates among other PCDD isomers. A study by Barkovskii and Adriaens (1998) indicates that, aside from direct deposition, the increase of

2,3,7,8-tetra-CDD concentrations in the environment may result from combined biotic and abiotic PCDD (penta-, hepta-, and octa-CDD) dechlorination, whereby HS act as catalysts (≅ electron shuttle, see below).

6.4 Biopolymers and Pheromones

HS also have the capacity to bind with biopolymers, such as enzymes, toxins, and DNA via hydrogen-, ionic- or covalent bonds. In many cases, these bonds result in protection from microbial decomposition. The biological activity is generally reduced but not totally eliminated (Crecchio and Stotzky 1998b); Ziechmann (1994) coined the name 'entropy buffer' for this property of HS.

6.4.1 Exoenzymes

There is substantial evidence that microbial extracellular enzymes (exoenzymes) provide the most efficient means of cleavage of biopolymers (Münster et al. 1992a,b; Münster and de Haan 1998; Wetzel 1991). These enzymes participate for example, in the C, N, P, and S, cycles. The inhibition of activities of exoenzymes, such as phosphatases, cellulases (Sinsabaugh and Linkins 1987), and proteases (Dudley and Churchill 1995; Jahnel and Frimmel 1994; Jahnel et al. 1994) are reported. Since the ecological effects of such inhibitions are well known for phosphatases and proteases, the following remarks will focus on these two enzyme families.

6.4.1.1 Phosphatases

Free, dissolved phosphatases can comprise 30 to 90% of all phosphatases. Münster et al. (1992a,b) report that 70 to 90% of the acidic phosphomonoesterases are not bound to particles, but are present as free, dissolved enzymes. The existence and activity of such enzymes is strongly influenced by HS. The HS-enzyme interactions are mainly based on HS-metal-protein interactions, dipole-dipole, and van der Waals forces via hydrogen bonds between NH_2- and/or carboxyl groups on the protein/enzyme surface, and the phenolic OH-groups of the polyphenols (Wetzel 1991). The exoenzymes are not competitively bound to HS, and so are non-permanently inactivated (entropy buffer sensu Ziechmann). Free, dissolved enzymes can be protected by the polyphenolic groups of HS, and consequently their catalytic properties and potential changes. It is assumed that this protection also acts as a shield against inhibitors and against proteolysis. This hypothesis is partially corroborated experimentally by Wetzel (1991), who shows that HS have a clear influence on the kinetic properties of alkaline phosphatases. A similar effect is described by Münster (1994), who reports that if polyvinylpyrrolidone (PVP), an effective absorber of HA, is added to the water sample, there is higher acPME activity in the treated sample than in the PVP-free control. Münster

(1994) assumes that the HS modifies the enzyme activity through changes in catalytic properties, such as in the catalytic center. Similar effects are known for soil enzymes, and established experimentally as summarized by Schinner and Sonnleitner (1996).

HS changes the capacity of plankton to exploit phosphomonoesters as an additional P-source, through hydrolysis by alkaline phosphatases. Stewart and Wetzel (1982) report that adding HS clearly increases alkaline phosphatase activity of a bacteria/algae system from Lawrence Lake (Michigan), particularly under low-light conditions. This study also gives evidence that low-molecular weight HS stimulates enzyme activity more than does high-molecular weight HS. This finding is not an exception, as the review of Jones et al. (1988) shows. Stewart and Wetzel (1982) formulate the hypothesis that HS inhibit extrecelluluar alkaline phosphatase, either through masking of the phosphomonoesters or through stimulation of bacterial or algal growth so that competition for external P resources is increased. This situation however, stimulates the synthesis of alkaline phosphatase in algae.

Wetzel (1993) further suggests that the HS-enzyme complexes remain in a preserved state for hours, days, or even longer, and that the enzymes can be released through photodegradation of HS. This proposal parallels the situation free of enzymes in soils (Schinner and Sonnleitner 1996). In the soil, the so-called 'abiontic' enzymes (enzyme-HS complexes) possess a clearly higher resistance to inactivation by environmental factors such as pH, heavy metals, proteolysis, temperature, complexation, and radiation, than do the freely dissolved enzymes. In addition, the enzyme substrates become stabilized by HS. Münster (pers. comm.) finds that, even after long exposure of phosphomonoesters in the dark at room temperature, the enzyme activity is still some 50–60% of initial activity.

In essence, it can be concluded that free, dissolved, and immobilized enzymes in humic waters can react more rapidly to substrate pulses and environmental changes than can enzymes on cell surfaces or within organisms. The former group of enzymes can have a key function in the initiation of biocatalysis and act as inducers of cell-surface enzymes.

In the euphotic zone, the exoenzymes are not only protected by absorption on HS, but are also subject to HS mediated photodegradation and consequent destruction. The concentrations and activities measured in lake waters are therefore in equilibrium with these contrasting processes. This situation is even more complex, since the HS protection shield can be destroyed by light, so that the enzymes can be rapidly released and become active. In the free, dissolved state, enzymes are in turn subject to enzymatic and microbial degradation.

From their studies, Boavida and Wetzel (1998) answer the question whether the HS bound enzymes are not only protected, but also immobilized for the long-term (Fig. 6.16).

Following destruction of the HS carriers, active enzyme are released again. The authors find that:

- phosphatases from different organisms bind differently to HS. The phosphatase from the green alga *Scenedesmus quadricauda* is only slightly inhibited by HS derived from cattail *Typha* sp., while that from *Escherichia coli* is inhibited by 40–50%. Similar results are obtained with Suwannee River HS;

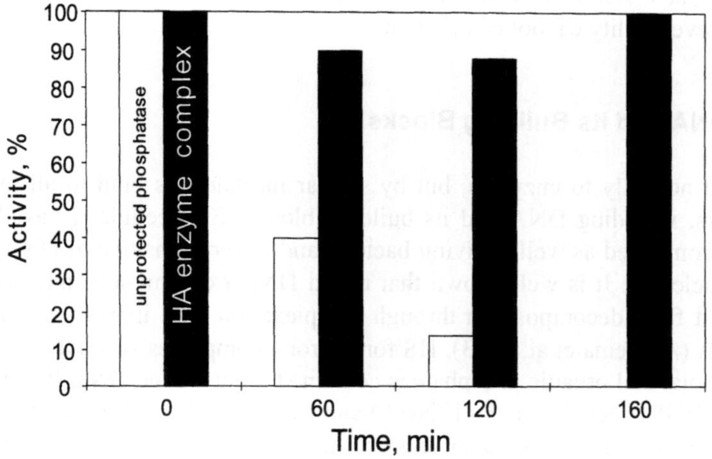

Fig. 6.16. Effect of natural sunlight exposure on unprotected *Escherichia coli* phosphatase and HS-enzyme complexes. HA: *Typha* HA, 5 mg/L C (after Boavida and Wetzel 1998, with kind permission of Blackwell Science)

- HS, to which exoenzymes are bound, generally act as a natural shield against UV radiation. The UV radiation of natural sunlight completely inactivates unprotected phosphatases within 160 min (Fig. 6.16).
  It appears that the formation of the protected enzyme-HS complexes is kinetically controlled. Significant degradation occurs only after 2 hours. The increase in phosphatase activity after 160 min may be due to the release of active enzyme, following the destruction of HS components in the HS-enzyme complex. The Suwannee River HA appear to be much more effective as a UV shield than is *Typha* derived HS;
- the phosphatase is protected from UV radiation when it is bound to HS, regardless of the source of the HS. Light quality has no effect on the formation of these complexes, since complexation occurs equally in darkness or light.

Enzyme inactivation can be reversed by UV radiation. The enzyme is preserved in complexes for later use, rather than being destroyed by UV radiation. This provides a lake with a reserve of potential active enzymes which can use the nutrient reserves in various lake compartments.

6.4.1.2 Proteases

The protection mechanism described for phosphatases is apparently not applicable to all enzymes. Dudley and Churchill (1995) report that of two proteases, serine protease and metalloprotease, only serine protease is inhibited by HA, but not by FA. The longevity of the inhibited protease clearly increased. Inhibition of serine protease can also occur due to elevation of the pH to 9.0, or through the increase in ionic strength (through addition of $CaCl_2$). It is concluded that the interactions between serine protease and HA are ionic in nature. The authors do not test UV radiation, so a comparable situation to that of the protected phosphatase inhibition and its reversibility cannot be ruled out.

6.4.2 DNA and its Building Blocks

HS bind not only to enzymes, but by similar mechanisms bind to all other biopolymers, including DNA and its building blocks. Extracellular ('naked') DNA comes from lysed as well as living bacteria and is very sensitive to DNAases and other nucleases. It is well known that naked DNA exists in soils and sediments, protected from decomposition through complexation with inorganic and organic materials (Aardema et al. 1983). HS form strong complexes with DNA, and up to 10% of total soil organic phosphate is contained in the bound DNA (Crecchio and Stotzky 1998a). Khairy et al. (1996a,b) and Davies et al. (1997) report on the underlying mechanism. The adsorption of aqueous nucleobases, nucleosides, and nucleotides on (compost-derived) HA occurs in detectable steps: adsorption on specific HA site, freeing up or 'creation' of more sites, adsorption at the newly available sites, and site hydration by water. It is interesting to note that HS building blocks have a selectivity for nucleobases, nucleosides, and nucleotides: for instance, carboxylic and phenolic groups interact strongly with purine adenine.

From an ecological perspective, Crecchio and Stotzky (1998a) describe the following details for binding of whole DNA to HS:

- adsorption and binding of bacterial DNA (from *Bacillus subtilis*) to HS,
- the transformability of bound DNA, that is to say its ability to act as 'cryptic genes' and
- the resistance of the bound DNA to nucleases (DNase I).

Crecchio and Stotzky (1998a) find that 70–80% of the adsorbed DNA is bound to HS. This bound DNA is able to transform competent auxotrophic and antibiotic-sensitive cells, although less effectively than free DNA. It is no surprise that the HS bound DNA are better protected from decomposition than free DNA. The concentration of DNAase required for inhibition of transformation of bound DNA is about 100 times as high as for the same quantity of free DNA. The described mechanism protects the DNA from degradation, but not from transformation. In other words, the 'cryptic genes' can eventually be expressed if sensitive hosts are present, and if the existing environmental conditions support transformation. Although to date, these findings stem only from soil and compost studies, we may expect that they are also applicable to aquatic ecosystems.

6.4.3 Pheromones

Among natural hydrophobic chemical compounds in aqueous systems, there are pheromones, particularly fish pheromones. Many of them are steroidal in nature, and therefore barely soluble in water. Hence, one would expect that the hydrophobic pheromones bind to the hydrophobic fraction of HS, as shown for xenobiotic chemicals in Chap. 7.

Only very recently, this aspect of freshwater ecology has been studied experimentally by Hubbard et al. (2002). The authors record electro-olfactogram response of the goldfish (*Carassius auratus*) olfactory epithelium to a steroid pheromone, 17α,20β-dihydroxy-4-pregnen-3-one (17,20β-P). This pheromone has a well characterized reproductive role in *C. auratus*. At all concentrations of HS tested, including the environmentally realistic concentration of 1 mg/L DOC, there is a significant attenuation of the amplitude of the initial response to 17,20β-P in the presence of HS compared to 17,20β-P alone. Furthermore, simultaneous recording of electro-encephalograms from the olfactory bulb demonstrate that the nervous activity evoked by the same concentration of 17,20β-P is less intense in the presence of HS than its absence. Interestingly, the data show that the electro-olfactogram response is dependent on both HS and 17,20β-P concentration, indicating an interaction between these two factors.

Prostaglandin $F_{2\alpha}$ ($PGF_{2\alpha}$) is also known to act as a pheromone in goldfish. However, it is non-steroidal and much more soluble in water 17,20β-P. Therefore, the effect of HS on the ability of goldfish to detect $PGF_{2\alpha}$ is expected to be much less marked than for 17,20β-P. Hubbard et al. (2002) show this to be the case; only the highest concentration of humic acid used significantly diminishes the electro-olfactogram amplitude in response to $PGF_{2\alpha}$. In fact, the lower concentrations of HS actually slightly enhanced the electro-olfactogram response to $PGF_{2\alpha}$, the underlying mechanism remains obscure.

In conclusion, Hubbard et al. (2002) suggest that HS may significantly reduce the concentration of 17,20β-P available for detection by *C. auratus* in natural environments, with consequent deleterious effects on the reproductive success of this species. Furthermore, as many teleost pheromones are steroid derivatives, this phenomenon is unlikely to be confined to this pheromone in this species, but may be applicable to chemical communication systems in teleosts in general.

6.5 Interactions with Electrons: HS as Redox Catalyst

It is well known that HS can participate in extracellular electron transport (Schindler et al. 1976). HS generally function as electron acceptors and donors. For instance, they carry out the electron transport from reduced material in sediments to electron acceptors which are thermodynamically not available. Microorganisms may benefit from this function of HS. Lovley et al. (1996) show that there are bacteria in soils and sediments which use HS as electron acceptors for the anaerobic oxidation of organic compounds and hydrogen. This electron transport requires energy to support growth. The microbial reduction of HS increases the ability of microorganisms to reduce less accessible electron acceptors, such as Fe(III) oxides. HS can apparently transfer electrons between the humus-reducing microorganisms, and the Fe(III) oxides. Lovley et al. (1996) refer to this as a 'shuttle'[10]. The former models for microbial Fe(III) reduction imply that the microorganisms must have direct physical contact with the Fe(III) oxide precipitate (Brock and Gustafson 1976). However, this concept is now revised by Lovley et al. (1996) (Fig. 6.17).

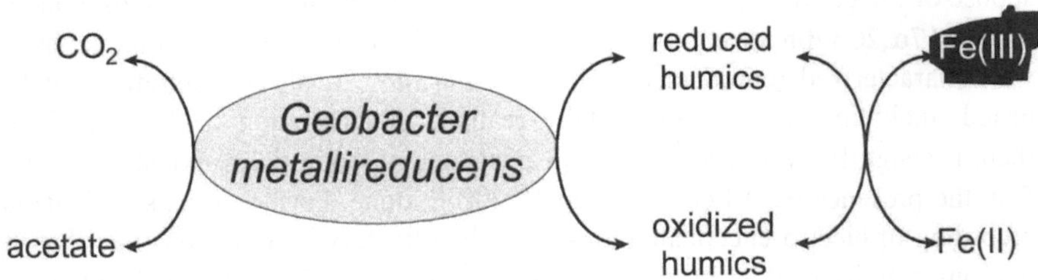

Fig. 6.17. Model concept of HS as electron shuttles in the microbial reduction of metals (after Lovley et al. 1996, with kind permission of Nature)

The new mechanism suggests a biological source of electrons for the HS-mediated reduction of metals and organic (micro)pollutants. As an example, Lovley et al. (1996) propose a model mechanism of how HA stimulate the Fe(III) reduction by the dissimilatory Fe(III) reducing bacterium *Geobacter metallireducens*. *G. metallireducens* transfers electrons to HS. Subsequently, the HS can stimulate Fe(III) reduction in two steps: (1) *G. metallireducens* oxidizes acetate for which HS acts as an electron acceptor, and (2) reduced HS transfers electrons to poorly crystalline Fe(III). In this way, the Fe(III) reducing bacterium gains energy for growth from the electron transfer to HS. The quinoide structures determine the reduction catalyst properties (Scott et al. 1998). Growth of *G. metallireducens* with poorly crystallized Fe(III) oxide as an electron acceptor can be stimulated by addition of the HS analogue anthrachinone-1,6-disulfonate.

The described ability of HS to act as redox catalysts is not limited to *G. metallireducens*. Many microorganisms representing a wide phylogenetic diversity can carry out Fe(III) reduction, and have the ability to transfer electrons to HS (Table 6.5). And *vice versa*, microorganisms which cannot reduce Fe(III) are also unable to reduce HS. The electron transfer shows that Fe(III) reducing bacteria need not necessarily live on Fe oxides in order to reduce them. The electron transfer accelerates the Fe(III) reduction. Furthermore, a wide phylogenetitc diversity of proteobacteria are, in turn, able to oxidize HS (Coates et al. 2002). Therefore HS are clearly redox catalysts.

In their more recent work, Lovley et al. (1998) also show other areas where this reductive mechanism plays a significant role. HS can stimulate the reduction of

[10] This process is really a catalysis. Lovley et al. (1996) do not speak of humic substances as redox catalysts, because the concentrations of HS are higher than in 'normal' catalysis reactions. By referring only to the redox-active structures, the quinones, which are only minor components in the humic molecules, the real catalyst concentration is rather low.

the crystalline forms of Fe oxide (goethite, haematite), and Fe species present in clay minerals. It is very likely that the reduced HS diffuse into clay minerals, where bacteria are excluded because of their size. The electron transfer between Fe(III) reducing bacteria and Fe(III) in the clay minerals can be mediated by such HS.

Table 6.5. Reduction of Fe(III), anthrachinone-2,6-disulfonat (AQDS), and HS, and oxidation of HS by various proteobacteria (from Coates et al. 2002; Lovley et al. 1998)

Organism	Reduction of Fe(III)	Reduction of AQDS	Reduction of HS	Oxidation of HS
δ-proteobacteria				
Desulfuromonas acetexigens	+	+		
Geobacter humireducens	+	+	+	+
Geobacter metallireducens	+	+	+	
Geobacter sulfurreducens	+	+	+	+
Stigmatella erecta				+
γ-proteobacteria				
Aeromonas hydrophila	+	+		
Escherichia coli				+
Marinobacter articus				+
Pseudomonas flavescens				+
Pseudomonas stutzeri				+
Shewanella alga	+	+	+	
Shewanella putrefaciens	+	+		
Shewanella sacchrophila	+	+		
β-proteobacteria				
Azoarcus evansii				+
Dechloromonas agitata				+
Dechloromonas aromatica				+
Geospirillum barnseii	+	+		
Rhodocyclus tenuis				+
Wolinella succinogenes	+	+	+	
α–proteobacteria				
Agrobacterium tumefaciens				+

The finding that microorganisms can transfer electrons to HS, has important consequences for the mechanism by which microorganisms can oxidize metals, as well as natural and xenobiotic organics in anoxic soils and sediments. This ability is also applicable to metal contamination in sediments and soils. It is known that Fe oxides can absorb many toxic heavy metals (Lovley 1991). As long as the Fe oxides are not reduced, it is difficult to remove the contaminants from sediments. The microbial reduction of Fe(III) can be a mechanism of toxic metal release. This process is increased through the addition of HS or similar compounds, particularly when a lack of pores in particles of sediments and soils makes direct access to the oxides impossible (Fig. 6.18).

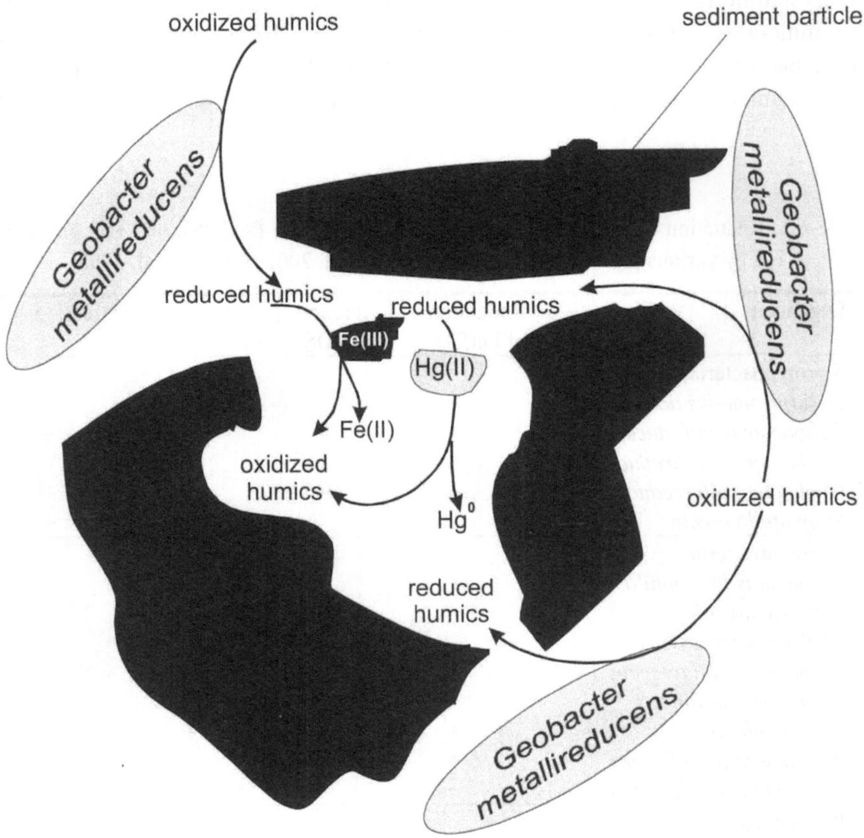

Fig. 6.18. Model of HS as electron shuttles in the microbial reduction of metals in fine pores in sediments and soils (after Lovley et al. 1998, with kind permission of Wiley-VCH)

Microbially reduced HS can generally transfer electrons to metallic contaminants such as Hg(II) (Alberts et al. 1974), and also U(VI) or Cr(VI) (Lovley et al. 1998; Nevin and Lovley 2000). This reduction makes the metals easier to remove or to be transformed to less poisonous chemical species (Lovley 1995a,b).

It is widely accepted that extracellular electron transfer is a general mechanism whereby microorganisms generate energy for cell growth and/or maintenance. Biofilm systems in particular are likely to be important environments for metabolisms that employ extracellular electron transfer (Hernandez and Newman 2001). The scheme in Fig. 6.18 shows that easily usable organic substances, such as acetate, are made available in two ways (Lovley et al. 1996):

- through excretion, death, and lysis of organisms, and
- through photodegradation of HS.

Meanwhile, Cervantes et al. (2000, 2001) demonstrate for the first time, that anaerobic degradation of phenolic compounds can be coupled to the reduction of quinones as terminal electron acceptors. In addition, when extracellular quinones are present in the environment only at low concentrations, this non-enzymatic mechanism of Fe(III) reduction is one of the most important of the many processes of cycling between reduced and oxidized states (Nevin and Lovley 2000). In all probability, the quinone structures of HS also function as inhibitors of photosynthesis of aquatic plants (Chap. 8.3).

A graphical summary of this chapter is shown in Fig. 6.19.

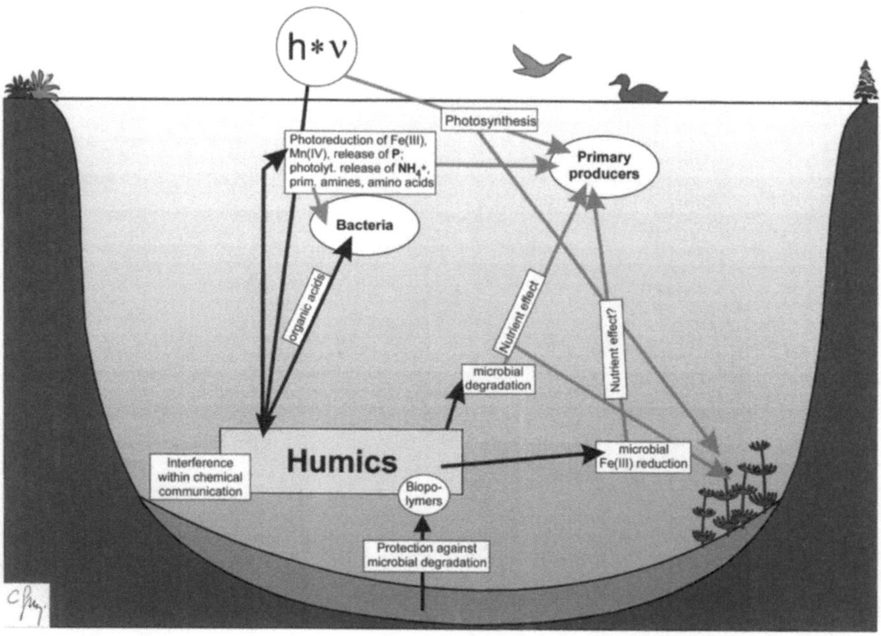

Fig. 6.19. Synopsis of humic substances as source of organic and inorganic nutrients (particularly phosphorus and nitrogen), as protecting agent of biopolymers, as interference within chemical communication, and as redox catalyst. h∗ν is solar energy

7 Indirect Effects on Organisms

Indirect effects of HS on organisms include changes in bioconcentrations and toxicity of xenobiotics and metals, thus reducing the chemical stress an aquatic organism may be exposed to. Compared to the current low level of knowledge of the direct effects of HS on aquatic organisms, the indirect effects are very well known. This different situation is no surprise because in the recent past, research on these substances is taken up in particular by soil and environmental chemists (Box 1.2). There are a number of studies on binding of xenobiotics and metals to HS, and subsequent changes in bioconcentration and toxicity of xenobiotics and metals, to aquatic organisms and thereby significantly altering the living conditions.

7.1 Binding of Xenobiotics to Humic Substances

In his rigorous review, Alexander (1995) presents information that (a) the availability of long-lived organic chemicals to microorganisms in soil in the field declines markedly with time, (b) some freshly added chemicals are readily available to microorganisms in soil in which the identical but aged compounds are not metabolized by indigenous microorganisms, (c) organic compounds incubated in sterile soil become increasingly less available to subsequently added microorganisms, (d) with time some compounds become increasingly resistant to extraction, and (e) sorption and desorption of hydrophobic compounds often require long time periods to reach equilibrium. These five lines of evidence are consistent with a sequestration of the molecules which diffuse into internal micropores or sites that are spatially remote. Sequestration appears to alter the hazard to microorganisms as well as to higher organisms.

What about dissolved HS in aquatic systems, where concentrations of both HS and xenobiotic chemicals are some orders of magnitude lower than in soil solutions? Dissolved HS can physically and/or chemically interact with organic pollutants in many ways, changing many properties of these compounds: the so-called water solubility (better expressed as partitioning in water) (Burnison 1994; Döring and Marschner 1998; Marschner 1998; Wershaw et al. 1969), hydrolysis kinetics (Schwarzenbach et al. 1993), volatilization (Gschwend and Wu 1985; Mackay et al. 1979), photolysis rates (Zepp et al. 1981), and bioconcentrations and toxicity of organic compounds (Bollag and Myers 1992; Cary et al. 1987; Day 1991; Hodge et al. 1993; Kadlec and Benson 1995; Kukkonen and Oikari 1987;

Oris et al. 1990; Steinberg et al. 1992, 1993). Our knowledge of the bonding forms for hydrophobic, and the very water soluble hydrophilic xenobiotics are very different.

7.1.1 Hydrophobic Chemicals

The following types of bond can be formed between HS and chemicals: occlusion, ion-exchange, hydrogen bonds, charge-transfer (π-π) bonds, covalent bonds, and hydrophobic adsorption and partitioning (Choudry 1984; Haider et al. 2000; Piccolo 1994). In soil organic matter, adsorption of hydrophobic compounds may follow a dual-mode sorption, with adsorption in the condensed domains and partitioning in the condensed as well as expanded domains (Xing 2001).

In general, one can state that the organic material or HS are the most important ligands, and the existing stable chemical bonds reduce the toxicity of the particular pollutant. The binding to HS can be initiated through an oxidative coupling reaction catalyzed biologically by polyphenoloxidases and peroxidases, or chemically by particular metals and clay minerals. This binding can also occur through autoxidation. A coupled reaction requires the release of free radicals, and ends in the formation of C–C–, C–O–, C–N–, and N–N bonds between HS and pollutants.

Bollag and Myers (1992) describe the catalytic effect of enzymes on phenols, aromatic amines, and some pesticides, through looking at the oxidative coupling products formed. They find that some humus constituents are very active against phenoloxidase, and can react with less reactive pollutants from which a greatly enhanced coupling activity results.

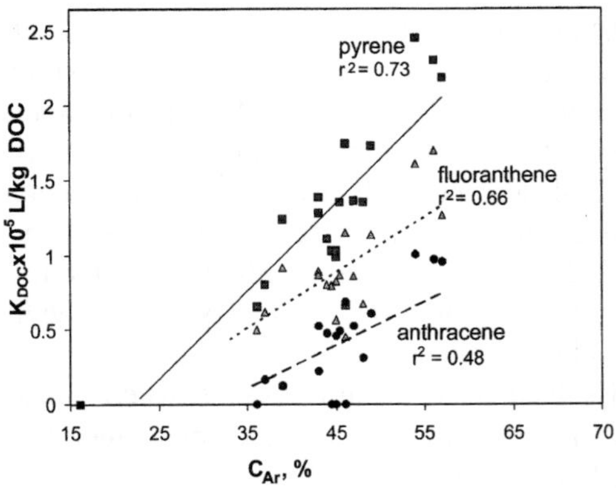

Fig. 7.1. DOC water partitioning coefficient of pyrene, fluoranthene, and anthracene as a function of lipophilicity of the humic isolates, expressed as share of aromatic moieties (from Steinberg et al. 2000, with kind permission of Wiley-VCH)

In most cases, the bonds are weak (Chien et al. 1997; Nanny et al. 1997). Schulten (1999a) shows by molecular modeling (Box 2.1) that the bonding of pentachlorophenol and DDT to HS molecules is predominantly through weak van der Waals forces, and hydrogen bonds which cause temporarily immobilization of xenobiotics. Welhouse and Bleam (1993a) describe that the commonly used triazine herbicide atrazine can act as both donor and acceptor of hydrogen bonds. Weak to mid-strength complexes held together by hydrogen bonds are formed by atrazine with amine, hydroxyl and carbonyl groups; strong complexes are held together with carboxyl groups (Welhouse and Bleam 1993b).

Other studies on biologically and physicochemically determined partition coefficients of polycyclic aromatic hydrocarbons (PAH) between water and organic C, show that such coefficients are a function of HS, but not of the NOM in general (Haitzer et al. 2000, 2001; Kopinke et al. 1995; Steinberg et al. 2000; and others) (Fig. 7.1, 7.2). Relatively weak bonds between HS and PAH are established by electron donor/acceptor complexes and hydrophobic interactions.

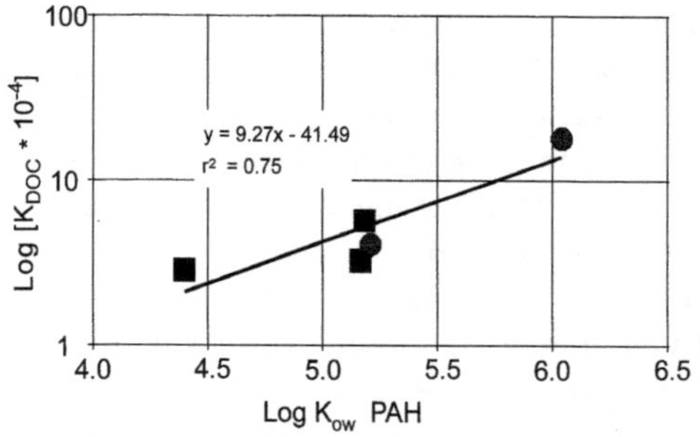

Fig. 7.2. K_{DOC} as a function of lipophilicity of the PAH. The function applies independent of the method (chemical: squares; biological: circles) (from Steinberg et al. 2000, with kind permission of Wiley-VCH). In this graph K_{DOC} is 0.1 K_{OW} and, therefore is almost identical with the value of 0.08 given by Burkhard (2000)

It should be mentioned here that K_{DOC} values of dissolved HS or DOC can only be compared with each other if they have been determined by the same method. Otherwise, large discrepancies, of up to 2 orders of magnitude, may be found (Burkhard 2000). In his review, Burkhard stresses that predictive relationships based solely upon the hydrophobicity of the chemical will have large uncertainties, because additional factors also affect the association of hydrophobic organic chemicals with DOC. To reduce uncertainties, substantial improvements will be required in characterizing the composition of DOC, and in understanding how the different components of the DOC interact with hydrophobic organic chemicals. Characterization of DOC by DOC measurement alone does not adequately describe this phase. More detailed evaluations, such as the ratio of the aromatic to aliphatic C content of HS appear to be very useful. The application of the ratio is shown in Fig. 7.1 which relates K_{DOC} to this molecular descriptor: between 50 and 70% of the K_{DOC} variability of the three PAH under study can be explained solely by this ratio.

Very recently, Laor and Rebhun (2002) report that linear partitioning, or site complexation in the presence of excess available sites, cannot fully describe the interactions of hydrophobic compounds with dissolved HS. Site-specific hydrophobic interactions at limited interior or external molecular surfaces may be considered.

The chemical interactions between natural organic C and chemicals are to a large extent determined by the lipophilicity of the HS on the one hand, and by the lipophilicity of the chemical compound on the other hand. For compartments such as soil and sediment, in which the organic C is present in high concentrations as colloids or particles, structural properties in addition to the aromicity are responsible for the accumulation of lipophilic environmental chemicals. From the realm of terrestrial ecological chemistry, it is known that the cuticular waxes of leaves and needles which are non-polar aliphatic compounds, are very good passive collectors of lipophilic environmental chemicals (Hinkel et al. 1989; Sabljic et al. 1990; Schramm et al. 1987). Recently, Chefetz et al. (2000) describe similar accumulators in particulate NOM isolates such as tomato cuticles, humins, HA, degraded lignin peat, and brown coal. A positive correlation is found between PAH adsorption and the aliphaticity of the structures, which determines the lipophilicity and the K_{OC}. The cuticles and humin from algal deposits have the highest K_{OC} values. The assumption that, in some instances, non-polar aliphatic compounds may be more effective than aromatic compounds in binding PAH, is supported by findings of Kopinke et al. (2001a,b).

The bonding of lipophilic chemicals to HS can also have a kinetic effect. Haider et al. (2000) describe a long-term lysimeter project investigating agrochemicals and their binding to, and release from, HS. The entrapment and sequestration may result in a gradient of increasingly stable fractions of xenobiotics, with a simultaneous decrease in bioavailability of the agrochemicals.

At the end of this bonding process, covalent bonds are formed with relatively water soluble xenobiotics, such as the triazine herbicide anilazine (Klaus et al. 1998a). These covalent bonds can form between ester and ether groups, via oxygen atoms originating from functional groups of the organic matter (Fig. 7.3). The depicted model applies not only for soil HS, but also for aquatic HS, and shows that anilazine is retained mainly by chemical interactions, charge transfer, and sequestration within the structural voids of the humic material.

A similar mode of action applies to diethyl phthalates (DEP). The sorption of DEP can be explained in terms of a two-step mechanism (Schulten et al. 2001). Absorption takes place as long as the host humic acid structure offers (a) enough internal docking space and (b) favorable interactions (energy release) with the guest molecule. This takes place for up to 7 DEP molecules. Further increase in the number to 30 DEP molecules will, due to the lack of free available internal voids, lead to surface controlled adsorption. The two-step sorption process apparently results in a linear increase in energy gain by DEP bonds, and similarly a constant incremental rise in molecular properties of the complexes, such as volume and surface area. Three outstanding observations emerge: (1) structural features at the atomic level (nanochemistry), such as partial atomic charges and high aromaticity of the HA, are observed to be dominating the intermolecular interactions in the complexes at the specific sorption sites; (2) torsional relief and favorable changes in bonding energy also prevail for the growing complex, the latter indicates both the structural flexibility of the HA host and the stabilizing effect of DEP on the complex, by filling of the voids within the HA molecule; (3) the intermolecular forces are described mainly by hydrogen bonds (electrostatic energy) and interactions between dipole-dipole, such as carboxylic functions and uncharged moieties such as aromatic rings (van der Waals energy).

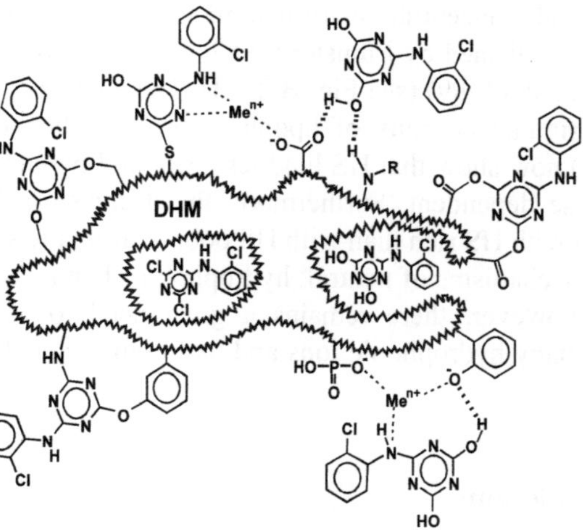

Fig. 7.3. Potential major binding interactions between anilazine molecules and aquatic HS (from Klaus et al. 1998a, with kind permission of Elsevier Publisher)

Dankwardt et al. (1996, 1998) also study triazines and the question of how these compounds are transformed to the bound residues, that is to say how they bind to HS. Using an enzyme immunoassay with antibodies with different binding properties to triazine functional groups, Dankwardt et al. (1996) show that atrazine is bound to HS by substitution of chlorine atoms rather than amino groups. In HA, the bound atrazine concentrations appear to be approximately one order of magnitude higher than in FA. Under laboratory conditions atrazine is more strongly bound than terbutylazine, another *s*-triazine (Dankwardt et al. 1998). Antibody recognition points to the availability of free ethyl and isopropyl groups exposed on the non-extractable *s*-triazine residue.

Recently, under the pressure of environmental politics, scientists begin to study the so-called endocrine disrupters, examples of molecules which affect non-target organisms. These chemicals are of concern because they exhibit hormone-like effects. For example, 17β estrodiol, a product of the microbial degradation of human contraceptives, no longer has effects on women but does have profound effects on aquatic organisms. Consequently, endocrine disrupters continue to be the subject of intensive study. With the formation of strong bonds between dissolved HS and these estrogen degradation products, ecotoxological potential of the latter is clearly reduced. In a recently published study, Loffredo et al. (2000) address this question and show that endocrine disrupting chemicals are only weakly bound to soil and sewage-sludge HA through hydrogen bonds, van der Waals forces, or hydrophobic bonds. As will be shown below for other chemicals, these bonds alone are sufficient to reduce both the bioconcentration and the ecotoxicity, and thereby diminish the ecotoxological risk.

One mechanism that has been very little studied in the binding process of lipophilic organic chemicals to HS is the interferrence with base cations, such as Ca^{2+}. From the classical study of Ghosh and Schnitzer (1980) we know that spherical colloids form at high concentrations of neutral salts (see Fig. B.2.1.4), which has been in principal confirmed by Münster (1982, 1985), Ephraim et al. (1995), and recently Myneni et al. (1999) (see Fig. B.2.1.3). The changes in HS conformation must affect the binding capacity of lipophilic organic chemicals. Akkanen and Kukkonen (2001) now show that HS bind less B*a*P and pyrene, if Ca^{2+} is present. This effect is dose dependent. Furthermore, the decrease in binding of B*a*P is more pronounced with HS rich than with HS poor in aromatic structures.

The bonding mechanisms of neutral, hydrophobic chemicals are comparatively well described; however, there remains a great need to explain the bonding mechanisms for many hydrophobic ions and hydrophilic xenobiotics (water soluble chemicals).

7.1.2 Hydrophobic Ions

From field studies, it is well known that even polar organic compounds, such as methylphenoxy-acetic acid, are retained by soil organic matter (Haberhauer et al. 2001). But, a decrease in C content of a soil does not necessarily imply a decrease of sorption capacity for these chemicals. In general, the well known types of bond also apply to hydrophobic ions. However, which type is important, and what structural properties of HS determine the type of bonding, are scarcely investigated. The studies of Arnold et al. (1998) on organotin compounds and their association with dissolved HS fill the void of knowledge on hydrophobic cations. These associations are found to be strongly dependent on pH, and the maximum binding occurs near to the particular acid dissociation constant (pK_a). The characteristic bond is the complexation of particular cations with negative ligands, such as carboxylate or phenolic compounds. The determining factors for the bonding of organotin cations are shown to be a) complex formation between the tin ions and the deprotonated ligands, and b) hydrophobic interactions. The quality of the HS also has an influence on the association, but the key characteristics are as yet not clear (Arnold et al. 1998).

Holten Lützhøft et al. (2000) describe the bonding of 4-quinolone, an antimicrobial used in aquaculture. They find unexpectedly high sorption coefficients (approximately 10^3–10^5), the same range as for highly hydrophobic chemicals, despite the low hydrophobicity of the 4-quinolones ($K_{OW} < 10^{1.7}$). The authors suggest that in addition to hydrophobicity, electrostatic interactions may play a role in adsorbing hydrophobic ions.

Sorption of aromatic amines to sediments and soils can occur by both reversible physical processes and irreversible chemical processes. Weber et al. (2001)

show that aniline and pyridine behave quite differently in sediment-water systems. The sorption kinetics of pyridine are quite fast, reaching equilibrium within 1–2 h. In contrast, the sorption kinetics of aniline are characterized by a rapid initial loss of aniline from the aqueous phase followed by a much slower rate of disappearance. Sequential extraction of a sediment suggests that pyridine is bound primarily through a reversible cation-exchange process, whereas aniline sorbs through both cation-exchange and covalent binding processes. At longer reaction periods sorption became increasingly dominated by covalent binding.

An additional mechanism by which hydrophilic chemicals, such as aniline, may form covalent bonds, are the photo-Fenton reactions. In these types of reactions, H_2O_2 reacts with Fe(II) to generate a hydroxyl radical ($HO^{\bullet}$). Chelating agents for iron are effective in facilitating Fenton reactions, because the ligands assist in the catalytic reactions with iron. HA are known to serve as chelating agents for Fe(II) and Fe(III), and as reducing agents for Fe(III) (Chap. 5). These characteristics permit HA to function in Fenton and photo-Fenton reactions. Fukushima et al. (2000 a, b, 2001) show that the degradation of aniline via photo-Fenton reactions leads to *p*-aminophenol, *p*-hydroquinone, maleic and fumatic acids, and the simultaneous release of NH_4^+ ions. However, the sum of the product concentrations is much smaller than the aniline concentrations added initially. This can be attributed to the majority of the aniline being incorporated into the polymeric structure in the HA after the reaction. NMR and pyrolysis studies indicate that aniline forms covalent bonds with quinone and the vinyl carbons in the HA, to form anilino-compounds, such as anilinoquinone and enominone.

7.1.3 Hydrophilic Chemicals

As with hydrophobic ions, the bonding of hydrophilic components of HS is also little studied. The same types of bond apply here as described above, with a dominance of ionic bonds. Schulten and Leinweber (2000) clearly demonstrate by molecular modeling that for water soluble compounds, such as hydroxyatrazine, peptides and saccharides, hydrogen bonds are the most important bond type (Fig. B.2.1.3).

Klaus et al. (1998b) and Oesterreich et al. (1999) describe the bonds of the fairly soluble amitrole, a triazole herbicide that inhibits photosynthesis. As with the less water soluble anilazine, amitrole forms many types of bond, including covalent and non-covalent bonds such as charge transfer and occlusion within the HS matrix. Hydrogen bonds are probably also formed, but not as yet directly shown. The latter is an important difference from anilazine. Covalent bonds occur mainly with acid amide bonds. The following evaluation of Klaus et al. (1998b) is most interesting: the total non-covalent bound amitrole fraction is bioavailable and can express its herbicidal effects as shown by the authors with the cress test.

In several instances, the hydrophilic components are fixed on/in HS only after degradation or metabolism of the educts. Achtnich et al. (1999) show this in a relatively unusual case, namely the covalent bonds of the comparatively hydrophilic xenobiotic from the formation of a biologically reduced metabolite of trinitrotoluene in soil HS. Following 33 days of anoxic incubation, nearly all trinitrotoluene is bound, of which 73% is in the humin fraction. The authors suggest that azoxy compounds are formed. This bonding probably completely eliminates the bioavailability of the xenobiotic, so that there is no bioconcentration and no toxic effects expressed. It is effectively bound to the HS after anaerobic microbial decomposition.

7.1.4 Synopsis of Binding Mechanisms

Based on experiments with xenobiotic chemicals with contrasting lipophilicity, Senesi et al. (2001) summarize and categorize the binding mechanisms: the nature of binding mechanism(s) and the adsorption capacity of HS for different chemicals appear to depend more on the molecular structure and chemical reactivity of the chemical than on the structural and functional properties of the HS. Chemicals which are moderately water soluble and rich in polar functional groups, are capable of forming hydrogen bonds and being protonated. They also possess electron-donor or electron-acceptor structural units, that can be adsorbed by HS functional groups by multiple-binding mechanisms including hydrogen bonds, ionic bonds and charge-transfer bonds. Chemicals which are characterized by low water solubility, low polarity, and high lipophilicity do not appear to form strong chemical bonds in the adsorption. These chemicals are expected to form hydrophobic bonding with aliphatic and aromatic moieties of which HS macromolecules are rich. In general, the adsorption capacity of HS for hydrophobic chemicals is two to three orders of magnitude higher than that for polar chemicals. The origin, composition, and chemical properties of HS appear to have a lower effect on the mechanism(s) and extent of adsorption than does the chemical type.

7.2 Decrease in Bioconcentration of Xenobiotics

Hydrophobic organic pollutants can reach much greater concentrations within aquatic organisms than in the surrounding water (Connell 1990; Nagel and Loskill 1991). Such enrichment or bioaccumulation can occur via nutritional routes (biomagnification), or via cell and body surfaces (bioconcentration). For water-breathing organisms, bioconcentration is the dominant uptake pathway (Connell 1998). The enrichment of hydrophobic materials in organisms, is due to an increase in entropy caused by the dissolution of hydrophobic chemicals in the lipid phase of the organism. In aqueous solutions, hydrophobic compounds have an organized hull of water molecules. On contact of the compound with the lipid phase of the organism, this hull collapses and the compound is dissolved within the lipid phase. With the collapse of the hydrate hull, there is an increase in entropy, which is the driving force for the enrichment of hydrophobic substances in organisms.

From an ecochemical point of view, substances which are bioconcentrated within organisms are considered dangerous, even when there is no acute toxic effect (Böhling and Loskill 1991). Even concentrations of aqueous pollutants which are shown to be non-toxic in acute or chronic toxicity tests, are subject to accumulation over a longer time period and can then be detrimental to the organisms and their consumers (Borgmann et al. 1990). Most bioconcentration and toxicity studies use artificial media, usually diluted solutions of inorganic salts in pure water. However, this approach does not take into account that natural waters always contain a multitude of dissolved and colloidal organic compounds (Steinberg and Münster 1985; Thurman 1985; Wetzel 2001). Such organic compounds have already been shown in the late 1960's, to be capable of binding organic contaminants (Ogner and Schnitzer 1970; Wershaw et al. 1969). Wershaw et al. (1969) point out that organic pollutants can be bound not only to particles, but also to dissolved organic material. In the late 1970's and early 1980's, various groups begin to study this phenomenon (Carter and Suffet 1982; Chiou et al. 1986; Gjessing and Berglind 1981; Hassett and Anderson 1979; Landrum et al. 1984; Means and Wijayratne 1982), and Gjessing and Berglind (1981) put forward the suggestion that the absorption of PAH on HS can influence the bioavailability of PAH. This suggestion is confirmed through a series of studies in the 1980's and 1990's.

It is now accepted that the molecular size of xenobiotics is increased through their association with HS. In this way, the permeability through membranes and the bioavailability of the chemicals is reduced, resulting in reduced bioconcentrations (Landrum et al. 1987; McCarthy 1989; Steinberg et al. 1993). This phenomenon can be shown for many chemicals. Only freely dissolved pollutants can pass through membranes and, thus the absorption of a toxicant onto DOC leads directly to a reduced bioavailability (Suffet et al. 1994). This applies to both organic chemicals (Twiss et al. 1999) and metals (Kim et al. 1999).

An extensive literature study on more or less lipophilic organics shows that in most studies, the presence of typical concentrations of DOC leads to a significant decrease in bioconcentrations (Haitzer et al. 1998, Figs. 7.4–7.8). The decrease in the bioconcentration in aquatic organisms per mg/L DOC, is more pronounced in the lower than in the higher DOC concentration range. For various PAH and DHAA, synthetic pyrethroids, superlipophilic substances, surface-active substances, and tributyl tin, the changes in bioconcentrations in the presence of dissolved HS are shown in Figs. 7.4–7.8. In this regard, PAH are the most commonly studied substances, and within this group most studies employ the carcinogenic B*a*P, shown separately in Fig. 7.5.

In the presence of DOC, the bioconcentration (BCF) is reduced by 2–98% compared to the DOC-free control (Figs. 7.4–7.8). In general, the decrease in bioconcentration is stronger with the same DOC, the more hydrophobic the compound is (MC, BaP: 50–98% decrease); compared to less hydrophobic PAH (NPH, ANTH: 2–48% decrease). The results further show that DOC of different origins, and therefore different qualities, leads to considerable differences in the BCFs of various DOC, with the same DOC concentration. The key question now is which structural characteristics of HS are responsible for the observed differences in BCF values. In the following section, recent studies on the influence of quantity and quality of HS are described, which can answer the above question.

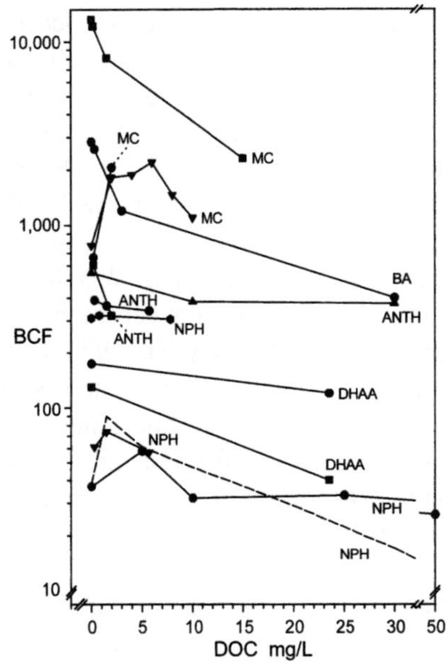

Fig. 7.4. Changes in bioconcentration factors (BCF) (on fresh weight basis) of polycyclic aromatic hydrocarbons and resin acid (dehydroabietic acid) in the presence of increasing DOC concentrations (from Haitzer et al. 1998, with kind permission of Elsevier Science). ANTH = anthracene, BA = benzanthracene, DHAA = resin acid, dehydroabietic acid, a pulp mill effluent constituent, with properties similar to those of polycyclic aromatic hydrocarbons, MC = methylcholanthrene, NPH = naphthalene

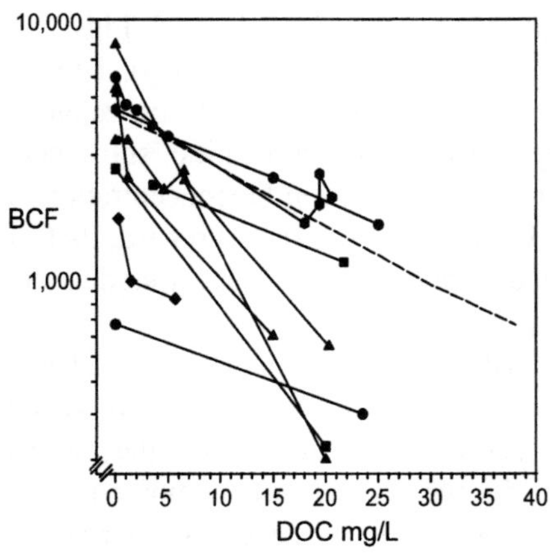

Fig. 7.5. Changes in bioconcentration factors (BCF) (on fresh weight basis) of benzo[*a*]pyrene (B*a*P) in the presence of increasing DOC concentrations (from Haitzer et al. 1998, with kind permission of Elsevier Science)

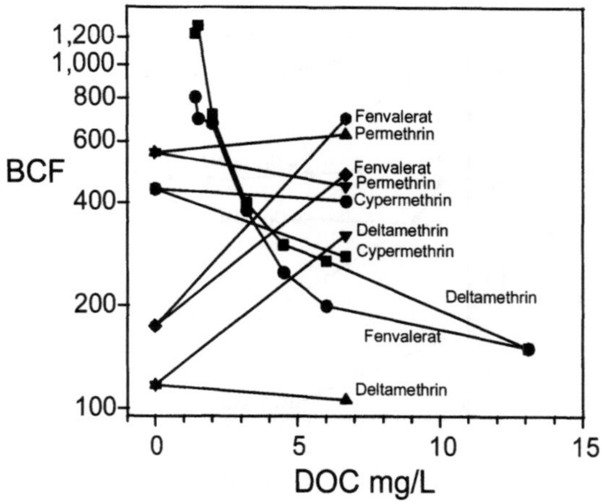

Fig. 7.6. Changes in bioconcentration factors (on fresh weight basis) of synthetic pyrethroids in the presence of increasing DOC concentrations (from Haitzer et al. 1998, with kind permission of Elsevier Science)

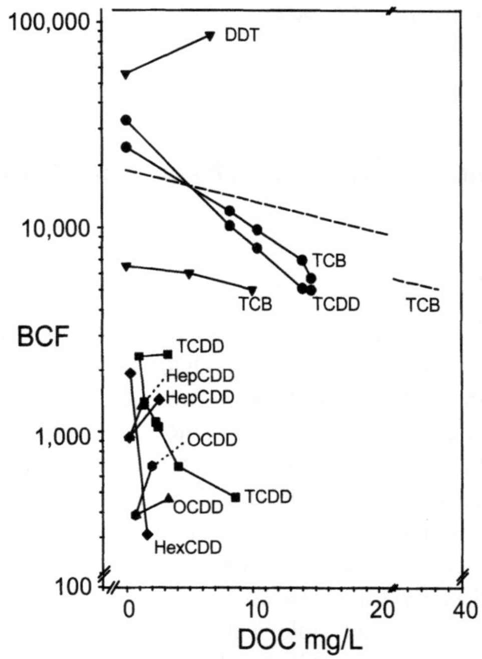

Fig. 7.7. Changes in bioconcentration factors (on fresh weight basis) of superlipophilic chlorinated hydrocarbons in the presence of increasing DOC concentrations (from Haitzer et al. 1998, with kind permission of Elsevier Science). DDT = dichlorodiphenyltrichloroethane, a (banned) contact insecticide, HepCDD = 1,2,3,4,6,7,8-heptachlorodibenzo-*p*-dioxin, HexCDD = 1,2,3,4,7,8-hexachlorodibenzo-*p*-dioxin, OCDD = 1,2,3,4,6,7,8,9-octachlorodibenzo-p-dioxin, TCB = 3,3',4,4'-tetrachlorobiphenyl, TCDD = 2,3,7,8-tetrachlorodibenzo-*p*-dioxin

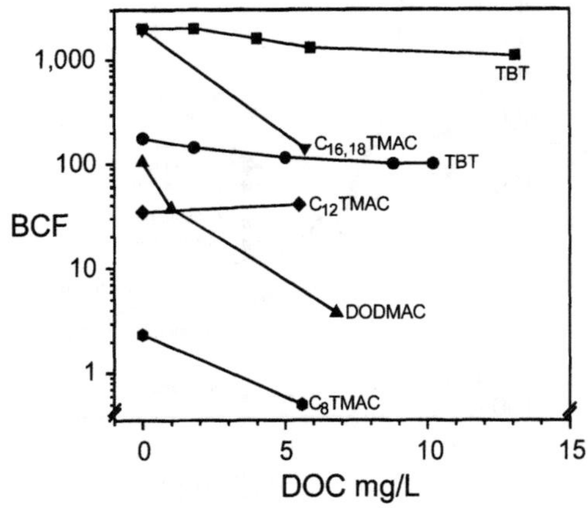

Fig. 7.8. Changes in bioconcentration factors (BCF) (on fresh weight basis) of surfactants and tributyl tin in the presence of increasing DOC concentrations (from Haitzer et al. 1998, with kind permission of Elsevier Science). C_8TMAC = octyl-trimethyl ammonium chloride, C_{12}TMAC = dodecyl-trimethyl ammonium chloride, $C_{16,18}$TMAC = monotallow-trimethyl ammonium chloride, DODMAC = dioctadecyl-trimethyl ammonium chloride, TBT = tributyl tin

7.2.1 Influence of Quantity and Quality of Humic Substances[1]

7.2.1.1 Quantity

The effect of DOC on the bioconcentration of hydrophobic contaminants increases with increasing concentration of DOC, that is to say bioconcentration factors become lower when more DOC is present. The quantitative relationship between concentration of DOC and BCF is not linear, however. At low concentrations of DOC, the decrease of BCF per mg/L DOC is much more pronounced than at higher concentrations of DOC (Fig. 7.9). This results in a dose (DOC concentration) – response (BCF) curve, similar to a logarithmic decrease curve (Day 1991; Kukkonen et al. 1989; Landrum et al. 1985).

To accurately describe this dose-response curve, a simple equation, based on the knowledge of the relevant biological and physicochemical processes, is derived (Haitzer et al. 1999a,b; McCarthy et al. 1994). This equation includes, as a precondition, the assumption that contaminants that are bound to DOC are not bioavailable, thus the bioconcentration of a contaminant is proportional to its freely dissolved concentration in water (Looser et al. 2000; McCarthy et al. 1994; Suffet et al. 1994). Therefore, the BCF in the presence of DOC can be expressed

[1] The following is based on Haitzer et al. (2002) with kind permission of Wiley-VCH.

as a function of the BCF in pure water (BCF_0), and the fraction of freely dissolved contaminant (f_{free})

$$BCF_{DOC} = BCF_0 f_{free} \tag{7.1}$$

$$f_{free} = \frac{1}{1 + K_{DOC}[DOC]}, \tag{7.2}$$

where K_{DOC} is the organic C based partition coefficient (in units of L/kg DOC) describing the distribution of the contaminant between DOC and water and (DOC) is the concentration of DOC measured in kg/L DOC (for a more detailed derivation of Eq. 7.2 see Haitzer et al. 1999b). These two equations (Eqs. 7.1, 7.2) can be combined to give an expression that describes the bioconcentration of a hydrophobic contaminant in the presence of DOC, as a function of the DOC concentration (Eq. 7.3)

$$BCF_{DOC} = BCF_0 \frac{1}{1 + K_{DOC}[DOC]}. \tag{7.3}$$

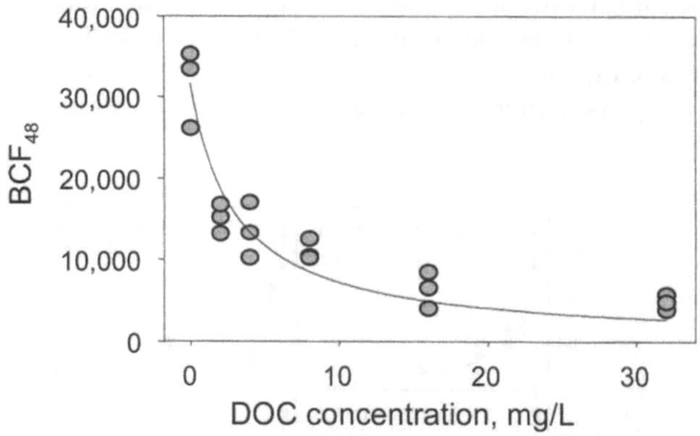

Fig. 7.9. Effect of different concentrations of a peat bog fulvic acid (HO13 FA) on the bioconcentration in 48 hours (BCF_{48}) of benzo[*a*]pyrene in the nematode *C. elegans* (from Haitzer et al. 2002, with kind permission of Wiley-VCH)

This simple model is used for fitting non-linear curves to experimental data that are obtained by measuring the bioconcentration of an organic contaminant in the presence of varying levels of DOC. Figs. 7.10 and 7.11 show that the effects of DOC from different sources on the bioconcentration of pyrene and B*a*P can be described reasonably well by this model (Haitzer et al. 1999a,b). Application of the model to data from a number of studies (Fent and Looser 1995; Kukkonen et al. 1989, 1990; McCarthy et al. 1985) give correlation coefficients between 0.85 and 1.00, suggesting that Eq. (7.3) is a universally valid model for characterizing the quantitative relationship between DOC concentration and BCF_{DOC}.

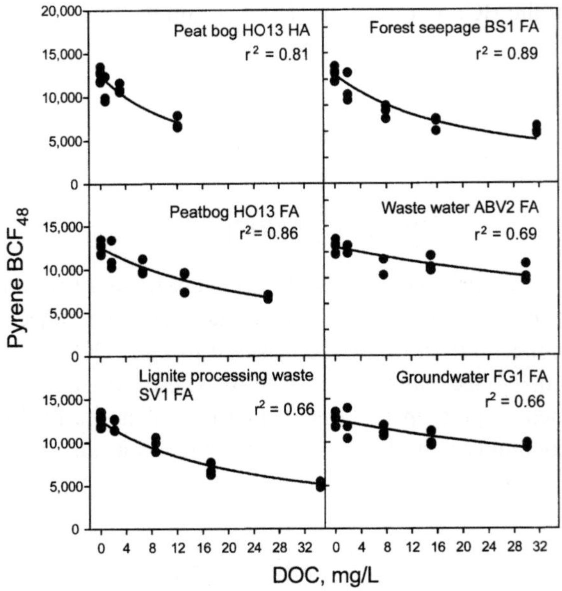

Fig. 7.10. Relationship between concentration of HA and FA from various sources and the bioconcentration (BCF) of pyrene in the nematode, *C. elegans*. BCF_{48} = bioconcentration factor after 48 h of exposure. r^2 = correlation coefficient for the curves fitted to the data using Eq. 7.3 as regression (from Haitzer et al. 2002, with kind permission of Wiley-VCH)

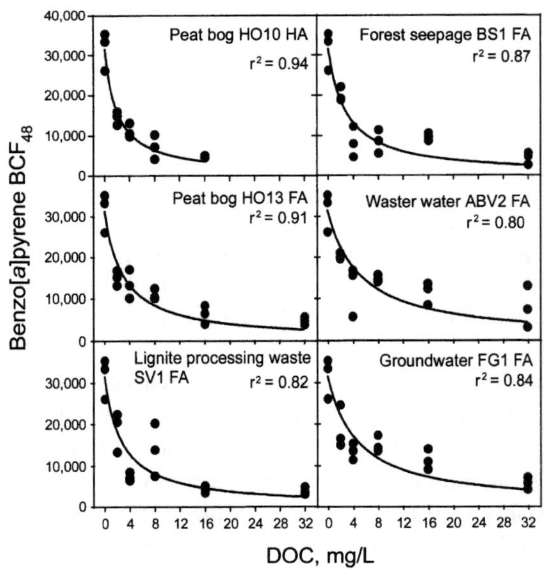

Fig. 7.11. Relationship between concentration of HA and FA from various sources and the bioconcentration (BCF) of benzo[*a*]pyrene in the nematode, *C. elegans* (from Haitzer et al. 2002, with kind permission of Wiley-VCH)

The regression procedure employed reduces each data set to a single sorption coefficient, by which it is possible to compare the capacity of various aspects of HS quality to decrease the bioconcentration of contaminants. The various origins of the HS are shown only as K_{DOC}, (this is acceptable, if the HS are isolated following the same protocol). The results are shown in Fig. 7.12.

It is obvious that the type of DOC can very clearly influence the bioconcentrations of organic contaminants. The effect of a high bonding capacity of the dissolved HS for benzo[*a*]pyrene (high K_{DOC} values) is similar to that of a high DOC concentration. The bioconcentration factor is approaching zero. If the dissolved HS used has, however, a low binding capacity (log K_{DOC} around 4), the BCF climbs about 25,000 fold (right edge of graph in Fig. 7.12) despite the same DOC concentration. It will now be elucidated which chemical structures in the HS may be responsible for their hydrophobic character and hence for the bonding of xenobiotics.

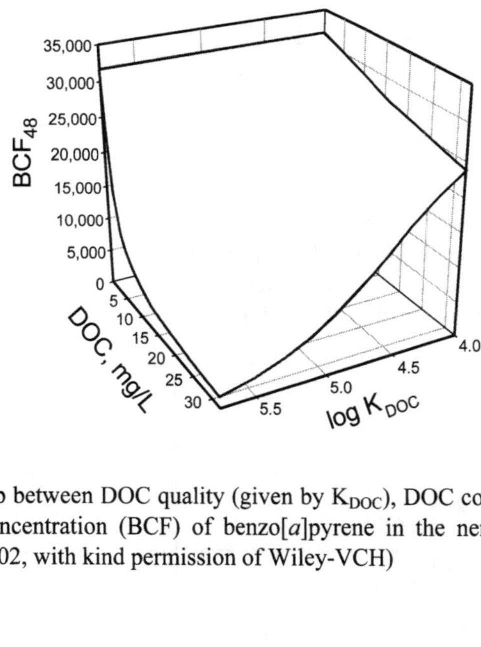

Fig. 7.12. Relationship between DOC quality (given by K_{DOC}), DOC concentrations (mg/L DOC), and the bioconcentration (BCF) of benzo[*a*]pyrene in the nematode *C. elegans* (from Haitzer et al. 2002, with kind permission of Wiley-VCH)

7.2.1.2 DOC Quality

Differences in the origin of DOC can clearly influence the effect of DOC on the bioconcentration of organic contaminants. This is illustrated by Fig. 7.13, which shows that comparable concentrations of DOC from different origins can have quite different effects on the bioconcentration of a hydrophobic contaminant.

Landrum et al. (1987) describe the effect of DOC from different sediment interstitial waters on the bioconcentration of B*a*P. By comparing the bioconcentration of B*a*P in the presence of DOC, to B*a*P bioconcentration in control treatments without DOC, Landrum et al. (1987) calculate biologically determined K_{DOC}s which are used as concentration-independent measures for the effect of DOC on the bioconcentration of B*a*P. The authors find that K_{DOC}s from the sediment interstitial waters vary by more than three orders of magnitude, depending on the sampling location of the waters. In a similar experiment, using B*a*P and DOC isolated from different humic surface waters, biologically determined K_{DOC}s range over more than one order of magnitude (Haitzer et al. 1999a,b). In agreement with these results, studies using physicochemical methods, such as equilibrium dialysis or reverse-phase separation, also reveal K_{DOC}s for B*a*P varying by one to two orders of magnitude, depending on the origin of DOC (Alberts et al. 1994; de Paolis and Kukkonen 1997; Kukkonen and Oikari 1991; McCarthy et al. 1989; Morehead et al. 1986).

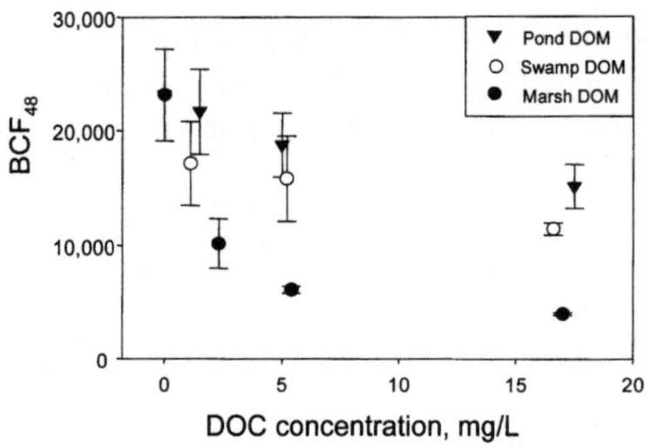

Fig. 7.13. Comparable concentrations of NOM from different origins caused different decreases in the bioconcentration of benzo[*a*]pyrene in the nematode, *C. elegans*. BCF_{48} = bioconcentration factor after 48 h of exposure, ± standard deviation (from Haitzer et al. 2002, with kind permission of Wiley-VCH)

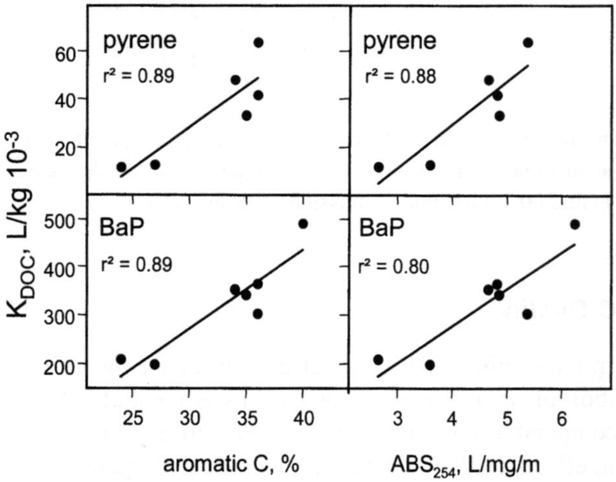

Fig. 7.14. Relationships between the content of aromatic carbon (determined by ^{13}C NMR spectroscopy) or the specific absorption at 254 nm of different HS/NOM isolates and K_{DOC} for pyrene and BaP (from Haitzer et al. 2002, with kind permission of Wiley-VCH)

To investigate if these variations in affinity of DOC for hydrophobic contaminants can be attributed to structural differences between the various organic materials, further studies relate the observed K_{DOC}s to certain DOC quality parameters. For non-functionalized hydrophobic contaminants (such as PAHs or PCB), a significant fraction of the variation in K_{DOC} can be attributed to variations in parameters describing the aromaticity of DOC, such as the content of aromatic C (determined by NMR spectroscopy) or the specific UV absorption (Akkanen et al. 2001; Chin et al. 1997; Gauthier et al. 1987; Haitzer et al. 1999c; Kukkonen and Oikari 1991; McCarthy et al. 1989; Perminova et al. 1999; Uhle et al. 1999). Examples of such relationships for the interactions of pyrene and B*a*P with various HA and FA are given in Fig. 7.14. In contrast to these findings, Akkanen et al. (2001) did not find that the bioconcentration of pyrene in *Daphnia magna* was significantly affected in the presence of DOC from 13 river waters, although equilibrium dialysis showed binding of pyrene to DOC. The underlying mechanism however, remains obscure. The results with the more hydrophobic B*a*P agree well with the results in Fig. 7.14.

Mathematically, these data can be expressed in the form of a linear relationship between K_{DOC} and a certain DOC quality parameter Q (for example the content of aromatic C or specific UV absorption) (Eq. 7.4)

$$K_{DOC} = a + bQ, \tag{7.4}$$

where a and b are empirically determined constants. The linear correlations between single parameters describing the aromaticity of DOC and K_{DOC} found in these studies may even be improved by using multiple regressions which additionally include the polarity of DOC (expressed as the H/O elemental ratio) (Georgi 1998).

Such empirical relationships (Eq. 7.4) can then be combined with Eq. 7.2, so that the fraction of freely dissolved contaminant can be estimated from a DOC quality parameter Q and the concentration of DOC (Eq. 7.5)

$$f_{free} = \frac{1}{1+(a+bQ)[DOC]}. \tag{7.5}$$

By substituting Eq. 7.5 into Eq. 7.1, the bioconcentration of a hydrophobic contaminant in the presence of DOC from a certain source can then be estimated from the bioconcentration in the DOC-free control, a DOC quality parameter, and the concentration of DOC (Haitzer et al. 1999c) (Eq. 7.6)

$$BCF_{DOC} = \frac{1}{1+(a+bQ)[DOC]}. \tag{7.6}$$

The ability of DOC to decrease the bioconcentration of an organic contaminant clearly depends on the type of contaminant. The more hydrophobic the contaminant, the stronger is the decrease of its bioconcentration by DOC (Kukkonen and Oikari 1991; Landrum et al. 1987; McCarthy et al. 1985; Steinberg et al. 2000). Significant effects of DOC on the bioconcentration of organic contaminants are only observed for contaminants with *n*-octanol/water partition coefficients greater than 10^4 (McCarthy 1989). The reason for this phenomenon is that more hydrophobic contaminants generally have a higher affinity for associating with DOC, so that larger fractions of these contaminants are bound by DOC, and are thus rendered unavailable for uptake by biota.

Quantitatively, this means that the DOC-water partition coefficient K_{DOC} of a contaminant can be related to its *n*-octanol-water partition coefficient (K_{OW}) (Eq. 7.7):

$$\log K_{DOC} = c + d \log K_{OW}, \tag{7.7}$$

where c and d are empirically determined constants. Such relationships track the behavior of compounds within one class of contaminants (PAHs) quite well (Landrum et al. 1984; McCarthy and Jimenez 1985; Rav-Acha and Rebhun 1992). However, the constants, especially d, can vary if different classes of contaminants are considered (Georgi 1998).

Equation 7.7 can be substituted into Eq. 7.3, so that the bioconcentration of different contaminants in the presence of a certain type of DOC can be expressed as a function of K_{OW} (Eq. 7.8):

$$BCF_{DOC} = BCF_0 \frac{1}{1 + (c + dK_{OW})[DOC]}. \tag{7.8}$$

It has to be taken into account, however, that this type of relationship has to be evaluated separately for each type of DOC, because differences in the quality of DOC are not considered in the model. In a recent study, Akkanen et al. (2001) show that best predictions for adsorption of hydrophobic chemicals on natural DOC from 13 river waters and one humic lake are obtained with compounds of higher hydrophobicity. The bioavailability of atrazine – with relatively low hydrophobicity – is not affected by DOC. The bioavailability of B*a*P, however, is significantly affected by both the quality and quantity of DOC, as shown above. There are several indications that the decreases in BCF lead also to reduced toxicity of xenobiotics (Perminova et al. 1999; Steinberg et al. 2000). This decrease of acute toxicity of organic chemicals in the presence of HS or NOM is referred to in Chapter 7.4.1.

7.2.2 Kinetic Effects on Bioavailability

The formation of xenobiotic-HS complexes is a kinetic process. In most studies, a standard incubation time is used, usually 48 or 96 hours, and it is assumed that complexes form their final configuration within this period. The sorption can usually be described as having two phase kinetics. A rapid sorption of free xenobiotics on the dissolved HS, is followed by a slower decrease of free xenobiotic concentration. The two phases are interpreted as a rapid adsorption on the dissolved HS surface, followed by gradual migration of the xenobiotic molecules to the interior of the dissolved HS molecules. According to published studies, the first phase is completed within minutes (Gauthier et al. 1986; McCarthy and Jimenez 1985;

Schlautman and Morgan 1993; Shaw et al. 2000). Some authors suggest that several effects such as modulation of bioconcentration and toxicity appear only after a longer contact time between the DOC and the xenobiotic chemical (Carlberg and Martinsen 1982; Johnsen 1987; Johnsen et al. 1987; Morehead et al. 1986).

To test the influence of isolate quality on bioconcentrations of xenobiotics, three NOM from wetlands and a pond in Ontario, Canada, are tested for their influence on decrease of bioconcentrations (Haitzer et al. 1999a). The bioconcentrations are determined after 1, 2, 4, 6, 8, and 12 days (Fig. 7.15).

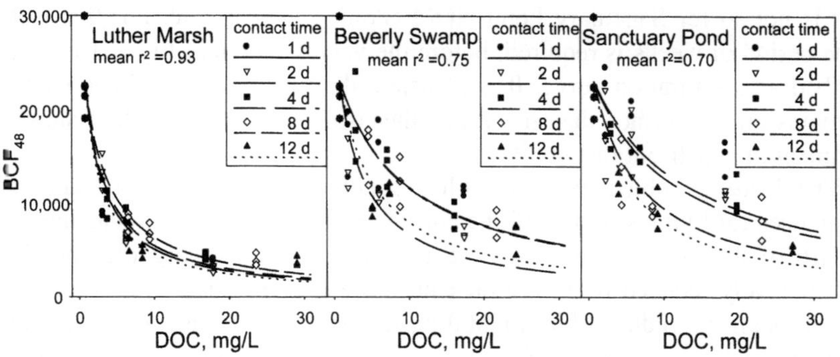

Fig. 7.15. Influence of NOM concentration from three sources and contact time on the bioconcentration of benzo[*a*]pyrene in the nematode *C. elegans* within 48 h (BCF_{48}) (from Haitzer et al. 1999a, with kind permission of the Society of Environmental Toxicology and Chemistry)

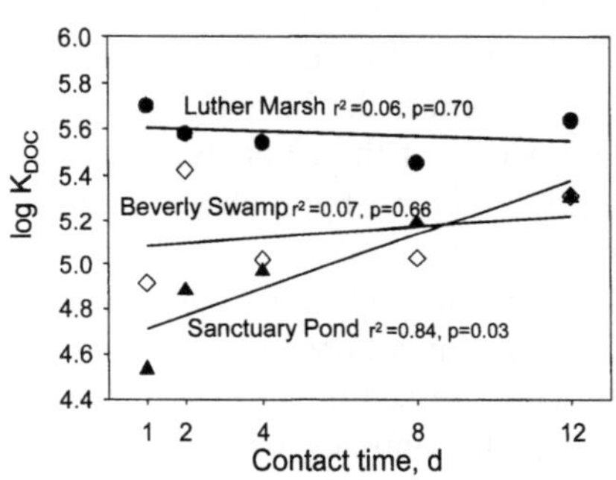

Fig. 7.16. Kinetic of bioconcentration of benzo[*a*]pyrene in the nematode *C. elegans* in the presence of 10 mg/L NOM from three sites in Ontario, Canada (from Haitzer et al. 1999a, with kind permission of the Society of Environmental Toxicology and Chemistry)

All NOM isolates show the expected concentration dependence for the decrease in bioconcentration for benzo[*a*]pyrene, but not all NOM isolates show a kinetic effect on xenobiotic-dissolved HS complex formation. For the NOM from Luther Marsh, the complexation process is completed within the first hour and there is no subsequent further decrease in bioavailability. The other two isolates exhibit a clear kinetic effect. To visualize the kinetic effect more clearly, time dependence of bioconcentration of B*a*P for 10 mg/L DOC as log K_{DOC} values, as a function of the contact time is displayed (Fig. 7.16).

It is obvious that a kinetic effect only occurs with the isolate from Sanctuary Pond. With this isolate, 84% of the variation in log K_{DOC} can be explained solely by contact time.

If one employs the model of Wershaw (1986, 1989), in which HS form micelles with a hydrophobic interior, the interactions between NOM and chemicals can be interpreted as distribution reactions (Chiou et al. 1983, 1986). Accordingly, the intramolecular hydrophobic domains of both wetland isolates (Luther Marsh and Beverly Swamp), are immediately available (<1 h) for benzo[*a*]pyrene, but this is not so for Sanctuary Pond NOM (Point Pelee) isolate. For this isolate, a time period of weeks is required, before the values shown in Fig. 7.16 can be described as a saturation curve. It is plausible that this pond DOC has more polar domains on the surface than have the other isolates, which lead to a slow establishment of a distribution equilibrium.

The decrease in bioavailability of hydrophobic chemicals has far-reaching consequences for living systems. If one considers the bioconcentration as internal exposure of the target organ, then the toxicity potential of organic chemicals is automatically reduced if HS are present in the system. Shaw et al. (2000) report that through the reduced bioavailability, the ability to biodegrade xenobiotics is also changed. In a decomposition study with 2,4-dichlorophenol, for example, the mineralization rate generally increases with culture age due to reduced acute toxicity. This experiment can also be explained by the model that only freely dissolved xenobiotics are decomposable: this pool is bioavailable. The pool is rapidly degraded and refilled from the xenobiotic-HS complex. The delivery rate of the xenobiotic is clearly not the rate-limiting step of the reaction. The increase in decomposition rate can be explained by decreases in toxicity through aging of the xenobiotic-HS complexes. Hence, the bioactivity (including biodegradation) increases. Overall, the formation of the complexes is very rapid and is completed within an hour. The authors suggest hydrophobic interactions or hydrogen bonds as probable bonding mechanisms.

Although the results of Shaw et al. (2000) are mechanistically well explained, these mechanisms cannot explain all observed kinetic effects. For example, Steinberg et al. (1992) describe increased rather than decreased toxicity of 2,4-dichlorophenol, and some substituted anilines, to *Daphnia magna* with increasing pretest contact times during which HS and the chemicals can react (Chap. 7.5.2).

Kinetic effects also occur with metals. In the presence of HS it takes a particular period of time until an equilibrium is reached between the various chemical species of the tested metals. For example, Kim et al. (1999) describe that the toxicity of Cu to *Ceriodaphnia dubia* is determined by the concentration of free Cu alone. The toxicity of Cu decreases, the longer it is able to react with, and to bind to, DOC. In addition, the poisoning effect is dependent on the DOC concentration. Exposure to Cu in the presence of 7 mg/L DOC for 24 hours, has the same toxic effect as exposure in the presence of 10 mg/L DOC for one hour. There is a linear relationship between the number of available binding sites on the DOC, and the decrease of potential Cu toxicity. In accordance with expectations, particular binding sites are more frequent on HS than in non-fractionated DOC.

7.3. Changes in Toxicity of Selected Heavy Metals

From the extensive literature on changes in speciation, bioavailability, and toxicity potential of non-organyl forming heavy metals in the presence of HS, one concludes that there is predominantly a decrease in bioavailability, and in the linked acute toxicity. For example:

- a variety of metals tested by single compound exposure or as mixed compounds exposure to fish, particularly rainbow trout, fathead minnows, and zebra fish (Bury et al. 1999a,b; Erickson et al. 1996, 1998; Hollis et al. 1996, 1997; Janes and Playle 1995; Karen et al. 1999; Macdonald et al. in press; Meinelt et al. 1995, 2001; Playle 1998; Playle et al. 1993a,b; Richards et al. 1999, 2001; Rose-Janes and Playle 2000; Welsh et al. 1993; Wood et al. 1999),
- precious metals affecting water fleas *D. magna* and *C. dubia* (Karen et al. 1999; Kim et al. 1999; Ma et al. 1999)
- various metals affecting amphipods (Amyot et al. 1994; Borgman and Norwood 1999).

The basic assumption of metal toxicity is that only freely dissolved metal species, that is to say those not bound to HS, are bioavailable and therefore can have a toxic effect (FIAM, Morel and Hering 1993, Chap. 6). In many cases, the free metal form is the metal ion[2] (Driscoll et al. 1980; Morel and Hering 1993; Wood et al. 1999). The validity of this concept is shown for many metals, most recently for Cu (Kim et al. 1999; Ma et al. 1999).

Details of metal uptake by aquatic organisms are not very well known. Active transport probably occurs via an enzyme-substrate complex and passive transport occurs by ion exchange, adsorption and diffusion (Thurman 1985). The role of HS in metal uptake is little known. The studies of Kozuch and Pempkowiak (1996, 1997) and Pempkowiak et al. (1994) on the bioaccumulation of Cd in the soft body of the edible mussels (*Mytilus edulis* and *M. trossulus*), show that uptake is up to twice as great in the presence of HS. Low-molecular weight HS have a greater influence on Cd accumulation than do high-molecular weight HS. The Cd complexation capacity increases with the aromaticity of the HS.

The effect of dissolved HS on metal toxicity to aquatic organisms is apparently contradictory, since there are both mitigating and adverse effects. For example, Gjessing (1981) describes a decrease of Cd toxicity for salmon (*Salmo salar*) and the green alga *Selenastrum capricornutum* (*Ankistrodesmus bibraianus*) in the presence of dissolved HS. In contrast, Winner (1984) finds that Cd toxicity for *Daphnia* increases, but Cu toxicity decreases in the presence of dissolved HS. It is known that HS have a higher affinity for Cd than for Cu. The apparent contradic-

[2] As discussed in Chapter 5.4, this does not fully apply to mercury: with this metal, chemical forms such as HgS^0_{aq} or $Hg(SH)(OH)^0$ are more bioavailable than are ions.

tion with the results of Gjessing can be related to the use of different HS in the various studies. Furthermore, there can be kinetic differences, as will be subsequently described for the Cd-Ca-HS system and its effects on fish eggs. Additionally, ambiguous effects on aquatic biocoenoses and guilds appear to be common with HS in aquatic environments (Klug 2002; Klug and Cottingham 2001; Steinberg and Brüggemann 2001, Chap. 9).

7.3.1 Iron

The draining of wetlands, often leads to high iron concentrations in freshwaters. This can result in exposure of fish and other lower vertebrates, such as amphibians, to toxic iron concentrations. The physiological pathology of iron in such animals is little known in comparison to that for mammals. It is, however, known that dissolved HS often have toxicity mitigating effects. Dissolved HS have a marked ability to bind Fe(II), or under light to reduce Fe(III) and bind the resulting Fe(II).

Lappivaara et al. (1999) describe that in natural iron-rich waters with natural DOC concentrations of 45 mg/L (approximately 480 mg/L Pt, typical for many Finnish lakes, but much higher than in humic lakes in most other regions), there is neither a bioaccumulation of iron nor a physiological reaction to iron. In the absence of HS, iron accumulates both in the gills and the liver of fish. This leads to decreased glycogen phosphorylase and mixed function oxygenase (7-ethoxyresorufin O-deethylase) activities in liver cells, and to reduced concentrations of sodium and potassium in blood plasma. When the natural iron complexation capacity of the HS is exceeded, for example through loading with Fe-rich drainage water to moderately humic freshwaters, in addition to the decreased enzyme activities further stress symptoms may appear in fish such as increased lactate, cortisol and 17β-estradiol concentrations.

7.3.2 Zinc

Like nearly all heavy metals, Zn is toxic to animals at high concentrations. Water hardness (Ca and Mg content), pH value, hydroxides, and dissolved HS can clearly change metal toxicity. Complexes can form between the heavy metals and HS via ion substitution and chelation (Sigg and Stumm 1991). In such complexes, cations lose their toxicity.

Meinelt et al. (1995) determine the changes in Zn toxicity in the 144 h zebrafish embryo larva test in the presence of dissolved HS, and with contrasting Ca concentrations. Exposure to 20 mg/L HS served as control. As toxicity endpoints, the survival of embryos and malformation of larvae after each 24 hours of the study period are applied. HS alone have no significant effect on the survival of embryos or vitality of larvae in either hard (Ca rich) or soft (Ca low) water. In the absence of HS, Zn is less toxic in hard than in soft water. Ca and Mg are two potential antidotes to heavy metals (Bradley and Sprague 1985). In the absence of Ca and Mg, that is in soft water, the toxicity of Zn is clearly enhanced. In hard-water and in the presence of HS, Ca and Mg compete with Zn for the binding sites of the HS. If Ca and Mg are present in high concentrations, Zn is displaced from the HS binding sites, and hence, the toxicity of Zn increases. Fig. 7.17 shows that in hard tap water, 20 mg/L HS are not sufficient to compensate for the toxicity of even 4 mg/L Zn.

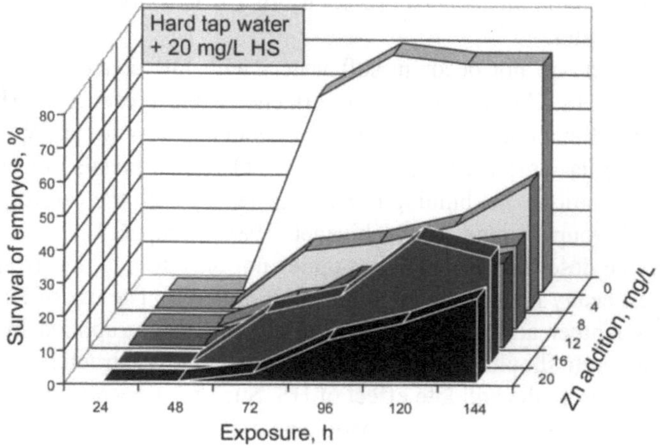

Fig. 7.17. In hard-water and in the presence of HS, Ca competes with Zn for binding sites, and hence, the toxicity of Zn is elevated, the survival of zebrafish embryos decreases (from Meinelt et al. 1995, with permission of ecomed Verlag)

In contrast, in soft water and with a DOC concentration of 2.5 mg/L, survival rate very clearly increases with 2.5 mg/L Zn present in the medium (Fig. 7.18).

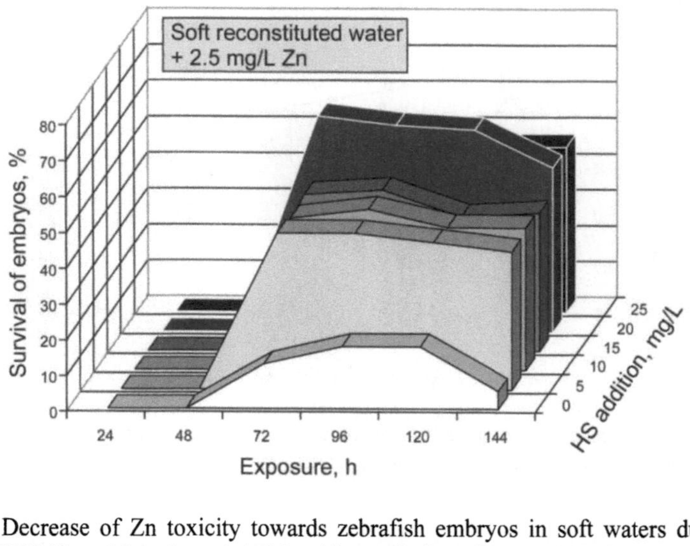

Fig. 7.18. Decrease of Zn toxicity towards zebrafish embryos in soft waters due to the presence of HS (from Meinelt et al. 1995, with permission of ecomed Verlag)

From various textbooks on aquaculture, it is known that water hardness and HS are the best protective agents against heavy metal toxicity (Gundersen et al. 1994). As shown, however, this protection mechanism tends to be eliminated, when Ca and Mg, as well as HS, are simultaneously present in the system. Although in hard-water, Zn is not toxic up to 20 mg/L, toxicity occurs at lower concentrations when HS are present. Comparable results are published, for example, by Winner (1986) who reports an increasing toxicity of Cd to *Daphnia* spp. when HS are added to the water.

These effects do not occur in soft waters with low concentrations of Ca and Mg. Without HS, Zn can exert toxic effects. In the presence of HS, the Zn is bound to the functional groups and is consequently less toxic. With bonding to HS, Zn loses its ionic character (Rashid 1971). When metals are present at elevated concentrations, the binding to HS is generally with the carboxyl and phenolic carboxyl groups (Linnik and Nabivanets 1984) (Chap. 6.3).

The above described mechanism applies not only for Zn, but is also generally valid for all heavy metals which do not form organyls. The use of HS as a natural antidote is neither adequately researched nor practically applied in aquaculture. Further research should be carried out on quantitative structure-effect relationships in relation to the antidote effect of HS. Such relationships have already been described for the bonding of hydrophobic neutral chemicals to HS (Chap. 7.1).

7.3.3 Cadmium

Toxic effects can only occur if Cd is bioaccumulated, and the target organs or tissues are internally exposed. This can be clearly shown for the uptake of Cd by eggs of the zebrafish (*Danio rerio*). After four hours of exposure, the uptake of Cd is significantly reduced in the presence of Luther Marsh NOM (Fig. 7.19).

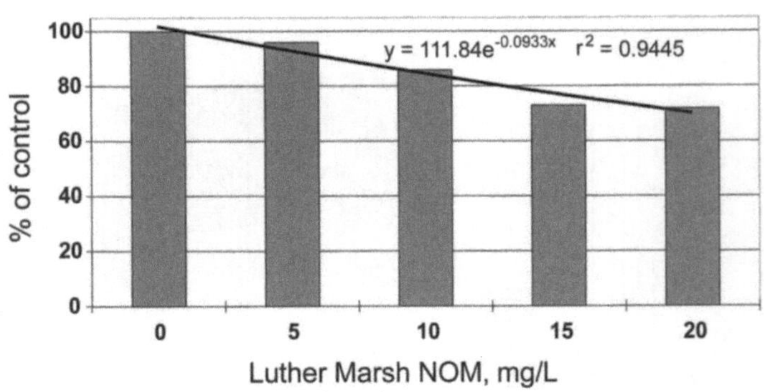

Fig. 7.19. Decrease of bioconcentration of Cd in zebrafish eggs in the presence of increasing concentrations of Luther Marsh NOM after 4-h exposure (from Burnison et al. 1999)

With a NOM concentration of 20 mg/L, only approximately 70% of the control levels of Cd is concentrated in the eggs. The HS-mediated decrease in bioconcentration can be described with a negative exponential function. Since an uptake via feeding in this experiment is excluded, uptake occurs only through bioaccumulation. Similar results are observed with other NOM isolates.

Interestingly, a large proportion (95%) of the Cd is retained by the chorion. The chorion maintains oxygen permeability of fish eggs and embryos, and also allows passage of organic molecules such as HS-like compounds (caffeic acid oxidation product, see Chap. 8) and cyanotoxins with a molecular mass up to 1.4 kDa (Wiegand et al. 1999). However, most Cd-NOM aggregates are obviously larger than cyanotoxins, hence a decrease in bioconcentration can occur.

The detrimental effect of Cd is caused by interactions with enzymes (Gill et al. 1991), and interference of water and ion balances (Hwang et al. 1995). Various previous studies indicate positive effects of HS on the survival of fish through a decrease in acute Cd toxicity effects. These studies include those of Gjessing (1981) on fish and algae, Hollis et al. (1996) on rainbow trout (*Oncorhynchus mykiss*) and Playle et al. (1993a,b) on the decrease of Cd binding in fish gills. With a simultaneous exposure of young rainbow trout to Cd and Cu, DOC reduces some acute toxic effects, particularly of Cu since only the Cu is excluded from the gills due to the greater bond strength of the Cu-DOC complexes. In contrast, Cd is found in higher concentrations in the gills. It is known that this metal interferes with ion regulation in the gills, particularly Ca regulation. Playle et al. (1993a,b) suggest that Cd binds to the high affinity Ca transport proteins, thus blocking Ca uptake and probably causing hypocalcaemia in the fish. NOM can therefore protect against acute respiratory and ion-regulatory effects of metals, but not against the particular chronic ion-regulatory effects of Cd (Richards et al. 1999).

The protective effect of HS against metal toxicity is, however, not universal. Negative effects of metals on aquatic organisms are observed in the presence of HS:

- the toxicity of Cd to *Daphnia* is higher in the presence of HS (Winner 1984). A possible mechanism is the photolytic break-up of metal-HS complexes, and the consequent release of easily bioavailable metal ions. Parkinson et al. (2000) report on this mechanism;
- the mortality of *Gammarus* sp. increases with increasing HS concentration when pH and Ca concentration are constant (Paarlberg 1984). Gammarids require Ca for building their exoskeleton, and it is concluded that the HS bind Ca until it becomes limiting,
- thus Cd toxicity to *Daphnia* is higher in hard than in soft water (Penttinen et al. 1998). The antidote function of DOC is reduced in the presence of Ca ions.

In a follow-up study, Meinelt et al. (2001), attempt to model Cd effects on zebrafish in a heavy metal-Ca-organism system. Fig. 7.20 shows the survival of developing embryos and larvae with exposure to increasing concentrations of Cd. With increasing Cd concentrations, the lowest survival rates are found with low Ca and HS concentrations. With a low Ca concentration but high HS concentration, an increasing protection occurs up to a Cd concentration of 6.2 mg/L. When concentrations of both Ca and HS are high, there is a complete protection against Cd up to 6.2 mg/L. Yet, at 9.3 mg/L Cd, there is only partial protection.

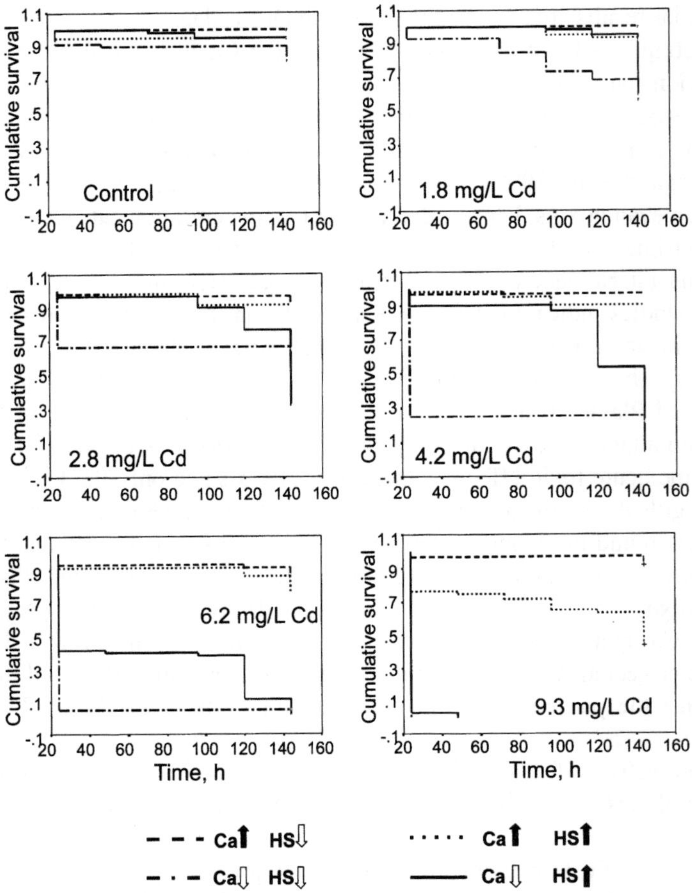

Fig. 7.20. Survival of zebrafish eggs at various concentrations of Ca and HS, if Cd is present (Kaplan-Meier plots) (from Meinelt et al. 2001, with kind permission of Elsevier Science). In the 9.3 mg/L Cd exposure, both graphs of low Ca content are almost identical. Low Ca concentration is 0.2, high is 2.0 mmol/L. Low HS concentration is 0 and high is 5 mg/L DOC. Low and high concentrations are indicated by downwards and upwards arrows, respectively

Ca and HS protect against toxic effects of Cd. Humic substances can also be deposited on the surfaces of eggs and embryos and alter membrane permeability, as observed by Campbell et al. (1997) as a detoxification mechanism. The protection, however, is more effective in the presence of both components. However, this effect declines with exposure to high Cd concentrations. It is notable that the survival of embryos at high Cd, Ca, and HS concentrations is significantly less than with high Cd, high Ca, and low HS concentrations (Fig. 7.20, bottom right).

These results lead to the following model for the effects of Cd (Fig. 7.21):

- when Ca and HS concentrations are both low, the binding sites on embryos are almost completely filled by Cd. This metal can have a highly toxic effect (Fig. 7.21a);
- in the case of low Ca, but high HS concentrations, some of the Cd is complexed by HS and consequently less Cd reaches the embryos. Toxicity is thus reduced and survival rate increased (Fig. 7.21b);
- with high Ca, but low HS concentrations, the Ca competes with Cd for embryo binding sites, thus reducing Cd binding and toxicity (Fig. 7.21c). This is comparable to the situation with Zn described above;
- with high Ca and high HS concentrations there are two possibilities (Fig. 7.21d and e): 1) there is a high proportion of Ca bound to HS, so that there is not enough to displace Cd from the embryos' surfaces and, hence reduce toxicity (Fig. 7.21d); 2) the competition for HS binding sites is won by Cd, consequently, the Cd toxicity is reduced and a higher proportion of the embryos survives (Fig. 7.21e).

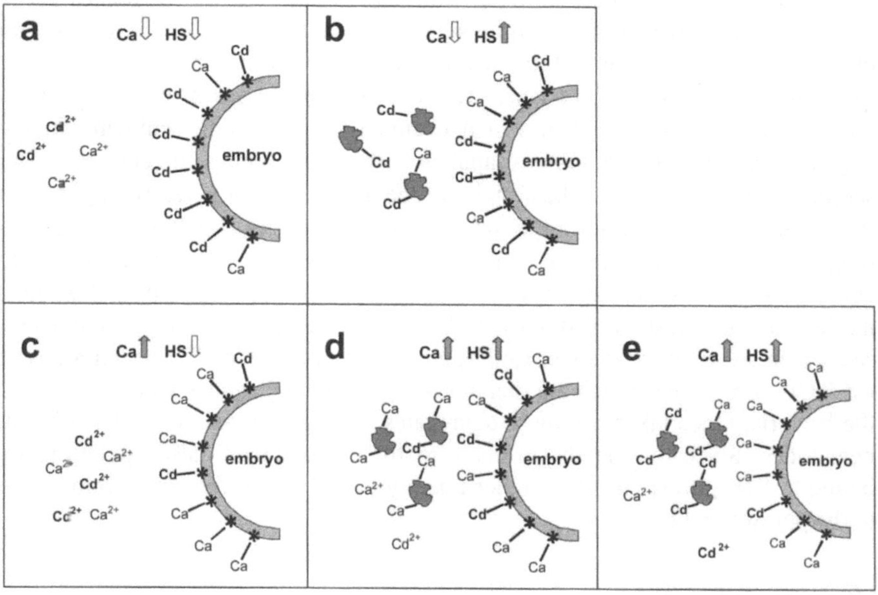

Fig. 7.21. Graphical structure of the binding sites model (from Meinelt et al. 2001, with kind permission of Elsevier Science). The grey flocs signify humic aggregates. For the sake of clarity, Cd is given in bold, Ca in normal font. Low Ca concentration is 0.2, high is 2.0 mmol/L. Low HS concentration is 0 and high is 5 mg/L DOC. Low and high concentrations are indicated by downwards and upwards arrows, respectively

7.3.3.1 Kinetic Effect

Under which conditions does each of the above mentioned cases occur? To answer this question, the Cd-gill model developed by Playle et al. (1993a,b) can be employed. The model predicts that with competition between embryo surfaces and HS for Cd, Cd will preferentially move to the HS, resulting in a toxicity decrease.

Negative results, as shown in Fig. 7.20, are not yet predicted by the model. Since the negative effect is dominant at the start of their experiment, Meinelt et al. (2001) conclude that there is a kinetic effect. HS bind excess Ca, which reaches an equilibrium over time if Cd is bound to the HS, and Ca is released into solution. In the meantime, the Cd ions can exert their toxic effects on the embryos. The Ca, which returns into solution, may be released too late to compensate for the toxic effect of Cd.

Such kinetic effects are relatively frequently reported (Kim et al. 1999; Ma et al. 1999 for Cu). Since there can be a kinetic effect with organic substances, this yields recommendations for use in cultured and ornamental fish practice. If organics are used therapeutically in waters containing HS, which are essentially all waters, such kinetic effects should be taken into account. Or, if HS are used as an antidote, a short-term undesirable contrary effect can occur.

Another indirect increase in toxicity through organic substances relates to dissolved and particulate chromophoric NOM. Höss et al. (1997, 2001b) describe the effects of Cd additions to natural sediments with NOM enrichment, on development of the nematode *C. elegans*. The Cd toxicity increases when the sulfide binding capacity is exceeded, which occurs with increasing sediment organic content. This increase in toxicity applies both to the total sediment and to the sediment porewater alone. The toxicity of the total sediment is clearly higher than that of the porewater. Relevant to the possible mechanism, is that the toxicity correlates well with porewater Cd concentration, independent of the porewater DOC concentration. This indicates that the potential complexation of Cd to DOC does not lead to a decrease in its bioavailability. Höss et al. (2001b) suggest that bacteria, which are food for the nematodes, act as vectors for Cd. The bacteria absorb Cd-DOC complexes, which are then taken up by the nematodes when they ingest the bacteria. Once taken up, these aggregates are digested and assimilated by the nematodes. Since this uptake mechanism for Cd and DOC also applies to Cd bound to POC, it explains the higher toxicity of the total sediment in comparison to the porewater alone.

7.3.4 Predicting Changes in Metal Toxicity

In a series of studies, Rick C. Playle and his colleagues at the Wilfrid Laurier University in Waterloo, Ontario, Canada, try to relate the decrease in metal accumulation on the gills of young fish and the subsequent toxicity to properties of the NOM isolates. Hence, they test various metals and metal mixtures, establish a speciation model for selected metals, and evaluate predictors for changes in metal behavior (Hollis et al. 1996, 1997; Janis and Playle 1995; Meinelt et al. 2001a,b; Playle 1998; Playle et al. 1993a,b; Richards et al. 1999, 2001; Rose-Janes and Playle 2000; Wood et al. 1999).

7.3.4.1 Silver and Lead

Silver is released by human activities into the aquatic environment via mining, industrial discharges, and sewage discharge. Ionic Ag interferes with sodium and chloride transport across the gills of freshwater fish, through the inhibition of the basolateral Na^+/K^+-ATPase responsible for providing the energy to transport sodium into the fish. The mechanism of acute Ag toxicity is ion loss, followed by decreased plasma volume, increased blood viscosity, and eventual cardiac failure. These ionoregulatory disruptions may be modified by water quality factors such as organic and inorganic complexing agents, acting through competition and complexation (Wood et al. 1996).

In a recent study, Rose-Janes and Playle (2000) determine the protective effects of DOC against the physiological effects of ionic Ag in ion-poor water. Rainbow trout exposed to $AgNO_3$ alone show large increases in plasma Ag, progressive losses of plasma Na and Cl, and have elevated concentrations of plasma glucose, with subsequent respiratory effects. In contrast, trout exposed to $AgNO_3$ plus DOC accumulate less Ag on their gills and in their plasma (Fig. 7.22), and show no adverse ionoregulatory or respiratory effects due to Ag. The study demonstrates the protective effects of natural complexing agents, through a decrease in the amount of ionic Ag available to bind at the gills. The authors assume that Ag^+ is bound first by the gills, then is transported actively across the gill epithelium into the plasma, possibly by incorporation into an active transport pump such as the basolateral Na^+/K^+-ATPase.

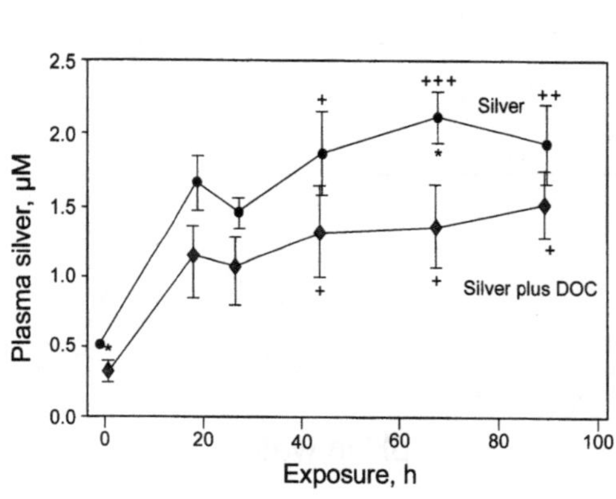

Fig. 7.22. Mean plasma silver concentration for rainbow trout exposed in ion-poor water to 0.1 µM $AgNO_3$ alone, and to 0.1 µM $AgNO_3$ plus 35 mg/L DOC. Data with standard deviation. + indicates significant differences from initial values within a treatment. * indicates means significantly different between the two exposures at a given sample time (from Rose-Janes and Playle 2000, with kind permission of Elsevier Science)

Pb is a common historical and contemporary metal contaminant throughout the world. Since Pb as Pb^{2+} acts as a Ca^{2+} analogue, and its uptake by gills of freshwater fish is presumably through high affinity Ca uptake mechanisms at the gills, the active uptake of Pb^{2+} at freshwater fish gills is therefore amenable to modeling using a biotic ligand model, described by Playle (1998). The behavior of Pb bound by DOC is illustrated in Fig. 7.23. As water pH increases, Pb complexation by DOC decreases as Pb carbonate complexes form, which frees sites on DOC for either Ca to bind or to remain unfilled ('-DOC'). The situation of high concentrations of Ca displacing some Pb from DOC can also be simulated in the Pb-gill binding model: Ca binds weakly to DOC compared to Pb ($\log_{Ca\text{-}OM}$=5.0 versus log $K_{Pb\text{-}OM}$=8.4), and any displaced Pb^{2+} going into solution is simultaneously kept from the gills by competition at the gill binding sites (log $K_{Ca\text{-}gillPb}$=4.0 versus log $K_{Pb\text{-}gillPb}$=6.0). The effect whereby high Ca concentrations slightly decrease the binding capacity of DOC for a metal by simultaneously competing for the binding sites, is demonstrated with Cd, DOC and *D. magna* by Penttinen et al. (1998). At least in these cases of a Ca analogue, Ca competition at the biotic ligand compensates for metal displaced from DOC by Ca.

Under acidic conditions, H^+ does not displace much Pb from DOC (in the model H^+ binds weakly to DOC compared to Pb). The Pb is only displaced from DOC at pH <2.5, with a pK of about 1.2 (Macdonald et al. in press). This result agrees with the Windermere Humic Aqueous Model (WHAM, Tipping 1994), where the pK for Pb and HA is 1.7, and the pK for Pb and FA is 0.9.

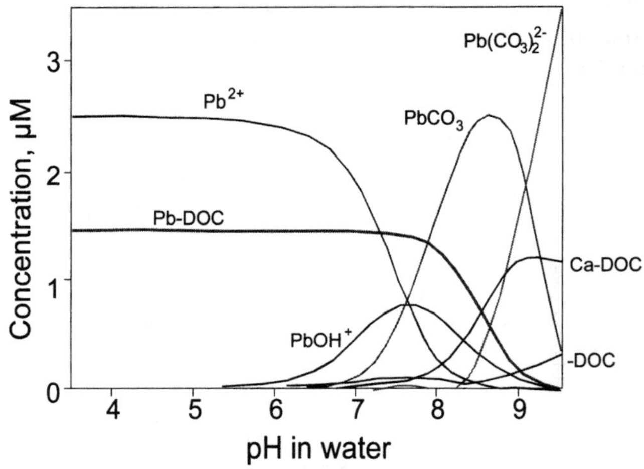

Fig. 7.23. Lead speciation as water pH increases from pH 3.5 to 9.5, as calculated by MINEQL+ and the Pb-gill binding model. Carbonate complexation at pH>7 decreases the concentration of Pb^{2+} in the water, eventually reducing the amount of Pb binding to DOC (-DOC) so that Ca can bind to the DOC (from Macdonald et al. in press, with kind permission of Elsevier Science)

Macdonald et al. (in press) assume that the gill ligand model can be extended to other organisms such as freshwater amphipods, in which water chemistry – especially Ca competition and DOC complexation – is shown to be important in modifying the accumulation of metals (Amyot et al. 1994; Borgman and Norwood 1999).

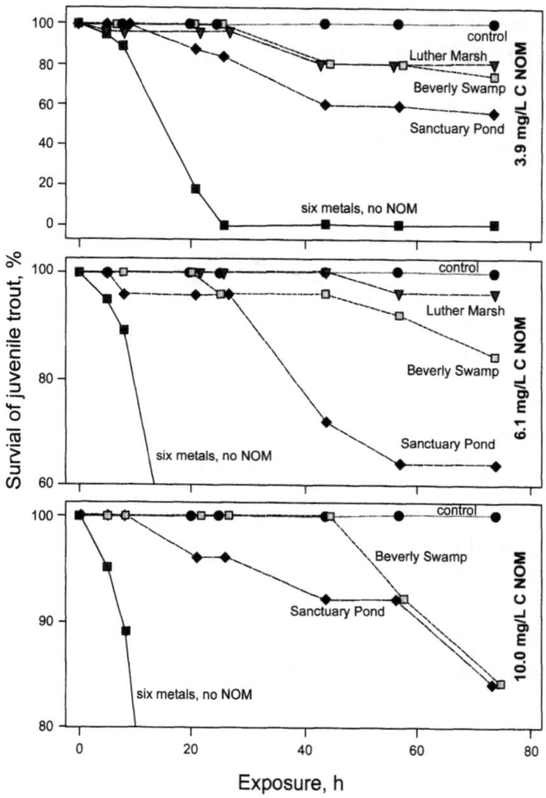

Fig. 7.24. Survival of juvenile trout exposed to six metals (Pb, Hg, Cd, Cu, Ag, Co) together plus 3.9, 6.1, and 10.0 mg/L NOM isolated from Luther Marsh, Beverly Swamp, and Sanctuary Pond. Control: metal free, soft water. Note the scale for survival rate has been changed for clarity (from Richards et al. 2001, with kind permission of NRC Research Press)

How does NOM quality affect the decrease of metal toxicity and metal binding to fish gills? In a recent study, Richards et al. (2001) address this question by applying three NOM isolated from three aquatic sources in southern Ontario, Canada, with contrasting chemical properties. Increasing concentration of each NOM increases trout survival, but the NOM having the most allochthonous properties, for instance highest specific absorbance (from Luther Marsh) increases fish survival most, while the NOM having the most autochthonous properties (from Sanctuary Pond, Point Pelee) increases fish survival least (Fig. 7.24). Specific absorbance is a substitute for aromaticity. That means that the aromaticity of the NOM isolates is involved in the decrease of metal toxicity, as it is in the decrease of xenobiotic concentration (Chap. 7.2), and toxicity (Chap. 7.4).

By testing several NOM isolates, Schultz (2002) address the question more specifically as to whether or not the aromaticity may be used as a predictor to estimate the decrease in metal toxicity towards juvenile rainbow trout. The study metal is again lead. Fig. 7.25 clearly shows that, with an $r^2 = 0.59$, approximately 60% of the decrease in toxicity (increase in LT50 values) can be explained by the specific absorption coefficient, a simple measure of aromaticity.

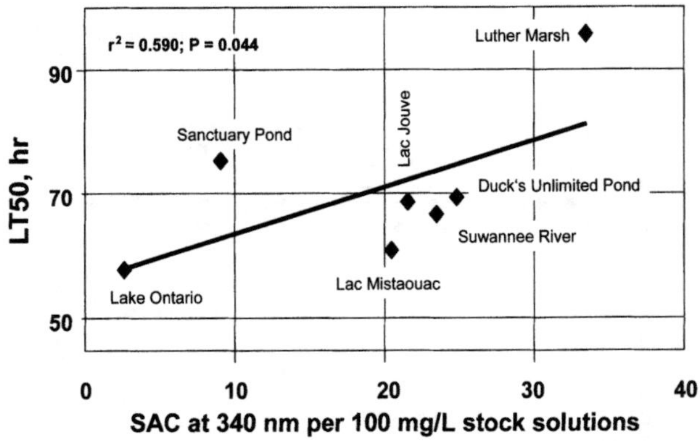

Fig. 7.25. Relationship between specific absorption coefficient (SAC) of various NOM and toxicity of Pb (indicated by the time to reach 50% fish mortality; LT50) of 2.5 μM Pb in the presence NOM (modified from Schultz 2002, with kind permission of the author)

A simple synoptic graphical summary of the interaction of HS and metals in an ecological context is shown in Fig. 7.26.

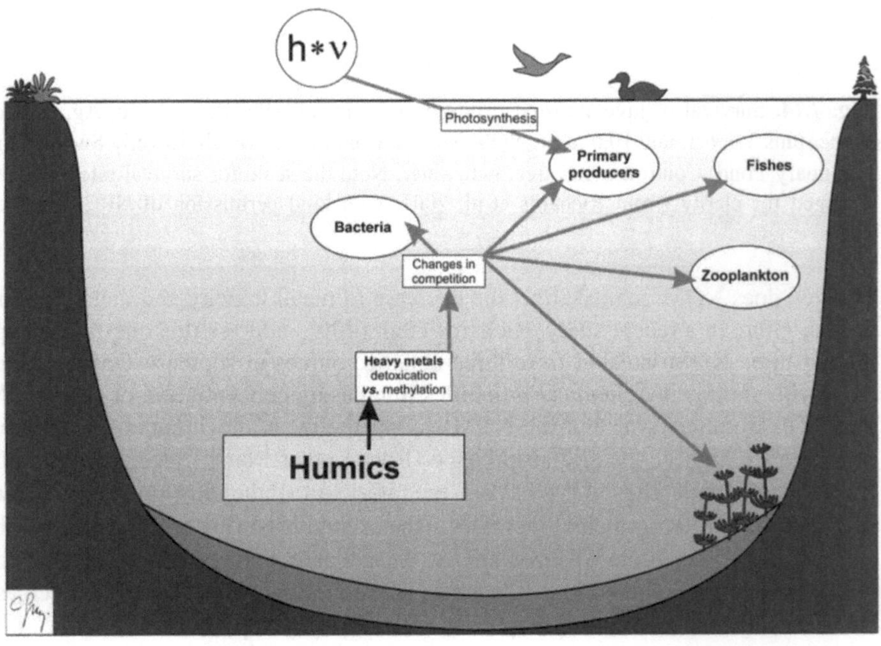

Fig. 7.26. Synopsis of the modulation of the toxicity of heavy metals by HS. Black arrow indicates direct effect, grey arrows indicate indirect effects. h∗ν is sunlight

7.4 Alteration of Xenobiotic Toxicity

There is information in the literature indicating that the presence of HS also alters the toxic effects of xenobiotics. The toxicity alteration also applies to situations where additional exposure is made more complex by UV radiation, or general radiation forms further potential toxins, or at least another modified environmental parameter. Toxicity alterations comprise either toxicity decreases [as exemplified by exposure of the water flea *D. magna* to organophosphate pesticide diazinon (Fig. 7.27), and selected PAH (Fig.7.28)], or toxicity increases (Chap. 7.5.2).

7.4.1 Humic Substances Mediated Decrease in Toxicity of Xenobiotics

While the mechanisms which lead to toxicity decrease can be largely explained, there are as yet only hypothetical explanations for the contrary situation of toxicity increase. There is still no certain theoretical background for this phenomenon as Kukkonen (1995) states; the evidence remains phenomenological.

7.4.1.1 Toxicity Decrease towards Daphnia magna

Phenomenological studies of decreasing toxicity in the presence of HS are frequent (Day 1991; Hodge et al. 1993; Kadlec and Benson 1995; Kukkonen 1995; Kukkonen and Oikari 1987, 1991; Lee et al. 1993; McCarthy 1989; Steinberg et al. 1992) and are exemplified with the water flea *D. magna*, exposed to diazinon, a herbicide (Fig. 7.27). The graph shows that with increasing HS concentrations, one has to apply increasing concentrations of diazinon to immobilize 50% of *D. magna*; the EC50 values are increasing.

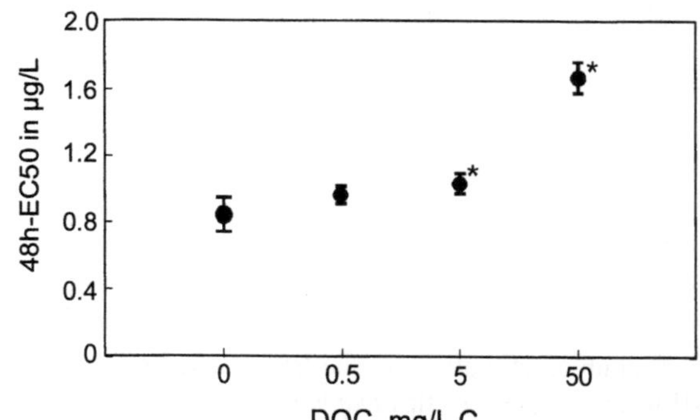

Fig. 7.27. Acute toxicity of diazinon towards *Daphnia magna* as a function of dissolved HS concentration. Note the decrease in toxicity. EC50 is the effective concentration which leads to an immobilization with 50% of *D. magna*. Data are means with standard deviation, * = significantly different from control (from Steinberg et al. 1993, with kind permission of Wiley-VCH)

Perminova et al. (1996) try to relate the decrease in toxicity to specific structures of the HS. They describe the effects of 19 various, chemically and spectroscopically well characterized HS on the toxic effects of PAH (pyrene, fluoranthene and anthracene) to the water flea *D. magna.* Grazing activity of the water fleas is employed as a sensitive measure of toxicity. The decrease in acute toxicity by HS (D) is calculated according to the following Eq. 7.9:

$$D = \left(1 - \frac{R_{HS} - R_{PAH+HS}}{R_{HS}} : \frac{R_0 - R_{PAH}}{R_0}\right) \cdot 100\%, \tag{7.9}$$

with R_0 = grazing activity of *D. magna* in the control without PAH and HS, R_{PAH} = grazing activity in the presence of PAH alone, R_{HS} = grazing activity in the presence of HS alone, R_{PAH+HS} = grazing activity in the presence of PAH and HS.

The toxic effects are correlated with the bonding of the PAH to the HS, in particular with the aromaticity of the HS (Chap. 7.1). Most (50–60%) of the variability of the decrease of toxicity of the PAH can be explained by the aromaticity of the HS (Fig. 7.28). The following simple mechanism is suggested: the interactions between HS, in particular their aromatic structures, with xenobiotics allows the formation of new associations, clearly increased in molecular size. Such associations are less bioavailable and, hence less bioconcentrated. The decreased bioconcentration means a reduced exposure of potential harmful substances for the target organs, so that less deleterious effects result.

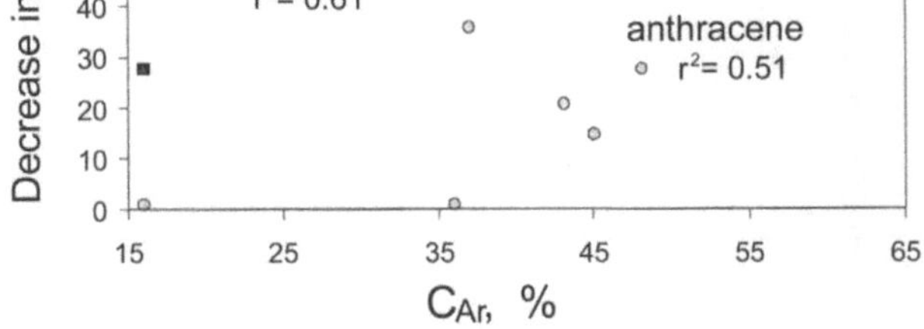

Fig. 7.28. Scatter diagram of aromaticity of the HS and their effect on reducing acute toxicity of three PAH towards the water flea *Daphnia magna.* D is decrease in acute toxicity in relation to the HS-free controls. ■ pyrene, △ fluoranthene, ● anthracene (from Steinberg et al. 2000, with kind permission of Wiley-VCH)

Perminova et al. (2001) plot the decrease in toxicity against the HS concentrations. The typical examples for pyrene, fluoranthene, and anthracene are given in Fig. 7.29. Comparably large magnitudes of decrease in toxicity are obtained for pyrene and fluoranthene, the lower decreases are observed for anthracene. This result can be related to the higher hydrophobicity of pyren and fluoranthene (log K_{OW} = 4.88 and 5.16, respectively), compared to the lower hydrophicity of anthracene (log K_{OW} = 4.45). This trend corresponds well with the decrease in bioconcentration of the PAH in the presence of HS.

The obtained structure–property relationship points out that humic materials rich in aromatic structures are the most efficient agents for reducing acute toxicity of PAH. Given that the governing mechanism of interaction between HS and PAH is hydrophobic binding, a similar action of HS on the other hydrophobic organic compounds can be expected (Perminova et al. 2001).

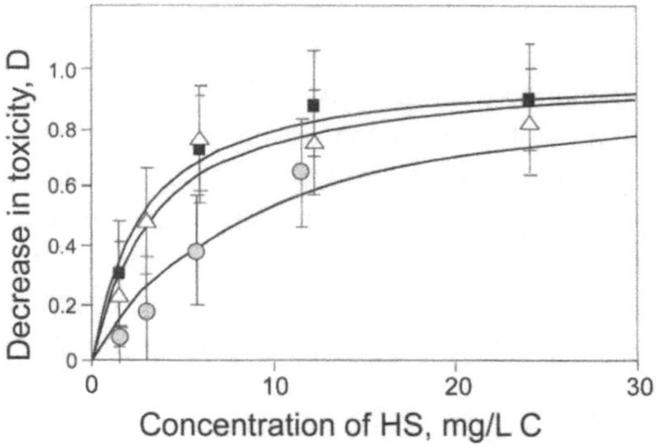

Fig. 7.29. Typical relationship of decrease in toxicity (D) versus concentration of HS for three model PAH: ■ pyrene, △ fluoranthene, ● anthracene. Data are means with standard deviation (from Perminova et al. 2001, with kind permission of the American Chemical Society)

7.4.2 Decrease in Xenobiotic Toxicity in the Presence of Dissolved HS and UV Radiation

The above-described situation is further complicated by UV radiation, which may result in further toxins, or at least altered toxicity parameters of the xenobiotics. Natural UV radiation in sunlight can, for example, increase the toxicity of PAH or PCB to aquatic animals (Holst and Giesy 1989; Oris et al. 1990) and plants (Gensemer et al. 1996; Greenberg et al. 1993). Mechanistic studies are much fewer in number than phenomenological descriptions. For instance, Oris et al. (1990) study the effect of dissolved HS on the light-induced toxicity of the PAH anthracene to the fathead minnow *Pimephales promelas* and the water flea *D. magna* in a laboratory system with simulated sunlight. It is known that in sunlight, in particular through the UV component, anthracene is photo-oxidized to more toxic products. The results in the presence of dissolved HS, show that this light-induced toxicity is primarily a phenomenon depending on the potential of a chemical to spread in the environment (fugacity, Mackay 1979). Without specifying the type of interaction or bonding between anthracene and dissolved HS, one can put forward the mechanism that in the presence of HS, anthracene diffusion in the aqueous system is hindered. In addition, particularly active UV wavelengths in sunlight are weakened in their effect by HS. This toxicity reducing effect occurs with low, environmentally relevant quantities of dissolved HS (<3 mg/L DOC).

In a detailed study, Gensemer et al. (1999) use biomarkers, such as chlorophyll *a* fluorescence to study the time-development of photo-induced toxicity of anthracene to the duckweed *Lemna gibba*, as well as the toxicity mitigating effect of HS. The expression of toxicity takes place in a definite sequence: chlorophyll fluorescence parameters react within one hour as the first biomarker, while chlorophyll breakdown requires at least two days. The presence of HS reduces both fluorescence interruption and chlorophyll degradation, if the toxic effects on photosystem II cannot be completely eliminated. This means that the tested humic acid slows down, but does not prevent the uptake of anthracene or more likely its photooxidation products. However, since the study is carried out only with a poorly defined commercial humic acid, no particular structure of the humic acid can be named as responsible for this slowing-down effect. The establishment of a quantitative structure property relationship, predicting which structural feature of the HS can lead to which weakening effect, is therefore not possible.

7.5 Humic Substances Mediated Increases of Adverse Effects

7.5.1 Controlled Release and Humic Substances-Mediated Transport of Xenobiotics and Metals

From an ecosystem perspective, adverse effects on aquatic organisms may occur via controlled release of xenobiotic chemicals and toxic metals bound to HS in the water column, or in the sediments, as well as via increased loading of chemicals transported into aquatic ecosystems via HS with adsorbed toxicants.

7.5.1.1 Controlled Release

Xenobiotic chemicals adsorbed to HS comprise a long-term risk of ecosystem contamination. Several studies address the question, how do HS modify the breakthrough of xenobiotic chemicals such as pesticides in soil columns. Interestingly, Celis et al. (2002) report that montmorillonites (expanding-lattice clay minerals) have a better adsorption capacity than do HS. Fernandez-Perez et al. (2000) describe an contrasting effect of HS, when studying the impact of bentonite (a clay mineral, similar to montmorillonite) and HA added to an alginate as a matrix for controlled release of the herbicides diuron and atrazine. The release of the active ingredient from alginate-based controlled release granules in water is more affected by the addition of HA, than by the addition of bentonite. In fact, HA have up to almost one order of magnitude higher adsorption capacities than clay minerals, as evaluated with the insecticide imidacloprid (Gonzalez-Prades et al. 1999).

Although trace element uptake on, and release from, solid soil and sediment phases are fundamental controls on the migration of the elements in the environment, the controls are still incompletely understood. By a one-dimensional migration model including kinetic factors, Braithwaite et al. (1997) show with uranium that the overall effect of HS is to retard, rather than completely prevent, migration of this metal in a peaty and a calcareous soil. Calculated half-times for uranium release are in the range of 30–60 days in acid, and around 10 years in neutral conditions, indicating that one major release mechanism is ion exchange.

Assessing the risk from release of humic-bound metals and xenobiotics, it appears that pH is a major control for metal retardation and release (also see Chap. 5). Under acid and particularly acidifying conditions, a severe secondary contamination by Al and heavy metals from HS-bound pools has to be taken into account. With xenobiotic chemicals, the risk of a secondary contamination appears to be smaller than the direct contamination via direct input into the aquatic ecosystems.

7.5.1.2 Humic Substances-Mediated Transport of Xenobiotics and Metals

The transport of ionic and non-ionic pollutants may be strongly impacted by the movement of naturally occurring DOC (see Chap. 3). Understanding chemical transport through soils is a requisite for forecasting environmental exposure and performing risk analysis for aquatic organisms.

Kaiser et al. (2001a) stress that the chemical composition of DOM in forest floor seepage water changes seasonally, with compounds in winter and spring showing greater mobility, greater biodegradability, and less interaction with metals and organic pollutants than those released during summer and autumn. Thus, the impact of DOM on transport processes varies significantly throughout the year due to changes in its composition. In addition, McCarthy (2001) stresses the diverse nature of NOM in groundwater and porewater, and how changes in hydrophobicity and molecular size dictate preferential interactions with contaminants and soil matrices. Larger, more hydrophobic organic compounds have a greater affinity for binding to mineral surfaces than smaller have. Small, hydrophilic components of NOM are preferentially transported. In natural systems, a steady state is established between the NOM in solution and on the immobile surfaces. This minimizes the retention of the NOM because sorption sites have been saturated. Under these circumstances, contaminants capable of binding to the native NOM will experience enhanced transport.

Studying transuranic radionuclides from a dumping site in Oak Ridge, Tennessee, USA, McCarthy et al. (1998a,b) report that NOM can dominate the mobilization, transport, and fate of contaminants, and that NOM-contaminant complexes are not retained, even by highly reactive mineral phases known to strongly adsorb actinides in the absence of NOM (McCarthy et al. 1998b). In the presence of NOM, radionuclides that associated with complexation sites on the NOM are transported tens of thousands of times faster than would be predicted in the absence of NOM. A similar phenomenon is reported from wastewater infiltration sites (sewage farms) near Berlin, Germany, even under acidifying conditions (Hoffmann et al. 1998).

In a recent paper, McCarthy (2001) draws a more detailed picture with respect to different NOM fractions. NOM transport through the soil involves a process of competitive sorption and potential displacement of previously adsorbed NOM. Depending on the relative competitiveness of the NOM containing bound contaminants and the previously adsorbed NOM, the migration of the contaminant may be either enhanced or retarded. One might predict, for example, that PAH or PCB can be immobilized. This is because the large, hydrophobic NOM moieties, which have the greatest affinity for binding hydrophobic organic contaminants, are preferentially retained on the immobile phase (McCarthy 2001). If metals are bound preferentially to the low-molecular weight, hydrophilic NOM, metal contaminants can migrate with little retardation. This is certainly the case with the transuranic radionuclides referred to above.

The quality of NOM in determining the co-mobility of chemicals is also stressed by Vereecken et al. (2001) who describe and numerically simulate the behavior of acenaphthene and dibenzofuran. The column studies show that the nature of NOM has a considerable influence on the mobility of acenaphthene. In contrast, no leaching of dibenzofuran is found. Vereecken et al. (2001) show that, in addition to the finding of McCarthy (2001), the transport of the bulk mass of hydrophobic organic contaminants is dominated by strong sorption between the contaminant and the soil matrix, resulting in slow migration. The sorption of the hydrophobic organic contaminant to the soil matrix is found to be reduced in the presence of NOM. The solubility of the contaminant increases linearly with NOM concentration.

Enfield et al. (2001) relate the co-transport of chemicals by DOC-mediated solubility enhancement to the partition coefficient. Once a partition coefficient is known, one can estimate the impact of the humic material on the transport of a chemical. West (1984) developed a correlation between the water solubility and naturally occurring HS-water partition coefficient as (Eq. 7.10):

$$\log K_{DOC} = -0.923 \log S(mg / L) + 3.294 \tag{7.10}$$

with S being water solubility of the hydrophobic chemical.

Enfield (1985) has estimated the relative enhancement that would be seen under normally occurring environmental conditions. Generally, the impact would be small, the most significant related to very hydrophobic conditions. Enfield et al. (2001) report that solubility enhancement of several orders of magnitude may occur, for instance 87,000 times for decane and 6,900 times for cyclodextrin. Because the authors consider the solubility enhancement of hydrophobic chemicals from the technical perspective of applying NOM macromolecules to remediation of contaminated sites, they do not consider the environmental fate of the co-transported chemicals in the aquatic environment. Thus, they do not discuss potential contamination of, and adverse effects on, aquatic organisms. This concern is an obvious gap in the environmental literature.

Increases in Bioconcentration of Hydrophobic Xenobiotics

On the organism level, the presence of dissolved HS can **increase** as well as **decrease** the bioconcentrations of hydrophobic xenobiotics, as shown in Figs. 7.4–7.8. Of 27 studies reviewed by Haitzer et al. (1998), 7 report an increase in bioconcentration. From these Figures and Table 7.1 it is clear that increases are only observed with DOC concentration below 10 mg/L. Although several hypotheses have been put forward and described by Haitzer (1998), none appears to be convincing and a recent experimental verification failed (Haitzer 2001). Very recently, Kukkonen (2002, pers. comm.) evolves an interesting hypothesis with relatively high plausibility: low concentrations of HS may lead to an increased activity of aquatic organisms (probably due to the moderate impact of ROS or moderate activation of the biotransformation system Chap. 8.3). A higher concentration of HS, however, may display too strong a stress, with reduced bioactivity. With exposure to low HS concentrations, the increased bioactivity may lead to increased active and/or passive uptake of particulate and dissolved substances, including inorganic and organic chemicals, from the water phase, with an ultimate increase in bioconcentration of chemicals from the water phase.

As described above (Chap. 7.1.1), in the presence of water hardness (Ca^{2+}), the binding of lipophilic organic chemicals to HS is affected by changing HS conformation. This must affect the bioconcentration of the lipophilic chemicals. Actually, Akkanen and Kukkonen (2001) present evidence that increasing Ca^{2+} concentrations lead to higher bioconcentrations of lipophilic chemicals in exposed organisms, as exemplified with B*a*P, pyrene, and tetrachlorobiphenyl. As a consequence, increased bioconcentrations lead to elevated toxicity of the organic chemical (see next Chap.).

7.5.3 Toxicity

The more risky scenario than pure increases in bioconcentration, is the increase of toxicity due to the presence of HS. First results indicating such effects are considered in relation to indirect photolysis of xenobiotics (Chap. 5.3.5). In the described studies, the toxicity potential is determined for small rodents, proxy models for humans.

Table 7.1. Data on the increase in bioconcentration of chemicals in the presence of dissolved HS (from Haitzer et al. 1998)

Increase absolute	Increase %	DOC mg/L	DOC source	Chemical	Reference
n.d.	54	2–6	A-HA*	C_{12}-LAS	Traina et al. (1996)
204	174	6.7	A-HA	deltamethrin	Muir et al. (1994)
528	303	6.7	A-HA	fenvalerate	Muir et al. (1994)
310	178	6.7	lake water	fenvalerate	Muir et al. (1994)
72	12	6.7	lake water	permethrin	Muir et al. (1994)
30,550	44	6.7	lake water	*p,p'*-DDT	Muir et al. (1994)
26	69	2–10	various natural waters	naphthalene	Kukkonen and Oikari (1991)
21	57	5	diluted river water	naphthalene	Kukkonen et al. (1990)
1,150	25	0.5	Nordic FA	benzo[*a*]pyrene	Kukkonen et al. (1989)
400	9	1–2.5	NIVA** concentrate	benzo[*a*]pyrene	Kukkonen et al. (1989)
2,300	31	1	diluted wetland water	benzo[*a*]pyrene	Kukkonen et al. (1989)
580	13	0.5–7	diluted lake water	benzo[*a*]pyrene	Kukkonen et al. (1989)
503	54	2.6	lake water	HepCDD	Servos et al. (1989)
397	42	1.2	A-HA	HepCDD	Servos et al. (1989)
78	20	3.3	lake water	OCDD	Servos et al. (1989)
289	75	2.0	A-HA	OCDD	Servos et al. (1989)
1,397	210	2	A-HA	3-methylcholanthrene	Leversee et al. (1983)
1,048	135	2	A-HA	3-methylcholanthrene	Leversee et al. (1983)
1,104	142	4	A-HA	3-methylcholanthrene	Leversee et al. (1983)
1,422	183	6	A-HA	3-methylcholanthrene	Leversee et al. (1983)
690	89	8	A-HA	3-methylcholanthrene	Leversee et al. (1983)
321	41	10	A-HA	3-methylcholanthrene	Leversee et al. (1983)

*Aldrich-HA, a commercial humic acid with changing composition
** NIVA = Norwegian Institute for Water Research, n.d. = not determined

7.5.3.1 Acute toxicity

The problem of increased toxicity is also intensively investigated in ecotoxicology. The first reliable results are published by Stewart (1984), who determined increased toxicity for substituted anilines. Similar studies are reported by Steinberg et al. (1992) with substituted anilines, substituted phenols, as well as the pesticide diazinon, and the flame retardant tetrabromophenol A.

Short-term toxicity to the waterflea, *D. magna,* is test with two differing pretest contact times. HS and chemicals can react together for 2 h or 60 h in the dark. The results are notably different (Fig. 7.30). Statistically significant increases in acute toxicity are observed, in particular with 2,4-dichlorophenol (with 60 h pretest contact time) and with some anilines. In contrast, 2,4-dichlorophenol exhibits no increased toxicity with a 2 h pre-test contact time. A photolytic mechanism can be excluded, since HS and the chemicals are kept dark.

Hence, at present, there is no known mechanism for this time-dependent (kinetic) effect. However, a good case can be made for detailed studies on the influence of contact time between dissolved HS and xenobiotics, because this clearly applies for both the changes in bioconcentrations and in toxicity. In general, for changes in acute toxicity of organic chemicals, three mechanisms can be put forward, including photoeffects:

- adsorption on dissolved HS
- dissolved HS-mediated production (for example by indirect photolysis) of compounds with toxic potentials which differ from the educts, and
- increased dissolved HS-mediated bioconcentrations (mechanism unknown) with subsequent increased toxic effect.

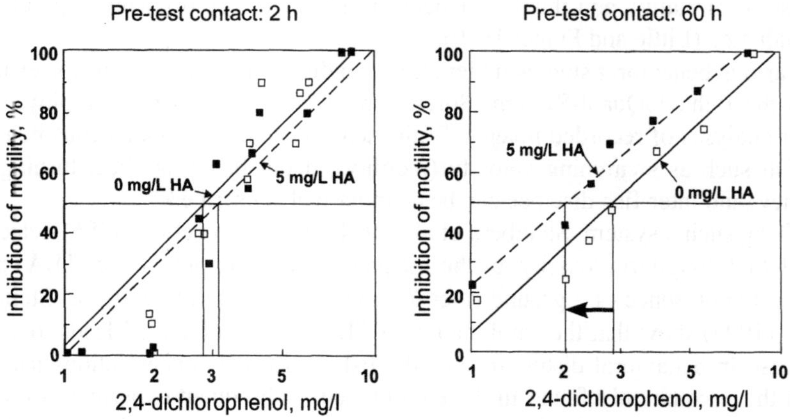

Fig. 7.30. Changes in acute toxicity of 2,4-dichlorophenol towards *Daphnia magna*, measured by inhibition of motility, in the presence of dissolved HS (commercial HS) exposed for two different pre-test contact times (from Steinberg et al. 1992, with kind permission of Wiley-VCH)

Acute toxic effects are, fortunately, relatively rarely seen in ecosystems, whereas sub-acute or chronic, (that is 'hidden' (subliminal)) effects are much more common. Therefore, in the following text, this aspect will be presented and discussed in detail.

In a ternary system (Ca-HS-toxicant), toxicity increases occur with HS present. Such a system is described by Körner et al. (in press) for the organo-phosphorus insecticide trichlorfon (TCF). In the absence of HS, Ca reduces the toxicity of TCF. There are two possible mechanisms. The positive charge of TCF-Ca complexes can inhibit the uptake of the toxicant. Alternatively, Ca can inhibit or prevent the degradation of TCF to the more toxic dichlorvos. If HS are added to the solution, the toxicity of TCF increases. This adverse HS effect is particularly marked at low Ca concentrations. The bonding of Ca to HS will result in less available sites for the process of reducing toxicity of TCF.

7.5.3.2 Subacute Toxicity: Behavioral Disturbances

A very sensitive parameter for subacute toxicity is the behavior of organisms (Baganz et al. 2001; Blübaum-Gronau 1994, 2001; Little 1990; Pluta et al. 1994; Lorenz et al. 1995). Behavior integrates many cellular processes, and is fundamental for the maintenance of organisms, populations, and communities in their habitats. Therefore, observations on behavior give a unique ecological perspective, in which the biochemical and ecological consequences of environmental pollution are linked. Changes in swimming behavior of fishes and other aquatic animals resulting from sub-lethal exposure to contaminants, is detrimental to the animals in terms of finding food and eating, avoiding predators, and reproduction. Hence, particular behavioral modifications provide information on sublethal toxicity, and indicate the potential for eventual mortality. In addition, modifications in swimming behavior caused by toxicants are generally perceived earlier than growth decrease or mortality, and this is independent of which type of swimming behavior is evaluated (Little and Finger 1990).

Various behavioral studies under stress conditions are carried out, for example with the BehavioQuant-System (Spieser and Scholz 1992) which employs computer analyses of recorded images. With such a system, the spontaneous behavior of fish such as swimming velocity (motility), depth below surface, turning frequency, and inter-fish distance can be quantitatively described.

Using such a system, Steinberg et al. (1994) and Lorenz et al. (1996) detect the sub-lethal, long-term toxicity, of the triazine herbicide terbutylazine (TBA) alone and in the presence of a groundwater FA to the zebrafish, *Danio rerio*. Steinberg et al. (1994) show that the combination of TBA and dissolved HS leads to a clear increase in behavioral disturbance of the fish. Since the concentrations used exceed those commonly found in the environment, subsequent experiments employ lower concentrations. Analysis of the vast quantity of data employed the Hasse diagram technique (Brüggemann and Halfon 1995; Brüggemann et al. 1995) from discrete mathematics. Motility (Fig. 7.31) and preference of light or dark habitats (Fig. 7.32) are chosen from the database, since they allow easy and informative interpretation. The Hasse diagrams are read from bottom to top, as the response intensity increases in this direction.[3]

For changes in motility, a clear dose-effect relationship is found, both in the absence and presence of HS. The fishes swim faster or more erratically with chemical or combined exposure, which may be attributable to attempts to escape exposure and find clean water. In closed aquaria, this goal cannot be reached and the increased swimming activity remains unsuccessful and never ceases. In the control without TBA and HS, the fishes display the least swimming activity. In the variant with 50 µg/L TBA and dissolved HS, motility exceeds that without dissolved HS, regardless of dissolved HS concentration. Dissolved HS alone increases the swimming activity of the fish.

[3] Circles linked to each other arise from each other with the analyzed criteria, those on the same level next to each other have the same intensity, but are not comparable to each other in the marked attributes.

Contrasting results are found with the preference for light/dark habitats. No clear dose-effect relationships are found. The seeking of dark habitats is an all or nothing phenomenon. With only 1 mg/L TBA exposure, the fish favor dark habitats. Further increases in exposure do not yield a more intense reaction. The seeking of dark habitats during exposure to chemicals makes ecological sense. The fish adopt their normal threat avoidance behavior. They move to darkness to avoid predators. With the exception of 5 µg/L TBA and dissolved HS control exposure, all responses to combined exposures lie below the corresponding responses to exposures with chemicals only. Apparently the HS act as a natural weak optical protection against predators.

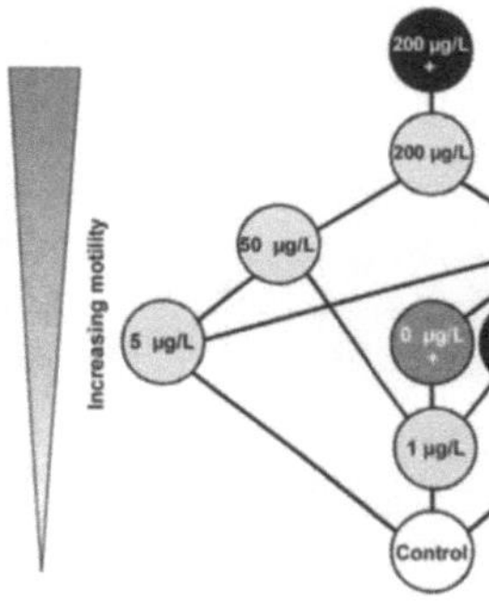

Fig. 7.31. Ranking of the changes in the motility of the zebrafish upon exposure to terbutylazine (grey circles), and to a mixture of terbutylazine and the groundwater fulvic acid (Fuhrberg, 2 mg/L DOC, black circles). The graph is Hasse diagram, which is read from bottom to top; the direction in which the intensity of the response increases. All results of the combined exposure (terbutylazine and fulvic acid) lie clearly above the corresponding exposure without fulvic acids (modified from Lorenz et al. 1996, with kind permission of Elsevier Science)

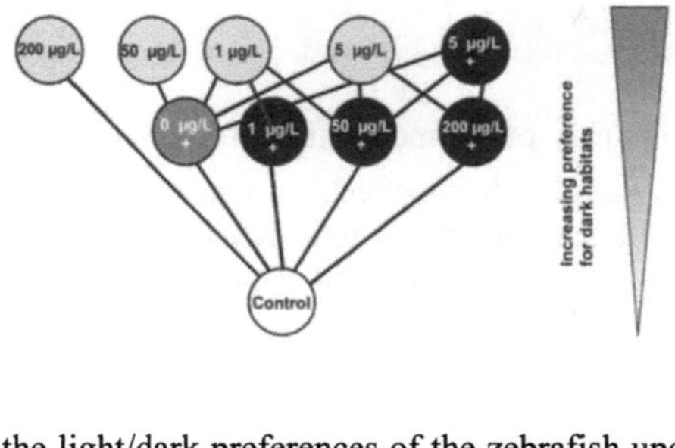

Fig. 7.32. Ranking of the light/dark preferences of the zebrafish upon exposure to terbutylazine (grey circles) and to a mixture of terbutylazine and the groundwater fulvic acid (Fuhrberg, 2 mg/L DOC, black circles). The graph is Hasse diagram, which is read from bottom to top; the direction in which the intensity of the response increases. All except one result of the combined exposure (terbutylazine and fulvic acid) lie clearly below the corresponding exposure without fulvic acids (modified from Lorenz et al. 1996, with kind permission of Elsevier Science)

As with other toxicological parameters, extrapolation of changes in behavior to effects which occur in natural populations is difficult. However, some hypothetical interpretations can be made. Hypoactivity and hyperactivity, as well as divergence from the daily rhythm, can interrupt feeding and increase susceptibility to predators. Furthermore, hypoactivity reduces the probability of finding prey, since the area of range is smaller. Consequently, the energy available for growth is reduced, so the fish remain smaller under such conditions (Little and Finger 1990). Sensitive individuals or species may be more affected than robust ones (Steinberg et al. 1994).

A general scheme summarizing how dissolved HS can change the toxicity of xenobiotics from an ecological point of view is shown in Fig. 7.33.

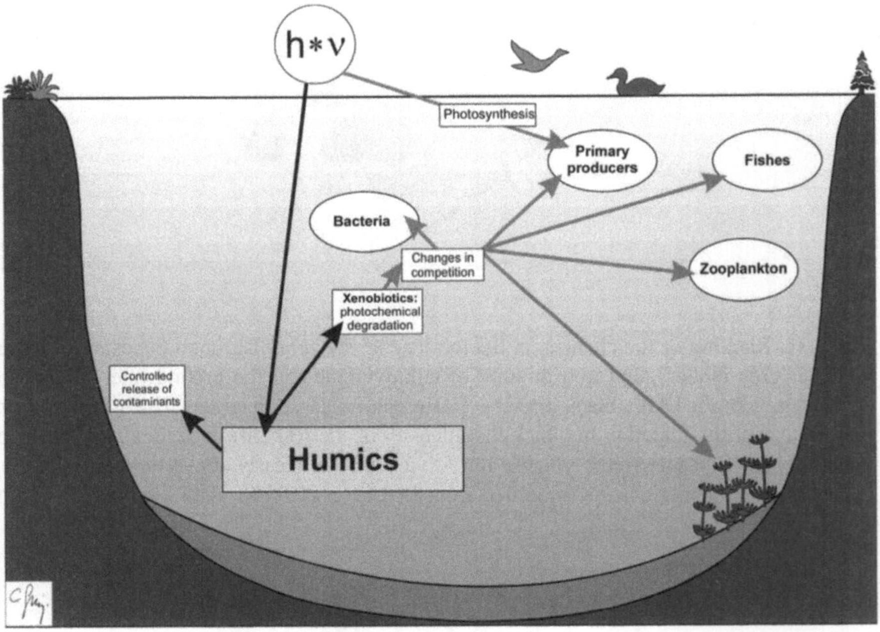

Fig. 7.33. Graphical summary of the modulation of toxicity of xenobiotics by HS. h∗ν is sunlight. Black arrows indicate direct effect, grey arrows indicate indirect effects. Controlled release of humic-bound organic chemicals into the aquatic environment has to be anticipated, but has not yet been in the focus of eco-chemical or ecological studies

8 Direct Effects on Organisms and Niche Differentiation

Various general properties of HS are involved in biological processes of surface waters:

- protonation/deprotonation: influences the acid-base status of freshwaters (Chap. 3)
- light absorption: acts as UV shield, light absorption limits the depth of the euphotic zone with strong thermal gradients during summer stratification (Chaps. 4, 5)
- particle formation: detritus formation, involvement in the detritus food chain, in some cases as direct trophic energy or carbon source
- complexation/binding: metals and nutrients (Chap. 6), xenobiotics (Chap. 7)
- photomineralization and both direct and indirect photolysis: release of substrates for microbial heterotrophy, release of carbon dioxide, release of reactive oxygen species (ROS, which are potentially toxic to organisms), photodegradation of potential toxins (Chap. 5)
- surface-active effects: potentially strongly stimulating at low concentrations, and inhibitory at higher concentrations. Inhibitory effects of FA on microorganisms and plant roots are possible, with a tenside-like effect on membranes and the linked increased permeability to apolar essential materials (Visser 1985). However, since the effect is strongest under weakly acidic conditions (Vigneault et al. 2000), an additional hydrophobic mechanism may also apply as postulated by Petersen (1991) (Chap. 8)
- accumulation on the surface of living cells whereby exchange of DOC with the membranes can occur and the permeability changes. An example: inorganic Al reduces the membrane permeability of the green alga *Chlorella pyrenoidosa*, whereby nutrient uptake is also reduced. If increasing quantities of DOC are then added, the membrane permeability is rapidly increased to the initial value (Campbell et al. 1997)
- metabolic interactions: modulation of plant photosynthesis and of the transformation systems of plants and animals (Chap. 8), and
- hormone-like actions: modulation of offspring numbers in the nematode *Caenorhabditis elegans*.

These effects of HS on aquatic organisms are of a direct or indirect nature. Direct effects also include pseudo-direct effects of HS on organisms, such as the effect of acids in the presence of HS. In many instances, the effect of protons from the acids and from the HS molecule cannot be separated. The basic question to be answered, however, is: can HS be taken up by aquatic organisms?

8.1 Uptake of HS and HS-like compounds

The question as to whether or not HS are taken up by organisms has been argued intensively in the literature. One cannot deny that beneficial, as well as adverse, effects can be observed when organisms come into contact with HS in their various forms: soil water solutions, peats, bogs, or HS-rich surface waters. Most soil scientists, for instance, attribute any effect to indirect modes of action, such as modulations of bioavailability of key nutrients, particularly iron (see Box 8.1). So do many freshwater ecologists. In addition to altered bioavailability of nutrients, further indirect modes of action are discussed, such as decreased light climates which affect primary producers and optical foragers alike, altered food web structures, and decreasing space for water breathing organisms due to increasing anoxia in the lower strata of the water column (see books of Hessen and Tranvik 1998 and Keskitalo and Eloranta 1999). Freshwater ecologists exclude direct interactions between HS and aquatic organisms, because uptake of HS is not considered to be feasible.

In contrast, biomedical scientists accept uptake of HS up to approximately 1.0 kDa (Beer et al. 2000; Brockow et al. 1998). Furthermore, they report interactions of HS with several receptors (Beer et al. 2000), and with blood coagulation (Klöcking et al. 2000).

Several early studies (Pflugmacher et al. 1999b; 2001) discuss the uptake as a decisive basic mechanism to explain surprising results, such as modulation of photosynthesis in macrophytes and algae (Chaps. 8.3.1, 8.3.2), induction of heat shock proteins in fish and invertebrates, modulation of transformations enzymes, and alteration of the endoplasmic reticulum (Chaps. 8.3.5, 8.3.8). Also, in one key German HS study, Ziechmann (1996) writes that HA precursors (HAP), namely from aqueous peat extracts and compressed peat, pass through the skin of pigs and mice and can accumulate subcutaneously. Recent studies (Wang et al. 1999)[1] also show that HS, or at least parts thereof, can be taken up by organisms, and at least parts of these molecules, can be found even in the DNA. Thus the uptake of HS by organisms can be counted as a direct effect, although mechanistic studies are rare.

In a very recent study, Pflugmacher et al. (in prep.) present evidence that ^{14}C labeled humic-like substances are taken up and bioconcentrated by several aquatic organisms. The applied oxidation product is synthesized by oxidation of caffeic acid with sodium *m*-periodate (Helbig and Klöcking 1983). Dunkelberg et al. (1997) attribute a molecular weight of 6.0 kDa to this oxidation product. Fig. 8.1 shows that, after 24 hours of exposure, a macrophyte (*Ceratophyllum demersum*),

[1] Because of methodological weaknesses, this study is not totally convincing: for example, the humic substances are not extensively radioactively labeled.

an invertebrate (*Gammarus pulex*), and a vertebrate (tadpoles of the moor frog *Rana arvalis*) are able to bioconcentrate ^{14}C in their bodies. The percentage uptake of the exposed ^{14}C caffeic acid oxidation product is: *C. demersum* 7.3 ±1.4%, *G. pulex* 6.9 ±1.0%, and *R. arvalis* 11.7 ±2.7%. It may still be argued that it is not the intact oxidation product of caffeic acid, but smaller photodegradation products, which are bioconcentrated. Nevertheless, it is evident that at least low-molecular weight (photodegradation) products of the humic-like substances are taken up and bioconcentrated by aquatic organisms, and that the bioconcentrated substances are responsible for several effects addressed below, which are feasible only if HS are themselves taken up by the aquatic organisms. If the humic-like products of the caffeic acid are in the molecular weight range of 1.0 kDa, they cover well the molecular weights of most FA in aquatic ecosystems (Thurman 1985).

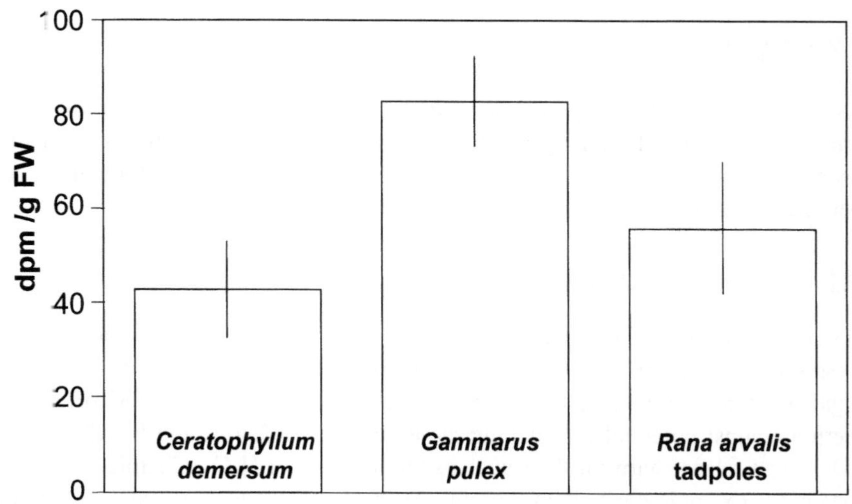

Fig. 8.1. Uptake of ^{14}C labeled caffeic acid oxidation product by three aquatic organisms within 24 hours (means ± standard deviation) (Pflugmacher, Greulich, Klöcking, Steinberg in prep)

8.2 Effects in Acidic Waters

In soft-water lakes, the predominant freshwater type in the boreal and tropical zones, HS determine the acid status. This process is so clear, that from the size of the catchment alone, one can determine the acidity status of the water bodies with a high degree of probability (Chap. 3, Fig. 3.19). It is also known that some acids are relatively strong, with dissociation constants $pK_a \approx 3$. In such waters the organisms are exposed to both acidity and to HS. This means that the effects of protons and HS on organisms in such waters cannot be separated from one another.

In addition, waters of the boreal zone can also be acidified through acidic depositions, so that the natural processes can be superimposed by inputs of mineral acids and cation acids (Al, Fe) which enter in catchment runoff. It is therefore fundamental to study how HS can change the effects of acids. It is known that HS can dampen the acidification effect through binding with protons and toxic metals such as Al, and yet can themselves have direct toxic effects.

The acid buffering effect of DOC in soft waters has already been discussed in Chap. 3. This effect naturally has effects on biocoenoses. As one of many examples, in the acidic Adirondack lakes (New York State, USA) increasing DOC content of the water is accompanied by increasing numbers of rotifers (Siegfried et al. 1989).

8.2.1 Algae

As a result of their short generation times, algal populations react rapidly to increased or decreased loading of nutrients or acids so they are often successfully used as bioindicators of water quality. This applies both to phytoplankton and benthic diatoms.

8.2.1.1 Phytoplankton

HS can affect algae in two ways: physically and chemically. Physical influence means that HS compete with the phytoplankton for light quanta, as cDOC and algal pigments are both light absorbers (Chaps. 5.1, 5.3). The cDOC offers protection against potentially damaging radiation such as UV-A and B, whereby the cDOC can release harmful ROS such as OH-radicals or H_2O_2. Chemical influence means for example changes in the bioavailability of nutrients or micro-pollutants (Chaps. 6.1, 8).

These physical and chemical influences of HS on phytoplankton are so fundamental that Thienemann (1925) adds a third category 'dystrophic' (famished), to the two trophic categories 'oligotrophic' and 'eutrophic' proposed by Naumann (1919) '*because the brown peaty waters of many of the lakes of the mountains of northern Europe seemed to support a different plankton association from that of the deep blue or blue-green oligotrophic lakes of central Europe*' (Hutchinson 1967). Already in the 1960s however, Hutchinson states that the phytoplankton of humic-rich lakes differs little from that of humic-poor (clear) waters, the former having only a slightly higher proportion of diatoms and chrysophytes. This has partly been confirmed in recent studies for example by Arvola et al. (1999a): in boreal lakes there is no clear dependence of phytoplankton biomass on the HS content, with the only exception that humic-poor lakes with a DOC content of less than 2 mg/L have the lowest biomass.

Jones (1998) correctly points out that the opposites 'humic' and 'clear' are artificial, since every water body, including the open sea, contains certain amounts of HS. There is a continuum from humic-poor to polyhumic. From this perspective, it is unlikely that humic lakes have fixed properties completely different from lakes with low HS contents. If one takes the number of phytoplankton species as a reference, then there is a similar picture: the lowest number of species are found in the clearest waters and in those with the highest HS content. This statement applies not only to phytoplankton but also to littoral diatoms. Since an increase in HS content is generally linked to an increase in acidity, the effects of these parameters cannot be separated from each other. Direct adverse effects of HS alone on phytoplankton cannot be ruled out, as the results presented in the next sections will show.

According to the studies of Arvola et al. (1999a), there are weak dependencies of particular strategies of phytoplankton on HS contents. In lakes with a high HS content, there is a tendency for motile forms to dominate, at least in particular seasons. The opposite applies for picoplankton: with increasing HS content, their proportion clearly decreases. This has, as yet, only been shown for small lakes. Overall, there is nothing like a 'humic community' of phytoplankton (Jones 1998), although the proportion of flagellates and particularly mixotrophic chrysophytes is very high (Jansson 1998).

8.2.1.2 Benthic diatoms

Diatoms express diverse sensitivity to protons and to HS. According to empirical studies in German middle range mountains, namely in the Fichtelgebirge and Frankenwald (Bavaria) and in the northern Black Forest and the Odenwald (Baden-Württemberg) (Schreiner 1990; Alles et al. 1991, respectively), diatoms differ in their responses to acidification and HS effects. While Schreiner (1990) lists the 22 most common species (Table 8.1), Alles et al. (1991) concentrate on the genus *Eunotia* (Table 8.2). Although these two studies differ in their empirical approach, they both provide valuable insight into the hitherto little studied direct effects of DOC (HS) on diatoms.

From the works of Schreiner (1990) and Alles et al. (1991), two situations are clear:

1. species react very differently to the three chosen environmental variables;
2. tolerance to the three environmental variables does not have the same meaning for the different diatoms. For example, *Eunotia exigua* is one of the most tolerant species to low pH, high Al and high DOC content. Another species in this genus, *Eunotia paludosa*, exhibits different tolerances: it is comparably acid tolerant, but has a lower Al tolerance, and only a minimal DOC tolerance.

In his dissertation, Alles (1999) makes the separation of DOC indicator diatoms more precise:

- oligodystrophic: 0–2 (3) mg/L DOC
- mesodystrophic: 3–10 mg/L DOC
- eudystrophic: >10 mg/L DOC.

Table 8.1. Tolerance levels for DOC, acidity, and aluminum of the 22 most common diatoms in streams of Bavarian middle range mountains, Germany (Schreiner 1990)

Species	DOC	pH	Al
Achnanthes austriaca (incl. var. *helvetica*)	3	2	3
Achnanthes lanceolata (incl. var. *rostrata*)	3	4	4
Achnanthes marginulata	3	2	1
Achnanthes minutissima	2	2	2
Achnanthes saxonica	3	3	3
Diatoma hiemale (incl. var. *mesodon*)	3	2	3
Eunotia bilunaris	1	1	1
Eunotia exigua	1	1	1
Eunotia incisa	3	2	3
Eunotia paludosa	4	1	2
Eunotia pectinalis (incl. var. *minor*)	3	3	4
Eunotia rhomboidea	1	1	1
Eunotia subarcuatoides	1	1	1
Fragilaria capucina	2	2	2
Fragilaria capucina var. *capucina*	3	3	4
Fragilaria construens (incl. var. *venter*)	3	4	4
Frustulia rhomboides (incl. var. *saxonica*)	3	3	4
Gomphonema parvulum	3	3	3
Navicula mediocris	2	1	1
Pinnularia appendiculata	2	1	1
Pinnularia microstauron	3	3	2
Pinnularia subcapitata var. *hilseana*	1	1	1

Tolerance level	DOC, mg/L	pH	Al, mg/L
1	≥9.0	<4.0	≥3.0
2	≥6.0 <9.0	≥4.0 <5.0	≥1.0 <3.0
3	≥3.0 <6.0	≥5.0 <6.0	≥0.2 <1.0
4	<3.0	≥6.0	<0.2

Table 8.2. Selected indicator properties of nine *Eunotia* species in streams of the Odenwald and Black Forest Mountains, Germany (Alles et al. 1991)

Species	DOC ('dystrophy')	pH	Al
Eunotia bilunaris	++	0+	++
Eunotia exigua	0-	0+	+++
Eunotia implicata	+	+/-	?
Eunotia meisteri	++	+++	++
Eunotia minor	0+	+/-	-
Eunotia paludosa	+++	+++	++
Eunotia silvahercynia	+	+/-	?
Eunotia subarcuatoides	---	0+	+++
Eunotia sudetica	-	+++	+

+, ++, +++: positive indication (three levels); -, --, ---: negative indication (three levels); 0: no clear indication; 0+: no clear indication but tending to higher values; 0-: no clear indication but tending to lower values; +/-: prefers mean values at neutral pH; ?: unknown

Bog-indicating species such as *Eunotia paludosa* (incl. var. *trinacria*), *Eu. meisteri*, *Eu. tenella*, *Frustulia rhomboides* var. *saxonica* and *Pinnularia subinterrupta* are found preferentially in eudystrophic flowing waters. In oligodystrophic environments, a marked decrease in species number occurs. In mesodystrophic conditions, the species number increases, whereas in eudystrophic waters, the species number again is reduced. In Black Forest streams, the presence of HS leads to a greater acid tolerance in the diatoms, which is probably explained by acidity reducing complexation by HS. At higher DOC concentrations (approximately >10 mg/L), HS themselves can have a direct toxic effect. Since no study has investigated the underlying mechanism, general mechanisms can only be hypothesized:

1. direct toxic effects of HS on the membranes, in particular of the hydrophobic fraction (Petersen's hypothesis, Chap. 8.1.2.1);
2. increased bioconcentration of HS and subsequent energy loss through additional metabolic performance of the biotransformation system (Chap. 8.3) and/or
3. changes in osmoregulation if one attributes a considerable osmotic potential to HS. Due to the character of polyelectrolytes, this is likely for HS. However studies on potential physiological mechanisms of the effects of HS on organisms are currently unavailable.

These possible mechanisms are not mutually exclusive, and all can be simultaneously effective.

8.2.2 Zooplankton

Zooplankton in HS rich freshwaters are mostly ubiquitous. Only a small number of species can be used as humus-indicators, and this only on a regional basis (see below). From a quantitative perspective, the zooplankton of HS rich water require allochthonous DOC as an additional food source. This is provided in the food chain through bacteria and ciliates (Sarvala et al. 1999), or directly by uptake by crustaceans (Hessen et al. 1990; Jones et al. 1998) (Chap. 9). Besides detritus (including HS), phytoplankton is only the second most important determinant for the zooplankton.

Studies on zooplankton in humic lakes have two features in common: they remain at the phenomenological level, and they are unable to differentiate between the effects of protons and of HS. Sarvala et al. (1999) state that many characteristics of zooplankton in humic-rich waters are much more dependent on the size of the lake and its wind-protected situation, than on the HS themselves. Many of the lakes are anoxic in the lower strata, strongly acidic, and contain few fish. However, one must ask: is there physiological evidence to exclude direct interactions of HS with zooplankton animals? Has anyone carried out physiological or biochemical measurements in the animals upon exposure to HS? The apparent recycling of paradigms somewhat resembles the beginning of zooplankton research in humic-rich lakes: since the ideas of Pütter (1909) on the potential role of dissolved organic materials in the nutrition of invertebrates, it is doggedly held that in humic-rich lakes, zooplankton species diversity is higher than in humic-poor lakes[2] (Naumann 1918; Nordqvist 1921). There is no proof of this claim, and it is often disproved, for example by Krogh (1931) (referred to in Jørgensen 1976). Järnefelt (1956) describes a very high abundance of zooplankton in some Finnish lakes, but finds no general instance to support the above hypothesis. In contrast, other authors state that humic lakes harbor an impoverished zooplankton fauna in Norway and Latvia (Druvietis et al. 1998; Eie 1974). Druvietis et al. (1998) also give a positive correlation between the content of the HS and the total number of zooplankton species: 40% of the variance of the zooplankton species composition can be explained by the HS content of the water.

How can one explain these diverse, even contradictory, observations?

Few countries have as many humic lakes as Finland, and thus this country is an outstanding location for the study of effects of humic-rich waters on diverse biocoenoses. A total of 138 studied lakes in southern Finland (Sarvala et al. 1999) provides an extensive database on the occurrence of zooplankton along a DOC gradient with particular reference to crustaceans and rotifers. We will see if the above contradictions can be resolved by the excellent Finnish studies.

8.2.2.1 Crustacea

Worldwide, the many reports on DOC content of water bodies as a determinant of crustacean diversity and biomass do not provide a unified picture. In southern Finland, the abundance of zooplankton crustaceans increases with TOC content (Fig. 8.2). Since the studied lakes with low HS contents are also anthropogenically acidified, no definitive effect can be assigned. The relationship between TOC and zooplankton abundance in Finnish lakes is, however, stronger than in Canadian lakes (Yan 1986) which are more dominated by their nutrient content and nutrient availability than by DOC content.

In the Finnish studies, there is no significant relationship between species number and TOC content. This also applies for Norwegian waters (Eie 1974). There too, ubiquists are dominant. Only *Daphnia longispina* senso stricto prefers small polyhumic over less humic lakes in Finland. This can also be an indirect effect, as in such lakes there are few planktivorous fish.

Berzins and Bertilsson (1990) report a number of HS-avoiding crustaceans in an extensive study of Swedish waters. These include *Bosmina coregoni, B. crassicornis, Cyclops scutifer, Daphnia cristata, D. cucullata, D. galeata, D. longispina* (!), *Holopedium gibberum,* and *Limnocalanus macrurus*.

Only weak indication of HS tolerance of zooplankton species is reported from

[2] The frequently used pair of terms 'brown water' and 'clear water' is linguistically incorrect. The opposite of 'brown' is 'colorless' or 'white' and the opposite of 'clear' is 'turbid'. Brown water lakes can be turbid, only if the DOC exceeds the threshold that allows colloid and particle formation. Throughout this book, these terms have been avoided, and 'humic rich' and 'humic poor' are used instead.

Coastal Plain blackwaters of the USA. In these blackwaters with 5–50 mg/L DOC, the zooplankton community consists of species of rotifers, copepods, and cladocerans, such as *Alona*, *Alonella,* and *Bosmina longirostris*. Harpacticoid copepods, including *Acanthocyclops vernalis* and *Attheyella illinoisensis*, can also be an important component of the invertebrates within the sandy substrates of blackwater streams (Smock and Gilinsky 1992, with further references). Brownlow and Bolen (1994) confirm that blackwater streams in eastern North Carolina harbor a somewhat richer fauna than is found in the alluvial streams.

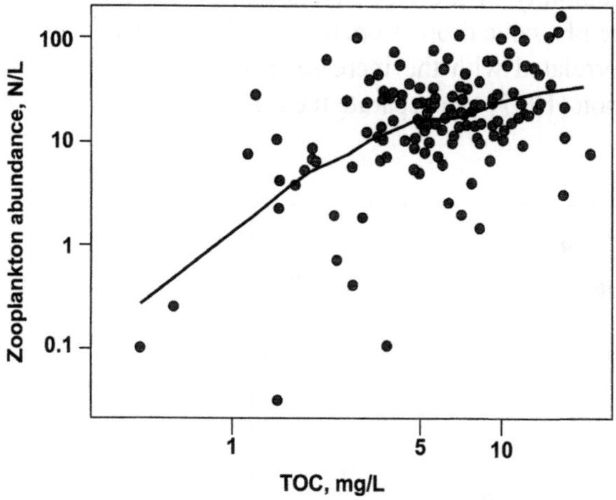

Fig. 8.2. Planktonic crustacean abundance as a function of TOC concentration in 138 southern Finnish lakes (after Sarvala et al. 1999, with kind permission of Backhuys Publishers)

The clearest relationship of species number and abundance to DOC content is reported for the Experimental Lakes Area in Canada. For these lakes, Patalas (1971) states that the copepod species number is negatively correlated with DOC content. However, abundance of some species increases with DOC content including that of *Diaptomus leptopus, D. oregonensis*, and *Tropocyclops prasinus mexicanus*. *Mesocyclops edax* and the cladocerans *Diaphanosoma brachyurum, D. leuchtenbergiana,* and *Holopedium gibberum* are also more common in humic lakes than in humic-poor lakes. The finding for *H. gibberum* is in disagreement with the Swedish study, for which there is no clear explanation. On the basis of some mechanistic studies, these seemingly contradictory findings will be revisited in Chap. 8.3.

Daphnia magna

One of the few studies on the influence of HS on acidification impact shows that natural dissolved HS can positively influence the survival and reproduction of acid-stressed water-fleas (*Daphnia magna*) (Petersen and Persson 1987; Petersen 1990) (Fig. 8.3).

Although there is only minimal data, the somewhat speculative interpretation of Petersen (1990) is repeated, as it remains plausible. Petersen describes the various acid-dependent effects of dissolved HS as changes of water or lipid solubility, which depend on the degree of protonation. The standard measure of bioavailability of an organic compound is the octanol/water partitioning coefficient (K_{ow}). Petersen and Kullberg (1985) define this coefficient for HS (K_{how}) as the relationship between the optical absorption of a HS at 430 nm in the octanol phase, to that in the water phase. With falling pH, the K_{how} of all studied HS increases. The potential bioavailability of the HS, expressed as lipid solubility, increases up to 5000% when the pH value drops from 6.5 to 3.5. The increasing octanol solubility of the HS is correlated with the increase in direct toxic effects with which Petersen and Persson (1987) can separate the effects of acidity from those of accompanying metals.

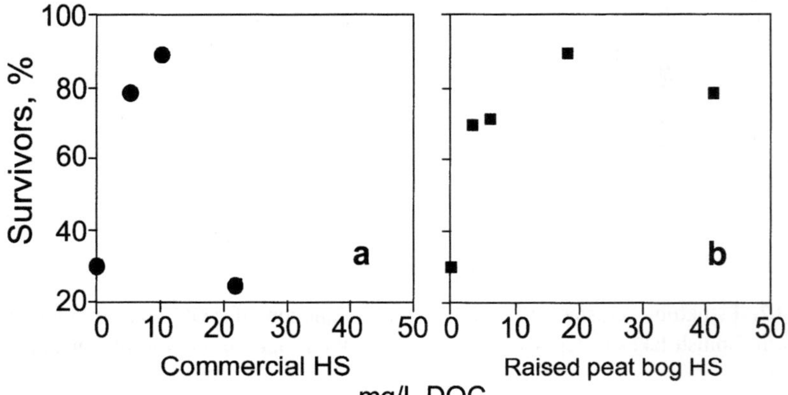

Fig. 8.3. Survival rate of *Daphnia magna* at pH 5.0 as a function of increasing concentrations of commercial (**a**) and raised peat bog HS (**b**) (modified from Petersen 1990, with kind permission of Wiley & Sons)

The differences in the toxic properties of HS from two sources at pH 5 to *Daphnia magna*, are shown in Fig. 8.3. At pH 5, the survival rate with the commercial HS is maximal at 10.7 mg/L. When a HS isolate from a bog is used, a maximum survival rate is found at 20 mg/L. With higher HS concentrations from both sources, adverse effects occur. From the results, one concludes that at a low pH and low HS concentrations, the direct effect of acid dominates the toxic effects, through a change in membrane permeability and the activity of heavy metals present in the culture medium. With increasing HS concentration, the test organism initially benefits from mechanisms such as the binding of heavy metals and/or reduction in membrane permeability. This means that the HS protect the organism from the direct negative effects of acid and reactive metals. The protection mechanism remains up to a specific maximum concentration. The maximum can be determined for example, by the amount of metals in the medium, the pH value, and the ability of the HS to form complexes.

Petersen (1990) concludes that the proportion of hydrophobic and lipid-soluble organic compounds in the dissolved HS plays an important role. It is probable that a small proportion of total available carbon is necessary for a strong decline in survival rate. In a study of some southern Swedish waters, the octanol-soluble carbon is generally 23–58 µg/L, representing only 0.12–0.37% of the total carbon. This should account for the toxic effect of low pH. The absolute amount of carbon is very low, but believed to be within the toxic concentrations of insecticides. The last statement is speculative, and should be confirmed in further studies.

Petersen's results do not necessarily indicate a toxic effect of the lipophilic fraction of HS. It is also possible that this fraction is more readily taken up by organisms due to its lipophilicity, and thus its greater ability to pass membranes. In the cell, HS are metabolized, while energy is consumed. Initial information that HS may be metabolized is presented by Pflugmacher et al. (1999b, 2001) on the aquatic macrophytes *Ceratophyllum demersum* and *Vesicularia dubyana.* Additional information will be presented for algae and cyanobacteria, amphipods, and clams in Chap. 8.3.

8.2.2.2 Rotifers

Sarvala et al. (1999) determine that in humic waters, there is a greater diversity of rotifers than of crustaceans. In northern European waters, the following species can be used as indicators of polyhumic conditions: *Anuraeopsis fissa, Ascomorpha agilis, Conochilus dossuarius coenobasis, Hexarthra mira, Keratella serrulata, Trichocerca similis.* There is a clear difference between the zooplankton composition in humic waters of North America and Europe: while rotifers dominate in North America, crustaceans dominate in Europe (Havens 1991, 1993; Sarvala et al. 1999). This difference is probably related to particular food webs. For example, the proportion of inedible algae is greater in North American than European lakes, thus favoring bacteria-feeding rotifers. However, this difference can also be related to the abundance of carnivorous invertebrates and planktivorous fish with high planktivory favoring small zooplankton form, such as rotifers.

8.2.3 Selected Benthic Invertebrates

8.2.3.1 Simuliids

Petersen (1991) describes the influence of HS, Al, and low pH (4.5) on the survival of the simuliids *Odagmia ornata* and *Simulium decorum.* In acidified waters, inorganic Al species (including Al^{3+}, AlF_3 and Al-sulfate complexes) play an important ecotoxicological role. HS are known in general to reduce the toxicity of these Al species. Fig. 8.4 shows that at pH 4.5, both HS and Al are toxic. At 56 mg/L TOC no simuliids survive. In the absence of HS the same is found at around 300 µg/L Al. When both HS and Al are present in the system, these factors are antagonistic. The toxic effect of 56 mg/L TOC and 300µg/L Al is such that only 50% of the larvae die.

Using the IHSS separation technique, Petersen determines that half the Al is in the hydrophobic fractions (divided equally between the FA and the HA fraction), and is therefore bound. If one presupposes Petersen's toxicity hypothesis (Petersen 1990), then the binding of Al to HS reduces the bioavailability of the toxic, hydrophobic DOC fraction. This explanation is plausible.

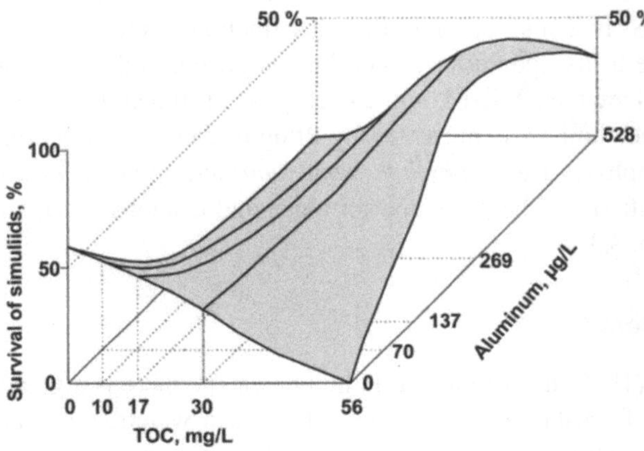

Fig. 8.4. Survival rates of simuliid larvae at pH 4.5 with increasing TOC concentration and Al concentrations. The original water color data are transformed into TOC concentration using Eq. 3.4 (modified from Petersen 1991, with kind permission of Springer Verlag)

8.2.3.2 *Amphipoda:* Gammarus pulex

Amphipods, in particular *Gammarus* spp., are acid sensitive organisms (Økland and Økland 1986), and exposure to pH 5.5 leads to mortality. Hargeby (1990) and Hargeby and Petersen (1988) report that during 21 days exposure of *Gammarus pulex* to acidity, with or without HS:

- the proton toxicity at pH 6 in the absence of HS leads to 92% mortality. If HS are added at a concentration of 7 and 20 mg/L, the mortality rates fall to 80% and 64% respectively (Fig. 8.5);
- animals surviving at pH 6 exhibit a reduced growth rate, less nutrient use, and a higher body water content than at pH 7.3 (control);
- the presence of HS significantly improves the survival rate in acidic conditions.

Adverse effects of dissolved HS, as described by Petersen and Persson (1987) for *D. magna*, are not apparent in the *G. pulex* study with the HS concentrations used. Yet, on day 10 of exposure, an exception from the beneficial effect of HS occurs in slightly acidic waters (Fig. 8.5). The higher HS concentration (marked with an arrow) reduces the survival rate significantly. Thus, in this study also, an adverse effect on animals cannot be excluded.

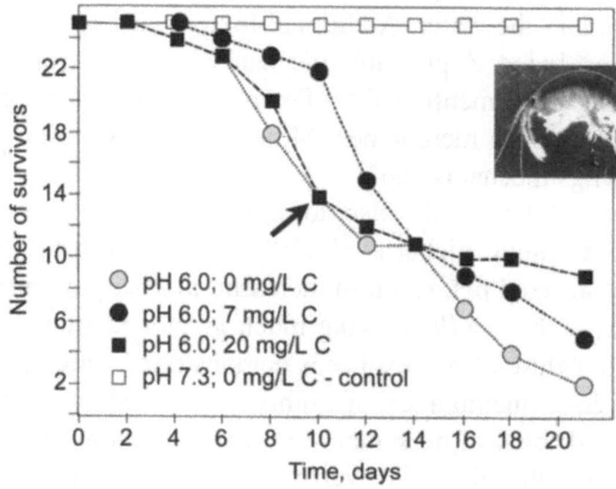

Fig. 8.5. Survival of *Gammarus pulex* during 21 days, exposed to pH 6.0 and 7.3, and HS concentrations of 7 and 20 mg/L C. The arrow marks the adverse effect of an elevated HS concentration (after Hargeby and Petersen 1988, with kind permission of Blackwell Science)

8.2.4 Amphibians

It has been known for a century, that in both temperate and tropical zones, there is less diversity in the amphibian fauna of waters rich in HS than where HS concentrations are low (when pH values are comparable). The literature is summarized by Saber and Dunson (1978). Janzen (1974) considers that HS themselves have a toxic influence, which may be attributed to phenolic compounds, although the mechanism is not elucidated. It is also possible that amphibians metabolize HS, as has recently been shown for some aquatic plants and macroinvertebrates (Chap. 8.3).

The first authors to establish an influence of DOC on amphibian survival are Saber and Dunson (1978) who describe that at pH 6.0, the mortality of tadpoles of the North American Bullfrog, *Rana catesbeiana,* is significantly higher in HS-rich water that in the control water with low HS content. This species of frog is therefore absent from black water rivers and lakes.

The hatching success of larvae of all amphibian species tested is less in HS-rich than in HS-poor waters. Clark and Hall (1985) find a strong negative correlation between DOC and hatching success. Freda (1986), Freda and Dunson (1986), and Dunson et al. (1992) suggest that a toxic component, in addition to low pH, is present in 'dark waters'. In a pond with pH 4.1 and 2.1 mg/L DOC (measured as tannic acid), 1/3 of the embryos of the Woodhouse toad *Bufo woodhousei* hatch, whereas in another pond with the same pH, but with 8.5 mg/L DOC (as tannic acid) all embryos die.

Similarly, the hatching success of the Jefferson's salamander *Ambystoma jeffersonianum* and the wood frog *Rana sylvatica* is linked to the concentration of organic components in the water. Additional toxicity due to Al in these waters is to a large degree excluded. A plausible mechanism, but far from proven, is that a toxic fraction of DOC, as mentioned by Petersen (1990), increases with decreasing pH and passes through membranes. More research is required in this area to explain these findings mechanistically.

Adverse effects of HS apply also to larvae of South African claw frogs, *Xenopus gilli* and *X. laevis*. Picker et al. (1993) show that solutions of increasing HS concentrations at fixed pH, result in increased mortality of embryos of both *X. gilli* and *X. laevis*, with *X. gilli* showing much greater resistance to this toxicity. The potential toxic nature of blackwater is demonstrated using an astringency test for leaf tannins. Jelly membranes of embryos exposed to blackwater become stained (tanned) dark brown, presumably by complexation with the humic compounds, probably via the phenolic hydroxyl groups. The toxicity of such high-molecular weight HS may result from their complexation with the glycoprotein of the jelly membranes, an event which would inhibit the formation of the perivitelline space, and thus prevent hatching (Picker et al. 1993).

Based on a field survey, Böhmer and Rahmann (1992) report on the amphibian fauna of lakes and ponds in the northern Black Forest (Germany). Since the most acidic lakes in this area also have the strongest UV absorption (an indirect measure of HS content), one cannot separate the acid (and Al) effect from the HS effect at low pH values. Yet, the result are similar to the ones reported above.

8.2.5 Fish

Following studies of Finnish lakes and using additional data from Sweden, Norway, and North America, Rask et al. (1999) come to the conclusion that, in general, there is no specific fish community in boreal humic lakes. The data does show, however, that some species thrive better than others in humic waters. If one considers the extreme conditions in strongly humic headwaters, one can often discern two groups of fish species. The first group comprises widespread species such as pike (*Esox lucius*) and perch (*Perca fluviatilis*) which are relatively tolerant of HS linked acidity. These species generally spend their entire life cycle in oxygen-rich epilimnetic waters. The second group of fish comprises small cyprinids, among others, which are tolerant of low oxygen content, but susceptible to predation. In Europe, the Crucian carp (*Carassius carassius*) is an example. In North America, this niche is occupied by the mudminnows (*Umbra* spp.) and minnows, which colonize small lakes and ponds in which species sensitive to low oxygen concentrations are absent.

In some Finnish lakes, there is a significant negative correlation between TOC content of the lakes and the body length of 5 year old perch and roach (Fig. 8.6). Rask et al. (1999) ascribe this principally to indirect HS effects, such as poor light conditions, that result in less efficient predation on fish and invertebrates. As a further mechanism, the authors mention density dependent factors. The growth of perch in particular in small Finnish lakes depends largely on population density, which in turn may determine the food availability. However, the authors do not explain why the population density effect is greater in lakes with a higher TOC content. When Rask et al. (1999) refer to effects of humic waters on fish, they always include the linked effects of acidity. Direct effects of HS themselves are not discussed, although they clearly exist.

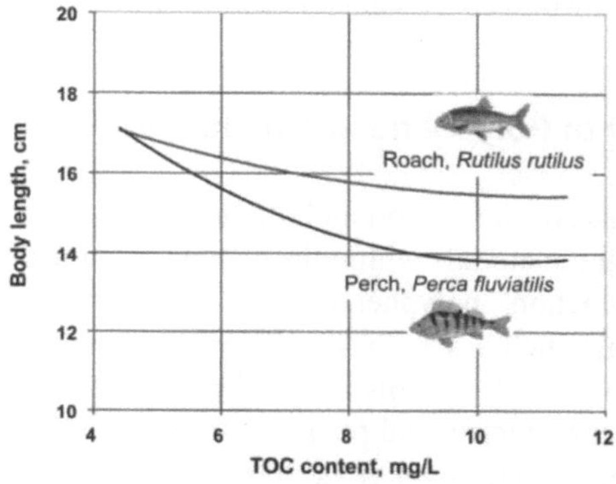

Fig. 8.6. Mean body length of perch and roach from lakes of different HS content (from Rask et al. 1999, with kind permission of Backhuys Publishers). For ease of presentation only the regression lines are taken. Original water color is transformed into TOC concentration using Eq. 3.4

Early life stages of fish such as eggs and embryos, which live on the sediment surface in close contact with HS, are more susceptible than the free swimming larvae and adults (Meinelt et al. 1997). The HS at the sediment surface are available at higher concentrations, via sediment pore water, than in the water phase. High exposure to HS occurs through direct contact of eggs and embryos with the HS particles, or the HS-covered sediment particles. If direct effects of HS on fish or their early developmental stages are expected, the effects on susceptible eggs and embryos will predominate over the effects on adults.

New results from studies using subacute sensitive biomarkers, such as heat shock proteins, show that juvenile fish also respond to HS exposure, although the physiological costs are not yet known (Chap. 8.3.7).

8.3 Effects in Non-Acidic Waters

The literature on the direct effects of HS comes mainly from agricultural production and human medicine. There is extensive knowledge for example on the role of HS in soils, soil processes, and the growth of plants and microorganisms (Chen and Aviad 1990; Kononova 1966; Lobartini et al. 1994; Muscolo and Nardi 1997; Nardi et al. 1994; Tan and Tantiwiramanond 1983; Box 8.1). The effects are not only stimulatory in nature, but can also be allelochemical as shown by Pellissier (1993) for two mycorrhizal fungi of spruce. There are also a number of reliable publications on the antiviral, bactericidal and other positive effects of HS on human cells and tissues (Riede et al. 1991; Schneider et al. 1994; Box 8.1). In contrast, there are comparatively few studies on the direct effects of HS on aquatic plants and animals. Recently, some studies are available using circum-neutral, often reconstituted, water to separate the effect of HS and HS-like substances from that of protons in acidic humic waters. These studies yield several unexpected results which will be reported and discussed below.

8.3.1 Allelopathy of Polyphenolic Substances

Polyphenols such as catechol, resorcinol, quinol, pyrogallol, gallic acid, and phloroglucinol are starting materials within the condensation pathway (Chap. 2.2.1). In addition to this function, polyphenols play a role as allelopathic chemicals, whereby allelopathic includes both stimulatory and inhibitory actions (Molisch 1937). Well known are polyphenols as a defense mechanism, most often as feeding deterrents, in various terrestrial plants (Harborne 1993; Haslam 1989), marine macroalgae (Harborne 1993; Paul et al. 2001), and some freshwater macrophytes (Gross 1999, 2000; Gross and Sütfeld 1994; Gross et al. 1996; Planas et al. 1982; Sütfeld 1998). Particular attention is given to the water milfoil *Myriophyllum* spp. (Haloragaceae) (Gross 1999; Nalewajko and Godmaire 1993; Planas et al. 1992; Saito et al. 1989).

Myriophyllum spp. are known to produce several phenolic compounds, including hydrolyzable polyphenols (tannins) such as tellimagrandin, a new class of allelochemicals in Haloragaceae (Gross 1999). The recent results of Walenciak et al. (2002) suggest that the food-derived tannins have an impact on gut microbiota in *Acentria ephemerella* (Lepidoptera: Pyralidae). Furthermore, it is evident that these tannins have potent algicidal properties (Gross 1986), inhibiting algal exoenzymes (Wetzel 1991, 1993) and reducing the photosynthetic oxygen production by inhibition of photosystem II (Leu et al. in press). This may be the same mode of action as proposed for the interference of plant litter leachates and HS within the photosynthesis of aquatic plants (Chap. 8.3.2).

8.3.2 Plants

It is very likely that the same active structure as in the above mentioned allelochemicals are responsible for adverse effects of plant litter leachates and HS to aquatic algae and macrophytes. However, most of the studies remain at the phenomenological level.

8.3.2.1 Algae

It is known that leachate from straw and leaf litter can inhibit algal development in circum-neutral waters (Welch et al. 1990), and this is used as a treatment against undesirable algae (Ridge et al. 1999). In particular, barley straw has been tested, and to a lesser extent leaf litter from oak (*Quercus robur*) and other broad-leaved species (Ridge et al. 1999). The inhibitory effects of such substances at concentrations of 3–50 mg/L are tested with green algae (*Chlorella* sp., *Cladophora* sp., *Selenastrum* sp.) and cyanobacteria (*Microcystis aeruginosa*) (Ridge and Pillinger 1996).

Only observations, and more or less plausible assumption on the mechanisms, are available:

- the milieu must be aerobic;
- the inhibitory effects on algae are first observed after 1–3 months;
- bacteria accelerate the release of materials inhibitory to algae, but do not produce these substances themselves.

The authors suggest that the active substances are oxidized polyphenols derived in part from lignin. As seen with allelochemical polyphenols, these substances are potent agents against algae. Furthermore, these chemicals can be precursors of HS and it may be assumed that, after being included into humic molecules, they retain their allelochemical potential, at least in parts. Since Ridge and Pillinger (1996) do not publish information on ecotoxicity-relevant parameters, the mechanism can only be suggested here. Either photosynthesis may be inhibited, or the transformation enzyme system may be activated, thus consuming energy that is no longer available for cell division and plant growth; or a combination of these two mechanisms may apply, as will be described for the effects of HS on the aquatic macrophytes and algae.

Not all algal taxa are equally inhibited by materials from straw and leaf litter. Benthic diatoms often found on the surfaces of rotting oak leaves (Ridge et al. 1999), seem to be less susceptible to the inhibiting compounds than planktonic algae.

Although the studies of Gjessing et al. (1998) are at the same phenomenological level as the report of Ridge and Pillinger (1996), it is informative to consider the former briefly (Fig. 8.7). Gjessing et al. (1998) describe the effect of NOM isolates from both sections (acidified and control) of the artificially acidified Lake Skjervatjern on which the European Humex project has been carried out. The cyanobacterium *Anabaena variabilis* and the coccal green algae *Selenastrum capricornutum* (syn. *Ankistrodesmus bibraianus*) are used as test organisms, and respond differently to exposure of NOM isolates. With the cyanobacterium, the NOM from the control part of the lake has an inhibitory effect, but the isolate from the acidified section has no clear effect. Both NOM isolates have an equally stimulatory effect on the green alga.

It is suggested that interactions of NOM with trace substances and/or potential pollutants are responsible for the described effects. That means HS modify bioavailability of nutrients and metals, a conventional explanation. There may also be direct effects, namely differing susceptibilities of the cyanobacterium and alga to the NOM, which can evoke responses similar to xenobiotics and activate the biotransformation systems.

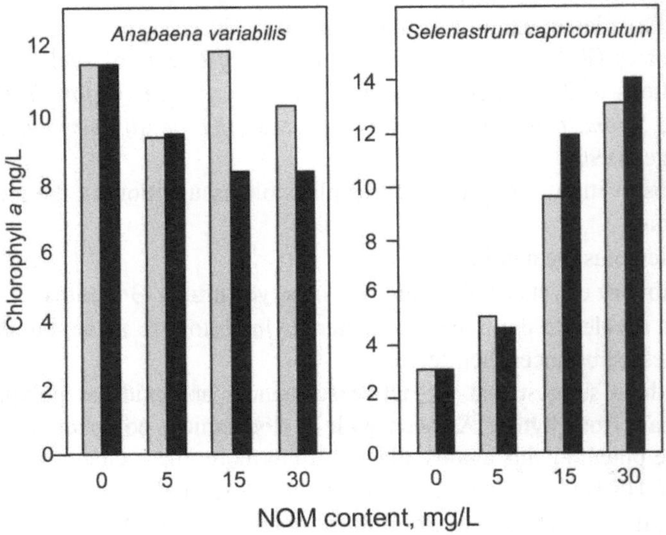

Fig. 8.7. Growth, measured as chlorophyll *a* content, of the cyanobacterium *Anabaena variabilis* and the coccal green alga *Selenastrum capricornutum* depending on the NOM content. Grey columns: NOM from the artificially acidified part of Lake Skjervatjern (Humex lake), black columns: NOM from the control part of the lake (after Gjessing et al. 1998, with kind permission of Elsevier Publishers)

Only recently, Amé et al. (in prep.) try to elucidate in more detail, the mechanism behind the HS- and NOM-induced modulation of photosynthetic oxygen release. The test alga is *Scenedesmus armatus*. Prior to photosynthesis measurement, the algae are exposed to HS and in order to avoid light quenching during measurement, are transferred into a HS-free synthetic medium. Suwannee River NOM, a forest soil leachate FA, and a synthetic HS (HS 1500) all significantly reduce the photosynthetic oxygen release (Fig. 8.8).[3] The reduction must be due to internal cell mechanisms, probably to interference of HS or their low-molecular weight fractions within the photosynthetic electron chain. Similar results, including hints on potential modes of action are obtained with two macrophytes, the

[3] Perminova et al. (2001) do not find any apparent effect of HS on the coccal green algal species *Chlorella vulgaris*, as determined by chlorophyll fluorescence measurements. This finding, however, does not contradict the statement above for two reasons: first, the applied toxicological endpoints are of different susceptibility (according to our experience with *S. armatus*, chlorophyll concentration is not a very sensitive toxicity endpoint); second, 1 h of exposure time may generally be too short to provoke detectable changes.

hornwort *C. demersum* (Fig. 8.9) and the tropical water moss *Vesicularia dubyana* (Fig. 8.10). Recent microbiological studies show that HS have the potential to act as electron acceptors for microbial respiration (Lovley et al. 1996, 1998), thus it appears most likely that this ability also applies to the effect described for algae, and is discussed in more detail in Chap. 8.3.2.2.

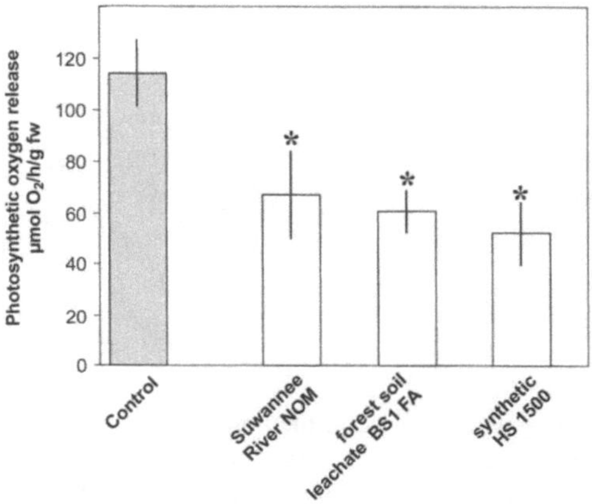

Fig. 8.8. Reduction of photosynthetic oxygen release in the coccal green alga *Scenedesmus armatus* after 18-h pre-exposure to 0.5 mg/L DOC of three HS and NOM isolates. Data are means of three replicates, ± standard deviation; * significantly different from control; fw = fresh weight (from Amé, Preuer, Nicklisch, Pflugmacher, Steinberg in prep.)

8.3.2.2 Macrophytes: Ceratophyllum demersum *and* Vesicularia dubyana

Nearly all submerged macrophytes can take up nutrients and pollutants both from the sediment through their roots, and from the surrounding water by shoots. In contrast, the hornwort *C. demersum*, as a macrophyte with a limited root system, can only take up nutrients and pollutants from the surrounding water. The same limitation applies to aquatic mosses. As a consequence of this, and because of their easy cultivation, the hornwort and the tropical aquatic moss *V. dubyana*, are popular and suitable species for ecophysiological and ecotoxicological studies. These two species are employed in studies on the direct effects of HS (Figs. 8.9, 8.10).

Pflugmacher et al. (1999b) describe the modulation of various physiological and biochemical parameters of *C. demersum* in the presence (at environmentally relevant concentrations) or absence of HS. In *C. demersum*, the photosynthetic oxygen release is significantly reduced by 8 out of 13 HS and NOM isolates, whereas in *V. dubyana* only 3 of 13 isolates reduce the oxygen release. The adverse effect of HS and NOM exposure can be seen even with the naked eye, for instance, *C. demersum* eventually turns yellow upon exposure to the forest soil leachate (BSI).

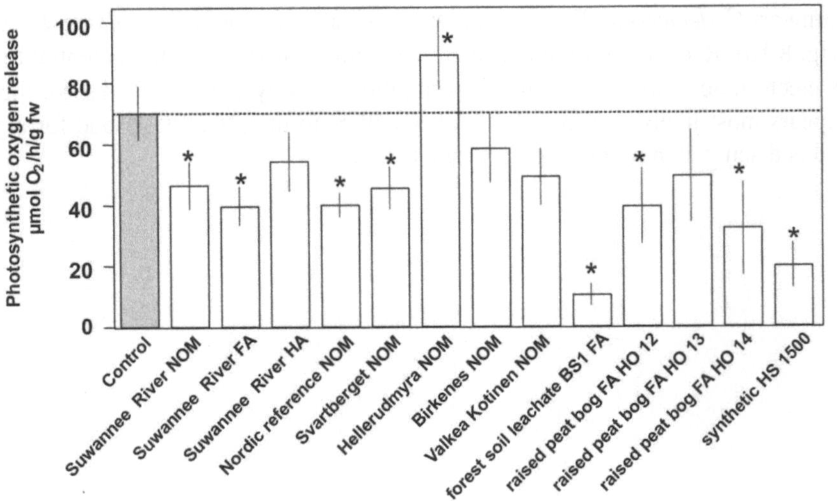

Fig. 8.9. Photosynthetic oxygen release in the hornwort *Ceratophyllum demersum* after 24 h exposure to different HS and NOM, 0.5 mg/ L C each. Prior to photosynthesis measurements, the plants are transferred into HS-free solutions. Most HS and NOM isolates significantly reduce the oxygen production, only one isolate significantly enhances it. Data are means of three replicates, ± standard deviation; * significantly different from the control; fw = fresh weight (from Pflugmacher et al. in press)

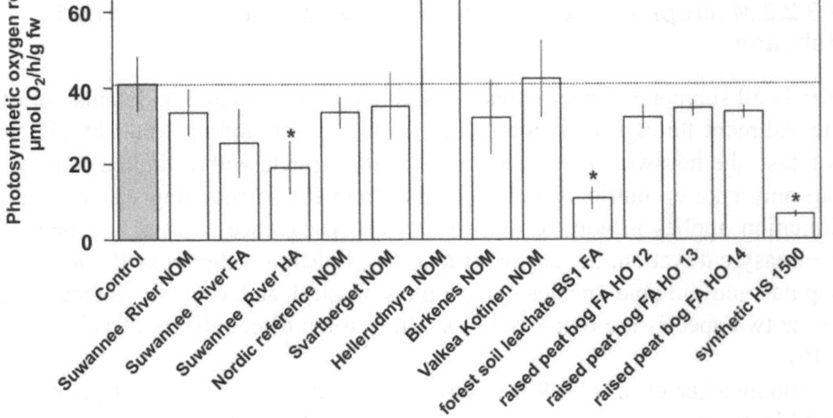

Fig. 8.10. Photosynthetic oxygen release in the tropical water moss *Vesicularia dubyana* after 24 h exposure to different HS and NOM, 0.5 mg/ L C each. Prior to photosynthesis measurements, the plants are transferred into HS-free solutions. Three HS isolates significantly reduce the oxygen production, only one isolate significantly enhances it. Data are means of three replicates, ± standard deviation; * significantly different from the control; fw = fresh weight (from Pflugmacher et al. in press)

So far, the evidence that HS or NOM are the causative agents of direct effects on aquatic organisms remains somewhat circumstantial, because one cannot exclude that the observed effects may be due to contaminants adsorbed onto the tested isolates. However, evidence would strongly increase, if the observed effects could be related to structural features of HS themselves, for instance by quantitative structure activity/effect relationships (QSAR). In fact, a QSAR can be established for the inhibition of photosynthetic oxygen release in aquatic plants. The electron trapping property can be attributed to the quinoide fraction of HS and NOM. Quinoide structures can rapidly form radicals which are electron acceptors, and can interfere with the electron flow in photosystem II. Taking electron spin resonance as an indirect, but significant measure for quinoide structures[4] (Rex 1960; Senesi and Steelink 1989; Steelink and Tollin 1962), it is evident that the reduction of photosynthetic oxygen release can be significantly related to quinoide structural units in the HS materials (Fig. 8.11). The spin content of the HS and NOM predicts approximately 90% (r^2=0.88) of the reduction of photosynthetic oxygen release in both macrophytes tested so far (Paul et al. in prep.). To date there is no mechanistic explanation for the different behavior of Hietajärvi NOM (Fig. 8.11).

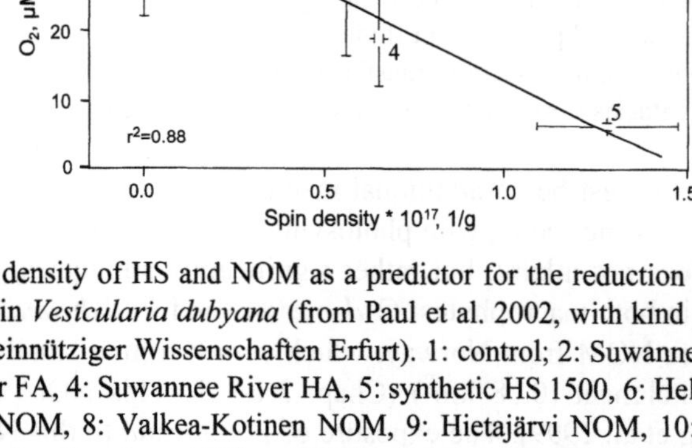

Fig. 8.11. Spin density of HS and NOM as a predictor for the reduction in photosynthetic oxygen release in *Vesicularia dubyana* (from Paul et al. 2002, with kind permission of the Akademie gemeinnütziger Wissenschaften Erfurt). 1: control; 2: Suwannee River NOM, 3: Suwannee River FA, 4: Suwannee River HA, 5: synthetic HS 1500, 6: Hellerudmyra NOM, 7: Svartberget NOM, 8: Valkea-Kotinen NOM, 9: Hietajärvi NOM, 10: Birkenes NOM. Note that several soil and peat HS and NOM isolates which have not been tested with aquatic plants so far, possess even a higher spin content than the synthetic HS 1500. Soil and peat HS have even higher spin contents than the displayed HS and NOM isolates. Hietajärvi NOM has been excluded from the regression

[4] Quinoide structures can form stable free radicals. Recently, Struyk and Sposito (2001) show that the redox capacity of HA is positively correlated with the stable free radical content, but the latter property can account for only a tiny fraction of the observed moles

The quinoide moieties have high reductive activities and can therefore act as electron traps and interrupt the electron transfer chain between photosystem II and photosystem I in plants (Fig. 8.12). The involvement of HS and NOM in the photosynthesis of aquatic organisms appears to be an intrinsic property of these substances.

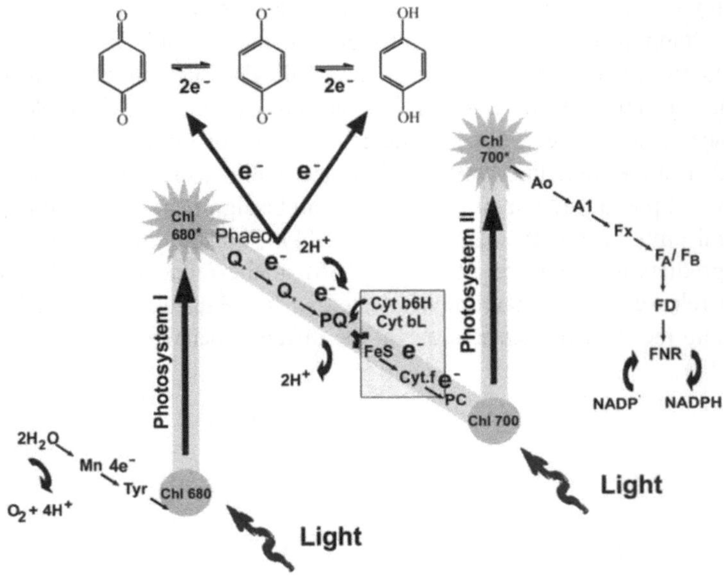

Fig. 8.12. Z-scheme of the photosynthesis in higher plants showing the electron transfer between photosystem I and photosystem II via a chain involving several plastochinones. In the proposed hypothesis, this electron transfer is interrupted by quinoide structures in HS, so that the photosynthesis is reduced or even blocked (from Pflugmacher et al. in press)

However, there must be an additional mechanism whereby HS and NOM can interfere directly, or indirectly, with photosynthesis of aquatic plants. There is one isolate which does not inhibit, but rather significantly stimulates, photosynthetic oxygen release in both macrophytes, *C. demersum* and *V. dubyana*. This isolate is the Hellerudmyra NOM from Norway, which comes from a small mountain lake with a small catchment (0.08 km^2) (Chap. 3.1.1.2). This NOM is assumed be young (Gjessing et al. 1999). The exposure of *V. dubyana* to this NOM leads to a near doubling in photosynthetic rate. The mechanism behind this effect is still obscure. Yet, a simple inorganic nutrient effect can be excluded, since the plants are kept in complete artificial nutrient-rich solutions. All macrophytes tests are performed non-axenically. Previous experiments with non-axenic coccal green algae show increases in algal yield upon dissolved HS exposure when bacteria are present in the system (Fig. 5.5). Hence, it may be concluded that some kind of metabolic support may also take place with the macrophytes. However, it still

of electrons transferred between HA and the oxidant.

remains unclear which dissolved HS/NOM structures might be responsible for such effects. Furthermore, it also is likely that both mechanisms (inhibition and enhancement of photosynthetic oxygen release) occur simultaneously, and that the net effect is the trade-off between these two modes of action (see the mathematical model for the reproduction of *C. elegans* in Chap. 8.3.4.1 and Fig. 8.25).

Clear changes also occur in other physiological parameters when aquatic plants are exposed to HS. For example, the chlorophyll pattern (the ratio of chlorophyll *a* to chlorophyll *b*) of *C. demersum* changes strikingly (Fig. 8.13). The reduction in chlorophyll *a* and the simultaneous increase in chlorophyll *b* can function as a reaction to xenobiotic stress, for which evidence is provided from studies on the terrestrial angiosperms *Petunia hybrida* and *Phaseolus vulgaris* (Debus and Schröder 1990), lichens of the genus *Cladonia* (Chettri et al. 1998), and the water fern *Salvinia minima* (Gardner and Al-Hamdani 1998).

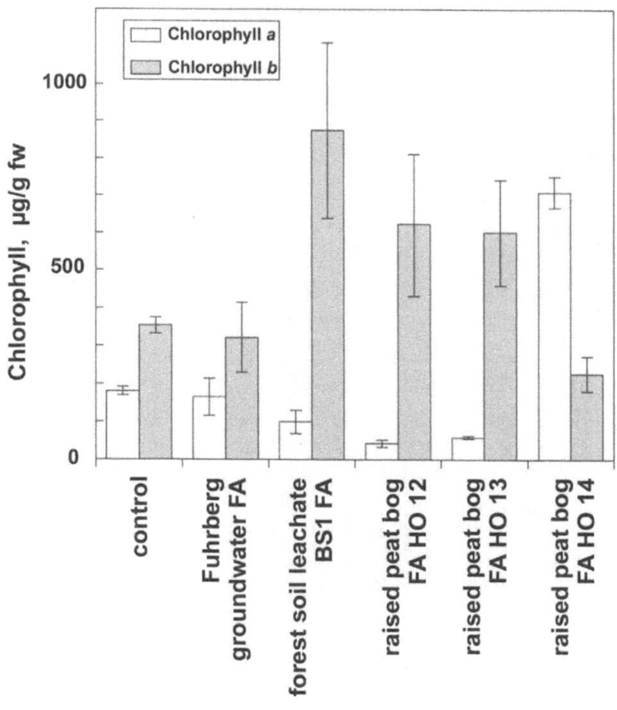

Fig. 8.13. Changes in concentrations of chlorophyll *a* and *b* in *Ceratophyllum demersum* after 7 d exposure to 0.5 mg/L DOC of different FA. Data are means, ± standard deviation. fw = fresh weight (from Pflugmacher et al. 1999b, with kind permission of the Institute of Plant Ecology, University Gießen, Germany)

The described effects on photosynthesis and pigment pattern in *C. demersum* indicate that the plant must take up the HS isolates. With these suppositions, other plant compartments such as biotransformation enzymes must also be considered as being affected by the taken up HS. Indeed, the activity of guaiacol peroxidase

(POD) climbs dramatically (to 1000 times that of the control) if the plants are exposed to only 0.5 mg/L C of various HS isolates (Fig. 8.14). This can be taken as further proof that HS put a xenobiotic stress upon plants (Roy et al. 1992). It can be suggested that the HS contain structures analogous to those of xenobiotics. Since the mentioned increase of POD is so large, Pflugmacher et al. (2001) conclude that the POD activity is not merely due to an external oxidative stress, that arises from the illumination of HS, and the subsequent release of H_2O_2, which can be destroyed by membrane-bound peroxidases and similar enzymes. POD is the only enzyme from phase I of the biotransformation system whose activity climbs in response to HS exposure. Other enzymes such as the mixed function oxygenases show either no increased activity or are slightly inhibited.

Pflugmacher et al. (2001) suggested that at least low-molecular weight HS enter intact plants and there they activate the biotransformation enzymes. That HS, HS-like substances, or parts of these molecules can penetrate cell membranes is not unreasonable as very recent evidence accumulates experimental evidence for this (Fig. 8.1).

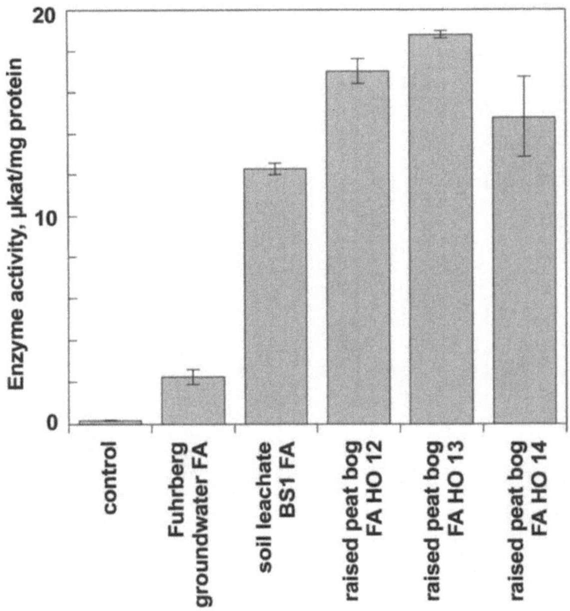

Fig. 8.14. Guaiacol peroxidase activity in *Ceratophyllum demersum* after 7 d exposure to 0.5 mg/L of various HS (from Pflugmacher et al. 1999b, with kind permission of the Institute of Plant Ecology, University Gießen, Germany)

How do the different size fractions of the organic carbon affect the physiological parameters of the macrophyte *C. demersum*? In a recent study, Wu et al. (in prep.) describe preliminary studies on the influence of two water stable aggregate size fractions, S1 (0.16–2.0 µm) and S2 (0.03–0.16 µm) from run-off from an agricultural field.

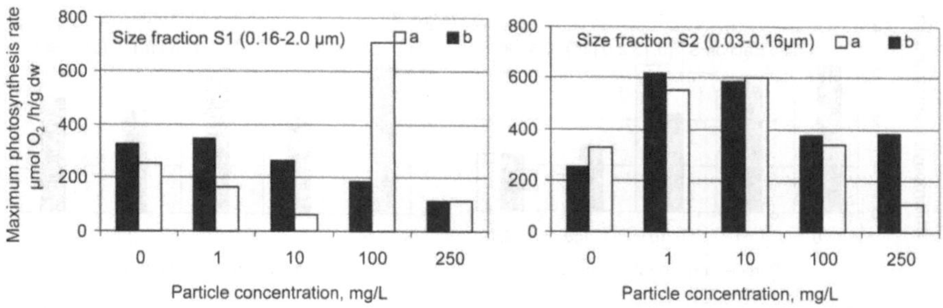

Fig. 8.15. Effects of particle concentration on photosynthesis of *Ceratophyllum demersum* after 24 h exposure to particle suspensions of two size fractions S1 (0.16–2.0µm) and S2 (0.03–0.16µm) from a run-off material, a and b are replicates (from Wu et al. submitted)

The aggregate size has significant effects on the photosynthesis of *C. demersum*, as shown in Fig. 8.15. After exposure to the coarse size fraction S1, the photosynthetic rate decreases with increasing aggregate concentration in suspension. Exposure to the colloidal fraction S2 stimulates photosynthesis at lower particle concentrations up to 10 mg/L, and inhibits the photosynthesis at higher concentrations. The content of chlorophyll *a* and *b* and the ratio of the two chlorophylls in *C. demersum* change significantly after exposure to the S1 fraction (Figs. 8.16 and 8.17), whereas the influence of the S2 fraction is not significant (Figs. 8.16 and 8.17). The stronger influence of the coarser size S1 fraction on photosynthesis, chlorophyll content, and chlorophyll pattern may be attributed to its higher turbidity compared to the colloidal S2 fraction at the same aggregate concentration. In addition, the relatively higher nitrogen content in S2 fraction may promote the photosynthesis and pigment production in *C. demersum*, and thereby mask a potential inhibitory effect on photosynthesis.

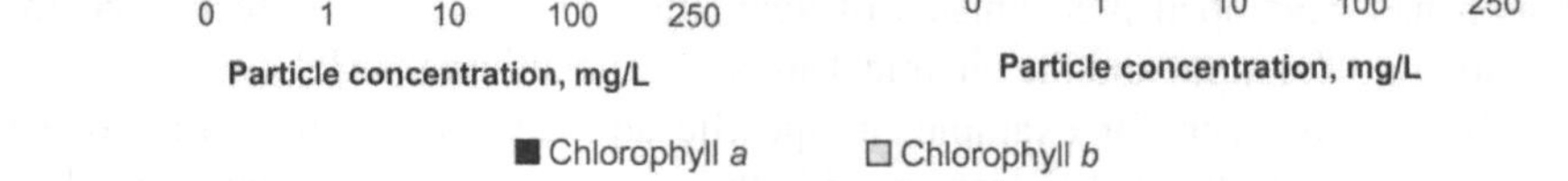

Fig. 8.16. Effects of particle concentration on chlorophyll concentrations of *Ceratophyllum demersum* after 24 h exposure to particle suspensions of two size fractions S1 (0.16–2.0µm) and S2 (0.03–0.16µm) from a run-off material (from Wu et al. submitted)

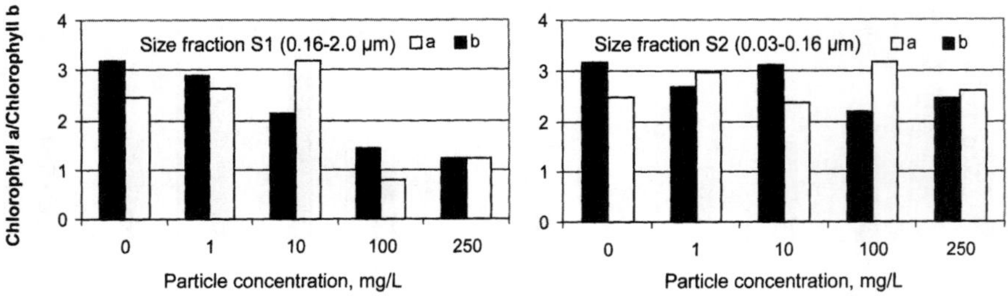

Fig. 8.17. Effects of particle concentration on the ratio of chlorophyll *a*/chlorophyll *b* of *Ceratophyllum demersum* after 24 h exposure to particle suspensions of two size fractions S1 (0.16–2.0µm) and S2 (0.03–0.16µm) from a run-off material, a and b are replicates (from Wu et al. submitted)

Comparisons of these results with those from uncontaminated forest soil samples from Ås, Norway, shows that the forest soil DOC also reduces photosynthetic activity of the macrophyte, indicating that the reduction in the agricultural run-off exposure experiment is probably not due to potential contaminants in these aggregates.

8.3.2.3 Niche Differentiation by Humic Substances for Aquatic Plants

Comparing Figs. 8.9 with 8.10, it becomes evident that the water moss *V. dubyana* appears less sensitive to several HS exposures than the angiosperm *C. demersum.* Taking this as a general difference between water mosses and angiosperms, a frequent field observation can be explained with an additional mechanism: in humic waters, in particular running waters, mosses are commonly the dominant macrophytic primary producers (Vuori and Muotka 1999)[5]. One cannot rule out that several mosses are less susceptible to HS-related effects than are angiosperms, despite the ability of mosses to use CO_2, the dominant inorganic carbon species in low pH conditions, as inorganic C source for photosynthesis. This statement is clearly at present hypothetical and implies the following: terrestrial vegetation has an effect on aquatic macrophytes via HS. HS have the potential to structure primary producer guilds or communities. If the coupled land/water system has sufficient time to develop concurrently, an aquatic flora will emerge which has developed mechanisms of adaptation to HS-related effects, and thus can tolerate the input and the chemical stress of allochthonous HS.

This can be seen for example, in specific aquatic plant communities of soft water and humic lakes. HS from outside the community or catchment, which are introduced through anthropogenic activities, such as the introduction of exotic terrestrial plant species, should lead to dramatic changes in the macrophyte com-

[5] Mosses are important primary producers in humic running waters, but their most important role is the accumulation and maintenance of allochthonous organic carbon which reaches higher trophic levels through the saprophytic food chain (Vuori and Muotka 1999).

munities. Various species react with differing sensitivity to 'foreign material', and in competition more sensitive species succumb.

When comparing all three aquatic plants tested so far (Fig. 8.18), however, it is evident that there is no 'most sensitive' species. With Suwannee River NOM, the most sensitive species is the coccal green alga *S. armatus*; with the soil FA it is the angiosperm *C. demersum*; and with the synthetic HS 1500 it is the water moss *V. dubyana*. That means that a specific region, with specific terrestrial plant cover resulting in HS with specific chemical features, will produce a specific aquatic community under non-eutrophicated conditions. Some general rules may apply, such as water mosses gain dominance in humic waters. But, even this statement has to be confirmed in future studies.

At present we are still clearly in the realm of information gathering, particularly since such effects of HS on photosynthetic oxygen release of aquatic plants are unexpected and contradict conventional paradigms on the inability of plants to take up HS or NOM.

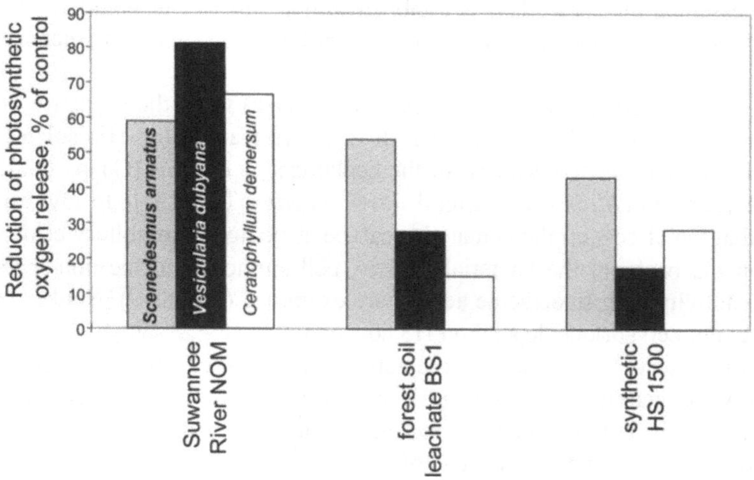

Fig. 8.18. Different susceptibilities of the three species of aquatic plants towards three different HS or NOM, as indicated by reduction of photosynthetic oxygen release

8.3.3 Fungi and Bacteria

8.3.3.1 Fungi

The growth of aquatic hyphomycetes is inhibited by aqueous leaf-litter extracts. Bärlocher (1992) describes a reduction in radial growth of four hyphomycetes on agar plates. This effect depends on the concentration of dissolved tannins (plant polyphenols, Chap. 8.3.4.2), as well as the leaching rate, and rarely lasts longer than 3–4 months. It is not certain if these substances are identical to the algae-inhibiting substances identified by Ridge et al. (1999). Nevertheless, there must be some kind of toxic action, since Mathur (1969) also describes a total inhibition of FA decomposing fungi in the presence of high HS (>10mg/L) concentrations.

8.3.3.2 Bacteria

How HS directly act on bacteria has very rarely been investigated; there is much more interest in the extent to which HS can support microbial metabolism following photolytic cleavage. However, since the antiviral, antibacterial, and antifungal effects of HS are known, for example from medicine and aquaculture (Boxes 8.1, 8.2), it is surprising that the aspect of potential adverse effects of HS on microorganisms is so little studied in an ecological context.

Visser (1985), for example, describes how low HS concentrations can strongly stimulate, and high HS concentrations inhibit, microorganisms. He explains this in terms of a tenside-like effect on membranes and consequent increased permeability to essential materials. This is established in an analogy technique in which the effect of FA is compared with that of known laboratory tensides (such as Tween 80).

In a completely different way, Höss et al. (2001a) study the influence of HS on the model bacterium *Escherichia coli*. It is known that various HS isolates clearly modify the reproductive success of the nematode, *C. elegans* (Chap. 8.3.4.1), and in the study, *E. coli* is used as food for *C. elegans*. To be able to investigate the idea that the effects on the nematodes can be attributed to inhibitory effects of the HS on the bacteria, the bacterial density, cell surfaces and the non-specific esterase activity with fluorescine acetate are examined (Fig. 8.19). Neither bacterial density nor cell surface depend on HS concentration (Fig. 8.19a, b). A weak effect of HS is observed with the esterase activity (Fig. 8.19c) when exposed to raised bog FA. Its activity remains significantly less than that of the control or other isolates at 72h. Yet, this reduced enzyme activity has no effect on bacterial density or cell surface within the time-span of the experiment. It is likely that the exposure time is too short. The reduction in enzyme activity can only be compensated for when excess energy is available. Under natural conditions, this is rarely the case.

In contrast to limnological literature, in the clinical microbiology literature there is a wealth of information about plant products as antimicrobial agents. The underlying antimicrobial mechanisms may also apply to HS, since the antimicrobial chemical compounds in plants are also constituents of HA and FA. These antimicrobial chemicals (Fig. 8.20) can be divided into several categories (Cowan 1999).

Phenolics and polyphenols Some of the simplest bioactive (phyto)-chemicals consist of a single substituted phenolic ring. Cinnamic and caffeic acids are common representatives of a wide group of phenylpropane-derived compounds which are in the highest oxidation state. Catechol and pyrogallol both are hydroxylated phenols, shown to be toxic to microorganisms. Catechol has two –OH groups, and pyrogallol has three.

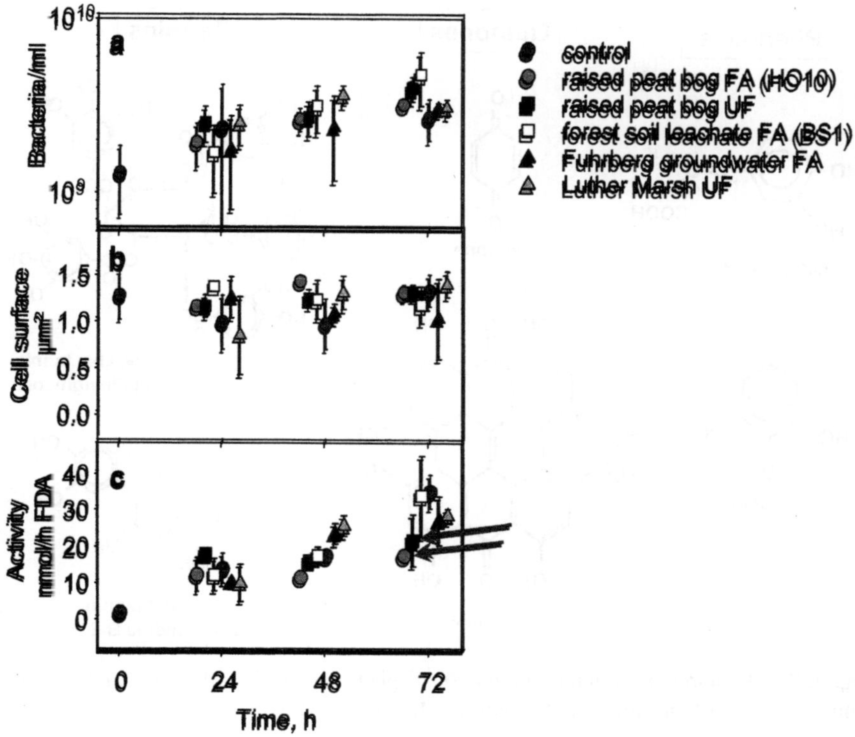

Fig. 8.19. Abundance (**a**), cell surface (**b**) and esterase activity (**c**) of *Escherichia coli* after 0, 24, 48 and 72 h exposure to various HS isolates (50 mg/L DOC) in the presence of the bacterivorous nematode *C. elegans* (from Höss et al. 2001a, with kind permission Blackwell Science). Data significantly different from control are indicated by arrows. UF = ultrafiltrate

The site(s) and number of hydroxyl groups on the benzene ring, are thought to be related to their relative toxicity to microorganisms, with evidence that increased hydroxylation results in increased toxicity. In addition, some authors report that more highly oxidized phenols are more inhibitory. The mechanisms thought to be responsible for phenolic toxicity to microorganisms include enzyme inhibition by the oxidized compounds, possibly through reaction with sulfhydryl groups, or through more non-specific interactions with the proteins (such as natural preservation of bodies in bogs, Box 8.1).

Quinones are aromatic rings with two ketone substitutions. They are ubiquitous in nature and are characteristically highly reactive. The switch between diphenol (or hydroquinone) and diketone (or quinone) occurs easily through oxidation and reduction reactions. The individual redox potential of the particular quinone-hydroquinone pair is very important in many biological systems. Hydroxylated amino acids may be made into quinones in the presence of suitable enzymes, such as polyphenoloxidase. The reaction for the conversion of tyrosine to quinone is shown in Fig. 8.21.

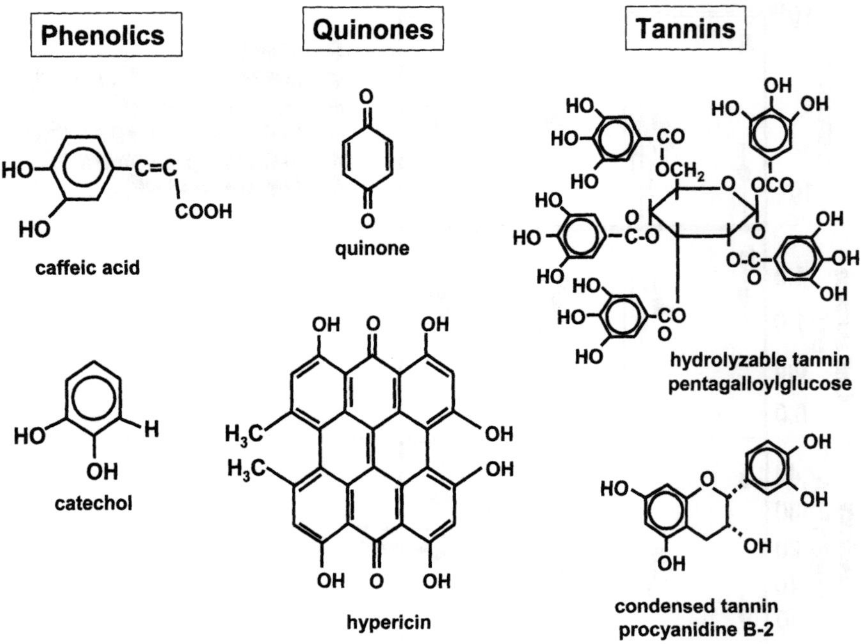

Fig. 8.20. Structures of common antimicrobial plant chemicals (after Cowan 1999, with kind permission of the American Society for Microbiology)

In addition to providing a source of stable free radicals, quinones are known to complex irreversibly with nucleophilic amino acids in proteins, often leading to inactivation of the protein and loss of function. For that reason, the potential range of quinone antimicrobial effects is great. Probable targets in the microbial cell are surface-exposed adhesins, cell wall polypeptides, and membrane-bound enzymes.

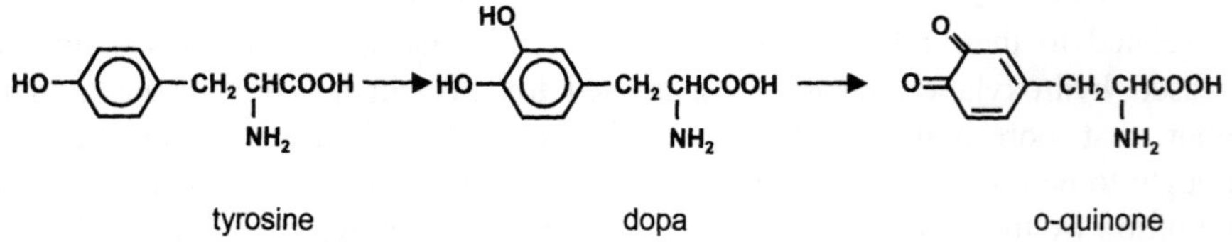

Fig. 8.21. Conversion of the protein amino acid tyrosine to quinone

Tannins 'Tannin' is a general descriptive name for a group of polymeric phenolic substances, with molecular weights ranging from 0.5–3.0 kDa. They are divided into two groups, hydrolyzable and condensed tannins. Hydrolyzable tannins are based on gallic acid, usually as multiple esters with d-glucose, while the more numerous condensed tannins are derived from flavonoid monomers (Fig. 8.20). Tannins may be formed by condensations of flavan derivatives, and by polymerization of quinone units.

From ecological biochemistry, tannins are known as feeding deterrents (Harborne 1993), as antifoulants, as chelators of metal ions, and as UV shields (Pautou et al. 2000). One of the molecular modes of action of tannins is to complex with proteins through so-called nonspecific forces such as hydrogen bonding and hydrophobic effects, as well as by covalent bond formation. Thus, their mode of antimicrobial action may be similar to that of quinones. They also complex with polysaccharide. Tannins in plants inhibit insect growth (Hammerschmitt and Schultz 1996; Schultz 1989). In freshwater, tannins determine ecological niches of crustaceans and mosquitoes (Pautou et al. 2000; Chap. 8.3.5.3).

8.3.4 Invertebrates

8.3.4.1 Nematodes

Höss et al. (2001a) test the effects of various HS on invertebrates, with the hypothesis that HS non-selectively influence growth and particularly reproduction. To test this, they use a nematode-bioassay with *C. elegans* (Traunspurger et al. 1997). Under the conditions used, these nematodes reproduce through self-fertilization, hence removing any genetic variability in the effects found. The HS and NOM used come from various sources including FA from waste water of a municipal treatment plant (ABV2), forest soil leachate (BS1), wastewater from lignite processing (SV1), a raised peat bog lake (HO10), a groundwater (Fuhrberg), and an ultrafiltrate from a marsh (Luther Marsh, Ontario, Canada).

None of the tested HS isolates influences the growth of *C. elegans*, but many have significant effects on reproduction (Fig. 8.22). The effects on reproduction are not uniform. While the Fuhrberg FA and the lignite waste water (SV1) FA have no influence on the number of offspring per worm; the soil leachate FA (BS1), the waste water FA (ABV2), and particularly the Luther Marsh ultrafiltrate significantly increase the number of offspring per worm at least at some concentrations. The enhancing effect of the FA from the municipal waste water may be due to human hormones and their degradation products, which are still present in measurable quantities in the effluent. However, since the hormones themselves do not adsorb very much to the resins during preparation unless they are not incorporated in the lipophilic HS matrix, it is most likely that other mechanisms cause the observed effects. We will try to elucidate potential mechanisms below.

At low concentrations, the raised peat bog lake FA (HO10) enhances the numbers of offspring, whereas concentrations above 5 mg/L DOC, environmentally relevant, completely inhibit reproduction. Since the food bacteria (*E. coli*) are less active in the Lake Hohlohsee isolate than in the others (Fig. 8.19), an adverse effect of food quality on the fertility cannot be ruled out. It is not yet known to what extent this negative effect is reversible.

In summary, HS can have both promotory and inhibitory effects on *C. elegans* reproduction. The effects depend both on the quality and quantity of the HS. The variability of the chemical building blocks from the various HS sources (aromatic fraction, aliphatic fraction, amino acids and carbohydrate-like structures, phenolic, and carboxylic groups) cannot statistically explain the variability of the effects.

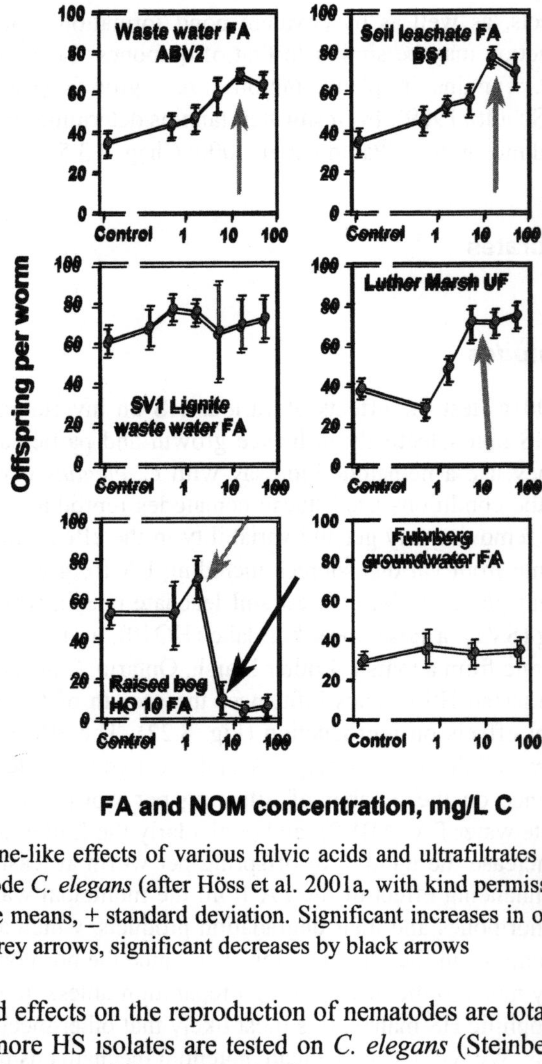

Fig. 8.22. Hormone-like effects of various fulvic acids and ultrafiltrates on the reproduction of the nematode *C. elegans* (after Höss et al. 2001a, with kind permission of Blackwell Science). Data are means, ± standard deviation. Significant increases in offspring numbers are indicated by grey arrows, significant decreases by black arrows

The described effects on the reproduction of nematodes are totally unexpected. Consequently, more HS isolates are tested on *C. elegans* (Steinberg et al. 2002). In detail the results are as follows (Fig. 8.23):

- Suwannee River FA. Above 8 mg/L DOC, there is a steady increase in offspring numbers with increasing exposure concentrations.
- Suwannee River HA. Concentrations above 8 mg/L DOC significantly increase the reproduction of *C. elegans*. The response seems to reach a plateau.
- Suwannee River NOM. No significant effect on reproduction at any concentration.
- Humex A NOM. Even the lowest exposure concentration leads to significantly elevated numbers of offspring. The highest concentrations increases

the numbers of offspring by more than a factor of four.

- Humex B NOM. Significant increases in reproduction are present only at the lower exposure concentrations.
- Hellerudmyra NOM. Significant increases in reproduction are present only at low exposure concentrations.
- Svartberget NOM. Significant promoting effects occurs with the higher exposure concentrations.
- Valkea-Kotinen NOM. A significant increase in offspring numbers is observed even with the lowest exposure concentration. The increased reproduction effect happens according to the 'all or nothing" principle. At the highest exposure concentration, a slight, but significant reduction in offspring numbers is obvious.
- Hietajärvi NOM. This NOM generally significantly enhances the number of offspring per worm, even at the lowest concentration.
- Nordic Reference NOM. The results are similar to the Valkea-Kotinen NOM exposure. Maximum numbers are observed at 4 and 64 mg/L DOC exposures.

Which kind of direct effect mechanisms may be valid? In this regard, the uptake of DOC for *C. elegans* seems to be important. The DOC can be taken up by nematodes via the cuticle (Lopez et al. 1979) as low-molecular weight DOC in dissolved form, or it can be ingested with the food, adsorbed on bacterial cells. Because the molecular weight of the isolates vary considerably from low-molecular weight moieties in the NOM isolates, to high-molecular weight moieties in HA, it is likely that both uptake routes may be operating. Once DOC is taken up by either route, it may serve as an additional food source, as is known for other invertebrates. For instance, Sherr (1988) showed that heterotrophic marine flagellates can utilize high-molecular weight polysaccharides as a carbon source. More recently, Ciborowski et al. (1997) and Wotton (1977) show that simuliid larvae can grow in a medium containing only allochthonous dissolved or colloid organic carbon. Although evidence is inconclusive, Ciborowski et al. (1997) speculate that DOC is precipitated on labral fans and ingested. In humic streams and rivers, DOC may be a direct and significant food source for simuliid larvae, particularly.

Another, even more promising explanation for the modulation of numbers of offspring is that specific structural units of the HS may act as pseudo-hormones. For instance, Klöcking et al. (1992) report that an isolate from a raised peat bog had a hormone-like effect on mice, roughly three orders of magnitude below the effect of estriol. Nevertheless, the authors conclude that the estrogenic potential of peat is considerably higher than previously anticipated. From pyrolysis mass spectrometry studies, it is well known that HS contain sterols, such as cholesterol or sitosterol, as minor constituents. These sterols mainly derive from plant material and usually do not exceed 5–8% of the structural units (Schmitt-Kopplin et al. 1998; Schulten 1999b).

However HS-bound sterols can be excluded as effective agents: in the nematode bioassay, cholesterol is a major constituent of the test medium. Consequently, it appears highly unlikely that additional HS-bound sterols, with presumably low bioavailability, will affect the nematodes any further. We therefore have to consider non-sterol structures in the HS which may have the potential of pseudo-hormones.

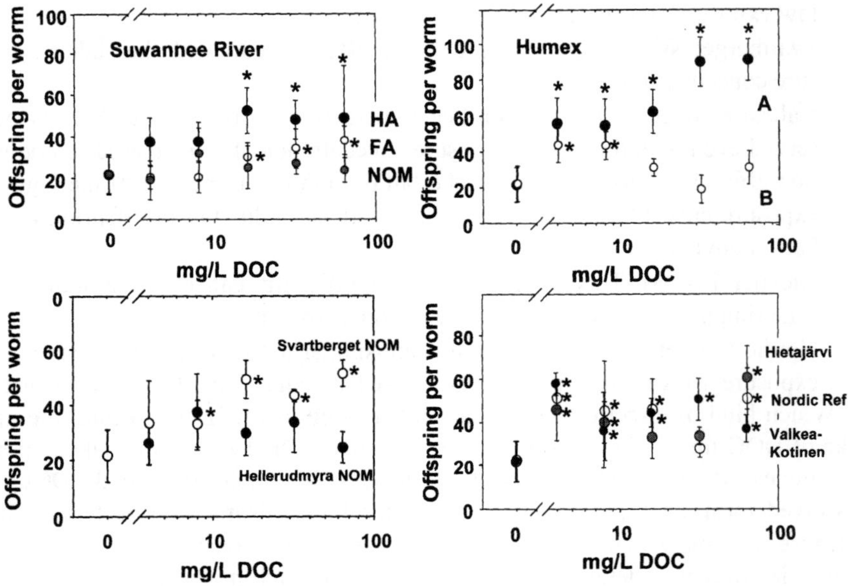

Fig. 8.23. Hormone-like effects of various HS and NOM on the reproduction of *C. elegans*. Data are means, ± standard deviation; * significantly different from control (from Steinberg et al. 2002, with kind permission of Wiley-VCH)

In contrast to sterols, alkylaromatics are major constituents of HS; and more importantly with respect to environmental relevance, they belong to the most photostable structures (Schmitt-Kopplin et al. 1998). They persist in the sunlit zones and are exposed in comparatively high concentrations to aquatic organisms. Actually, Höss et al. (2002) find that alkylphenols enhanced the reproduction of the nematode *C. elegans* in concentrations of 50 µg/L (lowest observed effect concentration, LOEC), as exemplified with 4-nonylphenol (Fig. 8.24).

Soil FA and HA from a site near Munich, Germany, are shown to contain alkylaromatics between 15 and 20% respectively, of the total of structural subunits (Schmitt-Kopplin et al. 1998). Schulten (1999b) reports that in FA and HA from a raised peat bog, alkylaromatics comprise 14% and 7%, respectively, of the total structural units. Thus alkylaromatics belong to the major building units of HS. If we assume 10% alkylaromatics as a model estimate of HS, and relate it simply to organic carbon, the following preliminary calculation can be made: a 4 mg/L DOC of a humic isolate contains up to approximately 400 µg/L alkylaromatic units. Even if most of these units are not available for receptors of the nematodes, an effective concentration of several tens of micrograms per liter can be predicted. This may explain the modulation of reproduction in general. It is interestingly to note that recently, estrogen receptors or binding proteins are demonstrated in *C. elegans* by radio immunoassay (Hood et al. 2000), and that environmental chemicals such as nonylphenol, bind to the receptor and thus alter the pathway of estrogen metabolism. The authors show that even relatively lipophilic chemicals, such as dieldrin and toxaphene, interfere with the estrogen receptor. This fact may serve as analogous evidence that low-molecular weight HS can also interact with the receptor.

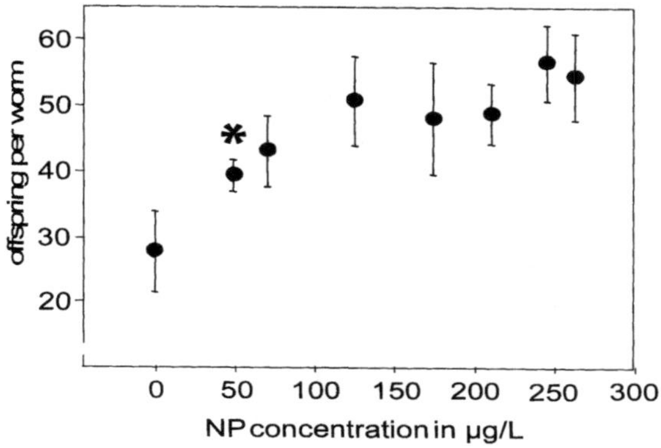

Fig. 8.24. Effect of 4-nonylphenol (NP) on the numbers of offspring of *C. elegans*. Data are means, ± standard deviation; * = LOEC (lowest observed effect concentration) (from Höss et al. 2002, with kind permission of Elsevier Publishers)

Two different results are obvious from the studies with *C. elegans*: HS may serve as a (minor) additional food source, and can modulate the reproduction of nematodes. How can the displayed responses be explained? This will be evaluated by a mathematical model. The basic assumptions of this model are:

1. nematodes follow (at least during the non-stationary phase) logistic growth
2. HS can non-specifically act as food (coefficient α)
3. HS acts specifically on the per capita growth rate by increasing or reducing the reproduction (coefficient β).

The parameter β will be positive, zero, or negative, according to whether:

- HS have an enhancing effect on growth,
- HS do not have any significant effect on the numbers of offspring (Suwannee River NOM in Fig. 8.23),
- HS have a negative effect on growth (FA from a raised peat bog, Fig. 8.22), and, in weak form, also Humex B, or Hellerudmyra NOM at higher concentrations). The most plausible mechanism appears to be that HS block the receptor structures and counteract the beneficial effect of cholesterol in the nutrient medium.

Scenarios with given conditions are going to be analyzed. For example, with concentrations of HS varying from 0 to 5 arbitrary units (a.u.) and the intensity of humic impact on numbers of offspring, β, from –0.08 to +0.08, the following diagram can be generated (Fig. 8.25).

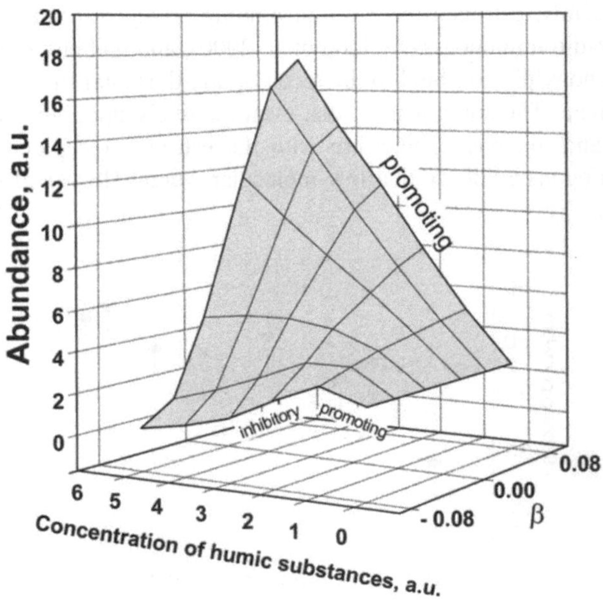

Fig. 8.25. Qualitative response of abundance of nematodes as an approximate measure for the consequences of the presence of HS. The dose-effect relationships can be identified by following parallels of concentration of HS along the abundance surface. β is the intensity of humic impact on numbers of offspring. For instance, with β = 0.08, the abundance increases with increasing concentrations of HS up to some 18 arbitrary units (a.u.) (from Steinberg et al. 2002). This model also to the ambiguous response of HS in aquatic ecosystems (Chap. 9; Steinberg and Brüggemann 2001, with kind permission of Wiley-VCH)

For instance, with β = 0.08, the abundance increases almost linearly with increasing concentrations of HS up to some 18 a.u.: the effect is predicted as promoting reproduction. Yet, with β of –0.03 there is a plateau of the abundance at higher humic concentrations: the effect becomes negligible. If β is <–0.03, the abundance of nematode will decrease, because the worms may have become infertile (as occurred with HO 10 FA above 5 mg/L DOC, Fig. 8.22). It is noteworthy that, with β = –0.08 and low HS concentrations, the model qualitatively also describes a small increase in abundance prior to a strong decrease.

Seemingly, there are two competing processes, and the resulting curve is a trade-off between them. It is intriguing to note that the ambiguous impact of HS on the numbers of offspring also occurred in the experiments of Steinberg et al. (2002), where an intermediate, or ultimate, decrease in reproduction has been observed, for instance with Humex B, Hellerudmyra or Hietajärvi, and Nordic Reference NOM (Fig. 8.23). To conclude: all experimental findings may be explained by one model. The inert response of dose-response relationships (negative β values, high HS concentrations) is due to a non-specific effect of enhancing the nutrient base, and a negative effect on the growth rate, and a trade-off between these two modes of action. This model does not only apply to the modulation of numbers of offspring of *C. elegans*, but also to the ecological behavior of HS in aquatic ecosystems, as described in Chap. 9 in more detail.

8.3.4.2 Crustacea

Tannins have a strong niche differentiating potential towards culicine mosquitoes, so they may be applied as a natural insecticide (Chap. 8.3.5.3) (Pautou et al. 2000). The effects of tannins on non-target organisms such as crustaceans have also been investigated. Bioassays indicate that exposure to tannic acid at concentrations of 0.06 to 2.0 mmol is more deleterious to *Chydrorus sphaericus, Diaptomus castor*, and *Eucypris fuscata*, than to *Acanthocyclops robustus*, *Daphnia pulex*, and *Eucypris virens*. Histophathological investigations after treatment with tannic acid, at concentrations from 0.125–0.50 mmol, reveal sequential degenerative patterns of the midgut epithelium depending on the taxon, duration of the treatment, and concentrations assayed.

Further effects of tannins on populations dynamics are described by Kautz et al. (2000) for the parthenogenic isopod *Trichoniscus pusillus*: condensed tannins in the food of the isopod negatively influence the reproductive success, while hydrolyzable tannins increase longevity of the isopod.

At present it is not well understood whether the impact of HS leads to reduced biodiversity in lakes and streams. For instance in the most prominent blackwater system of the world, the Rio Negro in South America, zooplankton does not show reduced species numbers compared to that in humic-poor systems (Brandorff 1978) In contrast, 'the profundal zone of the blackwater lakes has qualitatively and quantitatively the poorest benthic fauna of all lacustrine biotopes in the Central Amazon' (Reiss 1977).

It is interesting to note that crustaceans have developed specific life-history traits in blackwaters. Particular *Macrobrachium* sp. (Decapoda, Palaemonidae), from blackwaters typically very poor in phytoplankton, have abbreviated life cycles, as compared to *Macrobrachium amazonicum* from the floodplain lakes rich in phytoplankton, which are fed by the richer water of the Amazon River, and which maintains the marine pattern of development with numerous stages of filter-feeding, planktonic larvae (Walker 1992). This example may be taken as clear evidence that zooplankton and zoobenthos species in blackwaters have developed specific strategies to efficiently adapt to nutrient depleted and humic-rich habitats. However, with low species numbers of molluscs and the absence of mosquitoes, exceptions of the highly developed zoodiversity in Amazon blackwaters do exist (Fittkau 1981).

Daphnia magna

In an as yet unpublished study on *D. magna*, Meems et al. (in prep.) find that in pH neutral conditions, HS isolates from the Suwannee River, activate biotransformation enzymes such as glutathione-*S* transferases (GST) and glutathione peroxidases[6] (Fig. 8.26). Significant activation takes place of membrane-bound (microsomal) GST by Suwannee River HA, of soluble enzyme by Suwannee River NOM, and of guaiacol peroxidase, a biotransformation phase I enzyme, by both Suwannee River NOM and HA. In addition to biotransformation, elevated peroxidase activity may be caused by an oxidative stress, following radiation of the HS which leads to the release of ROS (Chap. 5.3). However, this is definitely not the case for the two forms of GST. It is much more likely that Suwannee River HS or fractions thereof are taken up by *D. magna*, and under energy consumption, are conjugated and metabolized. Thus the water fleas try to resist the HS stress and treat them as xenobiotic material.

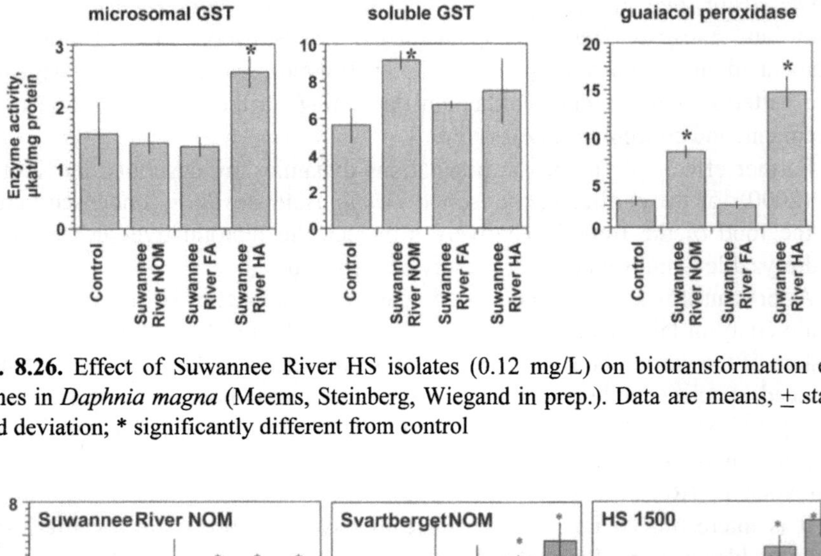

Fig. 8.26. Effect of Suwannee River HS isolates (0.12 mg/L) on biotransformation enzymes in *Daphnia magna* (Meems, Steinberg, Wiegand in prep.). Data are means, ± standard deviation; * significantly different from control

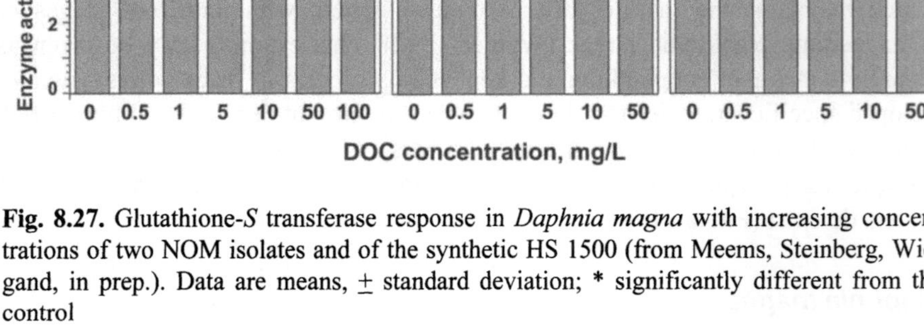

Fig. 8.27. Glutathione-*S* transferase response in *Daphnia magna* with increasing concentrations of two NOM isolates and of the synthetic HS 1500 (from Meems, Steinberg, Wiegand, in prep.). Data are means, ± standard deviation; * significantly different from the control

[6] glutathione peroxidase (GPX) reduces organic hydroperoxides, particularly lipid hydrop-

The question arises whether there is a dose-response relationship between HS and NOM exposure, and biotransformation enzyme activity. This is tested with three different HS/NOM isolates and the results are shown in Fig. 8.27. It is evident that *D. magna* responds significantly to increasing exposure concentrations, with increases in GST activity. This is most pronounced with the synthetic HS 1500 (slope 0.77) and least pronounced with Suwannee River NOM (slope 0.21). HS/NOM concentrations statistically account for as much as 83–95% of the variability of the enzymatic response (Table 8.3). Since all three isolates activate the phase II enzyme, a common mechanism might be assumed. Hence, the results serve as further support for the xenobiotic nature of HS and NOM.

Table 8.3. Statistics for linear regressions for the graphs in Figure 8.27: $y = ax + b$

HS/NOM	a	b	r^2
Suwannee River NOM	0.21	4.60	0.9563
Svartberget NOM	0.31	4.81	0.8307
HS 1500	0.77	2.34	0.8506

8.3.4.3 Diptera

Terrestrial vegetation may release natural chemicals, such as tannins, into freshwater ecosystems. Tannins are among the most important compounds known to mediate plant-insect relationships. The annual input of these organic compounds plays a prominent discriminating role in aquatic habitat selection, via ecological discrimination of mosquito breeding sites. Despite this interesting phenomenon, there are few field and laboratory studies address the discriminating role of tannins (or HS) in aquatic invertebrate habitat selection.

Ecological maps of the Alpine aquatic systems indicate specific correlation between vegetation zones and distribution of the mosquito fauna (Pautou et al. 1973). From this mapping, a guild structuring (niche differentiation) impact through tannins is hypothesized. Ecophysiological investigations determine that in mosquito breeding sites, tannins and, more generally, phenolic compounds from decaying leaves, ingested by Nematocera larvae, can have differing toxic effects and, thus, these compounds are involved in niche differentiation of Diptera (Rey et al. 1996, 1999b, 2000). Insensitive taxa, like *Aedes albopictus*, *Ae. rusticus*, *Anopheles claviger*, *Culiseta morsitans*, and *Chaoborus crystallinus*, can colonize waters with high levels of terrestrial tannins. Sensitive taxa, like *Culex pipiens*, *Simulium variegatum*, and *Chironomus annularius* are, however, confined to open habitats with low tannin levels. This implies that different adaptations of taxa, by acquisition of different detoxifying capacities, are correlated with their colonizing

eroxides that can be assembled in membranes. It also reacts directly with O_2 and H_2O_2.

ability (Rey et al. 1999a).

The above statement is supported by findings from the Rio Negro system. In blackwaters of the Amazon basin, mosquitoes are almost completely absent, due to obviously toxic effects of HS (Fittkau pers. comm.).

8.3.4.4 Zoobenthos

Several field studies from, for example, Finland, indicate that there are very few zoobenthos specific to HS-rich running waters. Many taxa which occur in HS-rich waters are also found in HS-poor waters such as *Asellus aquaticus* (Isopoda), *Helobdella stagnalis* (Hirudinea), *Glossosoma intermedia, Micrasema gelidum, Plectrocnemia conspersa* (Trichoptera), *Nemoura cinerea* and *N. flexuosa* (Plecoptera) (Vuori and Muotka 1999). However, the reverse is not the case; most species found in HS-poor running waters do not occur in HS-rich running waters. According to the Finnish studies, the latter waters harbor fewer species and are characterized by the predominance of a few taxa. An example is shown in Fig. 8.28.

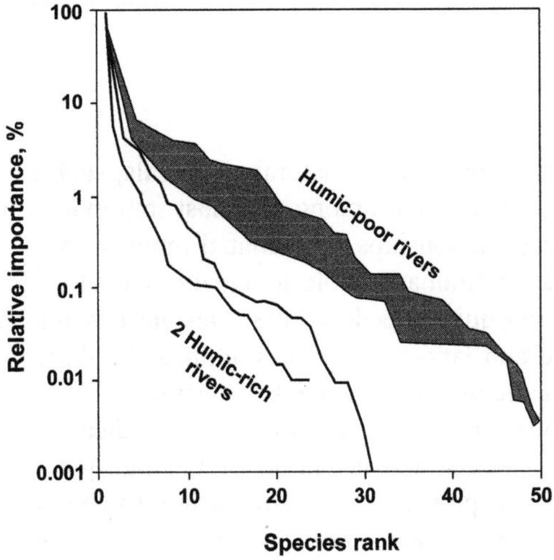

Fig. 8.28. Species abundance curves of 23 stream sites in northeastern Finland (from Vuori and Muotka 1999, with kind permission of Backhuys Publishers)

In the Finnish studies, it is obvious that in HS-rich sites, the species spectrum is clearly reduced. The fauna is dominated by species such as *Amphinemura sulcicollis, N. cinerea* (Plecoptera) and *Micrasema gelidum* (Trichoptera). These are species which belong to the guilds of shredders, seston collectors and detritus feeders. This and another study of boreal waters (Malmqvist and Mäki 1994) suggest that the guild of algae grazers is conspicuously absent from HS rich waters, probably due to short light attenuation.

In HS-rich streams, reduced biodiversity is also described from a small study in southern German habitats. In the HS-rich stream, only 58 macroinvertebrate taxa occur, and in contrast in the HS-poor stream, 73 macroinvertebrate taxa occur (Tham et al. 1997). Only 32 species are common to both streams. Diptera and Trichoptera represent the most diverse groups, contributing 35 and 17 taxa, respectively. Among Trichopteran species, only 13 species occur in the HS-rich stream, most of them belonging to the family Limnepilidae. This trend of reduced species diversity is even more pronounced in some other macroinvertebrate taxa. Only two species each of Gastropoda, Ephemeroptera and Plecoptera are described from the HS-rich stream; Turbellaria and Hirudinea are completely absent. In contrast, 7 Gastropoda, as well as 6 Ephemeroptera and 4 Plecoptera species are recorded from the HS-poor stream.

From an unnamed, first-order tributary of Quebrada Platanilla, Costa Rica, Benstead (1996) reports that benthic invertebrate density is extremely low when compared with similar streams in temperate latitudes. It is open to question, if the low biodiversity in the Costa Rican stream is due to the persistent high phenolic compound content after a period of rapid leaching. Absolute figures for the phenolic content, however, are not presented.

Reports on macroinvertebrate diversity are, however, by no means unequivocal. For instance, studies on macroinvertebrates in Coastal Plain blackwater streams of the USA contradict the classic but 'unwarranted' view of blackwater streams as having low macroinvertebrate taxonomic richness and productivity. Species richness can be very high, often being comparable to that found in streams located in other geophysical provinces in the southeast USA (Smock and Gilinsky 1992). Over 240 taxa are collected from a first-order stream in Virginia, 170 taxa are collected from second-order Cedar Creek in South Carolina, and over 550 taxa of aquatic insects are collected around fourth-order Upper Three Runs Creek (South Carolina). The latter number reflects the intensive work on adult insects at that site and also the more heterogeneous nature of the stream, due to the occurrence of large beds of macrophytes that are not found at the other sites. Diptera typically compose the greatest proportion of the taxa; Trichoptera are also well represented. Many of the insect species are endemic to the region (Smock and Gilinsky 1992).

Macroinvertebrate production in Coastal Plain blackwater streams is based on allochthonous detritus, in particular on fine particulate organic matter. Smock and Roeding (1986) report that fine particulate organic matter (detritus) supports 47–64% of macroinvertebrate production in Cedar Creek. Detritus and associated bacteria, much of which originates on the floodplain, are an important link between terrestrial primary production and invertebrate production in blackwater streams and rivers, for example, accounting for the majority of production by invertebrate microfilterers and collectors-gatherers of the Ogeechee River (Smock and Gilinsky 1992 with further references). The statement that detritus may serve as an energy and nutrient source (Smock and Roeding 1986) is noteworthy. Yet, direct evidence by stable isotope analysis, that detritus itself, rather than the attached bacteria, is the food of the macroinvertebrates is not presented. However, from zooplankton studies, we know that the detritus itself may be a major direct food source (Chap. 9, Fig. 9.3).

Molluscs

Only a few species of Mollusca are normally present in the smaller and more acidic Coastal Plain blackwater streams. The most abundant molluscs are the Sphaeriidae (fingernail clams), which have a greater tolerance to high acidity than do most other molluscs (Pennak 1989). Mollusc abundance increases as stream size increases, and is correlated with the decrease in acidity and higher calcium concentrations that often occur with increasing stream size. *Corbicula fluminea*, the introduced Asiatic clam, is abundant in some rivers such as the Ogeechee (Georgia, USA), but is not found in other rivers, probably because of low calcium concentrations.

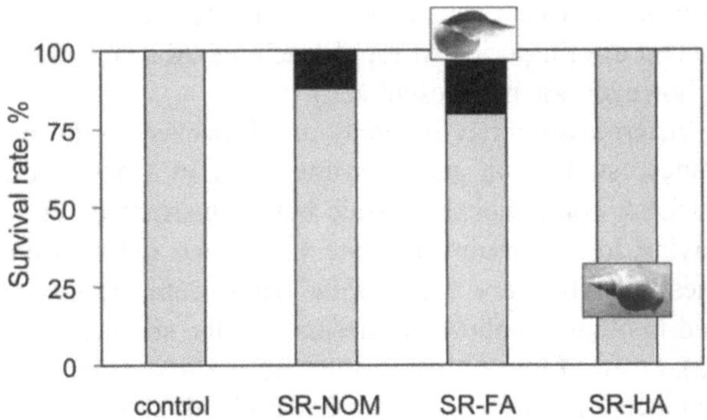

Fig. 8.29. Survival rates of the European freshwater snail, *Lymnea stagnalis*, exposed to 0.5 mg/L DOC of various HS from the Suwannee River (SR) for 24 hours. Grey columns: living snails; black columns: dead snails (from Hillmeister, Pflugmacher, Steinberg in prep.)

An additional mechanism affecting the distribution of molluscs may be attributable to HS. From recent cross experiments in which the European freshwater snail, *Lymnea stagnalis*, is exposed to different HS from the Suwannee River, there is clear evidence that under circum-neutral conditions, the snails respond to HS exposure as if exposed to man-made chemicals (Hillmeister et al. in prep.). After a 24 h-exposure to 0.5 mg/L DOC the activity of the transformation systems is extremely elevated, and Suwannee River NOM and FA, but not HA cause death of 10–20% of the animals (Fig. 8.29). This finding means that HS appear to have a toxic potential per se.

Future studies may determine whether the toxic potential of HS pertains only to exotic species or also to indigenous species. Probably, the indigenous flora and fauna are better adapted to 'their' HS, than are exotic species. Nevertheless, even in sixth order blackwater rivers of the Coastal Plain of Georgia, USA, individual size, growth rates, and production of the Asiatic clam, *C. fluminea*, indicate that this blackwater is a stressful environment, probably due not only to low alkalinity, as Stites et al. (1995) assume, but also to the chronic adverse effects of HS. It is interesting to note that in Amazon blackwaters such as the Rio Negro system, only very few mollusc species can exist (Fittkau 1981).

Trichoptera

From the paper by Tham et al. (1997) there is clear evidence that macroinvertebrates can be adversely affected if exposed to HS, even under circumneutral conditions. In their small study on ultrastructural changes, Tham et al. describe a direct, adverse effect of HS on larvae of the Trichoptera *Halesus digitatus*. Under the influence of HS, the chloride epithelia, which are the osmotically active cells, exhibit a smooth instead of a rough endoplasmic reticulum in the basal region (Fig. 8.30a,b). Furthermore, in individuals from the HS-rich stream, the number of polysomes in the same cellular area is generally lower. Since polysomes are predominantly formed from ribosomes, this finding indicates that HS reduce protein synthesis, at least in the chloride cells. It is worth noting that in this study, both the water with HS and the control are circum-neutral, thus the adverse effects are attributable to the HS themselves, and not to protons or metals.

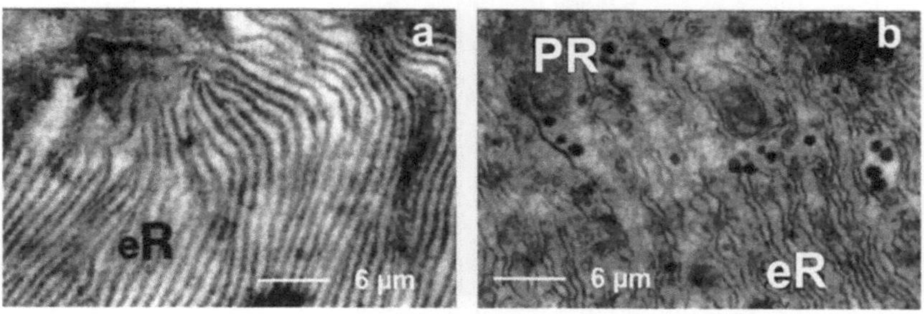

Fig. 8.30. Transmission electron micrograph of a cross-section from chloride epithelia from *Halesus digitatus* larvae. a: Subcellular structures from a larva from the HS-rich stream showing abnormal smooth endoplasmic reticulum (eR) in the basal cell part. Note also the scarcity of polysomes (PR). b: Subcellular structures from a larva from the HS-poor stream showing normal rough endoplasmic reticulum in the basal cell part. Note also the abundant polysomes (from Tham et al. 1997, with kind permission of the Polish Society of Humic Substances)

Gammarids

In the previous chapters, we demonstrate that aquatic plants, several invertebrates, and fish respond to exposure of pure HS isolates as if exposed to xenobiotic chemicals. Thus, it is not surprising that gammarids also show similar reactions upon exposure to HS isolates. In a screening test we exposed gammarids from Lake Müggelsee (Berlin, Germany) and Lake Baikal (Siberia, Russia) to Sanctuary Pond NOM (Ontario, Canada). Exposed individuals of *Gammarus tigrinus* and another undetermined *Gammarus* spp. have hsp 70 more strongly expressed than do control individuals (Fig. 8.31). As fish also respond to HS exposure by elevated hsp 70 concentrations, the induction of heat shock proteins appears to be a general reactions of aquatic organisms to exposure of HS and NOM.

Very recently Wiegand et al. (in press) describe the time-dependency of a HS-stress to a gammarid species, *Eulimnogammarus* sp., from Lake Baikal (Fig. 8.32). The toxicological endpoint is the guaiacol peroxidase activity. The activity of this enzyme increases from 30 minutes to 3 days and declines significantly before the ninth day. This may be due to two major reasons. First, the enzyme may be exhausted and/or inhibited or exhausted by the long-term exposure, and second the gammarids may get acclimatized to the stress or are able to compensate for the HS mediated stress by modes of action as yet unknown.

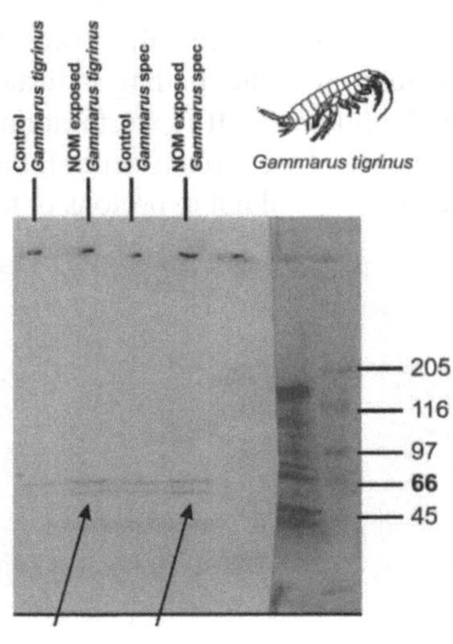

Fig. 8.31. Western blot to trace expression of hsp 70 in two gammarids from Lake Müggelsee, Germany, exposed to 0.5 mg/L NOM from Sanctuary Pond, Ontario, Canada. Exposed individuals show stronger bands of hsp 70 than control individuals (after Wiegand et al. in press)

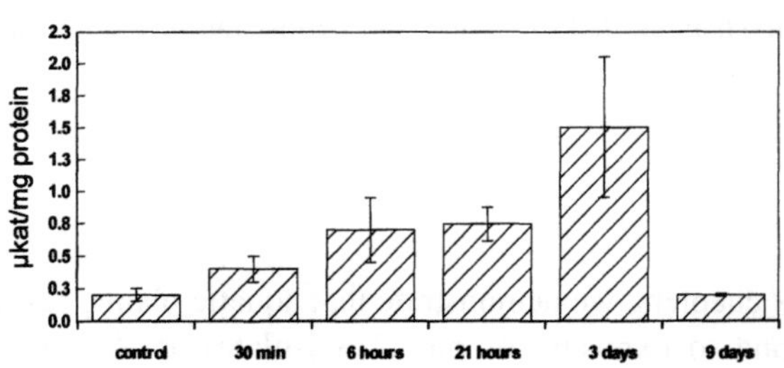

Fig. 8.32. Temporal changes in guaiacol peroxidase activity in *Eulimnogammarus* sp. from Lake Baikal exposed to Sanctuary Pond NOM (Ontario, Canada) at a concentration of 0.5 mg/L (after Wiegand et al. in press). Data are means, ± standard deviation.

8.3.5 Comparison of *Ceratophyllum, Dreissena,* and *Chaetogammarus*

Studies on the hornwort *C. demersum* with the caffeic acid oxidation product and various HS isolates suggest that HS can pass through cell walls and membranes, and can incite the activation of transformation enzymes. In a related study, Pflugmacher et al. (2001) address the main question of how molluscs and gammarids from hard-waters react to the HS isolates which interfere with the photosynthetic electron transport system in *C. demersum.* One of the most inhibitory isolates to *C. demersum* is the FA from the forest soil leachate (BS1 FA, Fig. 8.8). Therefore, the mussel *Dreissena polymorpha* and the gammarid *Chaetogammarus ischnus* are exposed to that FA at a concentration of 0.5 mg/L DOC for 7 days. For comparison, the plant is included in this study, since if different organisms respond similarly upon exposure of HS, the underlying mechanism may be generalized. The activities of microsomal and soluble GST are used as the toxicity endpoint. The results are shown in Fig. 8.33.

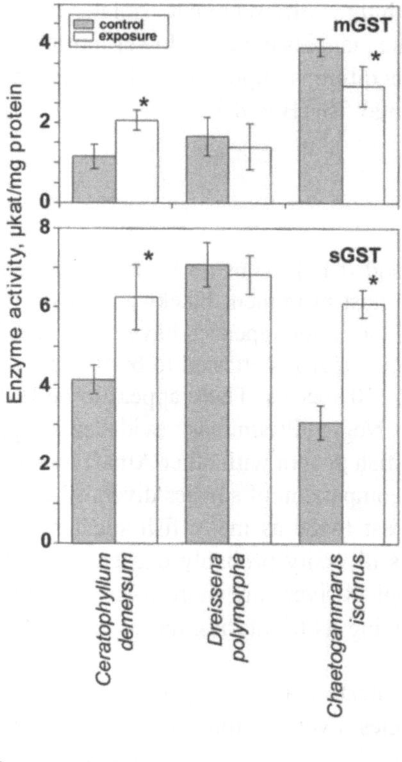

Fig. 8.33. Activity of microsomal glutathione-*S* transferase (mGST) and soluble glutathione-*S* transferase (sGST) in *Ceratophyllum demersum, Chaetogammarus ischnus,* and *Dreissena polymorpha,* exposed to FA of a forest soil leachate (BS1) for 7 days. Data are means, ± standard deviation; * significantly different from the control (from Pflugmacher et al. 2001, with kind permission of Wiley-VCH)

The results are not uniform. In the hornwort, both the soluble and microsomal GST are significantly activated. In the mussel *D. polymorpha*, neither of the GSTs differs in activity from the control, so it remains unclear whether the FA is taken up into the body of this species.[7] In the gammarid *C. ischnus*, the microsomal GST shows a significant inhibition. In contrast, the soluble GST shows a significant increase. The results indicate that this FA (BS1) must have been taken up by the macrophyte *C. demersum*, and by the gammarid *C. ischnus*, and that the FA activates the detoxification (transformation) system. Membrane-bound (microsomal) and soluble GST are differently activated or inhibited; there is, as yet, no clear explanation for this. This FA may not be taken up by the mussel *D. polymorpha*, or if it is taken up, the biotransformation system is not activated.

8.3.6 Amphibians

Despite the negative effects of HS on amphibia (Chap. 8.2.4), Ahlgren and Bowen (1991) report a positive effect. The authors feed tadpoles of the American Toad (*Bufo americanus*) with three different DOC isolates from a HS-rich river flowing into Lake Superior. The tadpoles exhibit a lower mortality rate than the unfed tadpoles. Their general condition is improved, a phenomenon which we shall also see with food and ornamental fish (Box 8.2).

8.3.7 Fish

It is not well understood if HS exposure leads to reduced biodiversity of fish in lakes and rivers. The most prominent blackwater river, the Rio Negro (Amazon basin, South America), does not appear to have a reduced diversity of genera. The high diversity of the Rio Negro is attested to by the fact that there are 200 genera out of a total of about 450 species. There appear to be few genera of fish that are found only in the Rio Negro. Preliminary evidence suggests that the Rio Negro shares nearly all of its fish genera with other Amazonian or Orinoco Rivers (Goulding et al. 1988). For comparison of species diversity, it may be noted that the Rio Negro alone has at least twice as many fish species as all the European rivers taken together, and its diversity probably exceeds that of all of North America. Comparisons with tropical rivers are more relevant, however. The Zaire system (Central Africa), including its tributaries, has at least 669 species (Goulding et al. 1988).

First impressions indicate that the Rio Negro alone may rival the entire Zaire Basin in total fish species diversity. Interestingly, with respect to the potential ad-

[7] The obvious lack of susceptibility of *D. polymorpha*, native to the Caspian Sea region of Asia, to HS may be one reason for its competitive success in invading exotic habitats, such as Europe (late 19th Century) and the North American Great Lakes. Since 1988, *D. polymorpha* spreads rapidly to all of the Great Lakes, and waterways in many states, as well as Ontario and Quebec (Nalepa et al. 1999).

verse effects of HS on fish communities, Goulding et al. (1988) hypothesis that extreme nutrient-poverty, low pH and high levels of humic compounds (which taken together are expressed in the aquatic environment as blackwaters) are not limiting factors to the diversity of fish families, genera, and species in the Rio Negro river system. One likes to add that the species composition of the ichthycoenoses developed over a long period of time, and the individual species had sufficient time to adapt physiologically to these conditions.

Fittkau (1973, 1997) offers a very interesting explanation for the extremely high biodiversity in tropical ecosystems in general, and in Rio Negro habitats in particular. He sees the magnificent richness as an adaptation to geochemically depleted habitats. The species richness is thus not an expression of the rich supply of food available in such habitats, but is obviously a form of adaptation to the continuous limitation of supply of nutrients and foodstuffs under otherwise permanently favorable conditions. This hypothesis has earlier been put forward by Fittkau (1973) and is inductively confirmed on the basis of numerous analyses of diversity (Fittkau 1997). Tropical, geochemically rich, and thus nutritionally rich, habitats are generally unstable both spatially and temporally; and they probably do not permit a large amount of diversity to develop in the ecosystems which they support. A deficient nutritional supply must always have played a decisive role in the development of the wealth of life forms, because lack of vitally required substances means selection, and selection determines adaptations, and adaptation forces changes, which ultimately leads to diversity and influences evolution. Surpluses weakens the selection process and restrict the development of biological diversity. Diversity decreases with eutrophication of a system. This statement means that the animal species inhabiting habitats of the Rio Negro system must successfully have developed mechanisms of adaptation not only to nutrient depletion, but also to chemical stresses from HS. Having this in mind and re-considering the findings from the above mentioned enzymatic studies, one likes to speculate that the decrease of the guaiacol peroxidase activity in *Eulimnogammarus* sp. (Fig. 8.32) is the start of an adaptation, or at least acclimatization, rather than the exhaustion of the enzyme. Furthermore, not only fish, but also potential pathogens and parasites are stressed by exposure to HS. If the pathogens and parasites are more susceptible to HS-mediated stress, the fish will benefit from the HS exposure. There are some indications that this may apply (Fig. 8.39).

In addition to Fittkau's evolutionary perspective, Tilman et al. (1996) offer a functional explanation. By using experimental grassland plots and native grassland with differently diverse communities, Tilman et al. (1996) show that ecosystem productivity increases significantly with plant biodiversity. Moreover, the main limiting nutrient, soil mineral nitrogen, is utilized more completely when there is a greater diversity of species, leading to lower leaching loss of nitrogen from these ecosystems. Similar effects apply for biodiversity of macrophytes and phosphorus retention in wetlands (Engelhardt and Ritchie 2001). These findings support the diversity–productivity and diversity–sustainability hypotheses and may functionally explain the species richness of several animal communities in Amazonian blackwaters.

In a recent paper, Saint-Paul et al. (2000) compare Amazonian blackwater and whitewater fish communities. A total of about 2,000 species are described so far, but the number may reach 2,500–3,000. The number of species in blackwater floodplains appears to be some 10% higher than in whitewater, and there is 54% similarity. The most abundant blackwater species are predators: *Plagioscion squamosissimus* (Sciaenidae: South American silver croaker), *Serrasalmus rhombeus* (Characidae: redeye piranha), and *S. manueli* (Characidae: a harmless piranha). The most abundant whitewater species are *Liposarcus pardalis,* (Loricariidae: armored catfishes), *Pygocentrus nattereri* (Characidae: red piranha), and *Pellona flavipinnis* (Clupeidae: yellowfin river pellona); the first is a detritus grazer, the two other species are predators. The number of species contributing to the main catch in the blackwater is much higher than in the whitewater site. This correlates with the lower dominance index for blackwater as compared to whitewater. Even seasonal differences are less pronounced, or do not even exist.

Saint-Paul et al. (2000) continue that there are very few data on biomass, growth, and production of Amazonian fish communities, and data from different studies are difficult to compare because of the different fishing methods used. But, it is very interesting to note that catches in blackwater are so low that no commercial fishery is carried out in this area. The data on fish biomass in black- and whitewater using the same gear and the same fishing effort show that in the Amazon floodplains, fish biomass is about five times higher than in the Rio Negro, indicating both a better food supply in the Várzea (floodplains of whitewater rivers) and a chronic chemical stress by HS, and/or a combination of HS and protons in the blackwater, which has to be compensated for by diverting energy, for instance, from growth.

Very recently, Gonzalez et al. (2002) describe some physiological strategies for ion regulation in fish from the Rio Negro, and detect a protective mechanism of HS. Rio Negro fish are generally fairly tolerant of low pH and extremely dilute waters. When fish are exposed to low pH in Rio Negro water, instead of deionized water with the same concentration of major ions, the effects of low pH are reduced. This suggests that high concentrations of dissolved organic molecules in the water may interact with the branchial epithelium in some protective manner.

In HS-rich boreal lakes, there are clear differences in the growth of various fish species in waters with increasing DOC content (Fig. 8.6). although there is generally no specific fish community. In general, these differences seem to attributable to indirect effects of HS, such as changed light regime and the consequent changes in hunting opportunities, as well as increased acidity. Initial results with carp yearlings (*Cyprinus carpio*) indicate that fish can also react directly to HS. In particular, in the gills which are in intensive contact with the media (5 mg/L Suwannee River FA, or 0.5 mg/L DOC for NOM), the expression of heat shock protein (hsp) 70 increases (Fig. 8.34, 8.35). In contrast, muscles show hardly any reaction to such exposure.

In the gills, the activity of soluble GST is significantly increased, following exposure to Suwannee River FA, but is significantlyreduced following exposure to NOM. In the liver, the activity of soluble GST is signifantly reduced by all Suwannee River isolates, whereas in muscle, soluble GST does not show significant differences from the control (Fig. 8.36). With one exception (gills, following Suwannee River FA exposure), microsomal GST activity is not significantly altered.

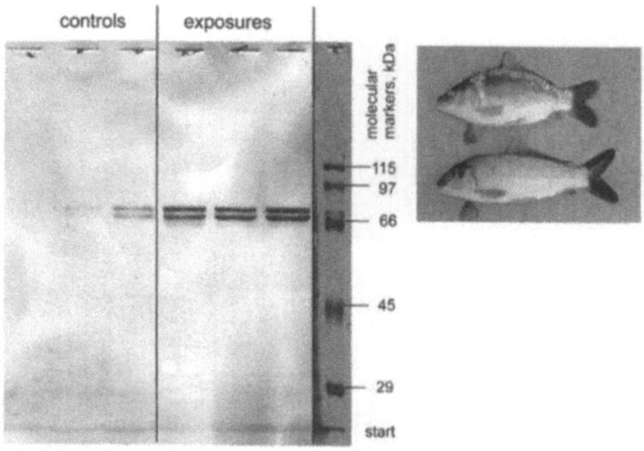

Fig. 8.34. Expression of heat shock protein (hsp) 70 in the gills of individual carp yearlings exposed to 5 mg/L DOC Suwannee River FA for 24 hours (after Wiegand et al. in press)

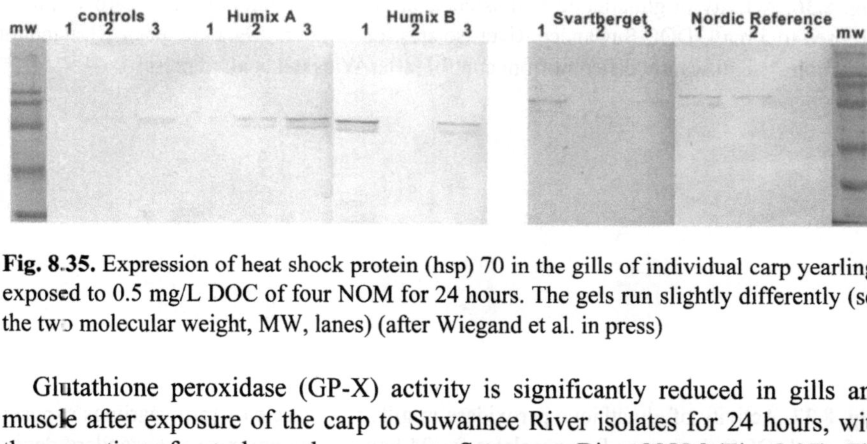

Fig. 8.35. Expression of heat shock protein (hsp) 70 in the gills of individual carp yearlings exposed to 0.5 mg/L DOC of four NOM for 24 hours. The gels run slightly differently (see the two molecular weight, MW, lanes) (after Wiegand et al. in press)

Glutathione peroxidase (GP-X) activity is significantly reduced in gills and muscle after exposure of the carp to Suwannee River isolates for 24 hours, with the exception of muscles and exposure to Suwannee River NOM (Fig. 8.37). The reactions are statistically significant, but difficult to interpret. High base-line GP-X activity in the gills is undoubtedly due to contact with the oxygen-rich external medium, and activity in the muscles may be attributed to transfer along the respiratory chain. The given DOC concentrations (5 mg/L Suwannee River HS,) are not so high that the GP-X must be depleted. The increased GST activity in the gills (Fig. 8.36) is not so strong that glutathione in the cells could be exhausted. One general explanation for the enzyme modulation may that the quinoide structures of the HS react, in the redox cycle, with flavin adenine dinucleotide and oxygen. In these structures, a large quantity of O_2^- radicals are produced which are converted to H_2O_2 by superoxide dismutase, and then the GP-X is depleted.

Further experiments, however, are necessary to confirm this potential mode of action.

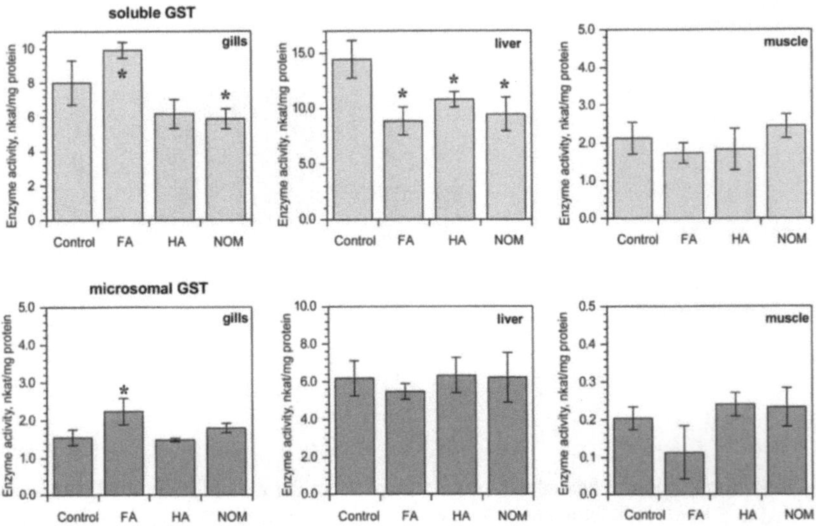

Fig. 8.36. Activity of glutathione-*S* transferase in gills, liver, and muscle of carp yearlings exposed to 5 mg/L DOC Suwannee River isolates for 24 hours. Data are means, ± standard deviation, * significantly different from control (after Wiegand et al. in press)

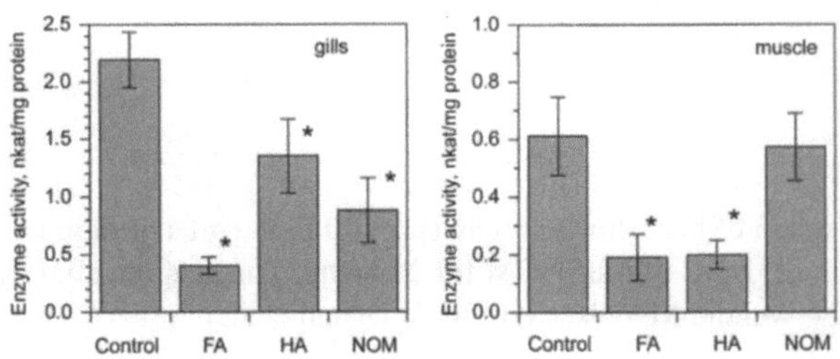

Fig. 8.37. Activity of glutathione peroxidase in gills and muscle of carp yearlings exposed to 5 mg/L DOC Suwannee River isolates for 24 hours. Data are means, ± standard deviation, * significantly different from control (after Wiegand et al. in press)

Elevated activity of biotransformation enzymes requires a lot of energy, which is presumably diverted from essential processes. The adverse effects of this switch can best be seen in studies with life stages, such as embryos that are not yet able to compensate for this energy loss. In a preliminary study, Meinelt et al. (in prep.) test the survival of embryos of a *r*-strategist (zebrafish, *Danio rerio*) exposed to the synthetic HS 1500. From a previous study, it is well known that zebrafish embryos possess transformation enzymes (Wiegand et al. in press). Fig. 8.38 shows that an elevated concentration of HS 1500 (500 mg/L, somewhat above environmental levels) significantly reduces the survival rate of the exposed embryos. Since even under natural conditions a great proportion of embryos die during ontogenesis, it cannot be judged whether the obvious increase of embryo survival is really beneficial or within the usual mortality of embryos.

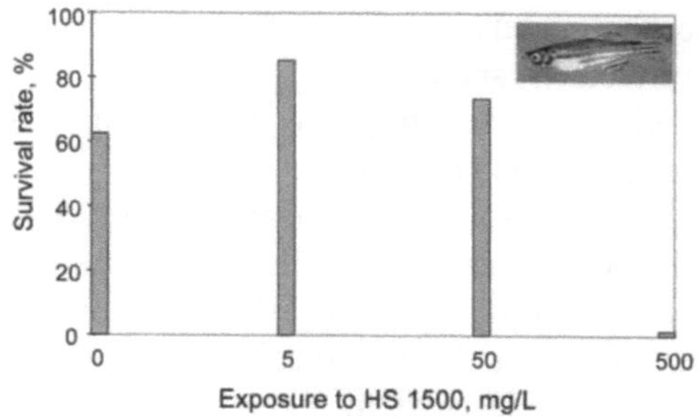

Fig. 8.38. Modulation of the survival rate of embryos of a *r*-strategist (zebrafish, *Danio rerio*) exposed to elevated concentrations of the synthetic HS 1500 (from Meinelt, Stüber, Steinberg in prep.)

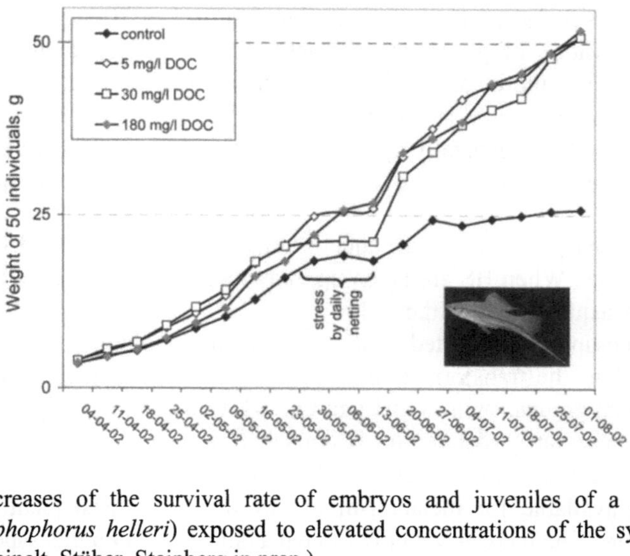

Fig. 8.39. Increases of the survival rate of embryos and juveniles of a *K*-strategist (swordtail, *Xiphophorus helleri*) exposed to elevated concentrations of the synthetic HS 1500 (from Meinelt, Stüber, Steinberg in prep.)

A different result is obtained with the embryos and juveniles with a *K*-strategist, the swordtail *Xiphophorus helleri* (Fig. 8.39). Any addition of HS leads to a weight gain by the young animals. However, if a specific adverse effect occurs, as shown in Figs. 8.37 and 8.38, it is obviously overcompensated by beneficial effects, such as reduction in parasite and pathogen infections. If this is true, the result is another trade-off between specific and non-specific HS-mediated effects (Steinberg and Brüggemann 2001, Fig. 8.25). This hypothesis has to be refined in future studies.

8.3.8 Potential Mode of Direct Action

The interactions of HS with aquatic organisms follow the basic rules of ecotoxicology:

1. effects apply to a variety of species. Hsp 70 is inducible in fish as well as invertebrates. Also, the modulation of transformation enzymes appears in some macrophytes, invertebrates, and fish;
2. effects may exhibit clear dose-response relationships, as exemplified with glutathione-*S* transferase and *D. magna*;
3. at least for the inhibition of photosynthetic oxygen release, a quantitative structure-effect relationship can be established. For the modulation of the fertility of *C. elegans*, there is a strong suggestion which humic structures may be responsible. Thus, effects in aquatic organisms start to become predictable, if the quality of the exposed HS or NOM is known.

Although not all parts of the puzzles are worked out yet, and some results remain contradictory, one may postulate the following: HS can be taken up by the cytosolic arylhydrocarbon (Ah) receptor. As a result of this bonding, hsp 70 or 90 dissociate. We assume that hsp 70 is released following loading of the Ah receptors in the same way as hsp 90, and starts the signal transduction. Only those particular HS isolates which cause an increase in hsp 70 synthesis, yield an increased activation of glutathione-*S* transferases. This hypothetical signal transduction which could be affected by HS is shown in Fig. 8.40. HS are handled by the cells in the same way as xenobiotics are.

If the hypothesis generally applies, a situation for which there is considerable evidence, one can conclude the following concerning the ecophysiological role of HS:

1. various, but possibly not all, aquatic organisms can take up HS through their membranes. When HS are taken up, they must somehow evoke a signal transduction sequence within the cell;
2. the taken-up HS are treated by the organism as xenobiotics, and metabolized in phase II of the transformation system. Since the HS already have sufficient functional groups, such as OH and COOH, to be conjugated, phase I enzymes are not necessarily required for HS metabolism;
3. the HS metabolism is 'expensive', because the organisms must have sufficient energy available for metabolism of HS which is consequently unavailable elsewhere. There are as yet only limited available data on this aspect;
4. at the ecosystem level, the findings indicate that through their direct effects on aquatic organisms, HS have effects on the structure of biocoenoses (guilds). For example, among nematodes there are species and/or individuals which can thrive by enhanced fertility, following HS exposure. As another example, the structure of aquatic plant communities is apparently affected by dissolved HS. The only species that survive are those for which HS do not inhibit photosynthesis, and do not provoke energy-requiring metabolism. Since in most freshwaters, HS are predominately of terrestrial origin, this implies that the plant communities in lakes must have developed together with the terrestrial vegeta-

tion; HS are mediators of this development;

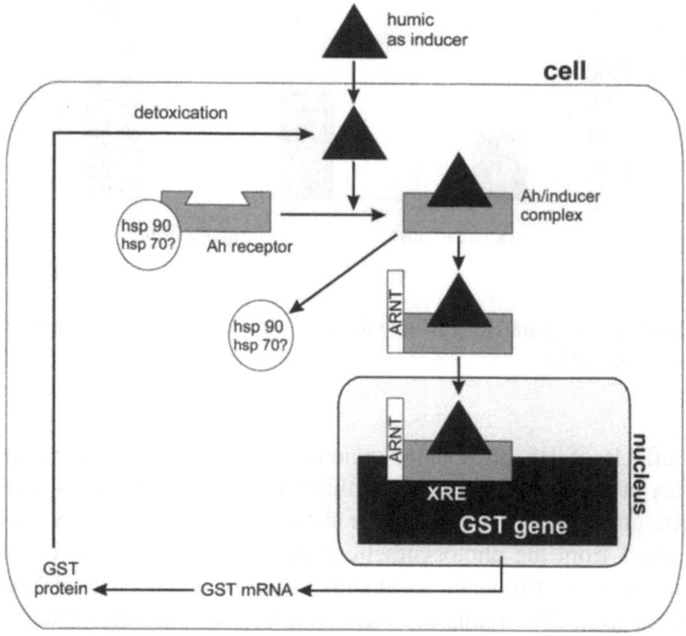

Fig. 8.40. Humic substances acting as potential internal cell regulators may explain the induction of glutathione-*S* transferase. Binding of HS may be mediated via the arylhydrocarbon (Ah) receptor. Upon binding, hsp 90, and probably also hsp 70, are released. The Ah/inducer complex binds to a translocator (ARNT) and is transported to a specific part of the DNA, for instance the xenobiotic regulating element (XRE). This activates the transcription of the structure genes and starts the production of biotransformation enzymes such as glutathione-*S* transferase (from Pflugmacher unpublished, with permission)

5. if one adopts the hypotheses and findings of Ziechmann's group (Chap. 2.3) based on the premise that HS or similar substances existed in the primitive atmosphere independent from, and parallel to, living and dead organisms, one can then speculate that the exposure of early life to HS or similar substances provokes the development of a protective mechanism, such as the transformation systems. This may be one reason why the transformation system is very conservative from bacteria to plants and to vertebrates;
6. transformation phase II enzymes metabolize HS. This means that these reactions not only convert HS to a soluble form which the particular cells or bodies can easily remove, but also to a form that must be more bioavailable for microorganisms. Interestingly, first indications are available that biotransformation enzymes are active even extracellularly. Pflugmacher et al. (1997) describe extra- and intracellular GST activation in sediments of two lakes (Fig. 8.41). It is therefore not inconceivable that these enzymes participate in the metabolism of HS and DOC in ecosystems.

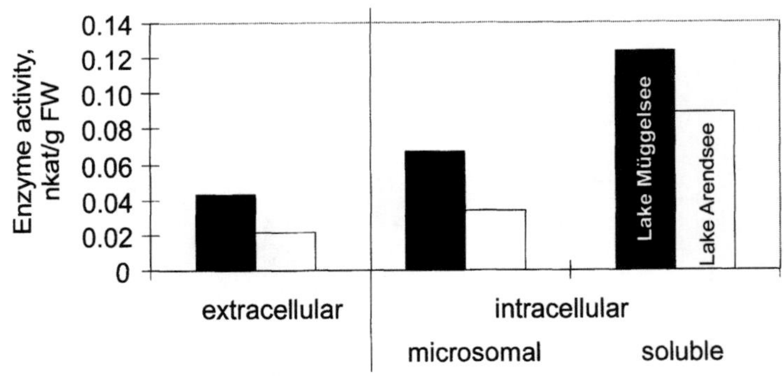

Fig. 8.41. Activity of glutathione-*S* transferase in sediments of two German lakes (from Pflugmacher et al. 1997)

Direct effects of HS on aquatic organisms are graphically presented in Fig. 8.42. HS are taken up by organisms. Within the organisms, these substances act as xenobiotics, and activate, or inhibit the detoxification system. HS can also turn electrons away from the photosynthetic electron transport chain. For these phenomena, mechanistic explanations already exist. In a few cases, HS can act directly as some kind of nutrients and increase the photosynthesis of macrophytes.

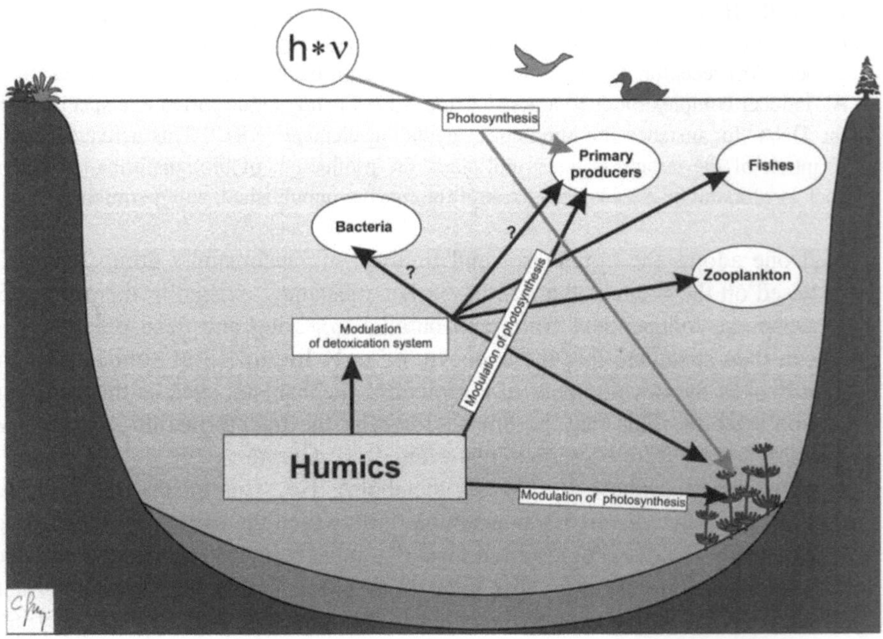

Fig. 8.42. Graphical summary of potential direct effects of HS: modulation of biotransformation enzyme systems and modulation of photosynthesis. h∗ν is solar energy

Box 8.1 Well Known Effects of Humic Substances on Terrestrial Plants and Vertebrates

The decay-inhibiting properties of HS, the beneficial effects of HS on terrestrial plants, and the beneficial or adverse effects on terrestrial vertebrates have been known for a longer time, particularly from applied science, archeology, agriculture, horticulture, and medicine.

Bog People

The decay-inhibiting, bacteriostatic properties of humic waters can particularly be seen with the 690 'bog people' found throughout northern Europe (Petersen et al. 1986). In the iron age, up to 2,000 years ago, people were buried in bogs, possible in ritual sacrifices associated with periods of famine. Determination of the contents of the last meal and time of execution are known, as for the Tollund man in the Silkeborg museum (Jutland, Denmark) (Fig. B.8.1.1).

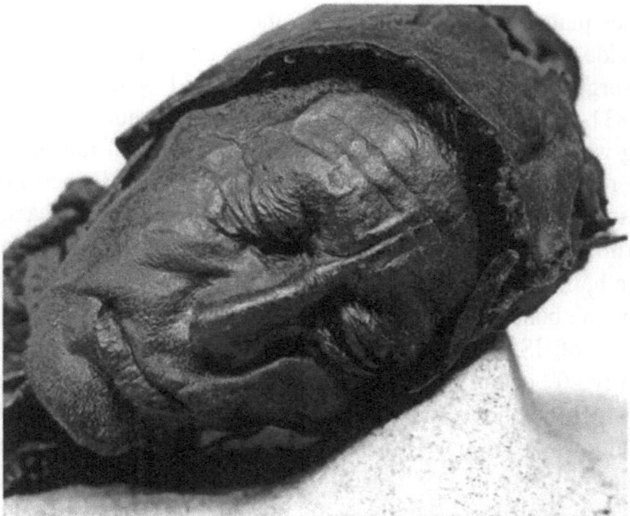

Fig. B.8.1.1. Head of a strangulated bog man, the Tollund man, Jutland, Denmark, buried approximately 2500 before present (with kind permission of the Silkeborg Museum)

The major reason for the decay-inhibiting, hence, preserving conditions is the inactivation of microbial proteases by the phenolic compounds of the HS. Petersen et al. (1986) write that the preservation can also be compared with a tanning by humic acids – the skin of the dead body slowly turns into leather. The tanning process is found to be so slow that after 2,000 years it is still incomplete. A minor reason for the preserving conditions may also be the absence of easily degradable carbon such as carbohydrates, so no, or only very little, microbial growth is feasible. The direct negative effects of HS on microorganisms, which have been described in the case of pathogens, may also apply to non-pathogenic, protein degrading microorganisms.

Terrestrial Plants

In their report, Chen and Aviad (1990) outline the positive effects of HS on useful plants, and note that for thousands of years people have noticed that dark-colored soils are more productive than light-colored soils and that productivity is tightly linked to rotten plant and animal materials. As early as 1627, Bacon suggests that plants take up soil juice as a source of nutrition. In the late 17th century, Woodward (1699) demonstrates that the reaction of plants to various sources of water follows the order: soils water extract >river water>well water, an effect correlated with the yellow color of the water. In the early 19th century, the direct role of HS in plant nutrition and growth is emphasized by Thaer (1808), who even suggests that humus alone is the material from which plants obtain nutrients. This suggestion is later encompassed in the 'humus theory' of Grandeau (1872) in which humus provides plants with major nutrients, including carbon. However, many scientists provide evidence against the humus theory and show that plants could synthesize organic matter from carbon dioxide and water (De Saussure 1804; Sprengel 1831-1832). The best known opponent of the humus theory is no lesser than Liebig who shows in many publications (1841, 1856) fundamental information on the role of minerals in plant nutrition. Lawes and Gilbert (1905) demonstrate that soil fertility may be maintained, for at least several years, by applying mineral fertilizers only. However, these experiments do not end the controversy between the humus and mineral theories. More exact experimentation is required to determine the benefit of humus to plant growth and to determine possible synergistic effects of HS and minerals.

In the early 20th century, Bottomly (1914a,b, 1917, 1920) publishes a series of studies in which he shows that HS can enhance the growth of plants in mineral nutrient solutions. If HS have a beneficial effect on plants, he attributes this to a hormone-like property of HS and calls them 'auximones'. In contrast, other researchers (Burk et al. 1932; Olsen 1930) attribute the beneficial effects of HS on plant growth to an improved solubility of some minerals ions such as iron. This controversy continues to this day.

With their studies, Rauthan and Schnitzer (1981) and Chen et al. (1994) contribute to the solution of this controversy. The true situation probably lies between the two extremes mentioned above as summarized by Chen and Aviad (1990):

- studies on the effects of HS on plant growth with sufficient mineral nutrient supply, show concurrently positive effects on plant biomass and composition;
- HS fertilization can enhance the nitrogen content of plants as shown for wheat (Gonet et al. 1996);
- the enhancement of root growth affects both growth in length and also the development of secondary roots. This is evident for wheat seedlings (*Triticum aestivum* L. var. *appulo*) as shown in Fig. B.8.1.2;

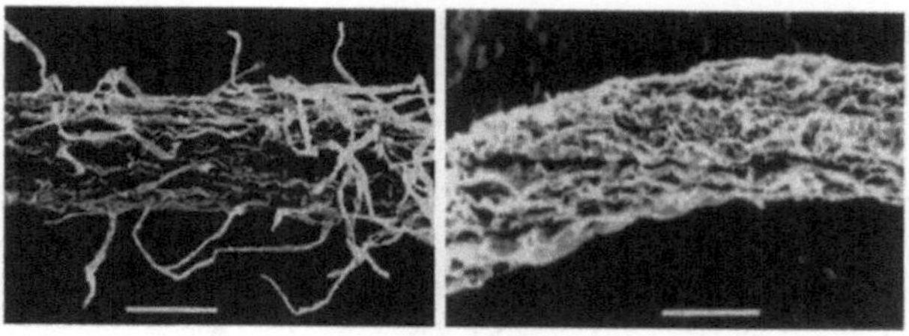

Fig. B.8.1.2. Scanning electron micrographs of roots of wheat seedlings grown in HS-containing media (left) and HS-free media (right). Scale bar: 100 μm (from Concheri et al. 1994, with kind permission of Elsevier Science)

- in general, HS stimulate root growth more than shoot growth. The effect on shoots is commonly only 1/10 of that on roots;
- the typical dose-response curve shows increasing growth with increasing HS concentrations in the nutrient solution followed by a decline at very high concentrations as displayed for cucumbers and melons in Fig. B.8.1.3. Such optima curves are also shown for the effects of acid on invertebrates (Fig. 8.3) in the presence of HS, or modulation in survival rates of zebrafish (Fig. 8.38);
- the stimulating effect of HS is correlated with increased uptake of macronutrients. HS can complex transition metal cations and, thus, improve their bioavailability. Sometimes, however, there is competition between HS and roots for the metals which can result in reduced bioavailability and plant uptake. Lastly, plants can release phytosiderophores which are stronger iron chelators than HS and can therefore lead to an exchange of ligands. Iron complexed in this way can be readily taken up by particular plants, for example grasses.

  A small fraction of the low molecular weight HS can be taken up by plants. These components apparently increase membrane permeability and can exert hormone-like effects. In later studies of Chen et al. (1994), neither auxins (such as indole acetic acid), cytokinins nor abscisic acid are found. The enhanced growth is attributed solely to improved bioavailability of iron and other trace metals (Chen et al. 1999);
- there are, however, also opinions which conflict with those of Chen. According to the pioneering studies of Vaughan et al. (1985), HS have direct and indirect

effects on plant growth and some metabolic processes. Modulating effects are found on the activities of phosphorylase and a peroxidase in wheat *T. aestivum*, and on indole-3-acetic acid oxidase in lentil *Lens culinaris* (Concheri et al. 1994 with references to the original literature). All these enzymes participate in biochemical processes which directly influence growth. It is concluded that particular HS have a hormone-like effect, as confirmed by Concheri et al. (1994), who describe a slight stimulation of the activity of such enzymes, while in older literature an inhibition is reported. The displayed curves for cucumbers of Rauthan and Schnitzer (1981) show strikingly high concentrations of FA (Fig. B.8.1.3, left). However, usual curves are in the concentration range shown for melons (Chen et al. 1994) (Fig. B.8.1.3 right).

- The positive effects of soil HS also apply to compost HS.

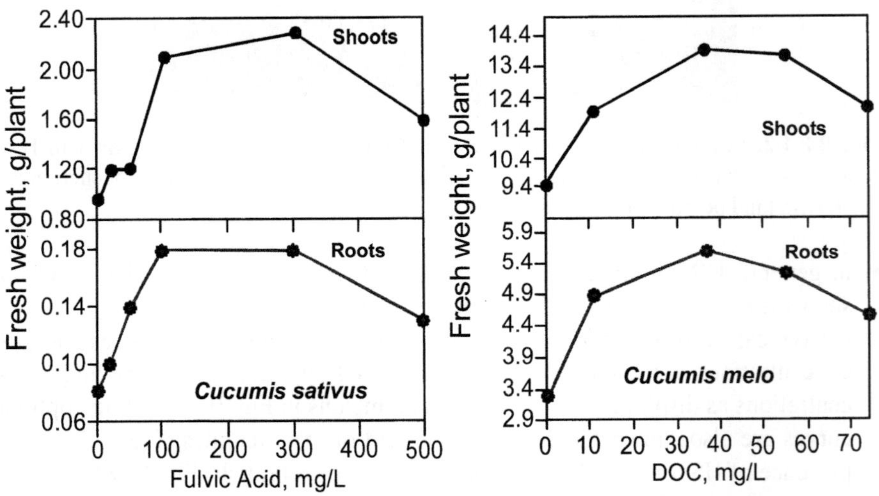

Fig. B.8.1.3. Influence of FA and DOC concentrations on dry weight of cucumber (*Cucumis sativus*) shoots and roots (based on Rauthan and Schnitzer 1981) and melon (*Cucumis melo*) shoots and roots (from Chen et al. 1994; with kind permission of Elsevier Science)

It is often shown that HS have an antidote effect on plants, which should mean that they compensate for the detrimental effects of chemicals or pathogen attacks (viruses, bacteria, fungi), although the exact mechanism is not known. Some examples will be given.

Terrestrial Plants: Humic Substances and Chemicals

The herbicide maleic hydrazide is known for its genotoxicity (clastogenic events). Micronucleated cells and aberrant anatelophases are described for treated peas (*Pisum sativum*), onions (*Allium cepa*), and broad beans (*Vicia faba*). However, the detrimental effects of the herbicide are significantly reduced by HS (De Simone et al. 1997; Ferrara et al. 2001), with those HS from peat being particularly effective. The HS themselves have no detrimental influence on the plants.

Loffredo et al. (1997) draw a more detailed picture by describing the effect of humic acids and herbicides, on the growth of tomato (*Lycopersicon esculentum*) seedlings in hydroponic culture. Morphological and physiological effects on the growth of young tomato plants caused by HA and herbicides, alone or in combination, are dependent on the origin and nature of both the HA and the herbicide. The presence in the nutrient medium of HA from non-amended or amended soils stimulates growth and attenuates the phytotoxic effects of herbicides. In contrast, a sewage-sludge-borne HA shows a marked phytotoxic activity by itself, and a negative synergistic effect when used in combination with any herbicide studied. The tested sludge-borne HA has comparatively low C/N, C/H, and O/C ratios, and carboxyl content, and has a comparatively high phenolic content.

Terrestrial Plants: Humic Substances and Pathogens[1]

In general, HS have the potential to influence plant health in ecological ways, that is without any application of chemicals. Although plants are in contact with HS only in the root zone, beneficial effect apply to both leaf and root pathogens. Organic matter from composts containing relatively young HS are well documented, with mature compost having beneficial effects on plants (Hoitink and Grebus 1997), and young compost being detrimental since it generally supports pathogenic infections. Mature compost can predictably control soil-borne diseases, such as those caused by the fungi *Fusarium* spp., *Phytophthora* spp., *Pythium* spp., and *Rhizoctonia solani*.

Two different modes of action are proposed:

- competition (general and specific suppression), antibiosis, and hyperparasitism or/and
- systematic strengthening of plant resistance.

Of these, the first mechanism is more thoroughly studied than the second.

The typically high microbial activity of the 'general microflora' of the soil hinders germination of *Phytophthora* and *Pythium* spp. spores and, thus, infection of the plants, probably through microbiostasis (Chen et al. 1988). Essential nutrients for the pathogens, derived from the seeds or roots, are metabolized by the 'general microflora', and the starved spores of the pathogens remain dormant, but are not killed. This mechanism is known as general suppression.

In contrast, *Rhizoctonia solani* is activated from the resting phase by many specific, mainly volatile substances. These substances are utilized by only a small number of microorganisms. This is the specific suppression which is more effective, the more diverse the microflora is.

What is the role of HS or organic material in defense against plant pathogens? The qualitative composition plays a key role. If the organic material slowly re-

[1] For the beneficial effects of HS on food and ornamental fish, see Box 8.2

leases organic nutrients (carbohydrates), the 'general soil microflora' will be stimulated, and thus provide and maintain a biological control against pathogens.

Animals and People[2]

Visser (1985) describes a general mechanism through in vitro and in vivo studies by which HS can improve the general condition of organisms. He shows that respiration of rat liver mitochondria is stimulated in the presence of either FA or HA at comparatively high concentrations (40–360 mg/L). Oxidative phosphorylation and supply of ATP is stimulated, and this energy store can, in turn, increase nutrient uptake and synthetic activities of the cells. The low-molecular weight fractions (approximately 5.5 kDa) are more effective than high-molecular weight fractions. At higher concentrations, the reverse – stimulation of mitochondria by high-molecular weight fractions – is sometimes noticed.

Ziechmann (1996) and Flaig (1997) report the chemical-biochemical properties of peloids and peats containing HS which are used in medicine. These include astringent and anti-phlogistic effects, disinfection through release of radicals, modulation of electrolyte status, prophylaxis and therapy of infectious and non-infectious dysenteries, chemotherapy for septicemia, cancer therapy, bacteriostatic, antimicrobial, and anti-viral effects in balneotherapy[3] (Brockow et al. 1998; Klöcking et al. 1997; Klöcking and Helbig 2001), local chemical effects, styptic effects, influence on smooth muscles (Kauffels 1988), trauma therapy for skin and tissue, and enhancement of skin resistance through improvement of physical properties of collagen (Riede et al. 1991).

Physicians clearly show that fulvic and humic acids with molecular weights of <1.0 kDa can penetrate the human skin. Furthermore, FA and HA have a partially agonistic effect on the α_2 adreno and D_2 dopamine receptors, but at the same time quite different effects on the spontaneous contractile activity of smooth muscles are noted (Beer et al. 2000).

The anti-adhesive, anti-inflammatory effects of HS are confirmed by Klöcking et al. (1968) and Ziechmann (1996) with experiments on animals, and humic acid precursors identified (HAP, see Fig. 2.5) as the active group. The observed anti-adhesive effect is attributed to an inhibition of coagulation through the binding of

[2] A lot of knowledge of beneficial effects of HS on diseases is buried in internal reports of research institutions and universities, particularly in eastern Europe, rather than being published in peer reviewed journals. This reduces public and scientific acceptance of HS as natural therapeutics.

[3] Raised peat bog is used therapeutically long ago in ancient Babylonia and the Roman Empire, where the inhabitants already recognize the healing effects of mud. As health clinic specialties, mud baths are offered in Europe in the early 19th century. Traditional indications for mud therapy are gynecological and rheumatic diseases (Klöcking and Helbig 2001).

thrombin to the HS. A very similar mechanism generally applies for the therapy of inflammatory diseases, an application for which HS have their most important use (Dunkelberg et al. 1997; Klöcking et al. 1968; Zeck-Kapp et al. 1991). Quecke and Loschen (1989) successfully use lignin degradation products to inhibit inflammatory diseases caused by the release of products of the arachidonic acid cycle. The release of arachidonic acid from membrane phospholipids is the rate-limiting key reaction for the synthesis of inflammatory agents on the basis of eicosanoic acid (similar to leucotriene, Schewe et al., 1991). The results of Schewe et al. (1991) and Dunkelberg et al. (1997) show a significant reduction in arachinodic acid release from membranes, and a subsequent reduced formation of leucotriene when HA and HA-like polymers are applied. HA from a natural bog reduces the release of arachinodic acid by as much as 80%. Hence, HS increase the integrity of membranes. It is very likely that the phenolic or quinoide monomer or polymer HAP play the key role, by interacting with the amino groups of proteinaceous compounds (Flaig 1997). Hence, the anti-inflammatory effect of HS is supported by a plausible biochemical explanation.

Published studies on bactericidal, antiviral, or virostatic effects of HS are limited in number (Ansorg and Rochus 1978; Brockow et al. 1998; Klöcking and Sprössig 1975; Schneider et al. 1994). In their study, Ansorg and Rochus (1978) give one of the first systematic surveys of the effectiveness of HA against human pathogenic bacteria. They conclude that many, but not all, tested HA isolates have anti-microbial properties at high concentration (600–2500 mg/L). A synthetic, HS-like polymer from hydroquinone, HS 1500, is effective at a relatively low concentration (39 mg/L). Based on the fact that the synthetic, HS-like polymers from brenzcatechin or hydrochinone are more effective than natural isolates, the authors conclude that the aromatic rings are the effective components.

Pro-inflammatory activity is associated with the synthetic low-molecular weight HS 1500, which activates human neutrophils (Zeck-Kapp et al. 1991). In addition, HS have the potential to act as electron donor-acceptor system and a 'buffering effect' is plausible: HS are able to produce, as well as bind to, activated oxygen species. This regulatory system is assumed to be important for the favorable influence of HS on wound healing and killing of cancer cells (from Klöcking and Helbig 2001).

Inspired by the finding that foot and mouth disease (a viral disease of livestock) is rapidly controlled by the simple application of peat (Schultz 1962), Klöcking and Sprössig (1975) study the growth of viruses and state that some, but not all, tested viruses are inhibited by HA from bog-water. They determine the following sequence in inhibition: herpes simplex virus Type 1>influenza virus A2>Coxsackie virus A9. This demonstrates that HS are effective against both naked and enveloped DNA viruses. The same is true for the synthetic HS 1500, which is, in part, more effective than natural HA (Klöcking and Helbig 2001). One of the most active anti-viral synthetic polymers is the oxidation product of caffeic acid. Further investigations (Klöcking and Helbig 2001) confirm the ability of HA-like polymers to selectively inhibit viruses for human immunodeficiency virus type 1 and type 2, cytomeglovirus, and vaccinia virus. No inhibition is found against poliovirus type 1, Semliki forest virus, parainfluenza virus type 3, reovirus and Sindbis virus. Adenovirus type 2 and ECHO virus type 6 show little or no response to natural HA. With most viruses, the inhibitory effect of HA and HA-like polymers is directed specifically against an early stage of virus replication, namely virus attachment to cells (Klöcking and Helbig 2001). The authors assume that the polyanionic HA occupy positively charged domains of the viral envelope glycoproteins, which are necessary for virus attachment to the cell surface.

The use of HS in living cells and in mammal blood leads to changes in the electrolyte equilibrium. In rat's blood, both synthetic and natural HS products affect measurable changes in Na^+, K^+, Mg^{++}, and Ca^{++} activity. With the synthetic HS 1500, Riede et al. (1991) find that this compound causes a receptor-mediated cell response in cell cultures of granulocytes and lymphocytes, indicative of a pro-inflammatory cell signal resulting in granulocyte bursting. The release of oxygen radicals and hydrogen peroxide is observed (Riede et al. 1991). Also, Schlickewei et al. (1991) give details of an astringent and tanning effect on collagen fibers upon exposure to the synthetic HS 1500. Subsequently, the resistance of the fibers increase.

Humic Substances as Agents for Diseases

HS-like substances such as tannins in tea and in several medicinal herbs have long been known also to have negative effects on human health, being linked to esophageal cancer. Tea is high in condensed catechin tannins, one of the two broad classes of tannins. However, thanks to an early warning by the British Medical Association, the British traditionally add milk to their tea to bind the tannins, thus reducing the risk of esophageal cancer. In contrast, the Dutch do not and about 100 year ago, when tea is the national drink of Holland, throat diseases are common. After the Dutch switch to coffee, throat diseases become rare (from Petersen et al. 1986).

Recently, a weak genotoxicity of HS is reported by Ribas et al. (1997) describing a small, but significant increase in sister chromatid exchange in human lymphocytes exposed to the commercially available Aldrich HA. The genotoxicity of this HA is less than that of herbicides. Modes of action are not mentioned. Such findings are rare, but not unique. For instance, genotoxicity of HA to the ovary of Chinese hamster is also reported by Cozzi et al. (1993).

Humic acids are also thought to be indirectly related to testicular atrophy and, hence, to reduced fertility of male mammals. Actually, Chen et al. (2001) describe with mice that the growth of Sertoli[4] cells is reduced upon HA exposure. The authors believe that HA-induced testicular atrophy is linked in part to an inhibitory effect on the growth of Sertoli cells.

---

[4] Sertoli cells are large cells in the vertebrate testis that support and nourish developing spermatozoa.

There are some illnesses which can be related to the uptake of HS. Three examples are given. The Kaschin-Beck disease, potentially caused by a lack of selenium affects some two million people in the People's Republic of China[5]. FA and other exogenous free radicals lead to membrane damage, through lipid oxidation under oxidative conditions. Selenium, as a coenzyme of particular anti-oxidative enzymes can counteract this disease (Peng et al. 1991; Sokoloff 1989).

Since the early sixties, there are scientific reports on the 'blackfoot' disease in various regions of China. This is an peripheral vascular disease, endemic in Taiwan, that affects the extremities, mainly the foot and occasionally the leg. The affected part becomes gangrenous with severe pain and often results in spontaneous amputation (Klöcking et al. 2000). For a long time, this disease has been attributed to a high arsenic content in drinking water. However, recent studies in a Taiwanese province show clearly that the drinking water has extremely high DOC concentrations (around 19 mg/L). In particular, the FA content of 7.5 mg/L is very high. These FA have a high fluorescence intensity caused by the high aromatic carbon content, and over 50% have a molecular weight <1.0 kDa.

Hung et al. (1994) suggest a connection between these FA characteristics and the incidence of blackfoot disease in Taiwan, and meanwhile it is now accepted that a particular HS fraction, the fluorescent FA, causes the blackfoot disease (Hseu et al. 2001; Yang et al. 1994). These FA intervene in blood coagulation and inhibit the activity of plasma protein C. Klöcking (1997) and Klöcking et al. (2000) describe a mode of action: protein C is a regulatory protein in hemostasis, activated by the thrombin/thrombomodulin complex. Activated protein C specifically inactivates two particular factors (Va and VIIIa) in the coagulation system, by limited proteolysis, thus, it has an anti-coagulant effect. The inhibition of active protein C by fluorescent FA is a strong potential mode of action for thromboembolic disorders via acceleration of blood coagulation.

In contrast to the activated protein C-inhibiting effect of the fluorescent FA from Taiwan drinking water, several HS activate protein C, and support the anti-coagulant effect of activated protein C leading to hypocoagulation in the blood. That is, HA inhibit the coagulation enzyme thrombin, thereby suppressing the formation of fibrin monomers from fibrinogen. With the exception of one tested bog HA, a dose-effect relationship can be established. The tested HS may be used as therapy for inhibited blood coagulation.

Another, probably not exclusive, mode of action is proposed by Gau et al. (2001). These authors find evidence that iron mediates HA-associated oxidative stress in endothelial cells, which may lead to atherothrombotic vascular injury observed for patients with blackfoot disease.

[5] Particularly in Qamdo, southwest China's Tibet Autonomous Region, a region with high incidence of Kaschin-Beck disease. Two-thirds of residents in Qamdo suffer from the disease, which currently has no effective treatment (http://www.china.org.cn)

Therapeutic Use of Humic Substances

HS are used medically in special cases, since they have only a very limited specificity due to their structural heterogeneity. Their only wide use is in balneology. For humans, the estrogen-like effect of HS has been known for a long time, yet without a good understanding of the mode of action. Early studies of Aschheim and Hohlweg (1933) identify bitumens in peat as the responsible agents. However in addition to these lipophilic materials, more water soluble substances are found, which also have steroid-like effects, penetrate the skin more easily and, are therefore now used in peat baths to treat gynecological diseases (Gierhake and Wehefritz 1936). The penetrability of HS, at least HAP, has been established for pig and mice skin (Ziechmann 1996).

With regard to therapeutic uses of HS, the effects on blood coagulation (Klöcking et al. 1996, 2000) and anti-inflammatory properties are the most studied. The mechanistic similarities of these processes are that the phenolic, or more probably quinoide, structures of HS are interacting with proteinaceous material (Flaig 1997). Although several descriptions of the mode of action with therapeutic usage of HS are still vague, the impact of HS on the quality of human health is increasingly recognized as an important subject for future work. Future studies have to aim at the molecular structure and mode of actions. To elucidate the chemical structure of synthetic HS, that originate from comparatively simple individual starting compounds, remains an important means to stimulate and facilitate the much more complicated exploration of natural HS (Klöcking and Helbig 2001).

Box 8.2 Application of Humic Substances to Food and Ornamental Fish

In the internet, several HS products are advertised: 'The product has a slight antiviral and antibacterial effect. Animals will be more vigorous and become more resistant to diseases. Anemia and anorexia syndromes can be avoided, and the already existing syndromes can quickly be cured.' Such HS may be administered by bath immersion or as food additives. How reliable are these claims?

Changing environmental factors, and anthropogenic stresses including grading and transportation, are acute or chronic stress situations, for which the fish requires substantial acclimatization (General Adaptation Syndrome). Prolonged stress and inadequate adaptation to stresses, lead to damage to the defense system, which is often followed by infection and disease. The medical treatment of these secondary infections is limited however, because:

1. there are only a few chemical substances approved for use in aquaculture, and these are not effective against all pathogens,
2. medical treatment may be linked to side effects such as toxic stress (Meinelt et al. 2001b, 2002).
3. all chemicals approved for use in aquaculture are suspected to be or known to be mutagenic[1] or carcinogenic[2] with a high toxic potential to non-target organ-

---

[1] Assessing malachite green, the British Committee on Mutagenicity concludes 'that although a limited negative in-vivo micronucleus test was available, the results of the recently conducted ^{32}P-post-labelling studies indicated that it would be prudent to assume that malachite green may be a potential in-vivo mutagen' (http://www.doh.gov.uk/com/com.htm).

'Methylene blue has been found to be mutagenic to *Micrococcus aureus*. In addition, methylene blue was mutagenic to *Salmonella typhimurium* in the presence and absence of metabolic activation. This compound was mutagenic to *Escherichia coli* in the microsuspension and DNA-cell binding assays. There are conflicting reports concerning the mutagenicity of methylene blue in *Bacillus subtilis*. One author reported that methylene blue was mutagenic to *B. subtilis* in the rec-assay. However, methylene blue was nonmutagenic to this bacteria in another study. Several reports were found concerning mutation induction via irradiation with visible light, in the presence of methylene blue. Photodynamic mutagenesis in the presence of methylene blue has been induced in bacteriophage *Serratiaphage X*, *Escherichia coli* and *Salmonella typhimurium*. Methylene blue was non-mutagenic to Chinese hamsters, *in vivo*. In addition, this compound was non-mutagenic in cultured Chinese hamster ovary cells and lung fibroblasts. Methylene isms, hence their application is restricted to closed culture systems.

The stimulation of defense mechanisms in fish, the consequent increase in disease resistance and use of less risky agents, are therefore of considerable interest. Such substances can include natural HS and synthetic, HS-like materials. The findings of Gamygin et al. (1991, 1992a,b)[3], Schreckenbach (1990), and Schreckenbach et al. (1991, 1994, 1996) indicate that, in general, HS exert a positive influence on the health status and the condition of food and ornamental fish, and strengthen their defense system, including their skin. Schreckenbach et al. (1996) generalize the health promoting effects of HS on fishes as follows:

1. general improvement of physiological processes,
2. stimulation of defense mechanisms, and
3. inhibition of adhesion and penetration of pathogens in the early life stages of fishes.

HS can heal skin lesions of fish caused by transport damage within 13 days. Normal medical treatment with FMM (mixture of formaldehyde, malachite green, and methylene blue) leads to total loss of damaged fish (Schreckenbach et al. 2002).

Schreckenbach et al. (1994) tested HS applications with later developmental stages of fish and discover interesting facts. Over a period of 5 weeks exposure to HS, larval biomass significantly increases, compared to unexposed controls. In addition, there is a significantly increased survival rate, compared to treatment with a conventional therapeutic chemical, such as malachite green. Schreckenbach et al. (1994) use the synthetic HS-like polymer HS 1500, which is synthesized from polyphenols and applied to freshwater invertebrates and plants (Chap. 8).

In total, the HS 1500 treatment has many positive effects with application both in the water phase and administration as a food additive (Schreckenbach 1994):

- increase in brood yield through prophylactic treatment of fish eggs and larvae,
- improvement in growth and food utilization,
- general improvement in condition and strength of resistance, as well as increase in vitality of fish during transportation and culture,
- healing of ectoparasite-damaged fish through therapeutic medication,
- suppression of secondary infections,
- limitation or even inhibition of outbreaks of primary infection through pro-

blue did not induce mutagenic effects in *Drosophila melanogaster*' (http://ntp-server.niehs.nih.gov/htdocs).

[2] 'In 1980, laboratory findings showed that exposure to formaldehyde could cause nasal cancer in rats. Since then, the question of whether exposure to formaldehyde increases a person's risk of cancer has been the subject of considerable controversy...' (http://cis.nci.nih.gov/fact/3_8.htm)

[3] Many obviously interesting results of HS application as pharmaceuticals lie in internal reports of research institutes or universities, particularly in Russia. It is hoped that these results will be published in peer-reviewed scientific journals, thus enhancing scientific and public acceptance of alternative pharmaceutical measures for food and ornamental fish.

phylactic application,

- detoxification of harmful metals and xenobiotics in water (Chapter 7).

One of the mechanisms of toxicity reduction can be the photolytic production of reactive oxygen species (ROS, Chapter 5.2) which can disinfect pathological infections, as recently shown by Liltved and Landfald (2000) for the fish pathogenic bacteria *Aeromonas salmonicida* and *Vibrio anguillarum.* In addition, the granulocyte activation, a non-specific cellular defense, is also stimulated in fish treated with HS 1500, as shown for carp (Schreckenbach et al. 1996). There is an increase in leukocyte number and activation in peripheral blood of carp following temperature stress and simultaneous HS application (Schreckenbach et al. 1991). The marked enrichment of hemosiderin[4] in the neutral granulocytes and monocytes accounts for an extensive phagocytosis of damaged erythrocytes with simultaneous HS exposure. It is suggested that the leukocyte activation promotes the subsequent antimicrobial defense as well as the breakdown of tissue damaged by stress. These effects of HS are also described by Isola and Mandrich (1991) and Gamygin et al. (1991, 1992a,b) for the successful healing of secondary infections in various food and ornamental fishes.

According to Gamygin et al. (1991, 1992 a, b), the exposure of young rainbow trout to synthetic HS 1500 leads to growth stimulation via increased food efficiency. Parasitic infections, and the associated loss of young fish are also reduced. These results are confirmed with carp by Schreckenbach et al. (1991). Additionally, Schreckenbach et al. (1991) find an elevated efficiency of the energy metabolism in carp by reduced temperature stress reactions. In various ornamental fish, treatments with HS concentrations of 50–90 mg/L lead to reduced sickness and mortality rate. The therapeutic success is greater than with usual medication such as a bath of a mixture of formaldehyde, malachite green, and methylene blue.

Schneider and Rieder (1992) observe antiviral effects of HS in carp. However, in contrast Schreckenbach et al. (1991) find with eggs, larvae, and juvenile fish, that applied HS have no therapeutic effect on the obligate fish pathogenic bacteria, *Aeromonas hydrophila, Pseudomonas fluorescens* and *P. putida.* The secondary effects of bacterial infections, such as liver infections, are clearly reduced through exposure to HS. Schreckenbach et al. (1996) accept these findings as another clue that the principal effect of HS lies not in weakening the defense of the pathogens, but more probably in strengthening the defense of the fish. This hypothesis is corroborated by the studies of Gamygin et al. (1991) and Hartung (1994). In Hartung's study, fish were infected by natural and artificial skin lesions. The subsequent HS applications brought about astonishing healing success. With infections of various ornamental fish, Burkart et al. (1994) also show that the HS dose has a greater healing effect on infections than does the usual medication of FMM. The HS treatment is significantly more effective than FMM for sensitive fish such as members of the Characidae (Fig. B.8.2.1).

[4] **Hemosiderin:** an intracellular storage form of iron; the granules consist of an ill-defined complex of ferric hydroxides, polysaccharides, and proteins having an iron content of about 33% by weight

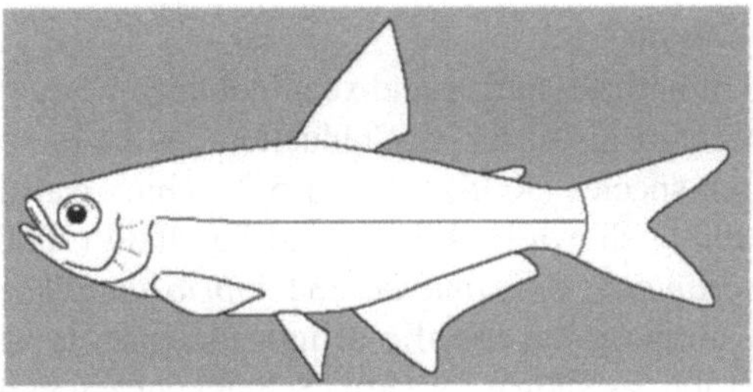

Fig. B.8.2.1. Sketch of a member of the Characidae (from Frose and Pauly 2002, with kind permission from FishBase)

Impact of Humic Substances on Fungal Infections

Fungal infections in fish farms are dominated by *Saprolegnia* sp. and *Achlya* sp. These ubiquitous fish pathogens lead to typical secondary infections of the skin, and the surface of fish eggs, with spores and fungal hyphae encroaching from damaged tissue to healthy fish and eggs. The economic losses caused by pathogenic fungi affecting food and ornamental fish are enormous.

Gamygin et al. (1992a) find no significant damage to rainbow trout eggs from mycosis over 4 days, following a one hour exposure to 5 mg/L HS. It is possible that HS deposited on egg surfaces intervene in the proteolytic enzyme system of the fungus, and thus reduce attachment of the pathogen. Applying HS 1500, Schreckenbach et al. (1991) find no direct inhibition of fungal growth, but a striking protective effect against *Saprolegnia,* when HS 1500 is applied prophylactically, even with artificially induced infection by *Saprolegnia*, compared to non HS 1500 controls. Schreckenbach et al. (1991) suggest a stimulation of the defense mechanisms of fish skin.

Schreckenbach et al. (1994) subsequently describe interesting experiments on eggs and larvae of rainbow trout, which are artificially infected with spores of the fungus *Saprolegnia* sp. The concurrent presence of the artificial FA significantly reduces mortality due to the fungal infection. In addition, a significant increase in the hatching rate is observed with FA concentrations of 15, 30, and 50 mg/L. These concentrations exceed water column concentrations, but can be found in sediment interstitial waters. Because eggs and embryos of many fish lie and develop on the sediment surface, where they come into contact with such high HS concentrations, the chosen concentrations are realistic, and the described effects are environmentally relevant.

For goldfish stressed by capture and transport, the application of a 1.5–2.0 hour bath with 10 mg/L of a sodium humate for 12 days leads to significant reductions of external mycosis, particularly of *Saprolegnia*. Only 6% of the HS treated fishes have mycosis, in contrast to 17% of the untreated controls (Schreckenbach et al. 2002).

Impact of Humic Substances on Parasite Infections

Although many parasites are found on and in fish in their normal state, parasitic infections commonly increase following weakening of hosts by stress factors. According to Gamygin et al. (1991), an application of HS can help reduce the number of infections over a long period, as well as healing secondary symptoms of infections such as ulcers and skin necroses. In contrast, Schreckenbach et al. (1994) find no direct anti-parasitic effects of HS applications. As long as the positive effects of HS against parasites cannot be related to specific structures of the HS, contradictory results with very different HS are not at all surprising.

The positive effects found by Gamygin et al. (1991) for HS treatment against *Ichthyophthirius*, *Trichodina* and *Costia* are attributed to an improvement in physiological condition of the fish, and to increased dermal resistance against invasive stages of the parasites. Burkart et al. (1994) also describe an improved 'healing' of infection symptoms in ornamental fish, and HS treated fish survive, while control fish and those receiving FMM have higher mortality. This strengthens the suggestion that normal treatment with FMM leads to additional stress to the fish, and thus increases mortality.

Impact of Humic Substances on Medications

The effectiveness of medications used in aquaculture and aquaria is strongly influenced by HS. The medication mixture FMM has a very small application range between effective treatment and detrimental effects for fish. Meinelt et al. (2001) show that the toxicity of FMM in soft waters is reduced in the presence of HS. The authors assume that ionic bonds can form complexes between the positively charged malachite green, and the negatively charged HS, and thus reduce the toxicity of malachite green. If calcium, however, is present in elevated concentrations in the aquarium water, it competes with malachite green for HS binding sites, causing an increase in the free concentration of malachite green and consequently increases toxicity to the fish. This finding contradicts some old paradigms of aquaculture which say that high calcium concentrations act as an antidote to potential medication toxicity. In the presence of HS, calcium ions may in fact act as toxicity promoters! It follows that the use of cationic medications must be very precisely dosed, the dosage depending on both the calcium and HS concentrations of the medium.

Very recently, Meinelt et al. (2002) also corroborate this mechanism for acriflavine. Acriflavine is used to treat external bacterial and protozoan diseases of fish and it is administered as a long-term bath. Meinelt et al. (2002) find that calcium ions are able to elevate the toxicity of acriflavine, while HS reduce the toxicity. The lowered LC50 values in the presence of HS and high calcium content can be seen in Table B.8.2.1.

Table B.8.2.1. Toxicity (6 d-LC50) of FMM (a mixture of formaldehyde, malachite green and methylene blue) and Acriflavine to *Danio rerio* (from Meinelt et al. 2001, 2002)

	↑Ca^{2+} / –HS	↓Ca^{2+} / –HS	↑Ca^{2+} / +HS	↓Ca^{2+} / +HS
FMM	17.3 µL/L	17.0 µL/L	14.0 µL/L	20.7 µL/L
Acriflavine	0.23 mg/L	0.47 mg/L	2.06 mg/L	2.34 mg/L

–HS: 0 mg DOC/L; +HS: 5 mg DOC/L; ↓Ca^{2+}: 0.2 mmol/L; ↑Ca^{2+}: 2 mmol/L

Conclusion

Common pharmaceuticals for food and ornamental fish are frequently harmful to fish, even at concentrations slightly above the therapeutical concentration range. Harmful effects range from toxic, to carcinogenic and teratogenic. The urgent need for alternative medication is obvious. Humic substances are a powerful alternative to chemical pharmaceuticals. However, in contrast to chemical pharmaceuticals, HS as natural products vary in their chemical composition due to source and isolation techniques. Even from a single source, the chemical composition will vary slightly from isolation to isolation. As a consequence of this variability, in many countries, HS are not legally permitted as pharmaceuticals. HS continue, however, to be considered as simple and very cheap 'additives'. They remain of little economic interest and appear to have little chance to enter the market as therapeutics. Thus a dilemma exists!

9 Ecological Significance

One cradle of limnology is located at the shore of Lake Geneva, Switzerland. Hence, the pioneering work in limnology is based on the particular type of water body that Lake Geneva represents, namely the summer-stable, stratified deep lake. However, on a world-wide scale is not the most common type of lake. For example, shallow lakes which do not develop stable thermal stratification are far more numerous than are HS-rich lakes. Because the paradigms developed in stratified humic-poor lakes are not questioned, limnologists have, for many years, tried to adapt these paradigms for HS-rich lakes.

One paradigm is that the pelagic primary production in lakes, at least in non-littoral dominated lakes, is the energetic basis of all heterotrophic processes in a lake. That means the various feeding types of zooplankton, of fish, and where applicable, the fish-eating birds and carnivorous mammals, the various types of zoobenthos and all dissimilatory micro-organisms in the water column and sediments – bacteria and fungi – depend on the production of phytoplankton. However, the energetic situation is by no means balanced, and much more energy is consumed by the heterotrophic processes as fixed by phytoplankton. Biomass production in sediments is overlooked or greatly underestimated, due partly to the fact that benthos has not been the very focus of limnology since its beginning.

Due to methodological progress, we now know that the energy fixed in algae is insufficient even to fuel only the pelagic food web. Little or almost no energy will be left to feed the benthic food webs. Hence, one becomes interested in the exudates excreted by the algae, or released during cell division, as food precursors for bacteria and zooplankton. In general, around 30–50% of the photosynthetically fixed organic carbon which released subsequently enters the microbial loop (Münster et al. 1999d). Even when almost 100% is exudated by the phytoplankton, as described by Sell and Overbeck (1992) for Lake Plusssee, the overall energy budget is still not balanced.

The first reports with a new paradigm come from the humus-dominated boreal lakes. Here, in many cases, bacterial production is clearly higher than the production of the planktonic algae. The supply of energy to the bacteria via HS photolysis products now comes into focus as an area of limnological interest.

Hence, three conceptual advances in freshwater ecology emerge:

1. freshwater lakes and their surrounding land are increasingly considered as being connected. There exists not only a material dependence (Chap. 3), but also a clear energetic dependence, of lakes on the adjacent catchment;
2. in addition to ecochemists, ecologists also begin to show an interest in dead or-

ganic matter as an ecological compartment. HS are an active component of ecosystems. They are not at all inert or refractory. Interesting and fascinating details are uncovered, especially HS mediated photolysis processes and the subsequent effects on the food web;

3. energetic budgets must include not only in-lake processes, but also inputs from the catchments. Hence, a complete budget comprises input, transformation, and output including sedimentation.

Finally, it becomes evident that even the large, thermally stratified lakes, and likewise parts of the oceans, are heterotrophic, that is to say, they are dependent on input of organic substances from terrestrial sources.

9.1 Net Heterotrophy

The contribution of bacteria to total planktonic respiration ranges from approximately 10 to 90%, with the highest contribution occurring in the most oligotrophic waters. Furthermore, in oligotrophic waters, most of the organic matter is dissolved, supporting a predominantly microbial food web, whereas in eutrophic waters there is an increased abundance of particulate organic matter supporting a food web consisting of larger autotrophs and phagotrophic heterotrophs (Biddanda et al. 2001). More surprisingly, Duarte et al. (2001) report that even in the subtropical northeast Atlantic, the community respiration significantly exceeds gross production. This applies for two-thirds of the stations investigated. The results of Duarte et al. (2001) provide evidence that even the subtropical northeast Atlantic is a net heterotrophic ecosystem, where planktonic communities respire more organic C than they produce by photosynthesis. Net heterotrophy, however, does obviously not apply to all regions of open oceans. Based on the rather conservative estimate of bacterial growth efficiency of 14%, Moran et al. (2001) believe that phytoplanktonically produced DOC will suffice to meet bacterial carbon demand in the Weddell and Scotia Seas. But, if the authors use a lower estimate of bacterial growth efficiency, phytoplanktonically produced DOC tends not to be sufficient for bacterial growth.

This means that a portion, and in boreal waters a large portion, of the organic C in the water body comes from the surrounding land or adjacent continent. If this new paradigm is generally applicable, then land-derived organic compounds support the aquatic food webs in non-eutrophicated systems. The connecting link into the aquatic food web is the bacterial activity within the water body. To estimate the quantity of organic C necessary for net heterotrophy of aquatic ecosystems, Cole (1999) set down a model calculation with the following assumptions: net heterotrophy is indicated by a CO_2 supersaturation for lakes with forested catchments when the supersaturation is around 1 matm (Cole et al. 1994). The net efflux of C as CO_2 will be about 40 $g/m^2/a$ with 250 ice-free days per year. If the lake's catchment area is 10-fold larger than the lake surface, the forest needs to export only 4 $g/m^2/a$ of C to meet this demand. Given a typical forest net primary production of approximately 500 g/m^2/a (Borman and Likens 1979), then only 1% of this terrestrial production is required to sustain net heterotrophy in the lake. If this net heterotrophy respires 50% of the terrestrial DOC input to the lake (Dillon and Molot 1997), 2% of terrestrial production is required for net heterotrophy.

HS composition changes characteristically during the annual hydrological cycle (Ivarsson and Jansson 1994; Visser 1984; Chap. 3.1.3). For instance, in boreal streams, peak DOC concentrations and their seasonality depend mainly on the leachate of deciduous litter (Hongve 1999), whereas base-flow DOC concentration and composition are determined by coniferous litter and soil organic matter release (Hongve 1999). Ivarsson and Jansson (1994) draw a similar picture with their study of the north Swedish River Lillån. The base-flow DOC concentration is 9–12 mg/L in summer and winter. The DOC concentration increases dramatically during high flow episodes during rain storms in summer and autumn, whereas concentrations climb only slowly during the snowmelt period in spring. DOC in the summer and autumn flood episodes comprises less decomposed, more acidic, organic material derived from soil surface layers. With snowmelt, the DOC comes from well-decomposed material mainly from deeper peat layers. With this qualitative variation, the direct or indirect bioavailability also varies annually. The HS provide C, N, and an energy source to bacteria (Carlsson et al. 1993; Hessen and Andersson 1990; Jones 1992; Tranvik and Sieburth 1989).

Although these qualitative facts are well known, quantitative data on the turnover of organic C compounds at the whole-lake level is little known. One of the few studies is for Lake Örträsket (7,3 km^2, DOC: 10 mg/L) in northern Sweden (Jonsson and Jansson 1997). Peak values for organic C retention following snowmelt are 75 t/d and exceed the autochthonous production by far. The yearly gross sedimentation of organic C is calculated as more than 300 t. For total N, 23 t/a is estimated. Interestingly, the particulate organic C of allochthonous origin appears to be more susceptible to microbial decomposition (32%), than is dissolved organic matter entering the lake from the same inlets (<10%). This is in good agreement with data from southern Sweden (Tranvik 1988). The relatively high bioavailability of sedimented allochthonous organic material implies that this material is the driving force of microbial activity in the sediment. The bacterial decomposition of HS, in particular in sediments, can result in anoxic hypolimnia (Hessen 1992), with decreased redox potential and high P release from the sediments, increased denitrification, and also fish kills under ice, particularly in small lakes (Rask et al. 1999).

9.2 Competition for Phosphorus

In humic lakes, dissolved P concentrations are very low and most P is released photochemically. In addition, the humic-Fe-P complex renders P available to the phytoplankton at very low concentrations. These facts have far reaching consequences for primary production by phytoplankton (Münster et al. 1999d), and for algae to acquire P. Low P concentrations are advantageous for the bacteria, since – due to the favorable surface area/volume ratio – they are the better competitors than algae. Hence, bacteria are one major P pool. Since zooplankton can be sustained by all living and dead particulate organic C, it is hardly surprising that they are another P reserve. To illustrate this, Hessen and Andersen (1991) determine that about 55% of the particulate P in a humic lake is found in bacteria, and the rest is bound in zooplankton. In extreme cases, up to 85% of the total P can be associated with the zooplankton (Salonen et al. 1994). Consequently, the P pool in phytoplankton is only marginal. This may lead to stoichiometric imbalances in zooplankters which may be re-balanced by feeding on a specific diet or by excreting excess nutrients.

The release of P from the sun lit humic-Fe-P complexes clearly changes the situation in favor of the phytoplankton. The daily fluctuations in the bioavailable P concentration induce particular phytoplankton strategies. Since the concentration of photolytically released P is highest in the topmost water layers, it is advantageous for the phytoplankton to stay there. However, in this layer potential damage from solar radiation is greatest, so the phytoplankton should avoid this zone. Therefore, a useful strategy is to spend only short periods in the P-rich strata of the water column. Conspicuously, in humic lakes there are many types of motile algae, obviously an evolutionary adaptation to the specific conditions in these waters.

An informative study on competition for P by Järvinen and Salonen (1998) was carried out in the polyhumic Lake Mekkojärvi, Finland, which is subject to biomanipulation to evaluate the effects of potential nutrient limitations on phytoplankton. Various combinations of nutrients, including P and glucose, are added to microcosms. The authors determine that primary production stays low in this bioassay. In general, the labile organic C supported the growth of pelagic bacteria, which then outcompete algae for P and, thus suppress their growth.

Another strategy of phytoplankton to successfully compete for P is to consume their competitors, such algae are mixotrophic, which means that they can grow both autotrophically and heterotrophically. Hessen (1998) coined the phrase 'If you can’t beat them, eat them'. Although the algae also gain additional organic C, the real reason for mixotrophy is the P shortage (Nygaard and Tobiesen 1993).

9.3 Bacterial Production

A number of studies (Hessen 1992; Salonen et al. 1992; Tranvik 1990) demonstrate that HS-rich waters have a higher bacterial production potential (bacterial carrying capacity) than HS-poor waters. As already mentioned, high-molecular weight DOC clearly contributes to this high bacterial carrying capacity (Tranvik 1990; Tulonen et al. 1992). The uptake of high-molecular weight DOC, however, requires an additional extracellular splitting before it can be assimilated by most bacteria (Billén 1991; Chróst 1990; Hoppe 1991).

These works contributes not only to the replacement of untenable old paradigms which consider HS as inert or refractory, but also explains the energetic gap in boreal and most non-eutrophicated waterbodies. There is hardly any non-eutrophicated ecosystem which is not net heterotrophic and depends on the allochthonous input of biochemically fixed energy. The autochthonous production and the view of a food pyramid sensu Lindeman (1942) is only a special case, and not the usual situation for energy cycling in a lake ecosystem (Hessen 1998).

In this superabundance of new knowledge and the subsequent change in paradigms, it is clear that the role of HS for freshwater organisms principally is seen as one-directional: as fuel and energy, which is transferred in the aquatic food web by heterotrophs and mixotrophic flagellates and ciliates. It is often overlooked, that HS can have direct adverse or even biocidal effects (Chap. 8.3).

In a recent study, Karlsson et al. (2001) examine changes in bacterioplankton biomass and production in subarctic lakes in northern Sweden to elucidate their coupling to the lake's physical, chemical, and biological characteristics. The studied lakes extend along an altitude gradient from the coniferous forest to the high-alpine belt. The results demonstrate that P probably restricts bacterial utilization of DOC in the coniferous forest lakes, while low DOC concentrations limit bacterial growth during the summer in the alpine lakes. The primary production of plankton is insufficient to support bacterial production in the lakes. High input of allochthonous DOC to the alpine lakes in spring is sufficient both to increase the bacterial production and to induce P-limitation. As a consequence, there is a tendency toward higher bacterial activity in the spring compared to the summer in the alpine lakes. The results indicate that most of the bacterial standing stock and production are supported by allochthonous DOC plus DOC from benthic production, and more or less limited by the P supply. Karlsson et al. (2001) therefore suggest that bacteria populations in subarctic lakes may be indirectly affected by climate variations through their impact on the input of DOC and nutrients from the lake catchments.

9.4 Food Webs in Humic-Rich Lakes as a Template for Non-Eutrophic Systems

The allochthonous input of organic C in most lakes is an important energy source used by bacteria. As a result of the dominance of heterotrophic processes, humic water bodies have quantitatively different food webs from humic-poor waters. Consideration of food webs shows that energy fixation is carried out by photoautotrophs and to some extent also chemolithoautotrophs. These primary producers supply biochemical energy and nutrients to the consumers (primary food web). The dead organic material of all trophic levels, the detritus, supports the so-called mineralizers, which through mineralization release inorganic nutrients. In addition, through filter feeders, the mineralizers enter the primary food chain. In this classical model, each member of the food chain has a clearly defined role and this naturally also applies to the planktonic community. The model must be updated following detailed studies on the role of bacteria in relation to the microbial loop (Currie and Kalff 1984a–c; Jansson 1998; Güde 1991; Rothhaupt and Güde 1992). As stated above, bacteria outcompete planktonic algae for inorganic nutrients. Growth of phytoplankton is nutrient or light limited, whereas growth of bacteria is limited by energy provided by phytoplankton (Currie 1990; Jansson 1998; Kirchman 1994).

A reversal of the food web model must be taken into account regarding the food web in HS-rich lakes. In such lakes, light and allochthonous HS provide two independent energy sources. The light is used by phytoplankton, the allochthonous HS are transformed by bacteria. Both groups of organisms mobilize energy and make it available to higher trophic levels (Jones 1992). As Jansson (1998) concludes, this should allow algae and bacteria to exist independently of each other. The experiments of Hessen et al. (1994) for Norwegian lakes, and of Jansson et al. (1996) for Swedish lakes, support this hypothesis. In the Norwegian lakes, the addition of N and P stimulates bacterial growth, while glucose has no effect on bacteria. In the Swedish experiments, only P addition gives a clear increase in bacterial biomass during summer, since bacteria are better competitors for nutrients than are algae.

The algae avoid the competitive pressure by using different nutrient resources, particularly for P (Rothhaupt 1992). It is striking that the phytoplankton of humic lakes is dominated by pigmented flagellates, mainly chrysophytes and cryptophytes. Many of these algae are mixotrophic. Heterotrophic growth can be achieved through both phagotrophy (uptake of organic particles) and osmotrophy (uptake of dissolved organic compounds) (Jones 1994). Phagotrophy through consumption of bacteria, pico- and nanoplankton is the best documented mixotrophic strategy of the phytoflagellates (Pringsheim 1963; Skuja 1948, 1956).[1] Jansson (1998) assumes that it appears to be energetically more efficient for the phytoplankton to consume bacteria, which concentrate nutrients from the medium, than to activate and maintain a specific uptake mechanism themselves. In addition, the bacteria can exploit the allochthonous detritus as a nutrient and energy source much more easily than can the phytoplankton. A frequent outcome for the mixotrophic phytoplankton, is that as a consequence of the high P content of the consumed bacteria, the phytoplankton becomes N limited (Jansson 1998)

The dominance of bacteria in humic lakes and the reaction of the various microplankton compartments (in terms of nutrient uptake) is summarized in Fig. 9.1. The enrichment of P and N alone, and in combination, supports the hypothesis above that the bacterial production is P limited, as bacterial biomass significantly

[1] This strategy of algae also occurs in other water bodies, when there is a lack of P. This can occur in eutrophic lakes after the spring bacterial maxima (Blomqvist et al. 1994), or in lakes acidified by deposition (Bleiwas et al. 1984; Earle et al. 1986; Schindler et al. 1985), or in extremely acidic flooded lignite pits (Nixdorf et al. 1998, in press). In some cases, phagotrophy can overcome lack of carbon and nitrogen limitation (Caron et al. 1990; Olaveson and Stokes 1989; Wehr et al. 1985, 1987).

increased only with P addition. Phytoplankton biomass and primary production usually increase after enrichment with N, but not with P. The mixotrophic organisms which comprise 75–90% during the ice-free period are, apart from the bacteria, the most important functional group of microorganisms. The uptake of P-rich bacteria leads to a relative N limitation for the mixotrophs which can be overcome through the addition of N alone or in combination with P. After addition of nutrients, the biomass of the mixotrophs (mainly dinoflagellates and chrysophytes), and potential mixotrophs (predominantly cryptophytes) is significantly greater than in the controls (Fig. 9.1). As potential mixotrophic organisms, the cryptophytes are generally more strongly limited by N or P than are the bacteria or the mixotrophs. In summary, bacterioplankton uses humic compounds as their principal energy source, and transfer energy and nutrients to potentially autotrophic organisms, with subsequent utilization by other components of the food web.

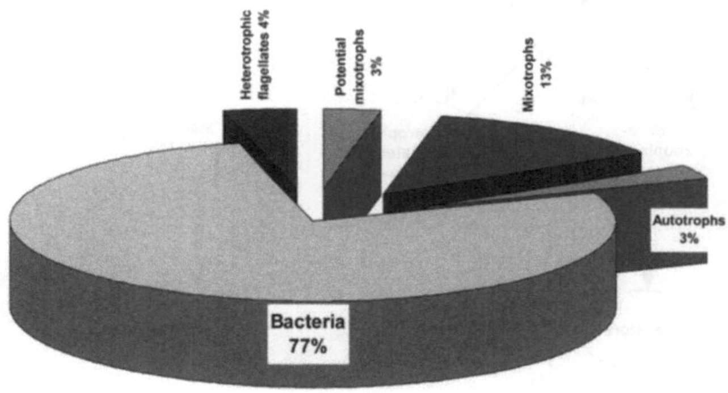

Fig. 9.1. Increases in biomass (µg/L C) of bacterioplankton, autotrophic, and mixotrophic phytoplankton, and heterotrophic nanoflagellates in nutrient enrichment experiments in Lake Örträsket, northern Sweden (after Jansson et al. 1996, with kind of the American Society of Limnology and Oceanography)

The same effect occurs in Coastal Plain, USA, blackwater streams too. Mallin et al. (2001) observe that the phytoplankton production is obviously N-limited, since any addition of inorganic or organic N-compound alone, or in combination with P, produces significant algal increases. However, neither inorganic nor organic P additions stimulate significant phytoplankton responses. In contrast, P addition stimulates ATP production without concurrent chlorophyll stimulation, indicating growth of heterotrophic microflora.

From these findings Arvola et al. (1999a) sketch the biomasses of the various compartments in non-eutrophic HS-poor and non-eutrophic HS-rich systems (Fig. 9.2). According to the classical view, plankton-dominated HS-poor water bodies are fueled by autochthonous primary production (Fig. 9.2a). However, this view is no longer tenable. HS-rich freshwaters receive their energy supply from the catchment in the form of HS, which is available to microorganisms mainly after photolytic degradation (Chap. 5). The planktonic primary production supplies only a small part of the required energy (Table 9.1). Yet, Chap. 9.7 will show that even in the non-eutrophicated HS-poor water bodies, the dominance of allochthonous C sources as the energy source (Fig. 9.2b) applies, and that the paradigm of plankton as the major energetic basis for the complete food web is only applicable to eutrophicated water bodies.

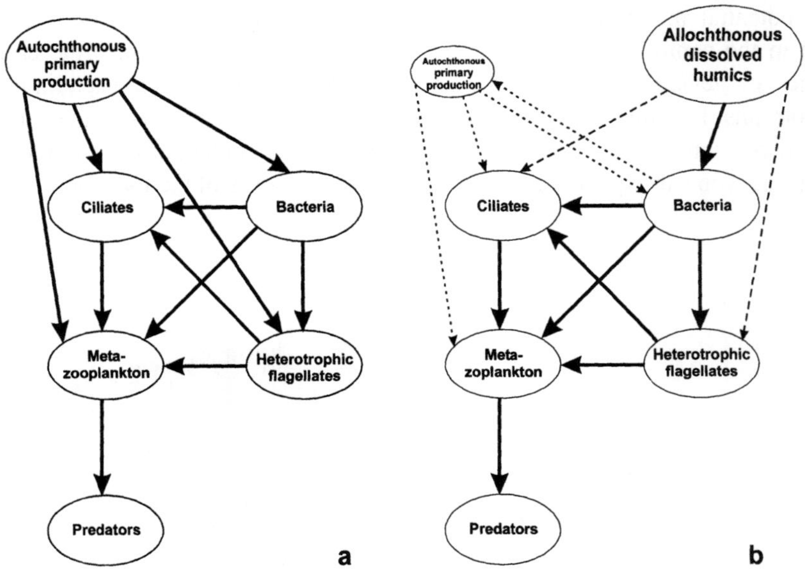

Fig. 9.2. Schematic carbon and energy flux in the pelagic systems of non-eutrophicated HS-poor lakes (**a**) and HS-rich lakes (**b**). Solid lines: main pathways, dotted lines: minor pathways of different degrees (after Arvola et al. 1999a, with kind permission of Backhuys). The size of the ovals is indicative of biomass. In (a) the pelagic food chain in the HS-poor water body represents the classical view of the functioning of a freshwater lake ecosystem, and appears to be somewhat outdated. In contrast, in (b) the food-chain appears to apply to most, if not all, non-eutrophic water bodies.

Strikingly, there is no balance in losses and inputs (Table 9.1). The inputs are some 1200 kJ/m^2/a too small to compensate for the losses. Since Arvola et al. (1999b) do not explain this discrepancy, it can only be suggested that the inputs are underestimated, probably due to extreme hydraulic situations such as heavy rain, snow melt, or the occurrence of diffuse input of percolating water. If this is the case, the allochthonous input would amount to 3663 (2466 + 1197) kJ/m^2/a and would account for 70% of all energy sources in Lammin Pääjärvi.

As Arvola et al. (1999b) write, the role of allochthonous DOC as an energy source is not nearly as clear as that of autochthonous DOC, and it remains controversial. The authors argue that the DOC pools are not static, and there is an urgent need for more studies on the details of decomposition, regeneration and subsequent supply. Analysis with stable isotopes is probably the method of choice as one can then determine not only the origin of organic carbon, but also the metabolic pathways, as in the impressive study of Kracht and Gleixner (2000).

Table 9.1. Energy balance for the Finnish humic Lake Lammin Pääjärvi (from Arvola et al. 1999b)

	$kJ/m^2/a$	%
Inputs		
Allochthonous material	2466	62.9
Primary production of phytoplankton	1096	28.0
Primary production of the littoral zone	358	9.1
Total	3920	100.0
Losses		
Respiration of bacterioplankton	1877	36.7
Respiration of zooplankton	698	13.6
Respiration of the benthos	762	14.9
Respiration of fish	65	1.3
Fishery output	7	0.1
Insect emergence	5	0.1
Outflow	1398	27.3
Sediment accumulation	305	6.0
Total	5117	100.0

9.5 Higher Trophic Levels

While dissolved HS can be a direct or – more probably – an indirect energy source for heterotrophic bacteria, the particulate organic carbon (POC), including colloidal and particulate dead organic material (detritus), is directly available to filter feeding zooplankton and zoobenthos (Haney 1973; Hessen et al. 1990; Hessen 1998; Wotton 1977). One can easily imagine that the aggregates illustrated in Fig. E.2.1.3d are available to filter feeders, such as daphnids or simuliids. The first investigator to mention the potential role of detritus as a food source was Naumann (1918), who describes the dominance of amorphous brown HS components in the guts of cladocerans in HS rich waters. Studying the zooplankton of a bog lake, Haney (1973) comes to a similar conclusion. '*... the high grazing and low algal productivity in Drowned Bog Lake indicates that allochthonous matter, for example from the* Sphagnum *mat, may provide an important input of nutrition into the bog's open water system.*'

A variety of studies show that in humic lakes, phytoplankton production is insufficient to support zooplankton production (Ojala and Salonen 2001). By use of two-compartment labeling, Hessen et al. (1990) explain the relative roles of detritus, bacteria, and phytoplankton as C sources for various crustaceans (Fig. 9.3): around 50% or even more of the organic C in the bodies of zooplankton originates from detrital sources, 5–10% from bacteria, and up to 40% from phytoplankton. The highest proportion of bacterial C is found in the two 'microfiltrators' *Daphnia* sp. and *Diaphanosoma* sp. Particularly surprising is the importance of detrital C for the copepod *Acanthodiaptomus denticornis*, in which generally more than 80% of body C is derived from detritus, because for selective feeding zooplankters as copepods, one would expect an even higher percentage of high-quality food (algae) than for the non-selective cladocerans (*Daphnia*). The data of Hessen et al. (1990) support Salonen and Hammar (1986) who find that unexpectedly high proportions of metazooplankton C in humic lakes are derived from non-living (detrital) C. Comparable data are also obtained by Jones et al. (1998) in Loch Ness (Scotland), a relatively large, moderately humic lake. Here, around 50% of the body C comes directly, or indirectly, via bacteria from detrital sources. These findings clearly show that allochthonous energy sources can support the food web in lakes, independent of their size.

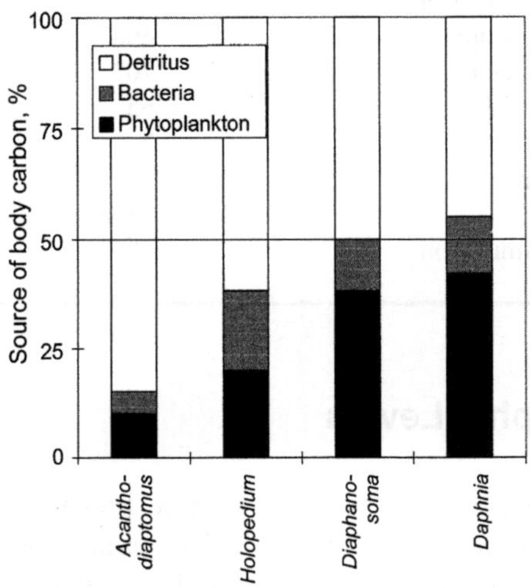

Fig. 9.3. Share of detritus, bacteria and phytoplankton in the diet of the dominating crustacean zooplankton in Lake Kjelsåsputten, based on separate labeling of food compartments. The shares are based on specific activity in sestonic fractions and animals at isotope equilibrium (after Hessen et al. 1990, with kind permission of the American Society of Limnology and Oceanography)

Meili et al. (1996) also find a clear difference in the ^{13}C and ^{15}N signatures of zooplankton attributable to different C sources. The signatures of the non-selective feeding cladocerans are very similar to those of terrestrial material whereas the algae-consuming *Eudiaptomus* sp. have a characteristic algal signature. Anderson and Benke (1994) find that *Ceriodaphnia dubia* fed natural forested floodplain swamp seston have somatic biomass growth rates as high as 61%/d, thus, the growth rates are within the range of literature estimates for other species of *Ceriodaphnia* under various food conditions. These studies again corroborate that allochthonous material is fundamental in fueling aquatic food webs.

Another link from dissolved organic material to zooplankton has recently been studied by Bastviken et al. (in press); the production of methanotrophic bacterial biomass. In three small lakes in southern Sweden, the methanotrophic bacterial production is equivalent up to 7% of the primary production, and as much as up to 120% of the total heterotrophic bacterial production. The methanotrophic biomass accounted for 3–11% of the total bacterial biomass on a depth-integrated basis, and as much as 41% of the total bacterial biomass at specific depths. Bastviken et al. (in press) conclude that methanotrophic bacteria have the potential to be an important food source for zooplankton in many lakes, particularly during winter, and that they can explain the depleted zooplankton ^{13}C signals frequently found. The finding about the extensive contribution of methane to the pelagic food web corresponds well with the statement of Kracht and Gleixner (2000) who report that methane is also an important of autochthonous production of humic materials (Chap. 2).

9.6 Seasonality of Production

Planktonic production in humic lakes depends, to a large extent, on two environmental factors: high concentrations of colored allochthonous organic compounds, and low concentrations of inorganic nutrients. In Lake Örträsket (Sweden), bacterial utilization of allochthonous organic C exceeds primary production in the epilimnion, and is largely determined by the input of organic C including HS, particularly during high water events (Jansson et al. 1999). The most important flooding occurs in spring, and this and later flooding events stimulate bacterial growth. The high water events later in the year are irregular, and generally smaller than those early in the year. The bacterial production is exploited mainly by mixotrophic flagellates which probably use bacteria as a source of C, P and N. The possibly extremely low availability of inorganic P during periods with high bacterial production, may allow the mixotrophs to outcompete obligate autotrophs and help the mixotrophs become dominant phytoplankters during most of the summer.

The C flux in the pelagic zone of Lake Örträsket is shown in Fig. 9.4. The results from Lake Örträsket indicate that the total production depends on bacterial energy mobilization from allochthonous organic C compounds, and that heterotrophic mobilized energy is linked via mixotrophs to higher levels in the food web. The photolysis of cDOC is an important prerequisite for microbial utilization of incoming humic C. The organic matter degradation, (bacterial and photolytic degradation), amounts to 1.5–2.5 mg C/L/a, where the photooxidation is responsible for approximately 10% of the organic C turnover (Jonsson et al. 2001; Pers et al. 2001). In the water column, organic matter is mineralized by bacteria (60%) and protozoan and metazoan zooplankton (30%). Metazoan zooplankton, in contrast to what may be the case in other lakes (Pace and Cole 1996; Riemann and Christoffersen 1993), are not important predators on bacteria in Lake Örträsket. Mixotrophs are by far the most important grazers.

The dominance of the mixotrophs is due to their high abundance relative to other grazers, and to their slightly higher clearance rates compared with heterotrophic flagellates. Primary production in the lake contributes at most 5% of the total organic carbon input and about 20% of the total organic carbon mineralization (Jonsson et al. 2001). Summarizing, by several independent methods, Jonsson et al. (2001) demonstrate that carbon mineralization during summer stratification in Lake Örträsket has three principal sources: 1) allochthonous particulate organic carbon entering predominantly during high flow in spring, 2) allochthonous dissolved organic carbon supplied continuously, and 3) fresh autochthonous organic carbon produced in the epilimnion during the summer. Of these sources, the large amounts of terrestrial carbon introduced in dissolved and particulate forms, is nearly always more important for the total respiration than in organic carbon produced in the lake, even though availability for degradation may be far higher for autochthonous carbon.

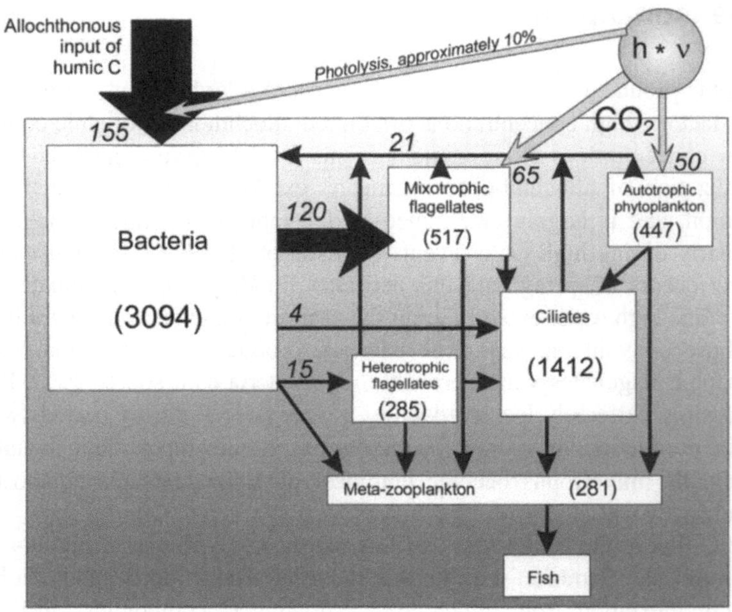

Fig. 9.4. Carbon fluxes (kg/d C) in the microbial food web of the epilimnion of Lake Örträsket during the period of summer stratification in 1995. Figures within brackets in boxes denote mean summer biomass (kg). h∗ν is solar energy influx (after Jansson et al. 1999 and amended from Jonsson et al. 2001, with kind permission of Schweizerbart Verlag)

It is interesting to note that heterotrophic flagellates and ciliates consume fewer bacteria than do mixotrophic phytoplankton. This might be surprising, since mixotrophs have to share bacterial prey resources with heterotrophic flagellates. Grazing efficiency is about the same, and the sensitivity to grazing by zooplankton should be similar in mixotrophic and heterotrophic flagellates. A difference between heterotrophs and mixotrophs is that the latter use light as an alternative energy source, and autotrophic growth may add to the competitive ability of mixotrophs relative to heterotrophs (Rothhaupt 1996).

9.7 Applicability of Net Heterotrophy

DOC from terrestrial sources clearly delivers energy and nutrients to aquatic bacteria. This energy is, in turn, transferred to zooplankton and to fish. The aquatic microorganisms therefore demonstrate the link between the terrestrial primary production and the aquatic secondary production (Cole 1999). The fixed C is made available to microorganisms in the aquatic system through the action of light on the chromophoric part of HS.

For which type of water is the hypothesis of an energy supply by terrestrial HS applicable? Where does the above mentioned net heterotrophy occur, and which systems are net autotrophic? All freshwaters receive HS from their catchments. Jones (1998) emphasizes that all freshwaters contain some HS of allochthonous origin. Therefore, humic-rich lakes with high water color must be seen as lying towards one end of a continuum between lakes with low concentrations of HS and lakes with high concentrations of HS. Support of microbial production by HS of terrestrial origin occurs more or less strongly in any freshwater body, even in eutrophicated ones. The first clear evidence of non-eutrophic lakes being net heterotrophic is presented by Salonen et al. (1992). In nutrient-rich waters, the autochthonous production dominates. Additional evidence for net heterophy is provided from Laurentian Great Lakes (Biddanda et al. 2001; Cotner et al. 2000), experimentally manipulated forest lakes (Cole 1999; Cole et al. 2000), and even from regions of oceans (del Giorgio and Peters 1994; del Giorgio et al. 1997; Duarte and Agustí 1998; Duarte et al. 2001).

The opposite, namely net autotrophy, is also reported very recently and convincingly. For instance, for oligotrophic Canadian Shield lakes, Carignan et al. (2000) report that gross photosynthesis is almost always larger than community respiration. Furthermore, HS appear to depress both photosynthesis and respiration. That means that heterotrophy, however, may be hidden or even hindered, due to several competing mechanisms, including humic-mediated adverse effects on bacteria, eutrophication-mediated changes in humification substrates, food web structures, and trade-offs between specific and non-specific effects of HS on primary producers.

One mechanism by which net autotrophy may be maintained, is the stimulation of algal photosynthesis by H_2O_2 concentrations significantly below toxic concentrations, as observed by Xenopoulos and Bird (1997). On the other hand, even small amounts of added H_2O_2 inhibit bacterial production; hence, the dominance of autotrophic over heterotrophic processes may appear.

To test the applicability of the net heterotrophy hypothesis, Blomqvist et al. (2001) add dissolved organic carbon, in the form of sucrose, to an oligotrophic humic-poor lake, initially characterized by a pronounced dominance of autotrophic phytoplankton (mostly by one species, the green alga *Botryococcus* sp.). The authors assume that it is organic carbon per se, and not other possible effects of HS, that determines the differences in structure of the planktonic ecosystem between humic-rich and humic-poor lakes. The additions of DOC result in a significant increase in bacterial biomass and a decrease in the biomass of autotrophic phytoplankton. The biomass of mixotrophic and heterotrophic flagellates increase significantly, whereas no effects are found to extend to higher trophic levels. As a result of the changes among biota, total planktonic biomass also decrease to a level typical of nearby humic lakes. Blomqvist et al. (2001) suggest that it is the carbon component of HS, and its utilization by bacterioplankton, that determines the structure and function of the pelagic food web in humic lakes.

9.7.1 Potential Changes in Humification Substrates During Eutrophication

Although evidence is growing that in non-eutrophicated systems, the majority of energy enters via allochthonous dissolved and particulate organic carbon, there are still some unsolved problems. For instance, Tranvik and Bertilsson (2001) describe the situation in more than 30 Swedish eutrophicated lakes. HS-mediated promotion of microbial growth occurs only under the following conditions: low electrical conductivity, pH values $\leq$8.0, and specific chlorophyll *a* concentrations $<$1 μg/mg DOC. A multiple regression shows that the bacterial carrying capacity is positively related to the DOC content, negatively related to the chlorophyll *a* content, and that both parameters account for about 75% of the variation in bacterial carrying capacity.

In oligotrophic to dystrophic soft-waters, the chemical conditions are as mentioned above. Calcium rich and eutrophic lakes are very different. According to the Swedish studies (Tranvik and Bertilsson 2001), the illumination of DOC in non-oligotrophic water bodies leads to an inhibition of microbial growth, similar to the findings of Benner and Biddanda (1998) and Obernosterer et al. (1999) for marine systems (Chap. 5.2.3). The mechanism of inhibition of microbial growth could be similar to the biogeochemical formation of marine HS, if lipids or other lipophilic substrates could serve as substrates for humification in freshwater systems, too. However, fatty acids do not play an important role in freshwater systems. This means that the inhibitory effect of DOC observed by Tranvik and Bertilsson (2001) for calcium-rich, eutrophic freshwater conditions remains unclear. A potential inhibitory effects of toxic cyanobacteria on microbial activity is not yet evaluated.

9.7.2 Food Web Structure

Cole et al. (2001) bring another aspect into the discussion: the food web structure. In a series of small lakes, in which the food web is manipulated and in part subjected to eutrophication, the gross primary production and total respiration increases over a period of up to seven years, as determined by three independent methods.

Without any treatment, the lakes are net heterotrophic, which means that production is smaller than respiration (P:R$<$1). Nutrient additions under weak planktivory do not change the predominance of respiration. However, nutrient additions under conditions of strong planktivory, result in respiration that is clearly smaller than production (P:R>1), with algal mass developments. These results are shown in Fig 9.5.

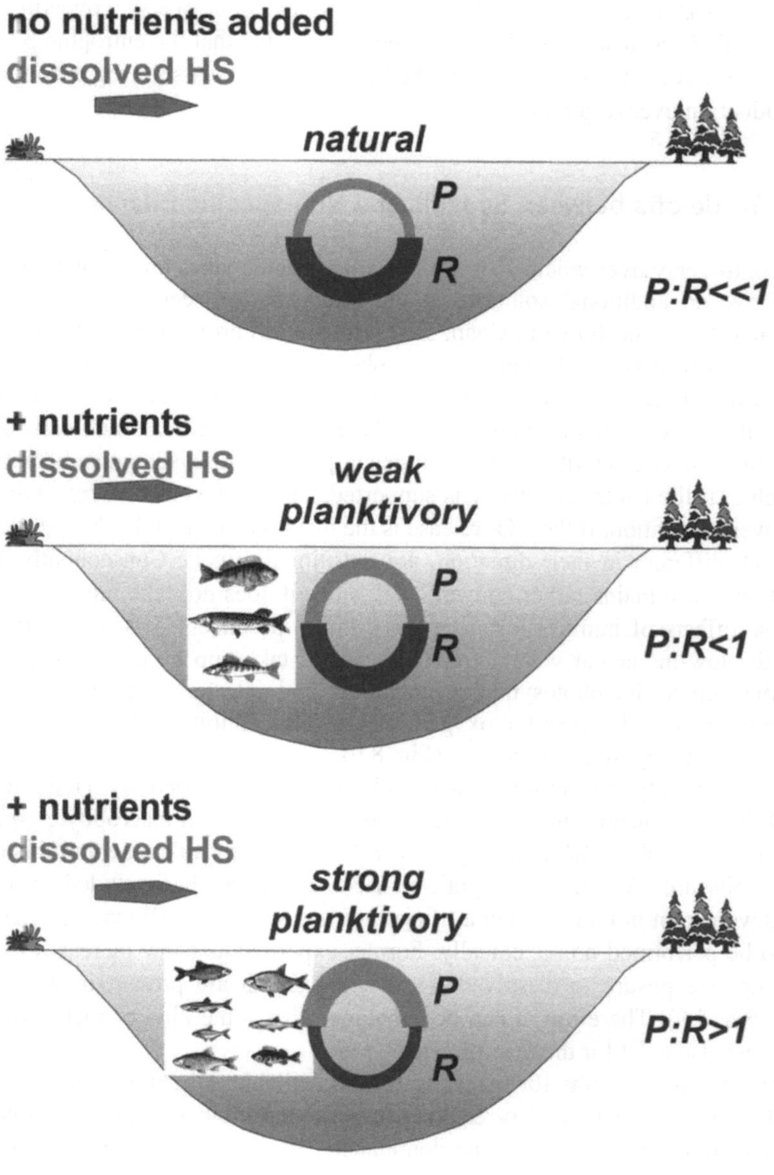

Fig. 9.5. Schematic presentation of gross production (**P**) in relation to total respiration (**R**) in lakes of contrasting nutrient conditions and contrasting food web structure (modified from Cole et al. 2001). The displayed fish species are representative of European food webs

Cole et al. (2001) report that for all studied lakes and all study years, a mean of approximately 26 mmol/m^2/d C of allochthonous origin is respired, which represents 13–43% of the total respiration. The respired proportion drops, when the gross production increases. These findings demonstrate that the food web structure has a much stronger influence on the production to respiration ratio, than does the trophic state itself. With respect to the controversy concerning the Swedish findings discussed above, it appears plausible that the eutrophic Swedish lakes may have a food web with marked planktivory, thus showing a dominance of production over respiration.

9.7.3 Trade-offs between Specific and Non-specific Effects

The controversy over whether or not non-oligotrophic lakes are net heterotrophic may have an additional solution, in the trade-offs between specific and non-specific effects. As shown in Chap. 5, several studies prove the nutritional potential of dissolved HS particularly after a photolytic attack of sunlight. This alone could explain the net heterotrophy of many non-eutrophic systems. Alternatively, the photolytic degradation may also produce carbon dioxide, which can support the photosynthetic activity of algae and macrophytes. The system may be net heterotrophic, if the bacterial activity is supported. The photosynthesis may predominate over respiration, if the CO_2 release is the main mechanism. Both mechanisms, although different in their direction, are nutritional effects. Consequently, interpretation still remains rather conventional, since it does not take into account the adverse effects of humic substances on primary producers. Dissolved HS or at least the low-molecular weight fractions, can be taken up by aquatic organisms, and may enter the photosynthetic apparatus and interfere within the electron transport chain and reduce the oxygen release. This inhibiting effect is mainly attributable to the quinoide structures (Fig. 8.9).

The ambiguity concerning photosynthesis is exemplified by Hellerudmyra NOM that significantly increases the photosynthesis of both macrophytes species, *C. demersum* and *V. dubyana* (Figs. 8.9, 8.10). The mechanism behind this effect is still obscure. A simple inorganic nutrient effect can be excluded, since the plants were kept in nutrient-rich artificial nutrient solutions. All macrophytes tests had to be performed non-axenically. Former experiments show increases in algal yield upon exposure to dissolved HS, when bacteria are present in the system (Figs. 5.8, 5.9). Therefore, it can be concluded that some kind of metabolic support must also exist for the macrophytes.

Corresponding results for bacteria are not available. Direct adverse effects of dissolved HS on bacteria also resulting in net autotrophic are most probable, but still remain to be determined. The dominance of heterotrophic processes may be caused by two mechanisms:

1. provision of microbial substrates, thus enhancing microbial growth and/or;
2. suppression of photosynthesis of the autotrophs. Both mechanisms apply to all aquatic systems, because all systems contain at least small quantities of DOC,

and humic substances are a major part of this.

The release of small fatty acids is based primarily on the chromophoric moieties that absorb light and transfer physical energy into potential (bio)chemical energy. In addition, suppression of photosynthesis is a function of HS and NOM quality: quinoide structures are responsible for the electron trapping effect, and also themselves comprise a moiety of the chromophoric DOC. Thus, HS and NOM, with their chromophoric (mainly aromatic) structures, display an ambiguous control on ecological processes in freshwaters (Klug 2002; Klug and Cottingham 2001; Steinberg and Brüggemann 2001): enhancement and inhibition may occur simultaneously.

10 Concluding Remarks

There are some old paradigms relating to HS which have to be replaced by new ones. There is clearly no process in freshwater systems that is not directly or indirectly influences by these substances. HS act as UV-shields and radical scavengers, and have done so previously. It is strongly hypothesized that without HS or HS-like substances, the formation of biomolecules during the development of life on Earth would not have taken place. Simultaneously the biomolecules would have been oxidized during the very formation process. In contrast, HS interact with biomolecules such as enzymes, hemoglobin, or DNA, such that the biomolecules must escape from or even repel their protectors against UV light and radicals. A harmless juxtaposition of the already existing biomolecules and the HS or HS-like substances was not possible. The biomolecules must immediately open metabolic routes which dispose of HS or HS-like substances. The biotransformation systems must probably have been established as one of the first metabolic pathways, and this could be one reason why the biotransformation system is extremely conservative in all organisms.

The most important presently known metabolic pathways involving HS in aquatic ecosystems are shown in graphical summary (Fig. 10.1). Some pathways may be highlighted again:

- HS reflect the state of the catchment, and control the water chemistry as long as no other proton buffering system dominates. HS are not only weak, but also relatively strong acids, and as such, they can cause pH values under 4.0, if a calcareous buffer system is absent;
- HS provide an effective natural shield against UV-radiation, with varying inputs of HS determining the extent of natural protection against UV radiation. The input depends on the vegetation in the catchment area, and on the climate. These influences can best be seen at the northern treeline (border boreal forest and tundra). During tundra periods, the lack of HS input results in an increase in UV penetration depth. The subsequent increased exposure of freshwater organisms to UV radiation, may reach such an extent that the ongoing anthropogenically caused damage to the global ozone layer, and even the consequent increase in UV-radiation, is inconsequential. Lack of HS is obviously a greater threat to aquatic organisms, than is the increasing UV radiation on Earth;
- through light exposure, HS are photolytically altered. These changes include the release of microbially available substrates, and can lead finally to mineralization (release of carbon monoxide and carbon dioxide). At least in non-eutrophic waters, the allochthonous HS feed heterotrophic processes more than

does autotrophic production. Most, if not all, non-eutrophic waters are net heterotrophic. Non-eutrophic aquatic systems are fed by their catchments. Terrestrial HS are the link between terrestrial and aquatic ecosystems. Which mechanisms hinder the microbial use of allochthonous dissolved HS in eutrophic waters, is not yet clear enough to establish paradigms;

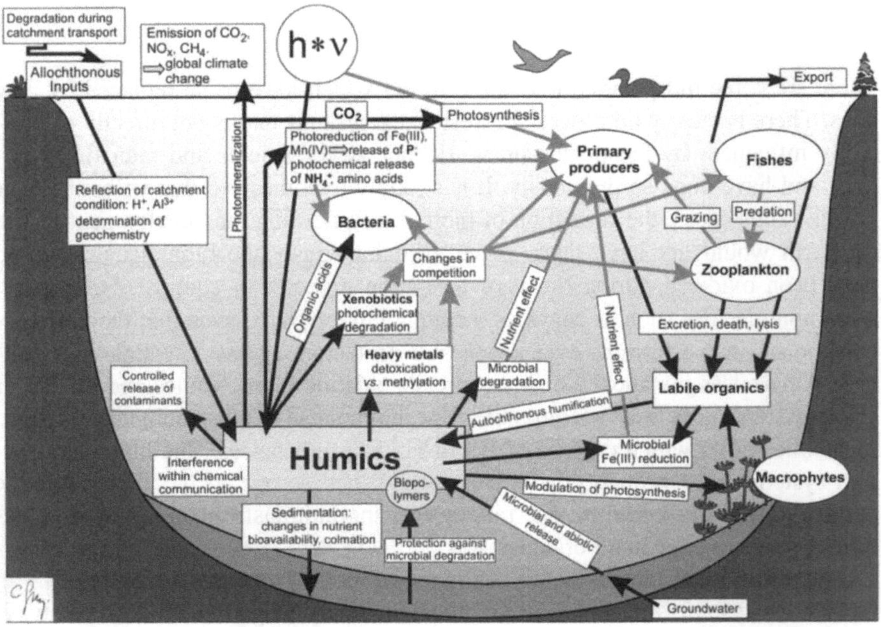

Fig. 10.1. Graphical summary of the ecological control HS display in aquatic ecosystems

- through exposure to light, a series of reactive oxygen species (ROS) is released from dissolved HS. These released species can oxidize HS, which leads to photomineralization. HS can also directly affect organisms, as indicated by the increased activity of enzymes such as peroxidases, to protect against the oxidative stress.
  Furthermore, the released ROS are natural purification tools in aquatic ecosystems, for natural toxic materials as well as chemical inputs from human activity. There is a variety of information indicating that the products of the oxidation processes are not harmless for aquatic organisms, but can increase the toxicity potential;
- HS can modify, in most instances decrease, the toxicity to organisms of xenobiotics and a series of metals, through the formation of various chemical complexes or compounds. For some metals, such as mercury, which can form organometal compounds, this toxicity reduction mechanism does not apply, and the toxicity potential significantly increases. In addition, the bioconcentration and toxicity of xenobiotics can be increased, but the underlying mechanisms are still obscure;

- HS have direct nutritive value. The idea that these high-molecular weight substances are not bioavailable, is no longer tenable. These materials can be broken down by microorganisms, so a size-reactivity model can be established, in which large molecules provoke high bioactivity, though not as high as that of small organic molecules (monosaccharides, amino acids, short-chain fatty acids).
- in the cycling of redox-sensitive metals, HS act as redox mediators (catalysts) and support growth of microorganisms. This applies, for instance, to Fe and Mn. In soils and sediments, such reactions take place independent of light. This process increases in significane, particularly when a lack of pores in soils and sediments makes direct access of bacteria to the oxides impossible;
- HS are entropy buffers. They can slow the degradation of complex substrates such as proteins or polysaccharide, by binding exoenzymes, and may even inhibit their activity. In addition, this binding protects exoenzymes against UV-radiation or against the attack of proteases and similar enzymes;
- HS contain, or at least control, information. They are generally able to bind extracellular DNA and can induce transformation in other organisms. These complex are the so-called cryptic genes;
- for the nematode *Caenorhabditis elegans*, it can be shown that HS modify the numbers of offspring per worm. Since, very recently, an estrogen receptor has been found in this species, a hormone-like mechanism may apply. Most probably, alkylaromatic structures, which form a major photostable building block of HS are the causative agents;
- HS have the potential to interfere with the chemical communication between aquatic organisms. At present, a strong binding of hydrophobic fish pheromones such as 4-androstene-3,17-dione to HS has been shown. However, the interference within the communication between aquatic organisms may be more generally applicable;
- antiviral, antibacterial, and antifungal properties of HS are known. For example, in aquaculture, HS can act as a natural pharmaceutical for fish fungal infections. A possible mechanism is via changes in the egg surface which can then no longer be recognized by the pathogens, or that inhibit pathogen adhesion and/or penetration. Another mechanism occurs via reactive oxygen species (ROS), oxidizing the membranes of the pathogens;
- HS, or low-molecular weight parts thereof, are generally able to block electron transport in photosystem II of water plants and algae, and thus inhibit photosynthetic oxygen release. The quinoide structures of HS are involved in this blocking process. In a few cases the opposite effect, namely increase in photosynthetic performance, has been observed with macrophytes and algae. This effect could be due to improved nutrition, but the precise mechanism is not yet known;
- HS, or low-molecular weight parts thereof, can be taken up by aquatic organisms This has been shown with labeled caffeic acid oxidation products with nominal molecular weights of 6 kDa. Within the cells, HS activate signal transduction chains, as has been shown for invertebrates, vertebrates, and macro-

phytes. However, whether this activation also occurs in microorganisms has yet to be determined. From medical studies, the adverse effects of humic-like substances, such as tannins, to pathogenic microorganisms are well documented, but comparable knowledge is lacking for aquatic microorganisms. In many ecological studies, HS are still seen only as an energy and nutrient source for microorganisms;

- several papers report contradictory biological and ecological effects of HS and NOM. For instance, promotion and inhibition of growth of algae and macrophytes, promotion and inhibition of net heterotrophy in non-eutrophic lakes, promotion and inhibition of reproduction of the nematode, *C. elegans*. These contradictions can be solved as trade-offs between specific and non-specific effects. For instance, net heterotrophy in non-eutrophic lakes can be considered as weak inhibition of photosynthesis of algae and macrophytes (specific effect), and simultaneous promotion of heterotrophic growth (non-specific effect). If net autotrophy predominates, a positive feed-back effect of bacteria on algae, based on growth promoting substances, can be observed.

These examples clearly show that HS are a major regulating force in freshwaters. They dominate the cycling of both substances and energy (Wetzel 1995). Thomas (1997) expands Wetzel's consideration and writes: '*The central dogma of the foodweb, and its implicit assumption that the energy flow in aquatic ecosystems can be quantified solely by measuring rates of photosynthesis, ingestion of solid food and its digestion by higher organisms, is invalid.*' In addition, there is much information emerging on the direct effects of HS on aquatic microorganisms, algae, macrophytes, zooplankton, and fish, the true direction of which is not yet completely understood. It can be expected that future research will yield further spectacular examples of interactions between HS and freshwater processes and organisms. It is evident that HS are a major biogeochemical regulator in freshwaters: no freshwater ecosystem can function without these substances.

References

Books

As a service to readers who are interested in humic substances and their scientific history, a compilation of reference books dealing with humic substances in soils, sediments, and natural waters will follow. Not all references are mentioned in the chapters of the book.

2001 and later

Clapp CE, Hayes MHB, Senesi N, Bloom PR, Jardine PM (eds) (2001) Humic substances and chemical contaminants. Soil Science Society of America, Madison

Frimmel FH, Abbt-Braun G, Heumann KG, Hock B, Lüdemann HD, Spiteller M (eds) (2002) Refractory organic substances in the environment. Wiley-VCH, Weinheim

Ghabbour EA, Davies G (eds) (2003) Humic substances: nature's most versatile materials. Taylor & Francis, New York (in press)

Ghabbour EA, Davies G (eds) (2001) Humic substances: structures, models, and functions. Royal Society of Chemistry, Cambridge

Hofrichter M, Steinbüchel A (eds) (2001) Biopolymers. Vol. 1: lignin, humic substances, and coal. Wiley-VCH, Weinheim

Swift RS, Spark KM (eds) (2001) Understanding and managing organic matter in soils, sediments and waters. Proc Intern Humic Subst Soc, St. Paul

1991–2000

Allard B, Borén H, Grimvall A (eds) (1991) Humic substances in the aquatic and terrestrial environment. Springer, Berlin, Lecture Notes in Earth Sciences 33

Beck AJ, Jones KC, Hayes MHB, Mingelgrin U (1993) Organic substances in soil and water: natural constituents and their influences on contaminant behaviour. Royal Society of Chemistry, Cambridge

Clapp CE, Hayes MHB, Senesi N, Griffith SM (eds) (1996) Humic substances and organic matter in soil and water environments. Proc Intern Humic Subst Soc, St. Paul

Davies G, Ghabbour EA (eds) (1998) Humic substances: structures, properties, and uses.

Royal Society of Chemistry, Cambridge
de Bertoldi M, Sequi P, Lemmes B, Papi T (eds) (1996) The science of composting. Blackie, London
Drozd J, Gonet SS, Senesi N, Weber J (eds) (1997) The role of humic substances in the ecosystems and in environmental protection.. Polish Soc Humic Substances, Wroclaw
Frimmel FH, Abbt-Braun G (eds) (1992) Refraktäre organische Säuren in Gewässern. Deutsche Forschungsgemeinschaft, Mitt XII Senatskomm Wasserforsch, VCH, Weinheim
Gaffney JS, Marley NA, Clark SB (eds) (1996) Humic and fulvic acids. Amer Chem Soc Symposium Series No. 651, Amer Chem Soc, Washington
Ghabbour EA, Davies G (eds) (1999) Understanding humic substances: advanced methods, properties and uses. Royal Society of Chemistry, Cambridge
Ghabbour EA, Davies G (eds) (2000) Humic substances: versatile components of plants, soil, and water. Royal Society of Chemistry, Cambridge
Hayes MHB, Wilson WR (eds) (1997) Humic substances, peats and sludges: health and environmental aspects. Royal Society of Chemistry, Cambridge
Hessen DO, Tranvik LJ (eds) (1998) Aquatic humic substances – ecology and biogeochemistry. Springer, Berlin
Keskitalo J, Eloranta P (eds) (1999) Limnology of humic waters. Backhuys, Leiden
Klavinš M (1997) Aquatic humic substances: characterization, structure and genesis. Riga University Press, Riga
McBride MB (1994) Environmental chemistry of soils. Oxford University Press, New York
Orlov DS (1995) Humic substances of soils and general theory of humification. Balkema, Brookfield
Piccolo A (ed) (1996) Humic substances in terrestrial ecosystems. Elsevier, Amsterdam
Salonen K, Kairesalo T, Jones RI (eds) (1992) Dissolved organic matter in lacustrine ecosystems: energy source and system regulator. Kluwer, Dordrecht
Senesi N, Miano TM (eds) (1992) Abstracts of oral and poster papers, 6th Meet Intern Humic Subst Soc, Bari
Senesi N, Miano TM (eds) (1994) Humic substances in the global environment and implications for human health. Elsevier, Amsterdam
Sparks DL (1995) Environmental soil chemistry. Academic Press, San Diego
Stevenson FJ (1994) Humus chemistry, 2nd edn Wiley, New York
Wilson WR (ed) (1991) Advances in soil organic matter research: the impact on agriculture and the environment. Royal Society of Chemistry, Cambridge
Zepp RG, Sonntag C (eds) (1995) The role of nonliving organic matter in the Earth's carbon cycle. Wiley, Chichester
Ziechmann W (1994) Humic substances. BI-Wissenschaftsverlag, Mannheim
Ziechmann W (1996) Huminstoffe und ihre Wirkungen. Spektrum, Heidelberg

1981–1990

Aiken GR, McKnight DM, Wershaw RL, MacCarthy P (eds) (1985) Humic substances in soil, sediment, and water. Wiley, New York
Averett RC, Leenheer JA, McKnight DM, Thorn KA (eds) (1989) Humic substances in the Suwannee River, Georgia: interactions, properties, and proposed structures, US Geological Survey Open-File Report 87–577, Denver

Baker EW (ed) (1985) Proceedings of the first meeting of the international humic substances society. Org Geochem 8

Becher G (ed) (1987) Special issue of humic substances research from the 3rd Internation Meet of Intern Humic Subst Soc, Sci Total Environ 62

Christmann RF, Gjessing ET (eds) (1983) Aquatic and terrestrial humic materials. Ann Arbor, Michigan

Frimmel FH, Christman RF (eds) (1988) Humic substances and their role in the environment. Wiley, Chichester

Hayes MHB, MacCarthy P, Malcolm RL, Swift RS (eds) (1989) Humic substances II: in search of structure. Wiley, New York

Hayes MHB, Swift RS (eds) (1984) Volunteered papers – 2nd international conference. University of Birmingham, Birmingham

Ishiwatari R (ed) (1990) Abstracts of oral and poster papers, 5th Intern Meet Intern Humic Subst Soc, Nagoya

MacCarthy P, Clapp CE, Malcolm RL, Bloom PR (eds) (1990) Humic substances in soil and crop sciences: selected readings. Soil Science Society of America and American Society of Agronomy, Madison

Orlov DS (1985) Humus acids of soils. Balkema, Rotterdam

Perdue EM, Gjessing ET (eds) (1990) Organic acids in aquatic ecosystems. Wiley, Chichester

Saiz-Jimenez C, Rosell RA, Albaiges J (eds) (1989) Special issue of humic substances research, 4th Intern Meet Intern Humic Subst Soc. Sci Total Environ 81/82

Sparks DL (1989) Kinetics of soil chemical processes. Academic Press, San Diego

Suffet IH, MacCarthy P (eds) (1989) Aquatic humic substances: influence of fate and treatment of pollutants. Amer Chem Soc, Washington

Thorn KA, Folan DW, MacCarthy P (1989) Characterization of the IHSS standard and reference fulvic and humic acids by solution state ^{13}C and ^{1}H NMR spectrometry. US Geological Survey, Water Resourc Investig Rep 89–4196

Thurman EM (1985) Organic geochemistry of natural waters. Martinus Nijhoff/Dr W. Junk, Dordrecht

1951–1980

Aleksandrova LN (1980) Soil organic matter and process of their transformation. LSHI, Leningrad (in Russian)

Gjessing ET (1976) Physical and chemical characteristics of aquatic humus. Ann Arbor, Michigan

Kononova MM (1966) Soil organic matter, its nature, its role in soil formation and in soil fertility. Pergamon, Oxford.

Maystrenko YuG (1965) Organic substances in waters and aquatic sediments of Ukraine. Naukova Dumka, Kiev (in Russian)

Scheffer F, Ulrich B (1960) Humus und Humusdüngung. Agrikulturchemie. Enke, Stuttgart

Schnitzer M, Khan SU (eds) (1972) Humic substances in the environment. Marcel Dekker, New York

Stevenson FJ (1982) Humus chemistry. Wiley, New York

Ziechmann W (1980) Huminstoffe. Probleme, Methoden, Ergebnisse. Verlag Chemie, Weinheim

Before 1950

Berzelius JJ (1839) Lehrbuch der Chemie. Arnold Dresden

Odén S (1922) Die Huminsäuren. Steinkopf, Leipzig

Skopintsev BA (1950) Organic substances in natural waters (aquatic humus). Gidrometeoizdat, Leningrad (in Russian)

Waksman S (1936) Humus, genesis, chemical composition and role in environment. Williams & Wilkins, Baltimore

Papers

Aardema BW, Lorenz MG, Krumbein WE (1983) Protection of sediment-adsorbed transforming DNA against enzymatic inactivation. Appl Environ Microbiol 46:417–420

Abbt-Braun G, Frimmel FH, Lipp P (1991) Isolation of organic substances from aquatic and terrestrial systems – comparison of some methods. Z Wasser Abwasserforsch 24:285–292

Abrahamsen G (1980) In: Drabløs D, Tollan, A (eds) Ecological Impact of Acid Precipitation, Proc Intern Conf, Sandefjord, Norway (SNSF Project), pp 58–63

Achard FK (1786) Chemische Untersuchung des Torfs. Cres's Chem Ann 2:391–403

Achtnich C, Fernandes E, Bollag JM, Knackmuss HJ, Lenke H (1999) Covalent binding of reduced metabolites of [$^{15}N_3$]TNT to soil organic matter during a bioremediation process analyzed by ^{15}N NMR spectroscopy. Environ Sci Technol 33:4448–4456

Ahlgren MO, Bowen SH (1991) Growth and survival of tadpoles (*Bufo americanus*) fed amorphous detritus derived from natural waters. Hydrobiologia 218:49–51

Aiken G, Cotsaris E (1995) Soil and hydrology: Their effect on NOM. J Amer Water Works Assoc 1995:36–45

Aiken GR (2000) Sources, transformations and reactivity of aquatic humic substances. Proc 10th Meet Intern Humic Subst Soc, 24–28 July 2000, Toulouse, pp 695–697

Aitkenhead JA, McDowell WH (2000) Soil C:N ratio as a predictor of annual riverine DOC flux at local and global scales. Glob Biogeochem Cycl 14:127–138

Aitkenhead JA, McDowell WH Sources of dissolved organic carbon and nitrogen in a hardwood forest floor. Soil Sci America J (in press)

Akkanen J, Kukkonen JVK (2001) Effects of water hardness and dissolved organic material on bioavailability of selected organic chemicals. Environ Toxicol Chem 20:2303–2308

Akkanen J, Penttinen S, Haitzer M, Kukkonen JVK (2001) Bioavailability of atrazine, pyrene and benzo[*a*]pyrene in European river waters. Chemosphere 45:453–462

Alberts JJ, Filip Z, Hertkorn N (1992) Fulvic and humic acids isolated from groundwater: Compositional characteristics and cation binding. J Contam Hydrol 11:317–330

Alberts JJ, Giesy JP (1983) Conditional stability of trace metals and naturally occurring humic materials: Application in equilibrium models and verification with field data. In: Christmann RF, Gjessing ET (eds) Aquatic and terrestrial humic materials. Ann Arbor, Michigan, pp 333–348

Alberts JJ, Griffin C, Gwynne K, Leversee GJ (1994) Binding of natural humic matter to polycyclic aromatic hydrocarbons in rivers of the southeastern United States. Water Sci Tech 30:199–205

Alberts JJ, Schindler JE, Miller RW, Nutter DE Jr (1974) Elemental mercury evolution mediated by humic acid. Science 184:895–897

Alexander M (1995) How toxic are toxic chemicals in soil? Environ Sci Technol 29:2713–2717

Allard B, Arsenie I (1991) Abiotic reduction of mercury by humic substances in aquatic systems – an important process for the mercury cycle. Water Air Soil Poll 56:457–464

Allard B, Borén H, Pettersson C, Zhang G (1994) Degradation of humic substances by UV irradiation. Environ Intern 20:97–101

Alles E (1999) Fließgewässerversauerung im Schwarzwald. Ökologische Bewertung auf der Basis benthischer Diatomeen. Dissertation, University Frankfurt. Ber Landesanst Umweltschutz BW, Karsruhe, Germany, Gewässerökologie 51, 507 pp

Alles E, Nörpel-Schrempp M, Lange-Bertalot H (1991) Zur Systematik und Ökologie charakteristischer *Eunotia*-Arten (Bacillariophyceae) in elektrolytarmen Bachoberläufen. Nova Hedwigia 53:171–213

Alomary A, Solouki T, Patterson HH, Cronan CS (2000) Elucidation of aluminum-fulvic acid interations by gas-phase hydrogen/deuterium (H/D) exchange and electrospray Fourier transform ion cyclotron resonance mass spectrometry (ESI FT-ICR). Environ Sci Technol 34:2830–2838

Amirbahman A, Reid AL, Haines TA, Kahl JS, Arnold C (2002) Association of methylmercury with dissolved humic acids. Environ Sci Technol 36:690–695

Amon RMW, Benner R (1994) Rapid cycling of high-molecular-weight dissolved organic matter in the ocean. Nature 369:549–552

Amon RMW, Benner R (1996a) Bacterial utilization of different size classes of dissolved organic matter. Limnol Oceanogr 41:41–51

Amon RMW, Benner R (1996b) Photochemical and microbial consumption of dissolved organic carbon and dissolved oxygen in the Amazon River system. Geochim Cosmochim Acta 60:1783–1792

Amon RMW, Fitznar HP, Benner R (2001) Linkages among the bioreactivity, chemical composition, and diagenetic state of marine dissolved organic matter. Limnol Oceanogr 46:287–297

Amyot M, Pinel-Alloul B, Campbell BGC (1994) Abiotic and seasonal factors influencing trace metal levels (Cd, Cu, Ni, Pb, and Zn) in the freshwater amphipod *Gammarus fasciatus* in two fluvial lakes of the St. Lawrence River. Can J Fish Aquat Sci 51:2003-2016

Anderson DH, Benke AC (1994) Growth and reproduction of the cladoceran *Ceriodaphnia dubia* from a forested floodplain swamp. Limnol Oceanogr 39:1517–1527

Anderson LJ, Johnson JD, Christman RF (1985) The reaction of ozone with isolated aquatic fulvic acid. Org Geochem 8:65–69

Andrews SS, Caron S, Zafiriou OC (2000) Photochemical oxygen consumption in marine waters: A major sink for colored dissolved organic matter? Limnol Oceanogr 45:267–277

Anonymous (2001): Synthesis and new conception of North Sea research (SYCON). Ber Zentr Meer Klimaforsch 14: 63pp

Ansorg R, Rochus W (1978) Untersuchungen zur antimikrobiellen Wirksamkeit von natürlichen und künstlichen Huminsäuren. Arzneim-Fosch/Drug Res 28:2195–2198

Arlauskas M, Slepetiene A (1997) Effects of conventional and minimum soil tillage systems, diverse fertilization and various crop rotations on the humic composition of loamy soils. In: Drozd J, Gonet SS, Senesi N, Weber J (eds) The role of humic sub-

stances in the ecosystems and in environmental protection.. Polish Soc Humic Substances, Wroclaw, pp 513–516
Arnold CG, Ciani A, Müller SR, Amirbahman A, Schwarzenbach RP (1998) Association of triorganotin compounds with dissolved humic acids. Environ Sci Technol 32:2976–2983
Arts MT, Robarts RD, Kasai F, Waiser MJ, Tumber VP, Plante AJ, Rai H, de Lange HJ (2000) The attenuation of ultraviolet radiation in high dissolved organic carbon waters of wetlands and lakes on the northern Great Plains. Limnol Oceanogr 45:292–299
Arvola L, Eloranta P, Järvinen M, Keskitalo J, Holopainen AL (1999a) Phytoplankton. In: Keskitalo J, Eloranta P (eds) Limnology of humic waters. Backhuys, Leiden, pp 137–171
Arvola L, Salonen K, Rask M (1999b) Trophic interactions. In: Keskitalo J, Eloranta P (eds) Limnology of humic waters. Backhuys, Leiden, pp 265–279
Aschan O (1907) Die Bedeutung der wasserlöslichen Humusstoffe für die Bildung der See- und Sumpfsee. Z Prakt Geol 15:56–68
Aschan O (1908) Die wasserlöslichen Humusstoffe der nordischen Süßgewässer. Z Prakt Chem 77:172–188
Aschheim S, Hohlweg W (1933) Über das Vorkommen östrogener Wirkstoffe in Bitumen. Dt med Wochenschr 59:12–14
Asplund G (1992) On the origin of organohalogens found in the environment. Dissertation University Linköping, Linköping Studies in Art and Science
Asplund G, Grimwal A (1991) Organohalogens in nature. Environ Sci Technol 25:1346–1350
Back RC, Watras CJ (1995) Mercury in zooplankton of northern Wisconsin lakes: Taxonomic and site-specific trends. Water Air Soil Pollut 80:931–938
Backes CA, Tipping E (1987) Aluminium complexation by an aquatic humic fraction under acidic conditions. Water Res 21:211–216
Backlund P (1992) Degradation of aquatic humic material by ultraviolet light. Chemosphere 25:1869–1878
Bacon F (1627) Sylva sylvarum. Printed by John Haviland for William Lee, London
Bada JL, Lazcano A (2000) Stanley Miller's 70th birthday. Origins Life Evol Biosphere 30:107–112
Baganz D, Staaks G, Spieser OH, Steinberg CEW (2001) How to use fish behaviour analysis to sensitively assess the hazard potentials of environmental chemicals. In: Butterworth FM, Gunatilaka A, Gonsebatt ME (eds) Biomonitors and biomarkers as indicators of environmental change, Vol. 2, Plenum, New York, pp 113–122
Bailey GW, Akim LG, Shevchenko SM (2001) Predicting chemical reactivity of humic substances for minerals and xenobiotics: use of computational chemistry, scanning probe microscopy, and virtual reality. In: Clapp CE, Hayes MHB, Senesi N, Bloom PR, Jardine PM (eds) Humic substances and chemical contaminants. Soil Science Society of America, Madison, pp 41–70 + 1 CD
Balogh SJ, Meyer ML, Johnson DK (1998) Transport of mercury in three contrasting river basins. Environ Sci Technol 32:456–462
Banerjee D, Nesbitt HW (2001) XPS study of dissolution of birnessite by humate with constraints on reaction mechanism. Geochim Cosmochim Acta 65:1703–1714
Barak P, Chen Y (1992) Equivalent radii of humic macromolecules from acid-base titration. Soil Sci 154:184–195
Barkay T, Gillman M, Turner RR (1997) Effects of dissolved organic carbon and salinity

on bioavailability of mercury. Appl Environ Microbiol 63:4267–4271

Barkovskii AL, Adriaens P (1998) Impact of humic constituents on microbial dechlorination of polychlorinated dioxins. Environ Toxicol Chem 17:1013–1020

Bärlocher F (1992) Effects of drying and freezing autumn leaves on leaching and colonization by aquatic hyphomycetes. Freshwater Biol 28:1–7

Barth EF, Acheson NH (1962) High-molecular weight material in tap water. J Amer Water Works Assoc 54:959–964

Bastviken D, Ejlertsson J, Sundh I, Tranvik L: Methane as a source of carbon and energy for lake pelagic food webs. Ecology (in press)

Bauer H, Frimmel FH (1987) Aufbereitungsorientierte Aspekte des photochemischen Abbaus aquatischer Huminstoffe. Z Wasser- Abwasser-Forsch 20:118–122

Bauer JE, Williams PM, Druffel ERM (1992) ^{14}C activity of dissolved organic carbon fractions in the north-central Pacific and Sargasso Sea. Nature 357:667–670

Baxter RM, Carey JH (1983) Evidence for photochemical generation of superoxide ion in humic waters. Nature 306:575–576

Becher G, Gjessing ET, Wright RF, Liltveld H (2001) Increasing colour in Norwegian surface waters – causes and consequences. Paper at 'Recent developments in hydrochemistry and hydrobiology', Techn Univ Karlsruhe, 23 Nov 2001

Beer AM, Lukanov J, Sagorchev P (2000) The influence of fulvic and humic acids from peat on the spontaneous contractile activity of smooth muscles. Phytomedicine 7:407–415

Behrendt H, Gelbrecht J, Huber P, Ley M, Uebe R, Fait M (1999) Geogen bedingte Grundbelastung der Fließgewässer Spree und Schwarze Elster und ihrer Einzugsgebiete. In: Landesumweltamt Brandenburg (ed), Studien Tagungsber 23:1–32.

Benedetti A, Alianiello F, Dell'orco S, Canali S (1994) A comparative characterization of organic matter in agrarian and forest soils from Italy. Environ Internat 20:419–424

Benner R, Biddanda B (1998) Photochemical transformations of surface and deep marine dissolved organic matter: Effects on bacterial growth. Limnol Oceanogr 43:1373–1378

Benoit JM, Gilmour CC, Mason RP (2001a) The influence of sulfide on solid-phase mercury bioavailability for methylation by pure cultures of *Desulfobulbus propionicus* (1pr3). Environ Sci Technol 35:127–132

Benoit JM, Gilmour CC, Mason RP, Heyes A (1999a) Sulfide controls on mercury speciation and bioavailability to methylating bacteria in sediment pore waters. Environ Sci Technol 33:951–957

Benoit JM, Mason RP, Gilmour CC (1999b) Estimation of mercury-sulfide speciation in sediment pore waters using octanol-water partitioning and implications for availability to methylating bacteria. Environ Toxicol Chem 18:2138–2141

Benoit JM, Mason RP, Gilmour CC, Aiken GR (2001b) Constants for mercury binding by dissolved organic matter isolates from the Florida Everglades. Geochim Cosmochim Acta 65:4445–4451

Benstead JP (1996) Macroinvertebrates and the processing of leaf litter in a tropical stream. Biotropica 28:367–375

Berg B (1986) The influence of experimental acidification on needle litter decomposition in a *Picea abies* L. forest. Scand J For Res 1:317–322

Bergamaschi B, Fram MS, Fujii R, Aiken, GR, Kendall C, Silva SR (2000) Using the reactivity, chemical composition, and $\delta\ ^{13}C$ isotopic composition of trihalomethanes to help understand the source of dissolved organic carbon found in the Sacramento-San Joaquin River Delta (CA USA). Proc 10th Meet Intern Humic Subst Soc, 24–28 July

2000, Toulouse, 1075–1079
Bertilsson S, Allard B (1996) Sequential photochemical and microbial degradation of refractory dissolved organic matter in a humic freshwater system. Arch Hydrobiol Spec Iss Adv Limnol 48:133–141
Bertilsson S, Tranvik LJ (1998) Photochemically produced carboxylic acids as substrates for freshwater bacterioplankton. Limnol Oceanogr 43:885–895
Bertilsson S, Tranvik LJ (2000) Photochemical transformation of dissolved organic matter in lakes. Limnol Oceanogr 45:753–762
Berzins B, Bertilsson J (1990) Occurrence of limnic micro-crustaceans in relation to pH and humic content in Swedish water bodies. Hydrobiologia 199:65–71
Bewley RJF, Parkinson D (1986) Sensitivity of certain soil microbial processes to acid precipitation. Pedobiologia 29:73–84
Beyer L, Blume HP, Sorge C, Schulten HR, Erlenkeuser H (1997) Humus composition and transformation in a pergelic cryohemist of coastal continental Antarctica. In: Drozd J, Gonet SS, Senesi N, Weber J (eds) The role of humic substances in the ecosystems and in environmental protection. Polish Soc Humic Substances, Wroclaw, pp 289–294
Biddanda B, Ogdahl M, Cotner J (2001) Dominance of bacterial metabolism in oligotrophic relative to eutrophic waters. Limnol Oceanogr 46:730–739
Billén G (1991) Protein degradation in aquatic environments. In: Chróst RJ (ed) Microbial enzymes in aquatic environments. Brock/Springer Ser Contemp Biosci, Springer Berlin, pp 123–143
Birge EA, Juday C (1926) Organic content of lake water. Bull US Bur Fish 42:185–205
Birge EA, Juday C (1934) Particulate and dissolved organic matter in inland lakes. Ecol Monogr 4:440–474
Black AP, Christman RF (1963) Characteristics of colored surface waters. J Amer Water Works Assoc 55:753–770
Bledsoe EL, Phlips EJ (2000) Relationship between phytoplankton standing crop and physical, chemical, and biological gradients in the Suwannee River and plume region, USA. Estuaries 23:458–473
Bleiwas ASH, Stokes PM, Olaveson MM (1984) Six years of plankton studies in the La-Cloche region of Ontario, with special reference to acidification. Verh Internat Verein Limnol 22:332–337
Blomqvist P, Jansson M, Drakare S, Bergstrom AK, Brydsten L (2001) Effects of additions of DOC on pelagic biota in a clearwater system: Results from a whole lake experiment in northern Sweden. Microbial Ecol 42: 383–394
Blomqvist P, Petterson A, Hyenstrand P (1994) Ammonium-nitrogen: A key regulatory factor causing dominance of non-nitrogen-fixing cyanobacteria in aquatic systems. Arch Hydrobiol 132:141–164
Bloom NS, Fitzgerald WF (1988) Determination of volatile mercury species at the picogram level by low temperature gas chromatography with cold vapor atomic fluorescence detection. Anal Chim Acta 208:151–161
Bloom NS, Gill GA, Cappelino S, Dobbs C, McShea L, Driscoll CT, Mason RP, Rudd JWM (1999) Speciation and cycling of mercury in Lavaca Bay, Texas, sediments. Environ Sci Technol 33:7–13
Blough NV, Zepp RG (1995) Reactive oxygen species in natural waters. In: Foote CS, Valentine JS, Greenberg A, Liebman JF (eds) Active oxygen in chemistry. Chapman and Hall, New York, pp 280–333
Blübaum-Gronau E, Hoffmann M, Spieser OH, Krebs F (1994) Der Koblenzer Verhaltens-

fischtest, ein auf dem Messsystem BehavioQuant® beruhender Biomonitor zur Gewässerüberwachung. Schriftenr Ver Wasser Boden Lufthyg 93:87–117

Blübaum-Gronau E, Hoffmann M, Spieser OH, Scholz W (2001) Continuous water monitoring: Changes of behavior patterns as indicators of pollutants. In: Butterworth FM, Gunatilaka A, Gonsebatt ME (eds) Biomonitors and biomarkers as indicators of environmental change, Vol. 2, Plenum, New York, pp 123–141

Boavida MJ, Wetzel RG (1998) Inhibition of phosphatase activity by dissolved humic substances and hydrolytic reactivation by natural ultraviolet light. Freshwater Biol 40:285–293

Boer MM, Koster EA, Lundberg H (1990) Greenhouse impact in Fennoscandia – preliminary findings of a European workshop on the effects of climatic changes. Ambio 19:2–10

Böhling S, Loskill R (1991) Purpose of the workshop. In: Nagel R, Loskill R (eds) Bioaccumulation in aquatic systems; contributions to the assessment. Proc Intern Workshop, Berlin 1990. VCH, Weinheim, pp 7–12

Böhmer J, Rahmann H (1992) Gewässerversauerung: Limnologische Untersuchungen zur Versauerung stehender Gewässer im Nordschwarzwald unter besonderer Berücksichtigung der Amphibienfauna. ecomed Landsberg/Lech, 231 pp

Bollag JM, Myers C (1992) Detoxification of aquatic and terrestrial sites through binding of pollutants to humic substances. Sci Total Environ 117/118:357–366

Borgmann U, Norwood WP (1999) Assessing the toxicity of lead in sediments to *Hyalella azteca*: the significance of bioaccumulation and dissolved metal. Can J Fish Aquat Sci 56:1492–1503

Borgmann U, Norwood WP, Ralph KM (1990) Chronic toxicity and bioaccumulation of 2,5,2',5'- and 3,4,3',4'-tetrachlorobiphenyl and Aroclor 1242 in the amphipod *Hyalella azteca*. Arch Environ Contam Toxicol 19:558

Borman FH, Likens GE (1979) Pattern and process in a forested ecosystem. Springer-Verlag, New York

Bottomly WB (1914a) Some accessory factors in plant growth and nutrition. Proc R Soc London, B 88:237–247

Bottomly WB (1914b) The significance of certain food substances for plant growth. Ann Bot (London) 28:531–540

Bottomly WB (1917) Some effects of organic growth-promotion substances (auximones) on the growth of *Lemna minor* in mineral cultural solutions. Proc R Soc London, B 89:481–505

Bottomly WB (1920) The effect of organic matter on the growth of various plants in culture solutions. Ann Bot (London) 34:353–365

Bourne DG, Jones GJ, Blakeley RL, Jones A, Negrie AP, Riddles P (1996) Enzymatic pathway for the bacterial degradation of the cyanobacterial cyclic peptide toxin microcystin LR. Appl Environ Microbiol 62:4086–4094

Bradley RW, Sprague JB (1985) The influence of pH, water hardness, and alkalinity on the acute lethality of zinc to rainbow trout (*Salmo gairdneri*). Can J Fish Aquat Sci 42:731–736

Braithwaite A, Livens FR, Richardson S, Howe MT, Goulding KWT (1997) Kinetically controlled release of uranium from soils. Eur J Soil Sci 48:661–673

Brakke DF, Henriksen A, Norton SA (1987) The relative importance of acidity sources for humic lakes in Norway. Nature 329:432–434

Brandorff GO (1978) Preliminary comparison of the crustacean plankton of a white water and a black water lake in Central Amazonia. Verh Intern Verein Limnol 20:1198–1202

Brassard P, Auclair JC (1984) Orthophosphate rate constants are mediated by the 10^3–10^4 molecular weight fraction in Shield lake waters. Can J Fish Aquat Sci 41:166–173

Breault RF, Colman JA, Aiken GR, McKnight D (1996) Copper speciation and binding by organic matter in copper-contaminated streamwater. Environ Sci Technol 30:3477–3486

Brezonik PL, King SO, Mach CE (1991) The influence of water chemistry on trace metal bioavailability and toxicity to aquatic organisms. In: Newman MC, McIntosh AW (eds) Metal ecotoxicology: Concepts and applications. Lewis, Chelsea, MI, pp 1–31

Brock TD, Gustafson J (1976) Ferric iron reduction by sulfur- and iron-oxidizing bacteria. Appl Environ Microbiol 32:567–571

Brockow T, Resch KL, Zacharias K, Franke A, Waldow R (1998) Therapeutic peat: Available evidence and future research priorities as seen by the example of gynaecological indications. (in German) Geburtsh Frauenheilk 58:459–463

Bronk DA, Sanderson MP, Koopmans DJ (2001) Contribution of photochemical ammonification to plankton nutrition in two rivers, and a coastal and open ocean environment. Meet Amer Soc Limnol Oceanog 2001, Albuquerque, New Mexico, February 12–16, 2001, Abstract SS 02–07

Browne KA, Tamburri MN, Zimmer-Faust RK (1998) Modelling quantitative structure-activity relationships between animal behaviour and environmental signal molecules. J Expl Biol 201:245–258

Brownlow CA, Bolen EG (1994) Fish and macroinvertebrate diversity in 1st order blackwater and alluvial streams in North Carolina. J Freshwater Ecol 9:261–270

Bruckmeier BFA (1994) Einträge von Säuren, PCB, PCDD und PCDF in den Großen Arbersee über eine Spanne von 130 Jahren. Diploma thesis, Munich University, pp 1–85

Bruckmeier BFA, Jüttner I, Schramm KW, Winkler R, Steinberg CEW, Kettrup A (1997) PCBs and BCDD/Fs in lake sediments of Grosser Arbersee, Bavarian Forest, South Germany. Environ Pollut 95:19–25

Brüggemann R, Halfon E (1995) Theoretical base of the program 'Hasse'. GSF-Report 20/95

Brüggemann R, Schwaiger J, Negele RD (1995) Applying Hasse diagram technique for evaluation of toxicological fish tests. Chemosphere 30:1767–1780

Buchan A, Neidle EL, Moran MA (2001) Diversity of the ring-cleaving dioxygenase gene pcaH in a salt marsh bacterial community. Appl Environ Microbiol 67:5801–5809

Buffle J, Deladoey P, Greter FL, Haerdi W (1980) Study of the complex formation of copper(II) by humic and fulvic substances. Anal Chim Acta 116:255–274

Buffle J, Leppard GG (1995a) Characterization of aquatic colloids and macromolecules.1. Structure and behavior of colloidal material. Environ Sci Technol 29: 2169–2175

Buffle J, Leppard GG (1995b) Characterization of aquatic colloids and macromolecules. 2. Key role of physical structures on analytical results. Environ Sci Technol 29:2176–2184

Bukaveckas PA, Robbins-Forbes M (2000) Role of dissolved organic carbon in the attenuation of photosynthetically active and ultraviolet radiation in Adirondack lakes. Freshwater Biol 43:339–354

Bunte C, Simon M (1999) Bacterioplankton turnover of dissolved free monosaccharides in a mesotrophic lake. Limnol Oceanogr 44:1862–1870

Burba P, Van den Bergh J, Klockow D (2001) On-site characterization of humic-rich hydrocolloids and their metal loading by means of mobile size-fractionation and exchange techniques. Fresenius J Anal Chem 371:660–669

Burk D, Lineweaver H, Horner CK (1932) Iron in relation to the stimulation of growth by humic acid. Soil Sci 33:413–435

Burkart D, Hönninger W, Rieder U (1994) Kalium-huminat (Sopar) bei der Behandlung von Fischkrankheiten. Res Rep to Sopar Pharma GmbH Mannheim (with permission)

Burkhard LP (2000) Estimating dissolved organic carbon partition coefficients for nonionic organic chemicals. Environ Sci Technol 34:4663–4668

Burnison K (1994) Solubility enhancement of fenvalerate by isolated DOC lakewater fractions. In: Senesi N, Miano TM (eds) Humic substances in the global environment and implications for human health. Elsevier, Amsterdam, pp 811–818

Burnison K, Meinelt T, Playle R, Pietrock M, Steinberg CEW (1999) Effect of humic substances on cadmium accumulation by zebrafish embryos and larvae. 26th Ann Aquat Toxicity Workshop, Edmonton, Alberta, Canada, 3.–6.10.1999 (Poster)

Burns A, Ryder DS (2001) Response of bacterial extracellular enzymes to inundation of floodplain sediments. Freshwater Biol 46: 1299–1307

Bury NR, Galvez F, Wood CM (1999b) Effects of chloride, calcium, and dissolved organic carbon on silver toxicity: Comparison between rainbow trout and fathead minnows. Environ Toxicol Chem 18:56–62

Bury NR, McGeer JC, Wood CM (1999a) Effects of altering freshwater chemistry on physiological responses of rainbow trout to silver exposure. Environ Toxicol Chem 18:49–55

Bushaw KL, Zepp RG, Tarr MA, Schulz-Jander D, Bourbonniere RA, Hodson RE, Miller WL, Bronk DA, Moran MA (1996) Photochemical release of biologically available nitrogen from aquatic dissolved organic matter. Nature 381:404–407

Buesing N (2002) Microbial productivity and organic matter flow in a littoral reed stand. Diss ETH Zurich No 14667, 136 pp

Cabaniss SE, Shuman MS (1988) Copper binding by dissolved organic matter: I. Suwannee River fulvic acids using a method for the determination of quickly reacting aluminium. Water Air Soil Pollut 84:103–116

Cameron RS, Thornton BK, Swift RS, Posner AM (1972) Molecular weight and shape of humic acid from sedimentation and diffusion measurements on fractionated extracts. J Soil Sci 23:395–408

Campbell PGC, Twiss MR, Wilkinson KJ (1997) Accumulation of natural organic matter on the surfaces of living cells: Implications for the interaction of toxic solutes with aquatic biota. Can J Fish Aquat Sci 54:2543–2554

Carignan R, Planas D, Vis C (2000) Planktonic production and respiration in oligotrophic Shield lakes. Limnol Oceanogr 45:189–199

Carlberg GE, Martinsen K (1982) Adsorption/complexation of organic micropollutants to aquatic humus: Influence of aquatic humus with time on organic pollutants and comparison of two analytical methods for analyzing organic pollutants in humus water. Sci Total Environ 25:245–254

Carlough (1994) Origins, structure, and trophic significance of amorphous seston in a blackwater river. Freshwater Biol 31:227–237

Carlsson P, Zuleika A, Granéli E (1993) Nitrogen bound to humic matter of terrestrial origin – a nitrogen pool for coastal phytoplankton? Mar Ecol Prog Ser 97:105–116

Caron DA, Goldman JC, Dennett MR (1990) Carbon utilization by the omnivorous flagellate *Paraphysomonas imperforata*. Limnol Oceanogr 35:192–201

Carpenter SR, Cole JJ, Kitchell JF, Pace ML (1998) Impact of dissolved organic carbon, phosphorus, and grazing on phytoplankton biomass and production in experimental lakes. Limnol Oceanogr 43:73–80

Carter CW, Suffet IH (1982) Binding of DDT to dissolved humic materials. Environ Sci Technol 16:735–740

Cary GA, McMahon JA, Kuc WJ (1987) The effect of suspended solids and naturally occurring dissolved organics in reducing the acute toxicity of cationic polyelectrolytes to aquatic organisms. Environ Toxicol Chem 6:469–474

Celis R, Hermosin MC, Carrizosa MJ, Cornejo J (2002) Inorganic and organic clay as carriers for controlled release of the herbicide hexazinone. J Agr Food Chem 50:2324–2330

Cervantes FJ, Dijksma W, Duong-Dac T, Ivanova A, Lettinga G, Field JA (2001) Anaerobic mineralization of toluene by enriched sediments with quinones and humus as terminal electron acceptors. Appl Environ Microbiol 67:4471–4478

Cervantes FJ, van der Velde S, Lettinga G, Field JA (2000) Quinones as terminal electron acceptors for anaerobic microbial oxidation of phenolic compounds. Biodegration 11:313–321

Cheam V, Gamble DS (1974) Metal-fulvic acid chelation equilibrium in aqueous $NaNO_3$ solution. Can J Soil Sci 54:413–417

Chefetz B, Deshmukh AP, Hatcher PG, Guthrie EA (2000) Pyrene sorption by natural organic matter. Environ Sci Technol 34:2925–2930

Chefetz B, Hadar Y, Chen Y (1998) Dissolved organic carbon fractions formed during composting of municipal solid waste: Properties and significance. Acta hydrochim hydrobiol 26:172–179

Chen Y, Aviad T (1990) Effects of humic substances on plant growth. Humic substances in soil and crop sciences; Selected readings, Amer Soc Agron, Soil Sci Soc Amer, pp 161–186

Chen Y, Clapp CE, Magen H, Cline VW (1999) Stimulation of plant growth by humic substances: Effects of iron availability. In: Ghabbour EA, Davies G (eds) Understanding humic substances: advanced methods, properties and uses. Royal Society of Chemistry, Cambridge, pp 255–263

Chen Y, Hoitink HAJ, Madden LV (1988) Microbial activity and biomass in container media predicting suppressiveness to damping-off caused by *Pythium ultimum*. Phytopathology 78:1447–1450

Chen Y, Khan SU, Schnitzer M (1978) Ultraviolet irradiation of dilute fulvic acid solutions. Soil Sci Amer J 42:292–296

Chen Y, Magen H, Riov J (1994) Humic substances originating from rapidly decomposing organic matter: properties and effects on plant growth. In: Senesi N, Miano TM (eds) Humic substances in the global environment and implications for human health. Elsevier, Amsterdam, pp 427–443

Chen Y, Schnitzer M (1976) Viscosity measurements on soil humic substances. Soil Sci Soc Am J 40:866–872

Chen YJ, Lin-Chao S, Huang TS, Yang ML, Lu FJ (2001) Humic acid induced growth retardation in a Sertoli cell line, TM4. Life Sci 69:1269–1284

Chettri MK, Cook CM, Vardaka E, Sawidis T, Lanaras T (1998) The effect of Cu, Zn, and Pb on the chlorophyll content in lichens *Cladonia convoluta* and *Cladonia rangiformis*. Environ Exp Bot 39:1–10

Chien YY, Kim EG, Bleam WF (1997) Paramagnetic relaxation of atrazine solubilized by humic micellar solution. Environ Sci Technol 31:3204–3208

Chin WC, Orellana MV, Verdugo P (1998) Spontaneous assembly of marine dissolved organic matter into polymer gels. Nature 391:568–572

Chin YP, Aiken GR, Danielsen KM (1997) Binding of pyrene to aquatic and commercial humic substances: The role of molecular weight and aromaticity. Environ Sci Technol 31:1630–1635

Chiou CT, Malcolm RL, Brinton TI, Kile DE (1986) Water solubility enhancement of some organic pollutants and pesticides by dissolved humic and fulvic acids. Environ Sci Technol 20:502–508

Chiou CT, Porter PE, Schmedding DW (1983) Partition equilibria of nonionic organic compounds between soil organic matter and water. Environ Sci Technol 17:227–231

Chiron S, Abian J, Ferrer M, Sanchez-Baeza F, Messeguer A, Barcelo D (1995) Comparative photodegradation rates of alachlor and bentazone in natural water and determination of breakdown products. Environ Toxicol Chem 14:1287–1298

Choi MH, Cech JJ Jr, Lagunas-Solar MC (1998) Bioavailability of methylmercury to sacramento blackfish (*Orthodon microlepidotus*): dissolved carbon effects. Environ Toxicol Chem 17:695–701

Choudry GG (1984) Interactions of humic substances with environmental chemicals. In: Hutzinger O (ed) The handbook of environmental chemistry. 2B Reactions and processes. Springer, Berlin, pp 103–128

Chróst RJ (1990) Microbial ectoenzymes in aquatic environments. In: Overbeck J, Chróst RJ (eds) Advanced biochemical and molecular approaches to aquatic microbial ecology. Brock/Springer Ser Contempor Biosci, Springer, Berlin, pp 47–78

Ciborowski JJH, Craig DA, Fray KM (1997) Dissolved organic matter as a food for black fly larvae (Diptera: Simuliidae). J N Am Benthol Soc 16:771–780

Clair TA, Bärlocher F, Brassard P, Kramer JR (1989) Chemical and microbial diagenesis of humic matter in freshwaters. Water Air Soil Poll 46:205–211

Clair TA, Ehrman JM (1996) Variations in discharge and dissolved carbon and nitrogen export from terrestrial basins with changes in climate: A neural network approach. Limnol Oceanogr 41:921–927

Clair TA, Sayer BG, Kramer JR, Eaton DR (1996) Seasonal variation in the composition of aquatic organic matter in some Nova Scotian brownwaters: nuclear magnetic resonance approach. Hydrobiologia 317:141–150

Clapp, CE, Hayes, MHB (1999) Size and shapes of humic substances. Soil Sci 164:777–789

Clark KL, Hall RJ (1985) Effects of elevated hydrogen ion and aluminum concentration on survival of amphibian embryos and larvae. Can J Zool 63:116–123

Clarkson TW (1994) The toxicology of mercury and its compounds. In: Watras CJ, Huckabee JW (eds) Mercury as a global pollutant: integration and synthesis. Lewis, Chelsea, pp 631–641

Claus H, Filip Z (1998) Degradation and transformation of aquatic humic substances by laccase-producing fungi *Cladosporium cladosporioides* and *Polyporus versicolor*. Acta hydrochim hydrobiol 26:180–185

Claus H, Gleixner G, Filip Z (1999) Formation of humic-like substances in mixed and pure cultures of aquatic microorganisms. Acta hydrochim hydrobiol 27:200–207

Coates JD, Cole KA, Chakraborty R, O'Connor SM, Achenbach LA (2002) Diversity and ubiquity of bacteria capable of utilizing humic substances as electron donors for anaerobic respiration. Appl Environ Microbiol 68:2445–2452

Cole JJ (1999) Aquatic microbiology for ecosystem scientists: New and recycled para-

digms in ecological microbiology. Ecosystems 2:215–225
Cole JJ, Caraco NF, Kling GW, Kratz TK (1994) Carbon dioxide supersaturation in the surface waters of lakes. Science 265:1568–1570
Cole JJ, Pace ML, Carpenter SR, Kitchell JF (2001) Persistence of net heterotrophy in lakes during nutrient addition and food web manipulation. Limnol Oceanogr 45:1718–1730
Concheri G, Nardi S, Piccolo A, Dell'Agnola G, Rascio N (1994) Effects of humic fractions on morphological changes related to invertase and peroxidase activities in wheat seedling roots. In: Senesi N, Miano TM (eds) Humic substances in the global environment and implications for human health. Elsevier, Amsterdam, pp 257–262
Connell DW (1990) Bioaccumulation of xenobiotic compounds. CRC Press, Boca Raton, FL
Connell DW (1998) Bioaccumulation of chemicals by aquatic organisms. In: Schüürmann G, Markert B (eds) Ecotoxicology. Wiley, New York, pp 439–450
Conrad R, Seiler W (1980) Photooxidative production and microbial consumption of carbon monoxide in seawater. FEMS Microbiol Lett 9:61–64
Cooper WJ, Zika RG (1983) Photochemical formation of hydrogen peroxide in surface and ground waters exposed to sunlight. Science 220:711–712
Cope WG, Wiener JG, Rada RG (1990) Mercury accumulation in yellow perch in Wisconsin seepage lakes: Relation to lake characteristics. Environ Toxicol Chem 9:931–940
Corin NS, Backlund PH, Kulovaara MAM (2000) Photolysis of the resin acid dehydroabietic acid in water. Environ Sci Toxicol 34:22-26
Corti S, Molteni F, Palmer TN (1999) Signature of recent climate change in frequencies of natural atmospheric circulation regimes. Nature 398:799–802
Cotner JB Jr, Heath RT (1990) Iron redox effects on photosensitive phosphorus release from dissolved humic materials. Limnol Oceanogr 35:1175–1181
Cotner JB, Johengen TH, Biddanda BA (2000) Intense winter heterotrophic production stimulated by benthic resuspension. Limnol Oceanogr 45:1672–1676
Cowan MM (1999) Plant products as antimicrobial agents. Clin Microbiol Rev 12:564–582
Cozzi R, Nicolai M, Perticone P, de Salvia R, Spuntarelli R (1993) Desmutagenic activity of natural humic acids: Inhibition of mitomycin C and maleic hydrazide mutagenicity. Mutat Res 299:37–44
Crecchio C, Stotzky G (1998a) Binding of DNA on humic acids: Effects on transformation of *Bacillus subtilis*. Soil Biol Biochem 30:1061–1067
Crecchio C, Stotzky G (1998b) Insecticidal activity and biodegradation of the toxins from *Bacillus thuringiensis* subsp. *kurstaki* bound to humic acids from soil. Soil Biol Biochem 30:463–470
Cronan CS (1990) Patterns of organic acid transport from forested watersheds to aquatic ecosystems. In: Perdue EM, Gjessing ET (eds) Organic acids in aquatic ecosystems. Wiley, Chichester, pp 245–260
Cronan CS, Aiken GR (1985) Chemistry and transport of soluble humic substances in forested watersheds of the Adirondack Park, New York. Geochim Cosmochim Acta 49:1697–1704
Currie DJ (1990) Large scale variability and interactions among phytoplankton, bacterioplankton and phosphorus. Limnol Oceanogr 35:298–310
Currie DJ, Kalff J (1984a) A comparison of the abilities of freshwater algae and bacteria to acquire and retain phosphorus. Limnol Oceanogr 29:298–310
Currie DJ, Kalff J (1984b) The relative importance of bacterioplankton and phytoplankton

in phosphorus uptake in freshwater. Limnol Oceanogr 29:311–321
Currie DJ, Kalff J (1984c) Can bacteria outcompete phytoplankton for phosphorus? A chemostat test. Microb Ecol 10:205–216
D'Arcy P, Carignan R (1997) Influence of catchment topography on water chemistry in southeastern Québec Shield lakes. Can J Fish Aquat Sci 54:2215–2227
Dahlén J, Bertilsson S, Petterson C (1996) Effects of UV-A irradiation on dissolved organic matter in humic surface waters. Environ Intern 22:501–506
Dankwardt A, Freitag D, Hock B (1998) Approaches to the immunochemical analysis of non-extractable triazine residues in refractory organic substances (ROS) and characterization of ROS. Acta hydrochim hydrobiol 26:145–151
Dankwardt A, Hock B, Simon R, Freitag D, Kettrup A (1996) Determination of non-extractable triazine residues by enzyme immunoassay: Investigation of model compounds and soil fulvic acids. Environ Sci Technol 30:3493–3500
Datsko VG (1939) Organic matter in waters of some seas. DAN SSSR 24:23–31 (in Russian)
Datsko VG (1940) About analysis of organic carbon in marine water. Tr. Azcherniro 12:56–67 (in Russian)
Datsko VG (1956) Fundamental aspects of organic matter investigations in natural waters. Gidrochim Mater 26:121–129 (in Russian)
Datsko VG (1959) Organic substances in waters of southern seas of USSR. Izd. An SSSR, Moskau (in Russian)
David MB, Vance GF, Rissing JM, Stevenson FJ (1989) Organic carbon fraction in extracts of O and B horizons from a New England spodozol. Effects of acid treatment. J Environ Qual 18:212–217
Davies G, Ghabbour EA, Khairy AH, Ibrahim HZ (1997) A 'site creation' model for specific adsorption of aqueous nucleobases, nucleosides, and nucleotides on compost-derived humic acid. J Phys Chem B 101:3228–3239
Davis RB, Anderson DS, Berge F (1985) Palaeolimnological evidence that lake acidification is accompanied by loss of organic matter. Nature 316:436–438
Day KE (1991) Effects of dissolved organic carbon on accumulation and acute toxicity of fenvalerate, deltamethrin and cyhalothrin to *Daphnia magna* (Straus). Environ Sci Technol 10:91–101
de Diego A, Tseng CM, Dimov N, Amouroux D, Donard OFX (2001) Adsorption of aqueous inorganic mercury and methylmercury on suspended kaolin: influence of sodium chloride, fulvic acid and particle content. Appl Organometal Chem 15:490–498
de Haan H (1974) Effect of a fulvic acid fraction on the growth of a *Pseudomonas* sp. from Tjeukemeer (The Netherlands). Freshwater Biol 4:301–310
de Haan H (1976) Evidence for the induction of catechol-1,2-oxygenase by fulvic acid. Plant Soil 45:129–136
de Haan H (1977) Effect of benzoate on microbial decomposition of fulvic acids in Tjeukemeer (The Netherlands). Limnol Oceanogr 22:38–44
de Haan H (1992) Impacts of environmental changes on the biogeochemistry of aquatic humic substances. Hydrobiologia 229:59–71
de Haan H (1993) Solar UV-light penetration and photodegradation of humic substances in peaty lake water. Limnol Oceanogr 38:1072–1076
de Haan H, Halma G, de Boer T, Haverkamp J (1981) Seasonal variations in the compositions of fulvic acids in Tjeukemeer, The Netherlands, as studied by Curie-point pyrolysis-mass spectrometry. Hydrobiologia 78:87–95

de Haan H, Jones RI, Salonen K (1990) Abiotic transformations of iron and phosphate in humic lake water revealed by double isotope labeling and gel filtration. Limnol Oceanogr 35:491–497

de Nobili M, Fornasier F, Bragato G (1997) Capillary electrophoresis of humic substances in polyethylenglycol solutions. In: Drozd J, Gonet SS, Senesi N, Weber J (eds) The role of humic substances in the ecosystems and in environmental protection. Polish Soc Humic Substances, Wroclaw, pp 97–101

de Paolis, F., Kukkonen, J. (1997) Binding of organic pollutants to humic and fulvic acids: Influence of pH and the structure of humic material. Chemosphere 34:1693–1704

De Saussure T (1804) Récherches chemiques sur la vegetation. Paris Ann 12:162–178

De Simone C, Piccolo A, De Marco A, D'Ambrosio C (1997) Antimutagenic activity of humic acids of different origin. In: Drozd J, Gonet SS, Senesi N, Weber J (eds) The role of humic substances in the ecosystems and in environmental protection. Polish Soc Humic Substances, Wroclaw, pp 945–950

de Wit JCM, van Riemsdijk, WH, Koopal LK (1993) Proton binding to humic substances. 1. Electrostatic effects. Environ Sci Technol 27:2005–2014

Debus R, Schröder P (1990) Responses of *Petunia hybrida* and *Phaseolus vulgaris* to fumigation with difluoro-chloro-bromo-methane (Halon 1211). Chemosphere 21:1499–1505

Dehorter B, Blondeau R (1992) Extracellular enzyme activities during humic acid degradation by the white rot fungi *Phanerochaete chrysosporium* and *Trametes versicolor*. FEMS Microbiol Lett 94:209–216

del Giorgio PA, Cole JJ, Cimbleris A (1997) Respiration rates in bacteria exceed phytoplankton production in unproductive aquatic systems. Nature 385:148–151

del Giorgio PA, Peters RH (1994) Patterns in planktonic P:R ratio in lakes: Influence of lake trophy and dissolved organic carbon. Limnol Oceanogr 39:772–787

Dickson W (1980) Properties of acidified water. In: Drabløs D, Tollan A (eds) Ecological impact of acid precipitation. Proc Intern Conf, Sandefjord, Norway (SNSF Project), pp 75–83

Dillon PJ, Molot LA (1997) Dissolved organic and inorganic carbon mass balances in central Ontario lakes. Biogeochemistry 36:29–42

Ding GW, Mao JD, Xing BS (2001) Characteristics of amino acids in soil humic substances. Commun Soil Sci Plant Anal 32:13–14

Donahue WF, Schindler DW, Page SJ, Stainton MP (1998) Acid-induced changes in DOC quality in an experimental whole-lake manipulation. Environ Sci Technol 32:2954–2960

Döring UM, Marschner B (1998) Water solubility enhancement of benzo(a)pyrene and 2,2',5,5'-tetrachlorobiphenyl by dissolved organic matter (DOM). Phys Chem Earth 23:193–197

Draper WM, Crosby DG (1983) The photochemical generation of hydrogen peroxide in natural waters. Arch Environ Contam Toxicol 12:121–126

Driscoll CT (1984) A procedure for the fractionation of aqueous aluminum in dilute acidic waters. Intern J Environ Anal Chem 16:257–283

Driscoll CT, Baker JJ, Bisogni JR, Schofield CL (1980) Effect of aluminium speciation on fish in dilute acidified waters. Nature 284:161–164

Driscoll CT, Blette V, Yan C, Schofield CL, Munson R, Holsapple J (1995) The role of dissolved organic carbon in the chemistry and bioavailability of mercury in remote Adirondack lakes. Water Air Soil Pollut 80:499–508

Driscoll CT, Yan C, Schofield CL, Munson R, Holsapple J (1994) The mercury cycle and fish in the Adirondack lakes. Environ Sci Technol 28:136A–143A

Druffel ERM, Williams PM (1990) Identification of a deep marine source of particulate organic carbon using bomb ^{14}C. Nature 347:172–174

Druvietis I, Springe G, Urtane L, Klavinš M (1998) Evaluation of plankton communities in small highly humic bog lakes in Latvia. Environ Internat 24:595–602

Duarte CM, Agustí S (1998) The CO_2 balance of unproductive aquatic ecosystems. Science 281:234–236

Duarte CM, Agustí S, Arístegui J, Gonzáles N, Anadón R (2001) Evidence for a heterotrophic subtropical northeast Atlantic. Limnol Oceanogr 46:425–428

Dudley RJ, Churchill PF (1995) Effect and potential ecological significance of the interaction of humic acids with two aquatic extracellular proteases. Freshwater Biol 34:485–494

Dunkelberg H, Klöcking HP, Klöcking R (1997) Suppression of heat-induced [^3H]arachidonic acid release in U937 cells by humic acid-like polymers. In: Drozd J, Gonet SS, Senesi N, Weber J (eds) The role of humic substances in the ecosystems and in environmental protection. Polish Soc Humic Substances, Wroclaw, pp 647–652

Dunnivant FM, Jardine PM, Taylor DL, McCarthy JF (1992) Transport of naturally occurring dissolved organic carbon in laboratory columns containing aquifer material. Soil Sci Soc Am J 56:437–444

Dunson WA, Wyman RL, Corbett ES (1992) A symposium on amphibian declines and habitat acidification. J Herpetol 16:349–352

Earle JC, Duthie HC, Scruton DA (1986) Analysis of the phytoplankton composition of 95 Labrador lakes, with special reference to natural and anthropogenic acidification. Can J Fish Aquat Sci 43:1804–1811

Eckhardt BW, Moore TR (1990) Controls of dissolved organic carbon concentrations in streams, southern Quebec. Can J Fish Aquat Sci 47:1537–1544

Egeberg PK, Christy AA, Eikenes M (2002) The molecular size of natural organic matter (NOM) determined by diffusivimetry and seven other methods. Water Res 36:925–932

Egeberg PK, Eikenes M, Gjessing ET (1999) Organic nitrogen distribution in NOM size classes. Environ Internat 25:225–236

Eie JA (1974) A comparative study of the crustacean communities in forest and mountain localities in the Vassfaret area (southern Norway). Norw J Zool 22:177–205

Ellerbrock RH, Höhn A, Rogasik J (1997) Influence of management practice on soil organic matter composition. In: Drozd J, Gonet SS, Senesi N, Weber J (eds) The role of humic substances in the ecosystems and in environmental protection. Polish Soc Humic Substances, Wroclaw, pp 233–238

Eloranta P (1999a) Light penetration in water. In: Keskitalo J, Eloranta P (eds) Limnology of humic waters. Backhuys, Leiden, pp 61–63

Eloranta P (1999b) Qualitative changes in light climate in water. In: Keskitalo J, Eloranta P (eds) Limnology of humic waters. Backhuys, Leiden, pp 64–66

Eloranta P (1999c) Light penetration and thermal stratification in lakes. In: Keskitalo J, Eloranta P (eds) Limnology of humic waters. Backhuys, Leiden, pp 72–74

Emmenegger L, Schönenberger R, Sigg L, Sulzberger B (2001) Light-induced redox cycling of iron in circumneutral lakes. Limnol Oceanogr 46:49–61

Enfield CG (1985) Chemical transport facilitated by multiphase flow systems. Water Sci Technol 17:1–12

Enfield CG, Lien BK, Wood AL (2001) Effect of non-target organics on organic chemical transport. In: Clapp CE, Hayes MHB, Senesi N, Bloom PR, Jardine PM (eds) Humic substances and chemical contaminants. Soil Science Society of America, Madison, pp 471–487

Engelhardt KAM, Ritchie ME (2001) Effects of macrophyte species richness on wetland ecosystem functioning and services. Nature 411:687–689

Engstrom DR (1987) Influence of vegetation and hydrology on the humus budgets of Labrador lakes. Can J Fish Aquat Sci 44:1306–1314

EPA (Environmental Protection Agency) (1997) Mercury study report to congress. Various Volumes, United States Environmental Protection Agency, Washington, DC, EPA-452/R-97-003

Ephraim JH, Pettersson C, Allard B (1996) Correlations between acidity and molecular size distributions of an aquatic fulvic acid. Environ Intern 22:475–483

Ephraim JH, Pettersson C, Nordén M, Allard B (1995) Potentiometric titrations of humic substances: do ionic strength effects depend on the the molecular weight? Environ Sci Technol 28:622–628

Erickson RJ, Benoit DA, Mattson VR, Nelson HP Jr, Leonard EN (1996) The effects of water chemistry on the toxicity of copper to fathead minnows. Environ Toxicol Chem 15:181–193

Erickson RJ, Brooke LT, Kahl MD, Venter FV, Hartin SL, Markee TP, Spehar RL (1998) Effects of laboratory test conditions on the toxicity of silver to aquatic organisms. Environ Toxicol Chem 17:572–578

Ertel JR, Caine JM, Thurman EM (1993) Biomarker compounds as source indicators for dissolved fulvic acids in a bog. Biogeochemistry 22:195–212

Ertel JR, Hedges JI (1983) Bulk chemical and spectroscopic properties of marine and terrestrial humic acids, melanoidins and catechol-based synthetic polymers. In: Christmann RF, Gjessing ET (eds) Aquatic and terrestrial humic materials. Ann Arbor, Michigan, pp 143–162

Ertel JR, Hedges JI, Perdue EM (1984) Lignin signature of aquatic humic substances. Science 223:485–487

Eurochlor: http://www.eurochlor.org/chlorine/science/htm

Faust BC, Hoigné J (1987) Sensitized photooxidation of phenols by fulvic acid in natural waters. Environ Sci Technol 21:957–964

Fee EJ, Hecky RE, Kasian SEM, Cruikshank DR (1996) Effects of lake size, water clarity, and climatic variability on mixing depths in Canadian Shield lakes. Limnol Oceanogr 41:912–920

Fent K, Looser PW (1995) Bioaccumulation and bioavailability of tributyltin chloride: Influence of pH and humic acids. Water Res 29:1631–1637

Fernandez-Perez M, Gonzalez-Pradas E, Villafranca-Sanchez M, Flores-Cespedes F, Urena-Amate MD (2000) Bentonite and humic acid as modifying agents in controlled release formulations of diuron and atrazine. J Environ Qual 29:304–310

Ferrara G, Loffredo E, Senesi N (2001) Aquatic humic substances inhibit clastogenic events in germinating seeds of herbaceous plants. J Agr Food Chem 49:1652–1657

Fettig J (1999) Characterization of NOM by adsorption parameters and effective diffusivities. Environ Intern 25:335–346

Fiebig DM (1995) Groundwater discharge and its contribution of dissolved organic carbon to an upland stream. Arch Hydrobiol 134:129–155

Field JA, Verhagen FJM, de Jong E (1995) Natural organohalogen production by basidomycetes. Trends Biotechnol 13:451–456

Filip Z, Alberts JJ (1995) Microbial utilization resulting in early diagenesis of salt-marsh humic acids. Sci Total Environ 144:121–135

Fischer H (2002) The role of biofilms in the uptake and transformation of dissolved organic matter. In: Findlay S, Sinsabaugh RL (eds) Dissolved organic matter in aquatic ecosystems. Academic Press, San Diego, pp 285–313

Fischer H, Sachse A, Steinberg CEW, Pusch M (2002): Differential retention and utilization of dissolved organic matter (DOC) by a microbial community in river sediments. Limnol Oceanogr Limnol Oceanogr 47:1702–1711

Fittkau EJ (1973) Artenmannigfaltigkeit amazonischer Lebensräume aus ökologischer Sicht. Amazoniana 4:321–340

Fittkau EJ (1981) Armut in der Vielfalt – Amazonien als Lebensraum für Weichtiere. Mitt Zool Ges Braunau 3:329–343

Fittkau EJ (1997) Structure, function and diversity of central Amazonian ecosystems. Nat Resourc Develop 45/46:28–41

Flaig W (1955) Zur Bildungsmöglichkeit von Huminsäuren aus Lignin. Holzforschung 9:1–4

Flaig W (1997) Aspects of the biochemistry of the healing effects of humic substances from peat. In: Hayes MHB, Wilson WR (eds) Humic substances, peats and sludges: health and environmental aspects. Royal Society of Chemistry, Cambridge, pp 346–356

Flaig W, Beutelspacher H, Rietz E (1975) Chemical composition and physical properties of humic substances. In: Gieseking JE (ed) Soil components. Vol. 1. Organic components. Wiley, New York, pp 1–48

Fooken U, Liebezeit G (2000) Distinction of marine and terrestrial origin of humic acids in North Sea surface sediments by absorption spectroscopy. Mar Geol 164:173–181

Forbes SA (1887) The lake as a microcosm. Bull Peoria (Ill) Sci Assoc, reprinted 1925 in Bull Ill Nat Hist Surv 15:537–550

Ford TE, Naiman RJ (1989) Groundwater-surface water relationships in boreal forest watersheds: Dissolved organic carbon and inorganic nutrient dynamics. Can J Fish Aquat Sci 46:41–49

Forsberg C (1992) Will an increased greenhouse impact in Fennoscandia give rise to more humic and coloured lakes? Hydrobiologia 229:51–58

Forsberg C, Petersen RC Jr (1990) A darkening of Swedish lakes due to increased humus inputs during the last 15 years. Verh Internat Verein Limnol 24:289–292

Fotijev A (1964) About the nature of humic substances in bog waters. Pochvovedenije 12:95–96 (in Russian)

Fotijev A (1970) The nature of aqueous humus. Dokl Akad Nauk SSSR 199:193–195 (in Russian)

Francko DA (1986) Epilimnetic phosphorus cycling: influence of humic materials and iron on coexisting major mechanisms. Can J Fish Aquat Sci 43:302–310

Francko DA (1990) Alteration of bioavailability and toxicity by phototransformation of organic acids. In: Perdue EM, Gjessing ET (eds) Organic acids in aquatic ecosystems. Wiley, Chichester, pp 167–177

Francko DA, Heath RT (1982) UV-sensitive complex phosphorus: association with dissolved humic material and iron in a bog lake. Limnol Oceanogr 27:564–569

Francko DA, Heath RT (1983) Abiotic uptake and photodependent release of phosphate from high-molecular-weight complexes in a bog lake. In: Christmann RF, Gjessing ET (eds) Aquatic and terrestrial humic materials. Ann Arbor, Michigan, pp 467–480

Freda J (1986) The influence of acidic pond water on amphibians: a review. Water Air Soil Poll 30:439–450
Freda J, Dunson WA (1986) Effects of low pH and other chemical variables on the local distribution of amphibians. Copeia 1986:454–466
Freeman C, Evans CE, Monteith DT, Reynolds B, Fenner N (2001b) Export of organic carbon from peat soils. Nature 412:784
Freeman C, Lock MA, Marxsen J, Jones SE (1990) Inhibitory effects of higher molecular weight dissolved organic matter upon metabolic processes in biofilms from contrasting rivers and streams. Freshwater Biol 24:159–166
Freeman C, Ostle N, Kang H (2001a) An enzymic 'latch' on a global carbon store. Nature 409:149
Frenkel AI, Korshin GV (1999) A study of non-uniformity of metal-binding sites in humic substances by X-ray absorption spectroscopy. In: Ghabbour EA, Davies G (eds) Understanding humic substances: advanced methods, properties and uses. Royal Society of Chemistry, Cambridge, pp 191–201
Frenkel AI, Korshin GV, Ankudinov AL (2000) XANES study of Cu^{2+}-binding sites in aquatic humic substances. Environ Sci Technol 34:2138–2142
Friese K, Herzsprung P, Witter B (2002) Photochemical degradation of organic carbon in acidic mining lakes. Acta hydrochim hydrobiol 30:141–148
Frimmel FH (1990) Characterization of organic acids in freshwater: a current status and limitations. In: Perdue EM, Gjessing ET (eds) Organic acids in aquatic ecosystems. Wiley, Chichester, pp 5–23
Frimmel FH (1994) Photochemical aspects related to humic substances. Environ Intern 20:373–385
Frimmel FH (1998) Impact of light on the properties of aquatic natural organic matter. Environ Intern 24:559–571
Frimmel FH, Bauer H, Putzien J, Murasecco P, Braun AM (1987) Laser flash photolysis of dissolved aquatic humic material and the sensitized production of singlet oxygen. Environ Sci Technol 21:4541–4545
Frimmel FH, Sattler D, Quentin KE (1980) Photochemical degradation of mercury-thio-compounds in oxygenated and deaerated water. Vom Wasser 55:111–120
Fritsche W (1998) Umweltmikrobiologie. Grundlagen und Anwendungen. Fischer, Jena
Froese R and Pauly D (eds) (2002) FishBase. World Wide Web electronic publication
Fukushima M, Tatsumi K (2001) Degradation pathways of pentachlorophenol by photo-Fenton systems in the presence of iron(III), humic acid, and hydrogen peroxide. Environ Sci Technol 35:1771–1778
Fukushima M, Tatsumi K, Morimoto K (2000a) The fate of aniline after a photo-Fenton reaction in an aqueous system containing iron(III), humic acid, and hydrogen peroxide. Environ Sci Technol 34:2006–2013
Fukushima M, Tatsumi K, Morimoto K (2000b) Influence of iron(III) and humic acid on the photodegradation of pentachlorophenol. Environ Toxicol Chem 19:1711–1716
Gagnon C, Fisher NS (1997) Bioavailability of sediment-bound methyl and inorganic mercury to a marine bivalve. Environ Sci Technol 31:993–998
Gamygin E, Ponomarew SW, Kanidjew AN, Sytschew GA, Schmakow NF, Marsanowa AG, Jelissenkowa NM (1991) Überprüfung der fischereibiologischen Effektivität des Präparates RHS 1500 der Firma WEYL GmbH Mannheim. Ministerium für Fischwirtschaft der UdSSR, Wissenschaftliche Produktionsvereinigung für Fischzucht, Wissenschaftlich-technisches Zentrum 'Aquakorm', Rybnoe, 29 pp (with permission)

Gamygin E, Ponomarew SW, Kanidjew AN, Sytschew GA, Schmakow NF, Marsanowa AG, Masloboischikow WS, Jelissenkowa NM, Nowoshenina DW (1992a) Überprüfung der fischereibiologischen Effektivität des Präparates RHS 1500 der Firma WEYL GmbH Mannheim. Komitee für Fischwirtschaft beim Ministerium für Landwirtschaft der Russischen Föderation, Allrussisches Forschungsinstitut für Teichwirtschaft, Wissenschaftliche Produktionsvereinigung für Fischzucht, Wissenschaftlich-technisches Zentrum 'Aquakorm', Rybnoe. IV. Präparatauswirkung auf die Fischeier während der Erbrütung, auf die Embryonen und Larven, pp 36–41 (with permission)

Gamygin E, Ponomarew SW, Kanidjew AN, Sytschew GA, Schmakow NF, Marsanowa AG, Masloboischikow WS, Jelissenkowa NM, Nowoshenina DW (1992b) Überprüfung der fischereibiologischen Effektivität des Präparates RHS 1500 der Firma WEYL GmbH Mannheim. Komitee für Fischwirtschaft beim Ministerium für Landwirtschaft der Russischen Föderation, Allrussisches Forschungsinstitut für Teichwirtschaft, Wissenschaftliche Produktionsvereinigung für Fischzucht, Wissenschaftlich-technisches Zentrum 'Aquakorm', Rybnoe. V. Prophylaktische und therapeutische Wirkung des Präparates bei den infektiösen und parasitären Fischerkrankungen sowie den traumatischen Haut- und Kiemenschädigungen, pp 41–50 (with permission)

Gao H, Zepp RG (1998) Factors influencing photoreactions of dissolved organic matter in a coastal river of the south-eastern United States. Environ Sci Technol 32:2940–2946

Garcia-Pichel F, Castenholtz RW (1991) Characterization and biological implications of scytonemin, a cyanobacterial sheath pigment. J Phycol 27:395–409

Gardner JL, Al-Hamdani SH (1998) Interactive effects of aluminium and humic substances on *Salvinia*. J Aquat Plant Manage 35:30–34

Gau RJ, Yang HL, Suen JL, Lu FJ (2001) Induction of oxidative stress by humic acid through increasing intracellular iron: a possible mechanism leading to atherothrombotic vascular disorder in blackfoot disease. Biochem Biophys Res Commun 283:743–749

Gaus C, Brunskill GJ, Weber R, Papke O, Muller JF (2001) Historical PCDD inputs and their source implications from dated sediment cores in Queensland (Australia). Environ Sci Technol 35:4597–4603

Gauthier TD, Seitz WR, Grant CL (1987) Effects of structural and compositional variations of dissolved humic materials on pyrene K_{OC} values. Environ Sci Technol 21:243–248

Gauthier TD, Shane EC, Guerin WF, Seitz WR, Grant CL (1986) Fluorescence quenching method for determining equilibrium constants for polycyclic aromatic hydrocarbons binding to dissolved humic materials. Environ Sci Technol 20:1162–1166

Geckeler KE, Eberhardt W (1995) Biogene Organochlorverbindungen – Vorkommen, Funktion, Umweltrelevanz. Naturwissenschaften 82:2–11

Geller A (1985a) Light-induced conversion of refractory, high molecular weight lake water constituents. Schweiz Z Hydrol 47:21–26

Geller A (1985b) Degradation and formation of refractory DOM by bacteria during simultaneous growth on labile substrates and persistent lake water constituents. Schweiz Z Hydrol 47:27–44

Geller A (1986) Comparison of mechanisms enhancing biodegradability of refractory lake water constituents. Limnol Oceanogr 31:755–764

Gensemer RW, Dixon DG, Greenberg BM (1999) Using chlorophyll *a* fluorescence to detect the onset of anthracene photoinduced toxicity in *Lemna gibba*, and the mitigating effects of a commercial humic acid. Limnol Oceanogr 44:878–888

Gensemer RW, Ren L, Day KE, Solomon KR, Greenberg BM (1996) Fluorescence induction as biomarker of creosote phototoxicity to the aquatic macrophyte *Lemna gibba*. In: Environmental toxicology and risk assessment. Amer Soc Test Mater STP 1306:163–176

Georgi A (1998) Sorption von hydrophoben organischen Verbindungen an gelösten Huminstoffen. Ph.D. Thesis University Leipzig, Germany.

Geyer S (1994) Isotopengeochemische Untersuchungen an Fraktionen von gelöstem organischem Kohlenstoff (DOC) zur Bestimmung der Herkunft und Evolution des DOC im Hinblick auf die Datierung von Grundwasser. GSF-Report 4/94:184 pp

Geyer S, Fischer M, Wolf M, Hertkorn N, Schmitt P (1996) Agriculture and its impacts on the isotope geochemistry and structural composition of dissolved organic carbon. In: Isotopes in water resources management, Internal Atomic Energy Agency, Vienna, Vol. 1:363–365

Geyer S, Wolf M, Fischer M, Fritz P, Buckau G, Kim JI (1992) Isotopendatierung von gelöstem organischem Kohlenstoff aus Grundwasser. In: Frimmel FH, Abbt-Braun G (eds) Refraktäre organische Säuren in Gewässern. Deutsche Forschungsgemeinschaft, Mitteilung XII der Senatskommission für Wasserforschung, VCH, Weinheim, pp 25–46

Ghosh K, Schnitzer M (1980) Macromolecular structure of humic substances. Soil Sci 129:266–276

Gibson JAE, Vincent WF, Pienitz R (2001) Hydrological control and diurnal photobleaching of CDOM in a subarctic lake. Arch Hydrobiol 152:143–159

Gierhake E, Wehefritz E. (1936) Chemische und balneologische Untersuchungen über das Vorkommen östrogener Wirkstoffe in deutschen Bademooren. Dt med Wochenschr 62:423–425

Giesy JP Jr (1976) Stimulation of growth in *Scenedesmus obliquus* (Chlorophyceae) by humic acids under iron limited conditions. J Phycol 12:172–179

Gilbert E (1980) Ozonation of humic acids in aqueous solutions. Vom Wasser 55:1–13

Gill GA, Bloom NS, Cappelino S, Driscoll CT, Dobbs C, McShea L, Mason RP, Rudd JWM (1999) Sediment – water fluxes of mercury in Lavaca Bay, Texas. Environ Sci Technol 33:663–669

Gill GA, Bruland KW (1990) Mercury speciation in surface freshwater systems in California and other areas. Environ Sci Technol 24:1392–1400

Gill T, Tewar H, Pande J (1991) *In vivo* and *in vitro* effects of cadmium on selected enzymes in different organs of the fish *Barbus conchonius* Ham. (Rosy barb). Comp Biochem Physiol 100C:501

Gjessing ET (1965) Use of Sephadex gels for the estimation of molecular weight of humic substances in natural waters. Nature 208:1091–1092

Gjessing ET (1971) Effect of pH on the filtration of aquatic humus using gels and membranes. Schweiz Z Hydrol 33:592–600

Gjessing ET (1981) The effect of aquatic humus on the biological availability of cadmium. Arch Hydrobiol 91:144–149

Gjessing ET (1992) The HUMEX project: experimental acidification of a catchment and its humic lake. Environ Internat 18:535–543

Gjessing ET (1994a) HUMEX (Humic Lake Acidification Experiment) Chemistry, hydrology, and meteorology. Environ Internat 20:267–276

Gjessing ET (1994b) The role of humic substances in the acidification response of soil and water – results of the humic lake acidification experiment (HUMEX). Environ Internat 20:363–368

Gjessing ET, Alberts JJ, Bruchet A, Egeberg PK, Lydersen E, McGown LB, Mobed JJ, Münster U, Pempkowika J, Perdue M, Ratnawerra H, Rybacki D, Takacs M, Abbt-Braun G (1998) Multi-method characterisation of natural organic matter isolated from water: characterisation of reverse osmosis-isolates from water of two semi-identical dystrophic lakes basins in Norway. Water Res 32:3108–3124

Gjessing ET, Berglind L (1981) Adsorption of PAH to aquatic humus. Arch Hydrobiol 92:24–30

Gjessing ET, Egeberg PK, Håkedal J (1999) Natural organic matter in drinking water – the 'NOM-Typing Project', background and basic characteristics of original water samples and NOM isolates. Environ Internat 25:145–159

Gjessing ET, Gjerdahl TC (1970) Influence of ultra-violet radiation on aquatic humus. Vatten 26:144–145

Gjessing ET, Källqvist T (1991) Algicidal and chemical effect of UV-radiation of water containing humic substances. Water Res 25:491–494

Goldstone JV, Pullin MJ, Bertilsson S, Voelker BM (2002) Reactions of hydroxyl radical with humic substances: bleaching, mineralization, and production of bioavailable carbon substrate. Environ Sci Technol 36:364–372

Goncharova IA, Starodomskaja AG, Datsko VG (1961) Determination of molecular weight of organic matter from natural waters. Gidrochim Mater 35:156–159 (in Russian)

Gonet SS, Dziamski A, Gonet E (1996) Application of humus preparations from oxyhumolite in crop production. Environ Internat 22:559–562

Gonzalez RJ, Wilson RW, Wood CM, Patrick ML, Val AL (2002) Diverse strategies for ion regulation in fish collected from the ion-poor, acidic Rio Negro. Physiol Biochem Zool 75:37–47

Gonzalez-Pradas E, Fernandez-Perez M, Villafranca-Sanchez M, Flores-Cespedes F (1999) Use of bentonite and humic acids as modifying agents in alginate-based controlled-release formulations of imidacloprid. Pestic Sci 55:546–552

Gorham E, Underwood JK, Martin FB, Ogden JG III (1986) Natural and anthropogenic causes of lake acidification in Nova Scotia. Nature 324:451–453

Goulding M, Leal Carvalho M, Ferreira EG (1988) Rio Negro, rich life in poor water. Amazonian diversity and foodchain ecology as seen through fish communities. SPB Academic Publishing, The Hague

Grandeau LN (1872) Recherches sur le role des matieres organiques du sol dans les phénomenes de la nutrition des vegetaux. Compt rend hebd academie sci. Paris

Granéli W, Lindell M, Barcal de Faria B, De Assis Esteves F (1998) Photoproduction of dissolved inorganic carbon in temperate and tropical lakes – dependence on wavelength band and dissolved organic carbon concentration. Biogeochemistry 43:175–195

Granéli W, Lindell M, Tranvik L (1996) Photo-oxidative production of dissolved inorganic carbon in lake of different humic content. Limnol Oceanogr 41:698–706

Greenberg BM, Huang XD, Dixon DG, Ren L, McConkey BJ, Duxbury CL (1993) Quantitative structure activity relationships for the photoinduced toxicity of polycyclic aromatic hydrocarbons to duckweed – a preliminary model. Environmental toxicology and risk assessment. Amer Soc Test Mater STP 1216:369–378

Gribble GW (1992) naturally occurring organohalogen compounds – a survey. J Natural Prod 55:1353–1395

Gribble GW (1994) Natural halogens, many more than you think! J Chem Educ 71:907–911

Grieb TM, Driscoll CT, Gloss SP, Schofield CL, Bowie GL, Porcella DB (1990) Factors affecting mercury accumulation in fish in the upper Michigan peninsula. Environ Toxicol Chem 9:919–930

Grøn C, Raben-Lange B (1992) Isolation and characterization of a haloorganic soil humic acid. Sci Total Environ 113:281–286

Gross EM (1999) Allelopathy in benthic and littoral areas: case studies on allochemicals from benthic cyanobacteria and submersed macrophytes. In: Inderjit S, Dakshini KMM, Foy CL (eds) Principles and practices in plant ecology: allelochemical interactions. CRC Press, Boca Raton, pp 179–199

Gross EM (2000) Seasonal and spatial dynamics of allelochemicals in the submersed macrophyte *Myriophyllum spicatum* L. Verh Internat Verein Limnol 27:2116–2119

Gross EM, Meyer H, Schilling G (1995) Release and ecological impact of algicidal hydrolysable polyphenols in *Myriophyllum spicatum*. Phytochemistry 41:133–138

Gross EM, Sütfeld R (1994) Polyphenols with algicidal activity in the submerged macrophyte *Myriophyllum spicatum* L. Acta hortic 381:710–716

Grossart HP, Plough H (2001) Microbial degradation of organic carbon and nitrogen on diatom aggregates. Limnol Oceanogr 46:267–277

Grossart HP, Simon M (1998a) The significance of limnetic organic aggregates (lake snow) for the sinking flux of particulate organic matter in a large lake. Aquat Microb Ecol 15:115–125

Grossart HP, Simon M (1998b) Bacterial colonization and microbial decomposition of limnetic organic aggregates (lake snow). Aquat Microb Ecol 15:127–140

Grunewald K, Korth A, Scheithauer J, Schmidt W (2002) Increase of DOM in Germany drinking water reservoirs. Wasser Boden 54 (in press)

Gschwend PM, Wu SC (1985) On the constancy of sediment-water partition coefficients of hydrophobic organic pollutants. Environ Sci Technol 19:90–96

Güde H (1991) Participation of bacterioplankton in epilimnetic phosphorus cycles of Lake Constance. Verh Internat Verein Limnol 24:816–820

Guetzloff TF, Rice JA (1994) Does humic acid form a micelle? Sci Total Environ 152:31–35

Guggenberger G, Zech W (1993a) Dissolved organic carbon control in acid forest soils of the Fichtelgebirge (Germany) as revealed by distribution patterns and structural composition analyses. Geoderma 59:109–129

Guggenberger G, Zech W (1993b) Zur Dynamik gelöster organischer Substanzen (DOM) in Fichtenökosystemen – Ergebnisse analytischer DOM-Fraktionierung. Z Pflanzenernähr Bodenk 156:341–347

Guggenberger G, Zech W (1994) Composition and dynamics of dissolved carbohydrates and lignin-degradation products in two coniferous forest, N. E. Bavaria, Germany. Soil Biol Biochem 26:19–27.

Guggenberger G, Zech W, Schulten HR (1994) Formation and mobilization pathways of dissolved organic matter: evidence from chemical structural studies of organic matter fractions in acid forest floor solutions. Org Geochem 21:51–66

Guildford SJ, Healey FP, Hecky RE (1987) Depression of primary production by humic matter and suspended sediment in limnocorral experiments at southern Indian Lake, northern Manitoba. Can J Fish Aquat Sci 44:1408–1417

Gümbel CW (1868) Geognostische Beschreibung des ostbayerischen Grenzgebirges. Verlag J Perthes, Gotha:968 pp

Gundersen DT, Bustaman S, Seim WK, Curtis LR (1994) pH, hardness and humic acid influence aluminium toxicity to rainbow trout (*Oncorhynchus mykiss*) in weakly alkaline waters. Can J Fish Aquat Sci 51:1345–1355

Haag I, Moritz K, Bittersohl J, Lischeid G (2001) Factors controlling total concentration and aqueous speciation of aluminium in an acidic headwater stream of the Bavarian Forest National Park: a modelling approach. Acta hydrochim hydrobiol 29:206–218

Haag WR, Hoigné J (1986) Singlet oxygen in surface waters. 3. Photochemical formation and steady-state concentrations in various types of waters. Environ Sci Technol 20:341–348

Haberhauer G, Pfeiffer L, Gerzabek MH, Kirchmann H, Aquino AJA, Tunega D, Lischka H (2001) Response of sorption processes of MCPA to the amount and origin of organic matter in a long-term field experiment. Eur J Soil Sci 52:279–286

Haemmerli SD, Leisola MSA, Sanglard D, Fiechter A (1986) Oxidation of benzo[*a*]pyrene by extracellular ligninases of *Phanerochaete chrysosporium*. J Biol Chem 261:6900–6903

Haiber S, Herzog H, Burba P, Gosciniak B, Labert J (2001a) Quantification of carbohydrate structures in size fractionated aquatic humic substances by two-dimensional nuclear magnetic resonance. Fresenius J Anal Chem 369:457–460

Haiber S, Herzog H, Burba P, Gosciniak B, Labert J (2001b) Two-dimensional NMR studies of size fractionated Suwannee River fulvic and humic acid reference. Environ Sci Technol 35:4289–4294

Haider KM, Martin JP (1988) Mineralization of ^{14}C labelled humic acids and of humic-acid bound ^{14}C-xenobiotics by *Phanerochaete chrysosporium*. Soil Biol Biochem 20:425–429

Haider KM, Spiteller M, Dec J, Schäffer A (2000) Silylation of soil organic matter. Extraction of humic compounds and soil-bound residues. In: Bollag JM, Stotzky G (eds) Soil biochemistry. Marcel Dekker, New York, pp 139–170

Haines TA, Komov VT, Jagoe CH (1994) Mercury concentration in perch (*Perca fluviatilis*) as influenced by lacustrine physical and chemical factors in two regions of Russia. In: Watras CJ, Huckabee JW (eds) Mercury as a global pollutant: integration and synthesis. Lewis, Chelsea, pp 397–407

Haines TA, Komov VT, Matey VE, Jagoe CH (1995) Perch mercury content is related to acidity and color of 26 Russian lakes. Water Air Soil Pollut 85:823–825

Haitzer M, Abbt-Braun G, Traunspurger W, Steinberg CEW (1999c) Effects of humic substances on the bioconcentration of polycyclic aromatic hydrocarbons: correlations with spectroscopic and chemical properties of humic substances. Environ Toxicol Chem 18:2782–2788

Haitzer M, Akkanen J, Steinberg CEW, Kukkonen JV (2001) No enhancement in bioconcentration of organic contaminants by low levels of DOM. Chemosphere 44:165–171

Haitzer M, Burnison BK, Höss S, Traunspurger W, Steinberg CEW (1999a) Effects of quantity, quality and contact time of dissolved organic matter on bioconcentration of benzo[*a*]pyrene in the nematode, *Caenorhabditis elegans*. Environ Toxicol Chem 18:459–465

Haitzer M, Höss S, Traunspurger W, Kukkonen J, Burnison BK, Steinberg CEW (2002) Biological Effects. In: Frimmel FH, Abbt-Braun G, Heumann KG, Hock B, Lüdemann HD, Spiteller M (eds) Refractory organic substances in the environment. Wiley-VCH, Weinheim, pp 361–381

Haitzer M, Höss S, Traunspurger W, Steinberg CEW (1998) Effects of dissolved organic matter (DOM) on the bioconcentration of organic chemicals in aquatic organisms – a review. Chemosphere 37:1335–1362

Haitzer M, Höss S, Traunspurger W, Steinberg CEW (1999b) Relationship between concentration of dissolved organic matter (DOM) and the effect of DOM on the bioconcentration of benzo[*a*]pyrene. Aquat Toxicol 45:147–158

Haitzer M, Löhmannsröben HG, Steinberg CEW, Zimmermann U (2000) *In vivo* laser-induced fluorescence detection of pyrene in nematodes and determination of pyrene binding constants for humic substances by fluorescence quenching and bioconcentration experiments. J Environ Monitor 2:145–149

Håkanson L (1996) A simple model to predict the duration of the mercury problem in Sweden. Ecol Model 93:251–262

Hall RJ, Driscoll CT, Likens GE (1987) Importance of hydrogen ions and aluminium in regulating the structure and function of stream ecosystems: an experimental test. Freshwater Biol 18:17–43

Hall RJ, Driscoll CT, Likens GE, Pratt JM (1985) Physical, chemical, and biological consequences of episodic aluminum additions to a stream. Limnol Oceanogr 30:212–220

Hammerschmitt R, Schultz JC (1996) Multiple defenses and signals in plant defense against pathogens and herbivores. Rec Adv Phytochem 30:121–154

Haney JF (1973) An *in situ* examination of the grazing activities of natural zooplankton communities. Arch Hydrobiol 72:87–132

Hapeman CJ, Bilboulian S, Anderson BG, Torrents A (1998) Structural influence of low-molecular-weight dissolved organic carbon mimics on the photolytic fate of atrazine. Environ Toxicol Chem 17:975–981

Harborne JB (1993) Introduction to ecological biochemistry, 4th edn, Academic Press, London

Hargeby A (1990) Effects of pH, humic substances and animal interactions on survival and physiological status of *Asellus aquaticus* L. and *Gammarus pulex* (L.) – A field experiment. Oecologia 82:348–354

Hargeby A, Petersen RC Jr (1988) Effects of low pH and humus on the survivorship, growth and feeding of *Gammarus pulex* (L.) (Amphipoda). Freshwater Biol 19:235–247

Hartmann H, Steinberg CEW (1986) Mallomonadacean (Chrysophyceae) scales: early biotic paleoindicators of lake acidification. Hydrobiologia 143:87–91

Hartmann H, Steinberg CEW (1989) The occurrence of scaled chrysophytes in some central European lakes and their relation to pH. Nova Hedwigia Beih 95:131–158

Hartung J (1994) Untersuchungsbericht: HS 1500 zur prophylaktischen und therapeutischen Erprobung bei Forellen. Sächsische Tierseuchenkasse Leipzig.

Harvey GR (1983) Dissolved carbohydrates in the New York bight and the variability of marine organic matter. Mar Chem 12:333–340

Harvey GR, Boran DA (1985) The geochemistry of humic substances in seawater. In: Aiken GR, McKnight DM, Wershaw RL, MacCarthy P (eds) Humic substances in soil, sediment, and water. Wiley, New York, pp 233–247

Harvey GR, Boran DA, Chesal LA, Tokar JM (1983) The structure of marine fulvic acid and humic acids. Mar Chem 12:119–132

Harvey GR, Boran DA, Piotrowicz SR, Weisel, CP (1984) Synthesis of marine humic substances from unsaturated lipids. Nature 309:244–246

Haslam E (1989) Plant polyphenols. Cambridge University Press, New York

Hassett JP, Anderson MA (1979) Association of hydrophobic organic compounds with dissolved organic matter in aquatic systems. Environ Sci Technol :1526–1529

Hatcher PG, Spiker EC (1988) Selective degradation of plant biomolecules. In: Frimmel FH, Christman RF (eds) Humic substances and their role in the environment. Wiley,

Chichester, pp 59–74

Havens KE (1991) Summer zooplankton dynamics in the limnetic and littoral zones of a humic acid lake. Hydrobiologia 215:21–29

Havens KE (1993) Pelagic food web structure in acidic Adirondack Mountain, New York, lakes of varying humic content. Can J Fish Aquat Sci 50:2688–2691

Hedges JI, Oades JM (1997) Comparative organic geochemistry of soils and marine sediments. Org Geochem 27:319–361

Hedin LO, Likens GE, Postek KM, Driscoll CT (1990) A field experiment to test whether organic acids buffer acid deposition. Nature 345:798–800

Hehmann A, Krienitz L (1996) The succession and vertical distribution of phytoplankton of the experimentally divided naturally acidic lake 'Große Fuchskuhle' (Brandenburg, Germany). Limnologica 26:301–309

Helbig B, Klöcking R (1983) Darstellung und Charakterisierung von Huminsäure-Modellsubstanzen. Z Physiother 33:31–37

Hemond HF (1994) Role of organic acids in acidification of fresh waters. In: Steinberg CEW, Wright RF (eds) Acidification of freshwater ecosystems: implications for the future. Wiley, Chichester, pp 103–115

Hendel B, Marxsen J (1997) Measurement of low-level extracellular enzyme activity in natural waters using fluorigenic model substrates. Acta hydrochim hydrobiol 25:253–258

Hendel B, Marxsen J (2000) Extracellular enzyme activity associated with degradation of beech wood in a central European stream. Internat Rev Hydrobiol 85:95–105

Henriksen A, Skogheim OK, Rosseland BO (1984) Episodic changes in pH and aluminium-speciation kill fish in a Norwegian salmon river. Vatten 40:255–260

Hering JG, Morel FMM (1988) Humic acid complexation of calcium and copper. Environ Sci Technol 20:349–354

Hernandez ME, Newman DK (2001) Extracellular electron transfer. Cell Mol Life Sci 58:1562–1571

Herndl GJ, Müller-Niklas G, Frick J (1993) Major role of UV-B in controlling bacterioplankton growth in the surface layer of the ocean. Nature 361:717–719

Herzsprung P, Friese K, Packroff G, Schimmele M, Wendt-Potthoff K, Winkler M (1998) Vertical and annual distribution of ferric and ferrous iron in acidic mining lakes. Acta hydrochim hydrobiol 26:253–262

Hesse S, Balz A, Frimmel FH (1997) Detaillierte Verfolgung des anthropogenen Kohlenstoffeintrags entlang des Schwarzwaldfließgewässers Forbach/Murg. Vom Wasser 88:103–117

Hessen DO (1992) Dissolved organic carbon in a small humic lake: effects on bacterial production an respiration. Hydrobiologia 229:115–123

Hessen DO (1998) Food webs and carbon cycling in humic lakes. In: Hessen DO, Tranvik LJ (eds) Aquatic humic substances – ecology and biogeochemistry. Springer, Berlin, pp 285–315

Hessen DO, Andersen T, Lyche A (1990) Carbon metabolism in a humic lake: pool sizes and cycling through zooplankton. Limnol Oceanogr 35:84–99

Hessen DO, Andersson T (1990) Bacteria as a source of phosphorus for zooplankton. Hydrobiologia 206:217–223

Hessen DO, Faerovig PJ (2001) The photoprotective role of humus DOC for *Selenastrum* and *Daphnia*. Plant Ecol 154:261–271

Hessen DO, Nygaard K, Salonen K, Vähätalo A (1994) The effect of substrate stoichiometry on microbial activity and carbon degradation in humic lakes. Environ Intern 20:67–

76
Hessen DO, Tranvik LJ (1998) Aquatic humic matter: from molecular structure to ecosystem stability. In: Hessen DO, Tranvik LJ (eds) Aquatic humic substances – ecology and biogeochemistry. Springer, Berlin, pp 333–342
Hessen DO, van Donk E (1994) Effects of UV-radiation of humic water on primary and secondary production. Water Air Soil Poll 75:325–338
Heumann KG, Rädlinger G, Erbes M, Heiber I, Obst U, Filip Z, Claus H (2000) Aging of dissolved halogenated humic substances and the microbiological influence on this process. Acta hydrochim hydrobiol 28:193–201
Hinkel M, Reischl A, Schramm KW, Trautner F, Reissinger M, Hutzinger O (1989) concentration levels of nitrated phenols in conifer needles. Chemosphere 18:2433–2439
Hintelmann H, Welbourn PM, Evans RD (1995) Binding of methylmercury compounds by humic and fulvic acids. Water Air Soil Pollut 80:1031–1034
Hintelmann H, Welbourn PM, Evans RD (1997) Measurement of complexation of methylmercury(II) compounds by freshwater humic substances using equilibrium dialysis. Environ Sci Technol 31:489–495
Hjelm O, Borén H, Öberg G (1996) Detection of halogenated organic compounds in soil from a *Lepista nuda* (wood blewitt) fairy ring. Chemosphere 32:1719–1728
Hjelm O, Johansson E, Öberg G (1999) Production of organically bound halogens by the litter-degrading fungus *Lepista nuda*. Soil Biol Biochem 31:1509–1515
Hocking PJ (1989) Seasonal dynamics of production, and nutrient accumulation and cycling by *Phragmites australis* (Cav.) Tin ex Steud. in a nutrient-enriched swamp in inland Australia. I. Whole plants. Aust J Mar Freshwat Res 40:421–444
Hodge VA, Fan GT, Solomon KR, Kaushik NK, Leppard GG, Burnison BK (1993) Effects of the presence and absence of various fractions of dissolved organic matter on the toxicity of fenvalerate to *Daphnia magna*. Environ Toxicol Chem 12:167–176
Hoekstra EJ, de Leer EWB, Brinkman UAT (1999) Findings supporting the natural formation of trichloroacetic acid in soil. Chemosphere 38:2875–2883
Hoekstra EJ, Lassen P, van Leeuwen JG, de Leer EWB, Carlsen L (1995) Formation of organic chlorine compounds of low molecular weight in the chloroperoxidase-mediated reaction between chloride and humic material. In: Grimvall A, de Leer EWB (eds) Naturally produced organohalogens. Kluwer, Dordrecht, pp 149–158
Hoekstra EJ, Verhagen FJM, Field JA, de Leer EWB, Brinkman UAT (1998) Natural production of chloroform by fungi. Chemosphere 49:91–97
Hoffmann C, Marschner B, Renger M (1998) Influence of DOM-quality, DOM-quantity and water regime on the transport of selected heavy metals. Phys Chem Earth 23:205–209
Hofrichter M, Scheibner K, Schneegass I, Fritsche W (1998) Enzymatic combustion of aromatic and aliphatic compounds by manganese peroxidase from *Nematoloma frowardii*. Appl Environ Microbiol 64:399–404
Hoitink HAJ, Grebus ME (1997) Composts and the control of plant diseases. In: Hayes MHB, Wilson WR (eds) Humic substances, peats and sludges: health and environmental aspects. Royal Society of Chemistry, Cambridge, pp 359–366
Hollis L, Burnison K, Playle RC (1996) Does the age of metal-dissolved organic carbon complexes influence binding of metals to fish gills? Aquat Toxicol 35:253–264
Hollis L, Muench L, Playle RC (1997) Influence of dissolved organic matter on copper binding, and calcium on cadmium binding, by gills of rainbow trout. J Fish Biol 50:703–720

Holst LL, Giesy JP (1989) Chronic effects of the photoinduced toxicity of anthracene on *Daphnia magna* reproduction. Environ Toxicol Chem 8:933–942

Holten Lützhøft HC, Vaes WHJ, Freidig AP, Halling-Sørensen, M, Hermens JLM (2000) Influence of pH and other modifying factors on the distribution behavior of 4-quinolones to solid phases and humic acids studied by 'negligible-depletion' SPME-HPLC. Environ Sci Technol 34:4989–4994

Hongve D (1999) Production of dissolved organic carbon in forested catchments. J Hydrol 224:91–99

Hongve D (2000) Will runoff from coniferous throughfall and lithomorphic soils contribute to DOC and colour in surface waters in southeast Norway? Proc 10th Meet Intern Humic Subst Soc, 24–28 July 2000, Toulouse, pp 203–204

Hood TE, Calabrese EJ, Zuckerman BM (2000) Detection of an estrogen receptor in two nematode species and inhibition of binding and development by environmental chemicals. Ecotox Environ Safe 47:74–81

Hoppe HG (1991) Microbial extracellular enzyme activity: a new key parameter in aquatic ecology. In: Chróst RJ (ed) Microbial enzymes in aquatic environments. Brock/Springer Ser Contemp Biosci, Springer, Berlin, pp 60–83

Horth H, Frimmel FH, Hargitai L, Hennes EC, Huc AY, Müller-Wegner U, Niemeyer J, Nissenbaum A, Sekoulov I, Tipping E, Weber JH, Zepp RG (1988) Environmental reactions and functions. In: Frimmel FH, Christman RF (eds) Humic substances and their role in the environment. Wiley, Chichester, pp 245–256

Höss S, Bergtold M, Haitzer M, Traunspurger W, Steinberg CEW (2001a) Refractory dissolved organic matter can influence the reproduction of *Caenorhabditis elegans* (Nematoda). Freshwater Biol 46:1–10

Höss S, Haitzer M, Traunspurger W, Gratzer H, Ahlf W, Steinberg CEW (1997) Influence of particle size distribution and content of organic matter on the toxicity of copper in bioassays using *Caenorhabditis elegans* (Nematoda). Water Air Soil Pollut 99:689–695

Höss S, Henschel T, Haitzer M, Traunspurger W, Steinberg CEW (2001b) Toxicity of cadmium on *Caenorhabditis elegans* (Nematoda) in whole sediment and porewater – the ambiguous role of organic matter. Environ Toxicol Chem 20:2794–2801

Höss S, Jüttner I, Traunspurger W, Pfister G, Schramm KW, Steinberg CEW (2002) Enhanced growth and reproduction of *Caenorhabditis elegans* (Nematoda) in the presence of 4-nonylphenol. Environ Pollution 120:169–172

Hseu YC, Chang WC, Yang HL (2001) Inhibition of human plasmin activity using humic acids with arsenic. Sci Total Envir 273:93–99

Hubbard PC, Barata EN, Canario AVM (2002) Possible disruption of pheromonal communication by humic acids in the goldfish, *Carassius auratus*. Aquat Toxicol 60:169–183

Huber SA, Frimmel FH (1994) Identification of diffuse and point sources of DOC in a small stream (Alb, Southwest Germany), using gel filtration chromatography with high sensitivity DOC-detection. Fresenius J Anal Chem 350:496–503

Huber SA, Frimmel FH (1996) Size-exclusion-chromatography with organic carbon detection (LC-OCD) A fast and reliable method for the characterisation of hydrophilic organic matter in natural waters. Vom Wasser 86:277–290

Hudson RJM, Gherini SA, Watras CJ, Porcella DB (1994) Modeling the biogeochemical cycle of mercury in lakes: the mercury cycling model (MCM) and its application to the MTL study lakes. In: Watras CJ Huckabee JW (eds) Mercury as a global pollutant: integration and synthesis. Lewis, Chelsea, pp 473–523

Hung TC, Kuo X, Jeng WL (1994) Characterization of dissolved organic carbon and humic substances in the well water of the blackfoot disease area in Taiwan. Toxicol Environ Chem 46:127–134

Hurley JP, Benoit JM, Babiarz CL, Shafer MM, Andren AW, Sullivan JR, Hammond R, Webb DA (1995) Influences of watershed characteristics on mercury levels in Wisconsin rivers. Environ Sci Technol 29:1867–1875

Hurley JP, Cowell SE, Shafer MM, Hughes PE (1998) Partitioning and transport of total and methyl mercury in the Lower Fox River, Wisconsin. Environ Sci Technol 32:1424–1432

Hutchinson GE (1957) A treatise on limnology. I: geography, physics, and chemistry. Wiley, New York, pp 557–652

Hutchinson GE (1967) A treatise on limnology. II: introduction to lake biology and the limnoplankton. Wiley, New York

Hwang PP, Lin SW, Lin HC (1995) Different sensitivities to cadmium in tilapia larvae (*Oreochromis mossambicus*, Teleostei). Arch Environ Contam Toxicol 29:1–7

Imai A, Fukushima T, Matsushige K, Kim YH (2001) Fractionation and characterisation of dissolved organic matter in a shallow eutrophic lake, its inflowing rivers and other organic matter sources. Water Res 35:4019–4028

Ishiwatari R (1985) Geochemistry of humic substances in lake sediments. In: Aiken GR, McKnight DM, Wershaw RL, MacCarthy P (eds) Humic substances in soil, sediment, and water. Wiley, New York, pp 147–180

Isola G, Mandrich A (1991) RHS 1500 in the prophylaxis of skin and branchial mycosis caused by capture stress in *Leuciscus souffia* Risso. CEntro Ricerche Sperimentazioni Acquacoltura Ligure, Genova 20.11.

Isosaari P, Pajunen H, Vartiainen T (2002) PCDD/F and PCB history in dated sediments of a rural lake. Chemosphere 47:575–583

Ivanovich M, Wolf M, Geyer S, Fritz P (1996) Isotopic characterization of humic colloids and other organic and inorganic dissolved species in selected groundwaters from sand aquifers at Gorleben, Germany. In: Gaffney JS, Marley NA, Clark SB (eds) Humic and fulvic acids. Amer Chem Soc Symp Ser No. 651, Washington. pp 220–243

Ivarsson H, Jansson M (1994) Temporal variations in the concentration and character of dissolved organic matter in a highly colored stream in the coastal zone of northern Sweden. Arch Hydrobiol 132:45–55

Jackson CR, Foreman CM, Sinsabaugh RL (1995) Microbial enzyme activities as indicators of organic matter processing rates in a Lake Erie coastal wetland. Freshwater Biol 34:329–342

Jackson TA, Hecky RE (1980) Depression of primary productivity by humic matter in lake and reservoir waters of the boreal forest zone. Can J Fish Aquat Sci 37:2300–2317

Jahnel JB, Frimmel FH (1994) Comparison of the enzyme inhibition effect of different humic substances in aqueous solution. Chem Engin Process 33:325–330

Jahnel JB, Frimmel FH (1995) Enzymatic release of amino acids from different humic substances. Acta hydrochim hydrobiol 23:31–35

Jahnel JB, Frimmel FH (1996) Detection of glucosamine in the acid hydrolysis solution of humic substances. Fresenius J Anal Chem 354:886–888

Jahnel JB, Ilieva P, Abbt-Braun G, Frimmel FH (1998) Aminosäuren und Kohlenhydrate als Strukturbestandteile von refraktären organischen Säuren. Vom Wasser 90:205–216

Jahnel JB, Mahlich B, Frimmel FH (1994) Beeinflussung der Enzymaktivität einer Protease durch Huminstoffe. Acta hydrochim hydrobiol 22:109–116

Jahnel JB, Schmiedel U, Abbt-Braun G, Frimmel FH (1993) Anwendung einer enzymatischen Methode zur Charakterisierung von Huminstoffen. Acta hydrochim hydrobiol 21:43–50

Jandl G, Schulten HR, Leinweber P (2002) Quantification of long-chain fatty acids in dissolved organic matter and soils. J Plant Nutr Soil Sci 165:133–139

Janes N, Playle RC (1995) Modeling silver binding to gills of rainbow trout (*Oncorhynchus mykiss*). Environ Toxicol Chem 14:1847–1858

Jansson M (1993) Uptake, exchange, and excretion of orthophosphate in phosphate-starved *Scenedesmus quadricauda* and *Pseudomonas* K7. Limnol Oceanogr 38:1162–1178

Jansson M (1998) Nutrient limitation and bacteria – phytoplankton interaction in humic lakes. In: Hessen DO, Tranvik LJ (eds) Aquatic humic substances – ecology and biogeochemistry. Springer, Berlin, pp 177–195

Jansson M, Bergström AK, Blomqvist P, Isaksson A, Jonsson A (1999) Impact of allochthonous organic carbon on microbial food web carbon dynamics and structure in Lake Örträsket. Arch Hydrobiol 144:409–428

Jansson M, Blomqvist P, Jonsson A, Bergström AK (1996) Nutrient limitation of bacterioplankton, autotrophic and mixotrophic phytoplankton, and heterotrophic nanoflagellates in Lake Örträsket, a large humic lake in northern Sweden. Limnol Oceanogr 41:1552–1559

Janzen DH (1974) Tropical blackwater rivers, animals, and mast fruiting by the Dipterocarpaceae. Biotropica 6:69–103

Järnefelt H (1956) Zooplankton und Humuswasser. Ann Acad Sci Fenn A IV Biol 31:1–14

Järvinen M, Salonen K (1998) Influence of changing food web structure on nutrient limitation of phytoplankton in a highly humic lake. Can J Fish Aquat Sci 55:2562–2571

Johnsen S (1987) Interactions between polycyclic aromatic hydrocarbons and natural aquatic humic substances: contact time relationship. Sci Total Environ 67:269–278

Johnsen S, Martinsen K, Carlberg GE, Gjessing ET, Becher G, Legreid M (1987) Seasonal variations in composition and properties of aquatic humic substances. Sci Total Environ 62:13–25

Johnson NM, Likens GE, Feller MC, Driscol CT (1984) Acid rain and soil chemistry. Science 225:1424–1425

Jones GJ, Bourne DG, Blakeley RL, Doelle H (1994) Degradation of cyanobacterial hepatotoxin by aquatic bacteria. Natural Toxins 2:228–235

Jones RI (1992) The influence of humic substances on lacustrine planktonic food chains. Hydrobiologia 229:73–91

Jones RI (1994) Mixotrophy in planktonic protists as a spectrum of nutritional strategies. Mar Microb Food Webs 8:87–96

Jones RI (1998) Phytoplankton, primary production and nutrient cycling. In: Hessen DO, Tranvik LJ (eds) Aquatic humic substances – ecology and biogeochemistry. Springer, Berlin, pp 145–175

Jones RI, Grey K, Sleep D, Quarmly C (1998) An assessment, using stable isotopes, of the importance of allochthonous organic carbon sources to the pelagic food web in Loch Ness. Proc R Soc Lond B 265:105–111

Jones RI, Salonen K, de Haan H (1988) Phosphorus transformations in the epilimnion of humic lakes: abiotic interactions between dissolved humic materials and phosphate. Freshwater Biol 19:357– 367

Jones RI, Shaw PJ, de Haan H (1993) Effects of dissolved humic substances on the speciation of iron and phosphate at different pH and ionic strength. Environ Sci Technol

27:1052–1059

Jonsson A, Jansson M (1997) Sedimentation and mineralisation of organic carbon, nitrogen and phosphorus in a large humic lake, northern Sweden. Arch Hydrobiol 141:45–65

Jonsson A, Meili M, Bergstrom AK, Jansson M (2001) Whole-lake mineralization of allochthonous and autochthonous organic carbon in a large humic lake (Örträsket, N. Sweden). Limnol Oceanogr 46:1691–1700

Jørgensen CB (1976) August Pütter, August Krogh, and modern ideas on the use of dissolved organic matter in aquatic environments. Biol Rev 51:291–328

Jüttner I, Henkelmann B, Schramm KW, Steinberg CEW, Winkler R, Kettrup A (1997a) Occurrence of PCDD/F in dated lake sediments of the Black Forest, southwestern Germany. Environ Sci Technol 31:806–812

Jüttner I, Lintelmann J, Michalke B, Winkler R, Steinberg CEW, Kettrup A (1997b) The acidification of the Herrenwieser See, Black Forest Germany, before and during industrialisation. Water Res 31:1194–1206

Kaczmarska I, Clair TA, Ehrman JM, MacDonald SL, Lean D, Day KE (2000) The effect of ultraviolet B on phytoplankton populations in clear and brown temperate Canadian lakes. Limnol Oceanogr 45:651–663

Kadlec MC, Benson WH (1995) Relationship of aquatic natural organic material characteristics to the toxicity of selected insecticides. Ecotox Environ Safe 31:84–97

Kaiser E, Herndl GJ (1997) Rapid recovery of marine bacterioplankton activity after inhibition by UV radiation in coastal waters. Appl Environ Microbiol 63:4026–4031

Kaiser K (2001) Dissolved organic phosphorus and sulphur as influenced by sorptive interactions with mineral subsoil horizons. Eur J Soil Sci 52:489–492

Kaiser K, Guggenberger G, Haumaier L, Zech W (2001a) Seasonal variations in the chemical composition of dissolved organic matter in organic forest floor layer leachates of old growth Scots pine (*Pinus sylvestris* L.) and European beech (*Fagus sylvatica* L.) stands in northeastern Bavaria, Germany. Biogeochemistry 55:103–142.

Kaiser K, Guggenberger G, Haumaier L, Zech W (2002) The composition of dissolved organic matter in forest soil solutions induced by seasons and passage through the mineral soil. Org Geochem 33:307-318

Kaiser K, Guggenberger G, Zech W (2001b) Isotopic fractionation of dissolved organic carbon in shallow forest soils as affected by sorption. Eur J Soil Sci 52:585–597

Kaiser K, Guggenberger G, Zech W (2001c) Organically bound nutrients in dissolved organic matter fractions in seepage and pore water of weakly developed forest soils. Acta hydrochim hydrobiol 28:411–419

Kaiser K, Zech W (1997) Structure-dependent sorption of dissolved organic matter on soils and related minerals. In: Drozd J, Gonet SS, Senesi N, Weber J (eds) The role of humic substances in the ecosystems and in environmental protection. Polish Soc Humic Substances, Wroclaw, pp 385–390

Kalbitz K (2001) Properties of organic matter in soil solution in a German fen area as dependent on land use and depth. Geoderma 104:203–214

Kalbitz K, Geyer S (2001) Humification indices of water-soluble fulvic acids from synchronous fluorescence spectra – effects of spectrometer type and concentration. J Plant Nutr Soil Sci 164:259–265

Kalbitz K, Geyer S (2002) Different effects of peat degradation on dissolved organic carbon and nitrogen. Org Geochem 33:319-326

Kalbitz K, Geyer S, Geyer W (2000a) A comparative characterization of dissolved organic matter by means of original aqueous samples and isolated humic substances. Chemos-

phere 40:1305–1312
Kalbitz K, Geyer S, Geyer W, Gehre M (2000c) Different dynamic of carbon and nitrogen as related to a different degree of humification of dissolved organic matter. Proc 10th Meet Intern Humic Subst Soc, 24–28 July 2000, Toulouse, pp 721–725
Kalbitz K, Geyer W, Geyer S (1999) Spectroscopic properties of dissolved humic substances – a reflection of land use history in a fen area. Biogeochemistry 47:219–238
Kalbitz K, Solinger S, Park JH, Michalzik B, Matzner E (2000b) Controls on the dynamics of dissolved organic matter in soils: a review. Soil Science 165:277–304
Kamiya M, Kameyama K (2001) Effects of selected metal ions on photodegradation of organophosphorus pesticides sensitized by humic acids. Chemosphere 45:231–235
Kang HJ, Freeman C, Ashendon TW (2001) Effects of elevated CO_2 on fen peat biogeochemistry. Sci Total Environ 27945–50
Karen DJ, Ownby DR, Forsythe BL, Bills TP, La Point TW, Cobb GB, Klaine SJ (1999) Influence of water quality on silver toxicity to rainbow trout (*Oncorhynchus mykiss*), fathead minnow (*Pimephales promelas*), and water fleas (*Daphnia magna*). Environ Toxicol Chem 18:63–70
Karentz D, Bothwell ML, Coffin RB, Hanson A, Herndl GJ, Kilham SS, Lesser MP, Lindell M, Moeller RE, Morris DP, Neale PJ, Sanders RW, Weiler CS, Wetzel RG (1994) Impact of UV-B radiation on pelagic freshwater ecosystems: report of working group on bacteria and phytoplankton. Arch Hydrobiol Beih Ergebn Limnol 43:31–69
Karlsson J, Jonsson A, Jansson M (2001) Bacterioplankton production in lakes along an altitude gradient in the subarctic north of Sweden. Microbial Ecol 42:372–382
Kauffels G (1988) Moorwirkung auf die glatte Muskulatur. In: Flaig W, Goecke C, Kauffels W (eds) Moortherapie - Grundlagen und Anwendung. Ueberreuter, Wien, pp 56-66
Kautz G, Zimmer M, Topp W (2000) Responses of the parthenogenetic isopod, *Trichoniscus pusillus* (Isopoda: Oniscidea), to changes in food quality. Pedobiologia 44:75–85
Kawahigashi M, Fujitake N, Tsurudome T, Suzuki S, Otsuka H (1997) Changes in configurations and surface active properties of humic acid with increasing concentration of NaCl. In: Drozd J, Gonet SS, Senesi N, Weber J (eds) The role of humic substances in the ecosystems and in environmental protection. Polish Soc Humic Substances, Wroclaw, pp 127–132
Kelly CA, Rudd JWM, Furutani A, Schindler DW (1984) Effects of lake acidification on rates of organic matter decomposition in sediments. Limnol Oceanogr 29:687–694
Keppler F, Eiden R, Niedan V, Pracht J, Schöler HF (2000) Halocarbons produced by natural oxidation processes during degradation of organic matter. Nature 403:298–299
Khairy AH, Davies G, Ibrahim HZ, Ghabbour EA (1996a) Adsorption of aqueous nucleobases, nucleosides, and nucleotides on compost-derived humic acid. 1. Naturally occurring pyrimidines. J Phys Chem 100:2410–2416
Khairy AH, Davies G, Ibrahim HZ, Ghabbour EA (1996b) Adsorption of aqueous nucleobases, nucleosides, and nucleotides on compost-derived humic acid. 2. Naturally occurring purines. J Phys Chem 100:2417–2421
Kieber DJ, McDaniel J, Mopper K (1989) Photochemical source of biological substrates in sea water: implications for carbon cycling. Nature 341:637–639
Kieber RJ, Li A, Seaton PM (1999) Production of nitrite from the photodegradation of dissolved organic matter in natural waters. Environ Sci Technol 33:993–998
Kieber RJ, Zhou X, Mopper K (1990) Formation of carbonyl compounds from UV-induced photodegradation of humic substances in natural waters: fate of riverine carbon to the

sea. Limnol Oceanogr 35:1503–1515

Killops SD, Killops VJ (1993) An introduction to organic geochemistry. Longman Scientific Technical, Essex, England

Kim SD, Ma H, Allen HE, Cha DK (1999) Influence of dissolved organic matter on the toxicity of copper to *Ceriodaphnia dubia*: effect of complexation kinetics. Environ Toxicol Chem 18:2433–2437

Kirchman DL (1994) The uptake of inorganic nutrients by heterotrophic bacteria. Microb Ecol 28:255–271

Kirk JTO (1994) Optics of UV-B radiation in natural waters. Arch Hydrobiol Beih Ergebn Limnol 43:1–16

Kisand V, Cuadros R, Wikner J (2002) Phylogeny of culturable estuarine bacteria catabolizing riverine organic matter in the northern Baltic sea. Appl Environ Microbiol 68:379–388

Klaus U, Mohamed S, Volk M, Spiteller M (1998a) Interaction of aquatic humic substances with anilazine and its derivatives: the nature of the bound residues. Chemosphere 37:341–361

Klaus U, Oesterreich T, Volk M, Spiteller M (1998b) Interaction of aquatic dissolved organic matter (DOM) with amitrole: the nature of bound residues. Acta hydrochim hydrobiol 26:311–317

Kleinhempel D (1970) Ein Beitrag zur Theorie des Huminstoffzustandes. Albrecht-Thaer-Arch 14:3–14

Klöcking HP (1997) Anticoagulatrory efficacy of poly(hydroxy)-carboxylates. In: Drozd J, Gonet SS, Senesi N, Weber J (eds) The role of humic substances in the ecosystems and in environmental protection. Polish Soc Humic Substances, Wroclaw, pp 951–953

Klöcking HP, Klöcking R, Helbig B (1996) Anti-factor IIa-activity of humic acid-like polymers derived from *p*-diphenolic compounds. In: Clapp CE, Hayes MHB, Senesi N, Griffith SM (eds) Humic substances and organic matter in soil and water environments. Intern Humic Subst Soc, St. Paul, pp 411–415

Klöcking HP, Mahr N, Klöcking R, Heise KH (2000) Influence of naturally occurring humic acids and synthetic humic acid like polymers on the protein anticoagulant pathway. Proc 10th Meet Intern Humic Subst Soc, 24–28 July 2000, Toulouse, 301–304

Klöcking R, Fernekorn A, Stölzner W (1992) Nachweis einer östrogenen Aktivität von Huminsäuren und huminsäureähnlichen Polymeren. Telma 22:187–197

Klöcking R, Helbig B, Schötz G, Wutzler P (1997) A comparative study of the antiviral activity of low-molecular phenolic compounds and their polymeric humic acid-like oxidation products. In: Drozd J, Gonet SS, Senesi N, Weber J (eds) The role of humic substances in the ecosystems and in environmental protection. Polish Soc Humic Substances, Wroclaw, pp 955–960

Klöcking R, Helbig R (2001) Medical aspects and applications of humic substances. In: Hofrichter M, Steinbüchel A (eds) Biopolymers. Vol. 1: lignin, humic substances, and coal. Wiley-VCH, Weinheim, pp379–392

Klöcking R, Hofmann R, Mücke D (1968) Tierexperimentelle Untersuchungen zur entzündungshemmenden Wirkung von Humaten. Arzneimittel-Forschung 18:941–941

Klöcking R, Sprössig M (1975) Wirkung von Ammoniumhuminat auf einige Viruszellsysteme. Z Allg Mikrobiol 15:25–30

Klug JL (2002) Positive and negative effects of allochthonous dissolved organic matter and inorganic nutrients on phytoplankton growth. Can J Fish Aquat Sci 59:85–95

Klug JL, Cottingham KL (2001) Interactions among environmental drivers: community re-

sponses to changing nutrients and dissolved organic carbon. Ecology 82:3390–3403
Köhler J (1994) Origin and succession of phytoplankton in a river-lake system (Spree, Germany). Hydrobiologia 289:73–83
Kok CJ, Haverkamp W, van der Aa HA (1992) Influence of pH on the growth and leaf-maceration ability of fungi involved in the decomposition of floating leaves of *Nymphaea alba* in an acid water. J Gen Microbiol 138:103–108
Kok CJ, van der Velde G (1994) Decomposition and macroinvertebrate colonization of aquatic and terrestrial leaf material in alkaline and acid still water. Freshwater Biol 31:65–75
Koopal LK, van Riemskijk WH, de Wit JCM, Benedetti MF (1994) Analytical isotherm equations for multicomponent adsorption to heterogeneous surfaces. J Colloid Interface Sci 166:51–60
Kopinke FD, Georgi A, Mackenzie K (2001a) Sorption and chemical reactions of PAHs with dissolved humic substances and related model polymers. Acta hydrochim hydrobiol 28:385–399
Kopinke FD, Georgi A, MacKenzie K (2001b) Sorption of pyrene to dissolved humic substances and related model polymers. 1. Structure-property correlation. Environ Sci Technol 35:2536–2542
Kopinke FD, Pörschmann J, Stottmeister U (1995) Sorption of organic pollutants on anthropogenic humic matter. Environ Sci Technol 29:941–950
Körner O, Meinelt T, Stüber A: Calcium and dissolved organic matter influencing the toxicity of the organophosphorus insecticide Trichlorfon. J Aquat Animal Health (in press)
Kortelainen P (1993) Contribution of organic acids to the acidity of Finnish lakes. Publ Water Envir Res Inst 13:1–48
Kortelainen P (1999a) Occurrence of humic waters. In: Keskitalo J, Eloranta P (eds) Limnology of humic waters. Backhuys, Leiden, pp 41–57
Kortelainen P (1999b) Acidity and buffer capacity. In: Keskitalo J, Eloranta P (eds) Limnology of humic waters. Backhuys, Leiden, pp 95–115
Kortelainen P, Mannio J (1988) Natural and anthropogenic acidity sources for Finnish lakes. Water Air Soil Pollut 42:341–352
Kortelainen P, Mannio J, Mäkinen I (1986) Strong and weak acids in lake waters – a methodological study. Aqua Fenn 16:221–229
Kortelainen P, Saukkonen S (1992) Leaching of organic carbon and nitrogen from peatland-dominated catchments. Suo 43:221–225
Kortelainen P, Saukkonen S (1998) Leaching of nutrients, organic carbon and iron from Finnish forestry land. Water Air Soil Poll 105:239–250
Kozuch J, Pempkowiak J (1996) Molecular weight of humic acids as a major property of the substrates influencing the accumulation rate of cadmium by a blue mussel (*Mytilus edulis*). Environ Intern 22:585–589
Kozuch J, Pempkowiak J (1997) Influences of marine humic substances of different properties on the accumulation of cadmium by the Baltic mussel *Mytilus trossulus*. In: Drozd J, Gonet SS, Senesi N, Weber J (eds) The role of humic substances in the ecosystems and in environmental protection. Polish Soc Humic Substances, Wroclaw, pp 937–944
Kracht O, Gleixner G (2000) Isotope analysis of pyrolysis products from *Sphagnum* peat and dissolved organic matter from bog water. Organ Geochem 31:645–654
Krog M, Grøn M (1995) Isolation of haloorganic groundwater humic substances. Sci Total

Environ 172:159–162

Krug EC, Frink CR (1983) Acid rain and acid soil: a new perspective. Science 221:520–525

Kukkonen J (1995) The role of natural organic material on the fate and toxicity of xenobiotics in the aquatic environment. In: Munawar M, Luotola M (eds) The contaminants in the nordic ecosystem: the dynamics, processes and fate. SPB Academic, Amsterdam, pp 95–108

Kukkonen J, McCarthy JF, Oikari A (1990) Effects of XAD-8 fractions of dissolved organic carbon on the sorption and bioavailability of organic micropollutants. Arch Environ Contam Toxicol 19:551–557

Kukkonen J, Oikari A (1987) Effects of aquatic humus on accumulation and acute toxicity of some organic micropollutants. Sci Total Environ 62:399–402

Kukkonen J, Oikari A (1991) Bioavailability of organic pollutants in boreal waters with varying levels of dissolved organic material. Water Res 25:455–463

Kukkonen J, Oikari A, Johnsen S, Gjessing E (1989) Effects of humus concentrations on benzo[*a*]pyrene accumulation from water to *Daphnia magna*: comparison of natural waters and standard preparations. Sci Total Environ 79:197–207

Kullberg A, Bishop KH, Hargeby A, Jansson M, Petersen RC Jr (1987) The ecological significance of dissolved organic carbon in acidified waters. Ambio 22:331–337

Kullberg A, Petersen RC Jr, Hargeby A, Svensson M (1992) Transport of octanol soluble carbon and dissolved organic carbon through the soil/water interface of the HUMEX lake. Environ Intern 18:631–636

Kulovaara M, Backlund P (1993) Effects of simulated sunlight on aquatic humic matter. Vatten 49:100–103.

Kulovaara M, Backlund P, Corin N (1995) Light-induced degradation of DDT in humic water. Sci Total Environ 170:185–191

Kvet J, Westlake DF (1998) Primary production in wetlands. In Westlake DF, Kvet J, Szczepanski A (eds) The production ecology of wetlands. Cambridge University Press, Cambridge, pp 78–139

Landrum PF, Nihart SR, Eadie BJ, Gardner WS (1984) Reverse-phase separation method for determining pollutant binding to Aldrich humic acid and dissolved organic carbon in natural waters. Environ Sci Technol 18:187–192

Landrum PF, Nihart SR, Eadie BJ, Herche LR (1987) Reduction in bioavailability of organic contaminants to the amphipod *Pontoporeia hoyi* by dissolved organic matter of sediment interstitial waters. Environ Toxicol Chem 6:11–20

Landrum PF, Reinhold MD, Nihart SR, Eadie BJ (1985) Predicting the bioavailability of organic xenobiotics to *Pontoporeia hoyi* in the presence of humic and fulvic materials and natural dissolved organic matter. Environ Toxicol Chem 4:459–467

Langhals H, Abbt-Braun G, Frimmel FH (2000) Association of humic substances: verification of Labert-Beer law. Acta hydrochim hydrobiol 28:329–332

Laor Y, Rebhun M (2002) Evidence for nonlinear binding of PAHs to dissolved humic acids. Environ Sci Technol 36:955–961

Lappivaara J, Kiviniemi A, Oikari A (1999) Bioaccumulation and subchronic physiological effects of waterborne iron overload on whitefish exposed in humic and nonhumic water. Arch Environ Contam Toxicol 37:196–204

Lara RJ, Thomas DN (1995) Formation of recalcitrant organic matter: humification dynamics of algal derived dissolved organic carbon and its hydrophobic fractions. Mar Chem 51:193–199

Larson RA, Hufnal JM Jr (1980) Oxidative polymerization of dissolved phenols by soluble and insoluble inorganic species. Limnol Oceanogr 25:505–512

Larson RA, Smykowski K, Hunt LL (1981) Occurrence and determination of organic oxidants in rivers and wastewaters. Chemosphere 10:1335–1338

Larsson P, Okla L, Tanvik LJ (1988) Microbial degradation of xenobiotic, aromatic pollutants in humic water. Appl Environ Microbiol 54:1864–1677

Laurion I, Ventura M, Catalan J, Psenner R, Sommaruga R (2000) Attenuation of ultraviolet radiation in mountain lakes: factors controlling the among- and within-lake variability. Limnol Oceanogr 45:1274–1288

Laurion I, Vincent WF, Lean DRS (1997) Underwater ultraviolet radiation: development of spectral models for northern high latitude lakes. Photochem Photobiol 65:107–114

Lawes JB, Gilbert JH (1905) Collected papers. In: Hall AS (ed) The book of the Rothamsted experiments. Murray, London

Lawson NM, Mason RP (1998) Accumulation of mercury in estuarine food chains. Biogeochem. 40:235–247

Lead JR, Balnois E, Hosse M, Menghetti R, Wilkinson KJ (1999) Characterization of Norwegian natural organic matter: size, diffusion coefficients, and electrophoretic mobilities. Environ Internat 25:245–258

Leavitt PR, Findlay DL, Hall RI, Smol JP (1999) Algal responses to dissolved organic carbon loss and pH decline during whole-lake acidification: evidence from paleolimnology. Limnol Oceanogr 44:757–773

Leavitt PR, Vinebrooke RD, Donald, DB, Smol JP, Schindler DW (1997) Past ultraviolet radiation environments in lakes derived from fossil pigments. Nature 388:457–459

Lee SK, Freitag D, Steinberg CEW, Kettrup A, Kim YH (1993) Effects of dissolved humic materials on acute toxicity of some organic chemicals to aquatic organisms. Water Res 27:199–204

Lee YH, Hultberg H (1990) Methylmercury in some Swedish surface waters. Environ Toxicol Chem 9:833–841

Leenheer JA (1981) Comprehensive approach to preparative isolation and fractionation of dissolved organic carbon from natural waters and wastewaters. Environ Sci Technol 15:558–587

Leenheer JA (1985) Fractionation techniques for aquatic humic substances. In: Aiken GR, McKnight DM, Wershaw RL, MacCarthy P (eds) Humic substances in soil, sediment, and water. Wiley, New York, pp 409–429

Leonard D, Reash R, Porcella DB, Paralkar A, Summers K, Gherini SA (1995) Use of the mercury cycling model (MCM) to predict the fate of mercury in the great lakes. Water Air Soil Pollut 80:519–528

Lettl A (1984) The effect of atmospheric SO_2 pollution on the microflora of forest soils. Folia Microbiol 29:455–475

Leu E, Krieger-Liszkay A, Goussias C, Gross E: Polyphenolic allelochemicals from the aquatic angiosperm *Myriophyllum spicatum* L. inhibit photosystem II. Plant Physiol (in press)

Leversee GJ, Landrum PF, Giesy JP, Fannin T (1983) Humic acids reduce bioaccumulation of some polycyclic aromatic hydrocarbons. Can J Fish Aquat Sci 40:63–69

Li D, Alic M, Gold MH (1994) Nitrogen regulation of lignin peroxidase gene transcription. Appl Environ Microbiol 60:3447–3449

Lichtfouse E, Chemu C, Baudin F, Leblond C, Da Silva M, Behar F, Derenne S, Largeau C, Wehrung P, Albrecht P (1998) A novel pathway of soil organic matter formation by selective preservation of resistant straight-chain biopolymers. Chemical and isotope evidence. Organ Geochem 28:411–415

Liebig J (1841) Organic chemistry in its application to agriculture and physiology. Tanslated by JW Webster, Owen, Cambridge

Liebig J (1856) On some points of agricultural chemistry. J Royal Agric Soc 17:284–326

Liltved H, Landfald B (2000) Effects of high intensity light on ultraviolet-irradiated and non-irradiated fish pathogenic bacteria. Water Res 34:481–486

Liltveld H, Wright RF, Gjessing ET (2001) Increasing colour in Norwegian surface waters - a consequence of climate change? 8th Symp Nordic Chapter Intern Humic Subst Soc, Copenhagen, 28–30 May, abstract

Lindell MJ, Granéli W, Tranvik LJ (1995) Enhanced bacterial growth in response to photochemical transformation of dissolved organic matter. Limnol Oceanogr 40:195–199

Lindell MJ, Granéli W, Tranvik LJ (1996) Impact of solar (UV)-radiation on bacterial growth in lakes. Aquatic Microb Ecol. 11:135–141

Lindell MJ, Rai H (1994) Photochemical oxygen consumption in humic waters. Arch Hydrobiol Beih Ergebn Limnol 43:145–155

Lindemann RI (1942) The trophic dynamic aspect of ecology. Ecology 23:399–418

Lindquist I (1967) Adsorption effects in gel filtration of humic acids. Acta Chem Scand 21:2564–2566

Lindqvist O, Johansson K, Aastrup M, Andersson A, Brinpauk I, Hovsenius G, Håkanson L, Meili M, Timm B (1991) Mercury in fish – geographical and temporal perspectives. Water Air Soil Pollut 55:159–177

Linnik PM (1996) Heavy metals in the rivers Dnjeper and Danube and the role of intrareservoir processes in the migration and transformation of their compounds. Arch Hydrobiol Suppl 113:541–547

Linnik PM (1998) The state of heavy metals in the intersitial solutions as an important characteristic of their migration mobility. Intern Rev Hydrobiol 83 Spec Iss:239–248

Linnik PM, Iskra IV (1996) Cadmium in the Dnjepr River: distribution, speciation and interaction with organic ligands. Arch Hydrobiol Suppl 113:559–564

Linnik PM, Nabivanets BI (1984) The state of metal ions in natural waters. Acta hydrochim hydrobiol 12:335–361

Little EE (1990) Behavioral toxicology: stimulating challenges for a growing discipline. Environ Toxicol Chem 9:1–2

Little EE, Finger SE (1990) Swimming behavior as an indicator of sublethal toxicity in fish. Environ Toxicol Chem 9:13–19

Liu C, Huang PM (2000) Role of hydroxy-aluminosilicate ions in the formation of humic substances. Proc 10th Meet Nordic, 24–28 July 2000, Toulouse, pp 61–64

Lobartini JC, Tan KH, Pape C (1994) The nature of humic acid-apatite interactions, products and their availability to plant growth. Commun Sci Plant Anal 25:2355–2369

Loffredo E, Pezzuto M, Senesi N (2000) Adsorption of the endocrine disruptor 17β estradiol onto humic acids from soils and sludges. Proc 10th Meet Nordic, 24–28 July 2000, Toulouse, pp 421–423

Loffredo E, Senesi N, D'Orazio V (1997) Effects of humic acids and herbicides, and their combinations on the growth of tomato seedlings in hydroponics. Z Pflanzenernähr Bodenk 160:455–461

Lönnerblad G (1931) Acta Univ Lund, NF Adv 2.27, No 14

Looser PW, Fent K, Berg M, Goudsmit GH, Schwarzenbach RP (2000) Uptake and elimination of triorganotin compounds by larval midge *Chironomus riparius* in the absence and presence of Aldrich humic acid. Environ Sci Technol 34:5165–5171

Lopez G, Riemann F, Schrage M (1979) Feeding biology of the brackish-water oncholaimid nematode *Adoncholaimus thalassophygas*. Mar Biol 54:311–318

Lorenz R, Brüggemann R, Steinberg CEW, Spieser OH (1996) Humic material changes effects of terbutylazine on behavior of zebrafish (*Brachydanio rerio*). Chemosphere 33:2145–2158

Lorenz R, Mayr C, Spieser OH, Steinberg CEW (1995) Neue Wege in die Ökotoxikologie: Quantitative Verhaltensmessungen an Fischen als Toxizitätsendpunkt. Acta hydrochim hydrobiol 23:197–201

Louchouarn P, Lucotte M, Mucci A, Pichet P (1993) Geochemistry of mercury in two hydroelectric reservoirs in Quebec, Canada. Can J Fish Aquat Sci 50:269–281

Loux NT (1998) An assessment of mercury-species-dependent binding with natural organic carbon. Chem Spec Bioavail 10:127–136

Lövgren L, Hedlung T, Öhman LO, Sjöberg S (1988) Equilibrium approaches to natural water systems. 6. Acid-base properties of a concentrated bog-water and its complexation reactions with aluminium. Water Res 11:1401–1407

Lövgren L, Sjöberg S (1989) Equilibrium approaches to natural water systems – 7. Complexation of copper(II), cadmium(II) and mercury(II) with dissolved organic matter in a concentrated bog-water. Water Res 23:327–332

Lovley DR (1991) Dissimilatory Fe(III) and Mn(IV) reduction. Microbial Rev 55:259–287

Lovley DR (1995a) Microbial reduction of iron, manganese, and other metals. Adv Agron 54:175–231

Lovley DR (1995b) Bioremediation of organic and metal contamination with dissimilatory metal reduction. J Industr Microbiol 14:85–93

Lovley DR, Coates JD, Blunt-Harris EL, Phillips EJP, Woodward JC (1996) Humic substances as electron acceptors for microbial respiration. Nature 382:445–448

Lovley DR, Fraga JL, Blunt-Harris EL, Hayes LA, Phillips EJP, Coates JD (1998) Humic substances as a mediator for microbially catalyzed metal reduction. Acta hydrochim hydrobiol 26:152–157

Lund V, Hongve D (1994) Ultraviolet irradiated water containing humic substances inhibits bacterial metabolism. Water Res 28:1111–1116

Luther GW III, Shellenbarger PA, Brendel PJ (1996) Dissolved organic Fe(III) and Fe(II) complexes in salt marsh porewaters. Geochim Cosmochim Acta 60:951–960

Lydersen E (1998) Humus and acidification. In: Hessen DO, Tranvik LJ (eds) Aquatic humic substances – ecology and biogeochemistry. Springer, Berlin, pp 63–92

Ma H, Kim SD, Cha DK, Allen HE (1999) Effect of kinetics of complexation by humic acid on toxicity of copper to *Ceriodaphnia dubia*. Environ Toxicol Chem 18:828–837

Macdonald A, Silk L, Schwartz M, Playle RC (2002) A lead-gill binding model to predict acute lead toxicity to rainbow trout (*Oncorhynchus mykiss*). Comp Biochem Physiol (in press)

Mackay D (1979) Finding fugacity feasible. Environ Sci Technol 13:1218–1223

Madari B, Michéli E, Johnston CT, Graveel JG, Czinkota I (1997) Spectroscoptic investigation of the effect of no-till and conventional tillage on soil organic matter. In: Drozd J, Gonet SS, Senesi N, Weber J (eds) The role of humic substances in the ecosystems and in environmental protection. Polish Soc Humic Substances, Wroclaw, pp 245–250

Makarov MI, Malysheva TI, Zech W, Haumaier L (1997a) Phosphorus compounds in humic and fulvic acids derived from various soils. In: Drozd J, Gonet SS, Senesi N, Weber J (eds) The role of humic substances in the ecosystems and in environmental protection. Polish Soc Humic Substances, Wroclaw, pp 407–412

Makarov MI, Malysheva TI, Zech W, Haumaier L (1997b) Interaction of inorganic phosphorus with soil humic substances. In: Drozd J, Gonet SS, Senesi N, Weber J (eds) The role of humic substances in the ecosystems and in environmental protection. Polish Soc Humic Substances, Wroclaw, pp 413–418

Malcolm RL (1985) Geochemistry of stream fulvic and humic substances. In: Aiken GR, McKnight DM, Wershaw RL, MacCarthy P (eds) Humic substances in soil, sediment, and water. Wiley, New York, pp 181–210

Malcolm RL (1991) Factors to be considered in the isolation and characterization of aquatic humic substances. In: Allard B, Borén H, Grimvall A (eds) (1991) Humic substances in the aquatic and terrestrial environment. Springer, Berlin, Lecture Notes in Earth Sciences 33:9–36

Malcolm RL, Hayes T (1994) Organic solute changes with acidification in Lake Skjervatjern as shown by ^{1}H-NMR spectroscopy. Environ Intern 20:299–305

Mallin MA, Cahoon LB, Parsons DC, Ensign SH (2001) Effect of nitrogen and phosphorus loading on plankton in Coastal Plain blackwater rivers. J Freshwater Ecol 16:455–466

Malmqvist B, Mäki M (1994) Benthic macroinvertebrate assemblages in north Swedish streams: environmental relationships. Ecography 17:9–16

Mantoura RFC, Dickson A, Riley JP (1978) The complexation of metals with humic materials in natural waters. Estuar Coast Mar Sci 6:378–408

Mao JD, Xing B, Schmidt-Rohr K (2001) New structural information on a humic acid from two-dimensional H-1-C-13 correlation solid-state nuclear magnetic resonance. Environ Sci Technol 35:1928–1934

Marmorek DR, Bernard DP, Jones ML, Rattie LP, Sullivan TJ (1987) The effects of mineral acid deposition on concentrations of dissolved organic acids in surface waters. Vancouver, Environmental and Social Systems Analysts Ltd., Report to US Environ Protect Agency, Envir Res Lab, Corvallis, Oregon, 110 pp

Marschner B (1998) DOM-enhanced mobilization of benzo(a)pyrene in a contaminated soil under different chemical conditions. Phys Chem Earth 23:199–203

Marschner B, Kalbitz K (2002) Controls of bioavailability and biodegradability of dissolved organic matter in soils. Geoderma (in press)

Martin JP, Haider K, Bondietti E (1975) Properties of model humic acids synthesized by phenoloxidase and autoxidation of phenols and other compounds formed by soil fungi. Proc Intern Meet Humic Substances, Nieuwersluis, 1972, Pudoc, Wageningen, pp 171–186

Marx G, Heumann KG (1999) Mass spectrometric investigations of the kinetic stability of chromium and copper complexes with humic substances by isotope-labelling experiments. Fresenius J Anal Chem 364:489–494

Marxsen J, Fiebig DM (1993) Use of perfused cores for evaluating extracellular enzyme activity in stream-bed sediments. FEMS Microbiol Ecol 13:1–12

Marxsen J, Schmidt HH (1993) Extracellular phosphatase activity in sediments of the Breitenbach, a central European mountain stream. Hydrobiologia 207–216

Masciandaro G, Marinari S, Grego S, Ceccanti B (1997) Use of pyrolysis technique to evaluate changes in soil organic matter quality caused by mineral and organic fertilization. In: Drozd J, Gonet SS, Senesi N, Weber J (eds) The role of humic substances in the ecosystems and in environmental protection. Polish Soc Humic Substances, Wroclaw, pp 425–430

Mason RP, Reinfelder JR, Morel FMM (1996) Uptake, toxicity, and trophic transfer of

mercury in a coastal diatom. Environ Sci Technol 30:1835–1845
Mason RP, Sullivan KA (1997) Mercury in Lake Michigan. Environ Sci Technol 31:942–947
Mason RP, Sullivan KA (1998) Mercury and methylmercury transport through an urban watershed. Water Res 32:321–330
Mathur SP (1969) Microbial use of podzol Bh fulvic acids. Can J Microbiol 15:677–680
Matilainen T, Verta M (1995) Mercury methylation and demethylation in aerobic surface waters. Can J Fish Aquat Sci 52:1597–1608
Matsunaga K, Ohyama T, Kuma K, Kudo I, Suzuki Y (1995) Photoreduction of manganese dioxide in seawater by organic substances under ultraviolet or sunlight. Water Res 29:757–759
Mattsson T, Kortelainen P, David MB (1995) Acid neutralizing capacity of solutions containing organic acids isolated from Finnish lakes. Water Air Soil Poll 85:505–510
Mattsson T, Kortelainen P, David MB (1998) Dissolved organic carbon fractions in Finnish and Maine (USA) lakes. Environ Intern 24:521–525
May R (1996) Mineralization and degradation of xenobiotics in soils and solutions through the white-rot fungus *Phanerochaete chrysosporium*. Dissertationsverlag NG Kopierladen, Munich, 277 pp
McCarthy JF (1989) Bioavailability and toxicity of metals and hydrophobic organic contaminants. In: Suffet IH, MacCarthy P (eds) Aquatic humic substances: influence of fate and treatment of pollutants. Amer Chem Soc, Washington, pp 263–277
McCarthy JF (2001) Subsurface transport of dissolved humic substances and associated contaminants. In: Clapp CE, Hayes MHB, Senesi N, Bloom PR, Jardine PM (eds) (2001) Humic substances and chemical contaminants. Soil Sci Soc Amer, Madison, pp 429–448
McCarthy JF, Czerwinski KR, Sanford WE, Jardine PM, Marsh JD (1998b) Mobilization of transuranic radionuclides from disposal trenches by natural organic matter. J Contam Hydrol 30:49–77
McCarthy JF, Jimenez BD (1985) Interactions between polycyclic aromatic hydrocarbons and dissolved humic material: binding and dissociation. Environ Sci Technol 19:1072–1076
McCarthy JF, Jimenez BD, Barbee T (1985) Effect of dissolved humic material on accumulation of polycyclic aromatic hydrocarbons: structure-activity relationships. Aquat Toxicol 7:15–24
McCarthy JF, Roberson LE, Burrus LW (1989) Association of benzo[*a*]pyrene with dissolved organic matter: prediction of K_{dom} from structural and chemical properties of the organic matter. Chemosphere 19:1911–1920
McCarthy JF, Sanford WE, Stafford PL (1998a) Lanthanide field tracers demonstrate enhanced transport of transuranic radionuclides by natural organic matter. Environ Sci Technol 32:3901–3906
McCarthy JF, Strong-Gunderson J, Palumbo AV (1994) The significance of the interactions of humic substances and organisms in the environment. In: Senesi N, Miano TM (eds) Humic substances in the global environment and implications for human health. Elsevier, Amsterdam, 981–997
McCarthy JF, Williams TM, Liang L, Jardine PM, Jolley LW, Taylor DL, Palumbo AV, Cooper LW (1993) Mobility of natural organic matter in a sandy aquifer. Environ Sci Technol 27:667–676
McDowell WH, Likens GE (1988) Origin, composition and flux of dissolved organic car-

bon in the Hubbard Brook valley. Ecol Monogr 58:177–195
McDowell WH, Wood T (1984) Podzolisation: soil processes control dissolved organic carbon in stream water. Soil Sci 137:23–32
McKnight DM, Aiken GR, Smith RL (1991) Aquatic fulvic acids in microbially based ecosystems: results from two desert lakes in Antarctica. Limnol Oceanogr 36:998–1006
McKnight DM, Andrews ED, Spaulding SA, Aiken GR (1994) Aquatic fulvic acids in algal-rich Antarctic ponds. Limnol Oceanogr 39:1972–1979
McKnight DM, Boyer EW, Westerhoff PK, Doran PT, Kulbe T, Andersen DT (2001) Spectrofluoreometric characterization of dissolved organic matter for indication of precursor organic material and aromaticity. Limnol Oceanogr 46:38–48
McKnight DM, Harnish R, Wershaw RL, Baron JS, Schiff S (1997) Chemical characteristics of particulate, colloidal, and dissolved organic material in Loch Vale watershed, Rocky Mountain National Park. Biogeochemistry 36:99–124
McKnight DM, Kimball BA, Bencala KE (1988) Iron photoreduction and oxidation in an acidic mountain stream. Science 240:637–640
McKnight DM, Wershaw RL (1989) Complexation of copper by fulvic acid from the Suwanne River – Effect of counter-ion concentration. In: Humic substances in the Suwannee River, Georgia: interactions, properties, and proposed structures. US Geol Survey, Open File Report 87–557, pp 63–69
McMurty MJ, Wales DL, Scheider WA, Beggs GL, Dimond PE (1989) Relationship of mercury concentrations in lake trout *(Salvelinus namaycush)* and smallmouth bass (*Micropterus dolomieui*) to the physical and chemical characteristics of Ontario lakes. Can J Fish Aquat Sci 46:426–434
Means JC, Wijayaratne R (1982) Role of natural colloids in the transport of hydrophobic pollutants. Science 215:968–970
Meili M (1991) The coupling of mercury and organic matter in the biogeochemical cycle – towards a mechanistic model for the boreal forest zone. Water Air Soil Pollut 56:333–347
Meili M (1992) Sources, concentrations and characteristics of organic matter in softwater lakes and streams of the Swedish forest region. In: Salonen K, Kairesalo T, Jones RI (eds) Dissolved organic matter in lacustrine ecosystems: energy source and system regulator. Kluwer, Dordrecht, pp 23–41
Meili M, Klein GW, Fry B, Bell RT, Ahlgren I (1996) Sources and partitioning of organic matter in a pelagic microbial food web inferred from the isotopic composition ($\delta^{13}C$ and $\delta^{15}N$) of zooplankton species. In: Simon M, Güde H, Weisse, T (eds) Aquatic microbial ecology. Schweizerbart, Stuttgart, pp 53–61
Meinelt T, Krüger R, Pietrock M, Stüber A, Osten R, Steinberg CEW (1997) Metal loads in sediments, pike (*Esox lucius*) and bream (*Abramis brama*) tissues in the River Oder (Germany/Poland). Arch Nat Lands 36:1–9
Meinelt T, Playle R, Schreckenbach K, Pietrock M (2001b) Interaction of the antiparasitic mixture FMC, humic substances and the water calcium content. Aquacult Res 32:405–410
Meinelt T, Playle RC, Pietrock M, Burnison BK, Wienke A, Steinberg CEW (2001a) Interaction of cadmium toxicity in embryos and larvae of zebrafish (*Danio rerio*) with calcium and humic substances. Aquat Toxicol 54:205–215
Meinelt T, Rose A, Pietrock M (2002) Effects of calcium content and humic substances on the toxicity of acriflavine to juvenile zebrafish *Danio rerio*. J Aquat Anim Health 14:35–38

Meinelt T, Staaks G, Stüber A (1995) Veränderung der Zinktoxizität in Wässern geringer und hoher Wasserhärte unter dem Einfluss eines synthetischen Huminstoffes. UWSF – Z Umweltchem Ökotox 7:155–158

Meyer JL, Tate CM (1983) The effects of watershed disturbance on dissolved organic carbon dynamics of a stream. Ecology 64:33–44

Meyer MW (1998) Ecological risk of mercury in the environment: the inadequacy of 'the best available science'. Environ Toxicol Chem 17:137–138

Micka J, Fiala L, Chalupa J (1985) Humic acids in the inflows to the Fláje Reservoir after deforestation of its watershed. Acta hydrochim hydrobiol 13:47–52

Mierle G, Ingram R (1991) The role of humic substances in the mobilization of mercury from watersheds. Water Air Soil Pollut 56:349–357

Miles CJ, Brezonik PL (1981) Oxygen consumption in humic-colored waters by a photochemical ferrous-ferric catalytic cycle. Environ Sci Technol 15:1098–1095

Mill T, Hendry DG, Richardson H (1980) Free-radical oxidants in natural waters. Science 207:886–888

Miller RW (1975) The role of humic acids in the uptake and release of mercury by freshwater sediments. Verh Intern Verein Limnol 19:2082–2086

Miller SL (1955) Production of some organic compounds under possible primitive earth conditions. J Am Chem Soc 77:2351–2361

Miller WL (1994) Recent advances in the photochemistry of natural dissolved organic matter. In: Crosby D, Helz GR, Zepp RG (eds) Aquatic and surface photochemistry. Amer Chem Soc Symp Ser, Lewis, Boca Raton, pp111–127

Miller WL (1998) Effects of UV radiation on aquatic humus: photochemical principles and experimental considerations. In: Hessen DO, Tranvik LJ (eds) Aquatic humic substances – ecology and biogeochemistry. Springer, Berlin, pp 125–143

Miller WL, Moran MA (1997) Interaction of photochemical and microbial processes in the degradation of refractory organic matter from a coastal marine environment. Limnol Oceanogr 42:1317–1324

Miller WL, Moran MA, Sheldon WM, Zepp RG, Opsahl S (2002) Determination of apparent quantum yield spectra for the formation of biologically labile photoproducts. Limnol Oceanogr 47:343–352

Miller WL, Zepp RG (1995) Photochemical production of dissolved inorganic carbon from terrestrial input: significance to the oceanic carbon cycle. Geophys Res Lett 22:417–420

Mills GL, Schwind D (1990) Photochemical degradation rates of tetraphenylborate and diphenylboric acid sensitized by dissolved organic matter in stream water. Environ Toxicol Chem 9:569–574

Milne CJ (2000) Measurement and modelling of ion binding by humic substances. Ph.D. Thesis, University of Reading, England, 131pp

Milne CJ, Kinnigurgh DG, Tipping E (2001) Generic NICA-Donnan model parameters for proton binding by humic substances. Environ Sci Technol 35:2049–2059

Miskimmin BM (1991) Effect of natural levels of dissolved organic carbon (DOC) on methyl mercury formation and sediment-water partitioning. Bull Environ Contam Toxicol 47:743–750

Mitsch WJ, Gosselink JG (2000) Wetlands. 3rd ed, Wiley & Sons, New York, 920 pp

Molisch H (1937) Der Einfluss einer Pflanze auf die Andere – Allelopathie. Fischer, Jena, Germany

Möller A, Kaiser K, Kanchanakool N, Anacksamphant C, Jirasuktaveekul W, Maglinao A,

Niamskul C, Zech W (2002) Sulfur form in bulk soils and alkaline soil extracts of tropical mountain ecosystems in northern Thailand. Aust J Soil Res 40:161–175

Monson BA, Brezonik PL (1999) Influence of food, aquatic humus, and alkalinity on methylmercury uptake by *Daphnia magna*. Environ Toxicol Chem 18:560–566

Moore TR (1987) Dissolved organic carbon in forested and cutover drainage basins, Westland, New Zealand. In: Swanson RH, Bernier PY, Woodard PD (eds) Forest hydrology and watershed management. Wallingford, NZ, Intern Ass Hydrol Sci Press, pp 481–487

Mopper K, Zhou X (1990) Hydroxyl radical photoproduction in the sea and its potential impact on marine processes. Science 250:661–663

Moran MA, Hodson RE (1990) Bacterial production on humic and nonhumic components of dissolved organic carbon. Limnol Oceanogr 35:1744–1756

Moran MA, Hodson RE (1994) Dissolved humic substance of vascular plant origin in a coastal marine environment. Limnol Oceanogr 39:762–771

Moran MA, Sheldon WM Jr, Zepp RG (2000) Carbon loss and optical property changes during long-term photochemical and biological degradation of estuarine dissolved organic matter. Limnol Oceanogr 45:1254–1264

Moran MA, Zepp RG (1997) Role of photoreactions in the formation of biologically labile compounds from dissolved organic matter. Limnol Oceanogr 42:1307–1316

Moran XAG, Gasol JM, Pedros-Alio C, Estrada M (2001) Dissolved and particulate primary production and bacterial production in offshore Antarctic waters during austral summer: coupled or uncoupled? Mar Ecol Progr Ser 222:25–39

Morehead NR, Eadie BJ, Lake B, Landrum PF, Berner D (1986) The sorption of PAH onto dissolved organic matter in Lake Michigan waters. Chemosphere 15:1655–1664

Morel FMM, Hering JG (1993) Principles and applications of aquatic chemistry. Wiley, New York

Morell JM, Corredor JE (2001) Photomineralization of fluorescent dissolved organic matter in the Orinoco River plume: estimation of ammonium release. J Geophy Res-Oceans 106:16807–16813

Morra MJ, Fendorf SE, Brown PD (1997) Speciation of sulfur in humic and fulvic acids using X-ray absorption near-edge structure (XANES) spectroscopy. Geochim Cosmochim Acta 61:683–688

Morris DP, Hargreaves BR (1997) The role of photochemical degradation of dissolved organic carbon in regulating the UV transparency of three lakes on the Pocono Plateau. Limnol Oceanogr 42:239–249

Mota AM, Rato A, Brazia C, Simões Gonçalves ML (1996) Competition of Al^{3+} in complexation of humic matter with Pb^{2+}: a comparative study with other ions. Environ Sci Technol 30:1970–1974

Mougin C, Laugero C, Asther M, Dubroca J, Frasse P, Aster M (1994) Biotransformation of the herbicide atrazine by the white rot fungus *Phanerochaete chrysosporium*. Appl Environ Microbiol 60:705–708

Moza PN, Hustert K (1997) Einfluss von Huminsäure auf die Photolyse von Pestiziden. GSF-Bericht 25/97:71–81

Muir DCG, Hobden BR, Servos MR (1994) Bioconcentration of pyrethroid insecticides and DTT by rainbow trout: uptake, depuration, and effect of dissolved organic carbon. Aquat Toxicol 29:230–240

Mulholland PJ, Dahm CN, David MB, Di Toro DM, Fisher TR, Hemond HF, Kögel-Knabner I, Meybeck MH, Meyer JL, Sedell JR (1990) What are the temporal and spatial variations of organic acids at the ecosystem level? In: Perdue EM, Gjessing ET (eds) Organic acids in aquatic ecosystems. Wiley, Chichester, pp 315–329

Mulholland PJ, Kuenzelr EJ (1979) Organic carbon export from upland and forested wetland watersheds. Limnol Oceanogr 24:960–966

Müller MB, Schmitt D, Frimmel FH (2000) Fractionation of natural organic matter by size exclusion chromatography – properties and stability of fractions. Environ Sci Technol 34:4867–4872

Münster U (1982) Physikalisch-chemische Untersuchungen zur Anreicherung, Fraktionierung und Strukturaufklärung monomerer und polymerer organischer Substrate im DOC vom Plusssee. PhD Thesis, University Kiel, Germany, 281 pp

Münster U (1984) Distribution, dynamic and structure of free dissolved carbohydrates in the Plusssee, a North German eutrophic lake. Verh Internat Verein Limnol 22:929–935

Münster U (1985) Investigation about structure, distribution and dynamics of different organic substrates in the DOM of lake Plusssee. Arch Hydrobiol Suppl 70:429–480

Münster U (1994) Studies on phosphatase activities in humic lakes. Environ Intern 20:49–59

Münster U (1999a) Bioavailability of nutrients. In: Keskitalo J, Eloranta P (eds) Limnology of humic waters. Backhuys, Leiden, pp 77–94

Münster U (1999b) Amino acid profiling in natural organic matter isolated by reverse osmosis from eight different boreal freshwaters. Environ Internation 25:209–224

Münster U, de Haan H (1998) The role of microbial extracellular enzymes (MEE) in the transformation of dissolved organic matter (DOM) in humic waters. In: Hessen DO, Tranvik LJ (eds) Aquatic humic substances – ecology and biogeochemistry. Springer, Berlin, pp 199–257

Münster U, Einiö P, Nurminen J, Overbeck J (1992a) Extracellular enzymes in a polyhumic lake: important regulators in detritus processing. Hydrobiologia 229:225–238

Münster U, Heikkinen E, Likolammi M, Järvinen M, Salonen K, de Haan H (1999a) Utilisation of polymeric and monomeric aromatic and amino acid carbon in a humic boreal forest lake. Arch Hydrobiol Spec Issues Advanc Limnol 54:105–134

Münster U, Heikkinen E, Salonen K, de Haan H (1998) Tracing of peroxidase activity in humic lake water. Acta hydrochim hydrobiol 26:158–166

Münster U, Nurminen J, Einiö P, Overbeck J (1992b) Extracellular enzymes in a small polyhumic lake: origin, distribution and activities. Hydrobiologia 243/244:47–59

Münster U, Salonen K, Tulonen T (1999b) Decomposition. In: Keskitalo J, Eloranta P (eds) Limnology of humic waters. Backhuys, Leiden, pp 225–264

Muscolo A, Nardi S (1997) Auxin or auxin-like activity of humic matter. In: Drozd J, Gonet SS, Senesi N, Weber J (eds) The role of humic substances in the ecosystems and in environmental protection. Polish Soc Humic Substances, Wroclaw, pp 987–992

Myneni SCB (2002) Formation of stable chlorinated hydrocarbons in weathering plant material. Science 295:1039–1041

Myneni SCB, Brown JT, Martinez GA, Meyer-Ilse W (1999) Imaging of humic substances macromolecular structures in water and soils. Science 286:1335–1337

Nagase H, Ose Y, Sato T, Ishikawa T (1982) Methylation of mercury by humic substances in an aquatic environment. Sci Tot Environ 24:133–142

Nagel R, Loskill R (1991) Bioaccumulation in aquatic systems; contributions to the assessment. Proc Intern Workshop, Berlin 1990. VCH, Weinheim

Nalepa TF, Fahnenstiel GL, Johengen TH (1999) Impacts of the zebra mussel (*Dreissena polymorpha*) on water quality: a case study in Saginaw Bay, Lake Huron. In: Claudi

R, Leach JH (eds) Non-indigenous freshwater organisms in North America; their biology and impact. CRC Press, Boca Raton, pp 255–271
Nalewajko C, Godmaire H (1993) Extracellular products of *Myriophyllum spicatum* L. as a function of growth phase and diel cycle. Arch Hydrobiol 127:345–356
Nanny MA, Bortiatynski JM, Hatcher PC (1997) Noncovalent interactions between acenaphtenone and dissolved fulvic acid as determined by ^{13}C NMT T_1 relaxation measurements. Environ Sci Techon 32:539–534
Nardi S, Pamuccio MR, Abenavolit MR, Muscola A (1994) Auxin-like effect of humic substances extracted from faeces of *Allolobophora caliginosa* and *Antennaria rosea.* Soil Biol Biochem 26:1341–1346
Naumann E (1918) Über die natürliche Nahrung des limnischen Zooplanktons. Lunds Univ Årsskr N F Avd 2, 14:1–47
Naumann E (1919) Några synpunkter angående plankton ökologi. Med särskild hänsyn till fytoplankton. Svensk Bot Tidskr 13:129–163
Naumann K (1993) Chlorchemie der Natur. Chem uns Zeit 27:33–41
Naumann K (1994) Natürlich vorkommende Organohalogene. Nachr Chem Tech Lab 42:389–392
Neidleman SL, Geigert J (1986) Biohalogenation – principles, basic roles, and applications. Ellis Horwood Publishers, Frome, Somerset
Nevin KP, Lovley DR (2000) Potential for nonenzymatic reduction of Fe(III) via electron shuttling in subsurface sediments. Environ Sci Technol 34:2472–2478
Niedan V, Pavasars I, Öberg G (2000) Chloroperoxidase-mediated chlorination of aromatic groups in fulvic acid. Chemosphere 41:779–785
Niedan V, Schöler HF (1997) Natural formation of chlorobenzoic acids (CBA) and distinction between PCB-degraded CBA. Chemosphere 35:1233–1241
Nikolaou AD, Lekkas TD (2001) The role of organic matter during formation of chlorinated by-products: a review. Acta hydrochim hydrobiol 29:63–77
Nilsson A, Håkanson L (1992) Relationships between mercury in lake water, water colour and mercury in fish. Hydrobiologia 235/236:675–683
Nixdorf B, Krumbeck H, Jander J, Beulker C: Comparison of bacterial and phytoplankton productivity in extremely acidic mining lakes and eutrophic hard water lakes. Acta Oecolog (in press)
Nixdorf B, Wollmann K, Deneke R (1998) Ecological potential for planktonic development and food web interactions in extremely acidic mining lakes in Lusatia. In: Geller W, Klapper H, Salomons W (eds) Acidic mining lakes. Springer, Berlin, pp 147–167
Nordqvist H (1921) Studien über das Teichzooplankton. Lunds Univ Årsskr NF Avd 2, 17:1–123
Norwood DL, Christman RF, Hatcher PG (1987) Structural characterization of aquatic humic materials. 2. Phenolic content and its relationship to chlorination mechanisms in an isolated aquatic fulvic acid. Environ Sci Technol 21:791–798
Nuutinen S, Kukkonen J (1998) The effect of selenium and organic material in lake sediments on the bioaccumulation of methylmercury by *Lumbriculus variegatus* (Oligochaeta). Biogeochemistry 40:267–278
Nygaard K, Tobiesen A (1993) Bacterivory in algae: a survival strategy during nutrient limitation. Limnol Oceanogr 38:273–279
O'Driscoll N, Evans RD (2000) Analysis of methyl mercury binding to freshwater humic and fulvic acids by gel permeation chromatography/hydride generation ICP-MS. Environ Sci Technol 34:4039–4043

Öberg G (1998) Chloride and organic chlorine in soil. Acta hydrochim hydrobiol 26:137–144

Öberg G (2001) The occurrence and origin of organic chlorine in soil. http://www.eurochlor.org/chlorine/science/chemistrySoil.htm

Obernosterer I, Reitner B, Herndl GJ (1999) Contrasting effects of solar radiation on dissolved organic matter and its bioavailability to marine bacterioplankton. Limnol Oceanogr 44:1645–1654

Odén S (1914) Zur Kolloidchemie der Humusstroffe. Kolloid Z 14:123–130

Odén S (1919) Die Huminsäuren. Kolloidchem Beih 11:75–260

Oesterreich T, Klaus U, Volk M, Neidhard B, Spiteller M (1999) Environmental fate of amitrole: influence of dissolved organic matter. Chemosphere 38:379–392

Ogawa H, Amagai Y, Koike I, Kaiser K, Benner R (2001) Production of refractory dissolved organic matter by bacteria. Science 292:917–920

Ogner G, Schnitzer M (1970) Humic substances: fulvic acid-dialkyl phthalate complexes and their role in pollution. Science 170:317–318

Ohle W (1934) Über organische Stoffe in Binnenseen. Verh Intern Verein Limnol 6:249–262

Ohle W (1935) Organische Kolloide in ihrer Wirkung auf den Stoffhaushalt der Gewässer. Naturwissenschaften 35:480–484

Ohle W (1937) Kooloidgele als Nährstoffgeneratoren der Gewässer. Naturwissenschaften 37:471–474

Ohno T, Doolan KL (2001) Effects of red clover decomposition on phytotoxicity to wild mustard seedling growth. App Soil Ecol 16:187–192

Ojala A, Salonen K (2001) Productivity of *Daphnia longispina* in a highly humic boreal lake J Plankton Res 23:1207–1215

Økland J, Økland KA (1986) The effects of acid deposition on benthic animals in lakes and streams. Experientia 42:471–486

Olaveson MM, Stokes PM (1989) Responses of the acidophilic alga, *Euglena mutabilis* (Euglenophyceae) to carbon enrichment at pH 3. J Phycol 25:529–539

Oliver BG, Thurman EM, Malcolm RL (1983) The contribution of humic substances to the acidity of colored natural waters. Geochim Cosmochim Acta 47:2031–2035

Olsen C (1930) On the influence of humus substances on the growth of green plants in water culture. Comp rend Lab Carlsberg 18:1–16

Opsahl S, Benner R (1998) Photochemical reactivity of dissolved lignin in river and ocean waters. Limnol Oceanogr 43:1297–1304

Oris JT, Hall T, Tylka JD (1990) Humic acids reduce the photo-induced toxicity of anthracene to fish and *Daphnia*. Environ Toxicol Chem 9:575–583

Orlov DS, Amosova IA, Glebova GI (1972) Molecular parameters of humic acids. Geoderma 13:211–229

Otto WH, Carper WR, Larive CK (2001) Measurement of cadmium(II) and calcium(II) complexation by fulvic acids using ^{113}Cd NMR. Environ Sci Technol 35:1463–1468

Ourisson G, Rohmer M (1992) Hopanoids. 2. Biohopanoids: a novel class of bacterial lipids. Acc Chem Res 25:403–407

Paarlberg A (1984) Effect of humic character, pH and lime on the growth and mortality of the freshwater shrimp *Gammarus pulex* (L.). MS Thesis, University Lund, Sweden, 56 pp

Pace ML, Cole JJ (1996) Regulation of bacteria by resources and predation tested in whole lake experiments. Limnol Oceanogr 41:1448–1460

Pace ML, Cole JJ (2002) Synchronous variation of dissolved organic carbon and color in lakes. Limnol Oceanogr 47:333–342

Paciolla MD, Davies G, Jansen SA (1999) Generation of hydroxyl radicals from metal-loaded humic acids. Environ Sci Technol 33:1814–1818

Patalas K (1971) Crustacean plankton communities in forty-five lakes in the Experimental Lakes Area, northwestern Ontario. J Fish Res Bd Can 28:231–244

Paul A, Pflugmacher S, Pietsch C, Stösser R, Steinberg CEW: Correlation of spin concentration in humic substances with inhibitory effects on photosynthetic oxygen evolution of aquatic macrophytes. (submitted)

Paul A, Pflugmacher S, Steinberg C.E.W. (2002) Korrelation der Spinkonzentration von Huminstoffen mit inhibitorischen Effekten bei der Photosynthese aquatischer Makrophyten. Sitzungsber Akad gemeinnr Wissenschaft Erfurt, Mathem-Naturwiss Klasse (in press)

Paul VJ, Cruz-Rivera E, Thacker RW (2001) Chemical mediation of macroalgal–herbivore interactions: ecological and evolutionary perspectives. In: McClintock JB and Baker BJ (eds) Marine chemical ecology. CRC Press, Boca Raton, pp 227–265

Pautou G, Aïn G, Gilot B, Cousserans J, Gabinaud A, Simonneau P (1973) Cartographie écologique appliquèe à la démoustication. Doc Cartogr Ecol Univ Scientif Médic (Grenoble, France) 11:1–16

Pautou MP, Rey D, David JP, Meyran JC (2000) Toxicity of vegetable tannins on crustaceae associated with alpine mosquito breeding sites. Ecotox Environ Safe 47:323–332

Peijnenburg WJGM, Hart MJ, den Hollander HA, van de Meent D, Verboom HH, Wolfe NL (1992) Reductive transformation of halogenated aromatic hydrocarbons in anaerobic water–sediment systems: kinetics, mechanisms, and products. Environ Toxicol Chem 11:301–314

Pellissier F (1993) Allelopathic effect of phenolic-acids from humic solutions on two spruce mycorrhizal fungi – *Genoccum graniforme* and *Accaria laccata*. J Chem Ecol 19:2015–2114

Pempkowiak J, Kozuch J, Southon T (1994) The influence of structural features of marine humic substances on the accumulation rates of cadmium by a blue mussel (*Mytilus edulis*). Environ Internat 20:391–395

Peng A, Wang Z, Wang W, Yang C (1991) A new hypothesis for the etiology of Kaschin-Beck disease. J Environ Sci (China) 3:5–14

Pennak RW (1989) Freshwater invertebrates of the United States, Protozoa to Mollusca. Wiley, New York, 3rd edn

Penttinen S, Kostamo A, Kukkonen JVK (1998) Combined effects of dissolved organic material and water hardness on toxicity of cadmium to *Daphnia magna*. Environ Toxicol Chem 17:2498–2503

Perdue EM (1998) Chemical composition, structure and metal binding properties. In: Hessen DO, Tranvik LJ (eds) Aquatic humic substances – ecology and biogeochemistry. Springer, Berlin, pp 41–61

Perdue EM (2001) Modeling concepts in metal-humic complexation. In: Clapp CE, Hayes MHB, Senesi N, Bloom PR, Jardine PM (eds) Humic substances and chemical contaminants. Soil Science Society of America, Madison, pp 305–316

Perminova IV, Grechishcheva NYu, Kovalesvskii DV, Kudryavtsev AV, Petrosyan VS, Matorin DN (2001) Quantification and prediction of the detoxifying properties of humic substances related to their chemical binding to polycyclic aromatic hydrocarbons. Environ Sci Technol 35:3841–3848

Perminova IV, Kovalevsky DV, Yashchenko NYu, Danchenko NN, Kudryavtsev AV, Zhilin DM, Petrosyan VS, Kulikova NA, Philippova OI, Lebedeva GF (1996) Humic substances as natural detoxifying agents. In: Clapp CE, Hayes MHB, Senesi N, Griffith SM (eds) Humic substances and organic matter in soil and water environments. Intern Humic Subst Soc, St. Paul, pp 399–406

Perminova IV, Yashchenko NYu, Petrosyan VS (1999) Relationships between structure and binding affinity of humic substances for polyaromatic hydrocarbons: relevance of molecular descriptors. Environ Sci Technol 33:3781–3787

Pers C, Rahm L, Jonsson A, Bergström AK, Jansson M (2001) Modelling dissolved organic carbon turnover in humic Lake Örträsket, Sweden. Environ Model Assess 6:159–172

Persson G, Broberg O (1985) Nutrient concentrations in the acidified Lake Gårdsjön: the role of transport and retention of phosphorus, nitrogen and DOC in watershed and lake. Ecol Bull 37:158–175

Persson L, Alsberg T, Odham G, Kiss K (2000) On-line size-exclusion chromatography/electrospray ionisation mass spectrometry of aquatic humic and fulvic acids. Proc 10th Meet Intern Humic Subst Soc, 24–28 July 2000, Toulouse, pp 1113–1116

Petasne RG, Zika RG (1987) Fate of superoxide in coastal sea water. Nature 325:516–518

Peters AJ, Hamilton-Taylor J, Tipping E (2001) Americium binding to humic acid. Environ Sci Technol 35:3495–3500

Petersen RC Jr (1990) Effects of ecosystem changes (e. g., acid status) on formation and biotransformation of organic acids. In: Perdue EM, Gjessing ET (eds) Organic acids in aquatic ecosystems. Wiley, Chichester, pp 151–166

Petersen RC Jr (1991) The contradictory biological behaviour of humic substances in the aquatic environment. In: Allard B, Borén H, Grimvall A (eds) (1991) Humic substances in the aquatic and terrestrial environment. Springer, Berlin, Lecture Notes in Earth Sciences 33:369–390

Petersen RC Jr, Kullberg A (1985) The octanol/water partition coefficient of humic material and its dependence on hydrogen ion activity. Vatten 41:236–239

Petersen RC Jr, Persson U (1987) Comparison of the biological effects of humic materials under acidified conditions. Sci Tot Environ 62:387–398

Petersen RC Jr, Petersen LMM, Persson U, Kullberg A, Hargeby A, Paarlberg A (1986) Health aspects of humic compounds in acid environments. Water Qual Bull 11:44–49

Peuravuori J (1992) Isolation, fractionation and characterization of aquatic humic substances. Does a distinct humic molecule exist? Acad Diss Univ Turku, Finland. Finnish Humic New 4:1–99

Peuravuori J, Pihlaja K (1999) Structural characterization of humic substances. In: Keskitalo J, Eloranta P (eds) Limnology of humic waters. Backhuys, Leiden, pp 22–34

Pflugmacher S, Geissler K, Steinberg CEW (1999a) Activity of phase I and phase II detoxication enzymes in different cormus parts of *Phragmites australis*. Ecotoxicol Environ Safety 42:62–66

Pflugmacher S, Hupfer M, Steinberg CEW (1997) Detoxifizierende Enzymsysteme im Sediment – Erste Ansätze zur Isolation und Akvitätsbestimmung von Endo- und Ectoenzymen aus limnischen Sedimente. Soc Environ Toxical And Chem German Branch, 1997, Aachen, Germany (Poster)

Pflugmacher S, Pietsch C, Rieger W, Paul A, Preuer T, Zwirnmann E, Steinberg CEW (2003) Humic substances and their direct effects on the physiology of aquatic plants. In: Ghabbour EA, Davies G (eds) Humic substances: nature's most versatile materials. Taylor & Francis, New York (in press)

Pflugmacher S, Spangenberg M, Steinberg CEW (1999b) Dissolved organic matter (DOM) and effects on the aquatic macrophyte *Ceratophyllum demersum* in relation to photosynthesis, pigment pattern and activity of detoxication enzymes. J Appl Bot 73:184–190

Pflugmacher S, Steinberg CEW (1997) Activity of phase I and phase II detoxication enzymes in aquatic macrophytes. J Appl Bot 71:144–146

Pflugmacher S, Tidwell LF, Steinberg CEW (2001) Dissolved humic substances can directly affect freshwater organisms. Acta hydrochim hydrobiol 29:34–40

Phlips EJ, Cichra M, Aldridge FJ, Jembeck J, Hendrickson J, Brody R (2000) Light availability and variations in phytoplankton standing crops in a nutrient-rich blackwater river. Limnol Oceanogr 45:916–929

Piccolo A (1994) Interactions between organic pollutants and humic substances in the environment. In: Senesi N, Miano TM (eds) Humic substances in the global environment and implications for human health. Elsevier, Amsterdam, pp 961–979

Piccolo A (1997) New insights on the conformational structure of humic substances as revealed by size exclusion chromatography. In: Drozd J, Gonet SS, Senesi N, Weber J (eds) The role of humic substances in the ecosystems and in environmental protection. Polish Soc Humic Substances, Wroclaw, pp 19–35

Piccolo A, Conte P, Cozzolino A, Spaccini R (2001) Molecular sizes and association forces of humic substances in solution. In: Clapp CE, Hayes MHB, Senesi N, Bloom PR, Jardine PM (eds) Humic substances and chemical contaminants. Soil Science Society of America, Madison, pp 89–118

Piccolo A, Nardi S, Concheri G (1996) Macromolecular changes in humic substances induced by interaction with organic acids. Eur J Soil Sci 47:319–328

Picker MD, McKenzie CJ, Fielding P (1993) Embryonic tolerance of *Xenopus* (Anura) to acidic blackwater. Copeia 1993:1072–1081

Pienitz R, Smol JP (1993) Diatom assemblages and their relationship to environmental variables in lakes from the boreal forest-tundra ecotone near Yellowknife, Northwest Territories, Canada. Hydrobiologia 269/270:391–404

Pienitz R, Smol JP, Lean DRS (1997a) Physical and chemical limnology of 24 lakes located between the Yellowknife and Contwoyto Lakes, Northwest Territories (Canada). Can J Fish Aquat Sci 54:347–358

Pienitz R, Smol JP, Lean DRS (1997b) Physical and chemical limnology of 59 lakes located between the southern Yukon and the Tuktoyaktuk Peninsula, Northwest Territories (Canada). Can J Fish Aquat Sci 54:330–346

Pienitz R, Smol JP, MacDonald GM (1999) Paleolimnological reconstruction of Holocene climatic trends from two boreal treeline lakes, Northwest Territories, Canada. Arct Antarct Alp Res 31:82–93

Pienitz R, Vincent WF (2000) Effect of climate change relative to ozone depletion on UV exposure in subarctic lakes. Nature 404:484–487

Planas D, Sarhan F, Dube L, Godmaire H, Cadieux C (1982) Ecological significance of phenolic compounds of *Myriophyllum spicatum*. Verh Internat Verein Limnol 21:1492–1496

Playle RC (1998) Modelling metal interactions at fish gills. Sci Total Environ 219:147–163

Playle RC, Dixon DG, Burnison K (1993a) Copper and cadmium binding to fish gills: estimates of metal-gill stability constants and modelling of metal accumulation. Can J Fish Aquat Sci 50:2678–2687

Playle RC, Dixon DG, Burnison K (1993b) Copper and cadmium binding to fish gills:

modification by dissolved organic carbon and synthetic ligands. Can J Fish Aquat Sci 50:2667–2677

Pluta HJ, Knie J, Leschber R (eds) (1994) Biomonitore in der Gewässerüberwachung. Fischer, Jena

Poléo ABS, Lydersen E, Rosseland BO, Kroglund F, Salbu B, Vogt RD, Kvellestad A (1994) Increased mortality of fish due to changing Al-chemistry of mixing zones between limed streams and acidic tributaries. Water Air Soil Poll 75:339–351

Poltz J (1972) Untersuchungen über das Vorkommen und den Abbau von Feten und Fettsäuren in Seen. Arch Hydrobiol/Suppl 40:315–399

Ponomareva VV, Etlinger AI (1954) About characteristics of organic matter in waters of Newaa. Zh Prikl Him 27:774–781 (in Russian)

Porvari P, Verta M (1995) Methylmercury production in flooded soils: a laboratory study. Water Air Soil Poll 80:789–798

Pracht J, Boenigk J, Isenbeck-Schröter M, Keppler F, Schöler, HF (2001) Abiotic Fe(III) induced mineralization of phenolic substances. Chemosphere 44:613–619

Prescott CE, Parkinson D (1985) Effects of sulphur pollution on rates of litter decomposition in a pine forest. Can J Bot 63:1436–1443

Pringsheim E G (1963) Farblose Algen: Ein Beitrag zur Evolutionsforschung. Fischer, Stuttgart

Putschew A, Wischnack S, Jekel M (2001) Bildung organischer Bromverbindungen in Oberflächengewässern. Wasser Boden 53:21–23

Pütter A (1909) Die Ernährung der Wassertiere und der Stoffhaushalt der Gewässer. Fischer, Jena

Qualls RG, Haines B L (1992a) Measuring adsorption isotherms using continuous, unsaturated flow through intact soil cores. Soil Sci Soc Am J 56:456–460

Qualls RG, Haines BL (1992b) Biodegradability of dissolved organic matter in forest throughfall, soil solution, and stream water. Soil Sci Soc Am J 56:578–586

Quecke K, Loschen C (1989) Beeinflussung der Arachidonsäure-Kaskade durch pflanzliche Inhaltsstoffe und Torfbestandteile. In: M Hornig (ed) Gynäkologische Balneotherapie. Bad Waldsee, pp 166–174

Quémerais B, Cossa D, Rondeau B, Pham TT, Gagnon P, Fortin B (1999) Sources and fluxes of mercury in the St. Lawrence River. Environ Sci Technol 33:840–849

Quigley MS, Santschi PH, Hung CC, Guo L, Honeyman BD (2002) Importance of acid polysaccharides for ^{234}Th complexation to marine organic matter. Limnol Oceanogr 47:367–377

Rädlinger G, Heumann KG (1997) Determination of halogen species of humic substances using HPLC/ICP-MS coupling. Fresenius J Anal Chem 359:430–433

Rädlinger G, Heumann KG (2000) Transformation of iodide in natural and wastewater systems by fixation on humic substances. Environ Sci Technol 34:3932–3936

Rashid MA (1971) Role of humic acids of marine origin and their different molecular weight fractions in complexing di- and tri-valent metals. Soil Sci 111:289–306

Rask M, Viljanen M, Sarvala J (1999) Humic lakes as fish habitats. In: Keskitalo J, Eloranta P (eds) Limnology of humic waters. Backhuys, Leiden, pp 209–224

Rasmussen JB, Godbout L, Shallenberg M (1989) The humic content of lake water and its relationship to watershed and lake morphometry. Limnol Oceanogr 34:1336–1343

Rasyid U, Johnson WD, Wilson MA, Hanna JV (1992) Changes in organic structural group composition of humic and fulvic acids in sediments from similar geographical but different depositional environments. Org Geochem 18:521–529

Rauthan BS, Schnitzer M (1981) Effects of soil fulvic acid on the growth and nutrient content of cucumber (*Cucumis sativus*) plants. Plant Soil 63:491–495

Rav-Acha C, Rebhun M (1992) Binding of organic solutes to dissolved humic substances and its effects on adsorption and transport in the aquatic environment. Water Res 26:1645–1654

Ravichandran M (1999) Interactions between mercury and dissolved organic matter in the Florida Everglades. Ph.D. Thesis, University of Colorado, Boulder

Ravichandran M, Aiken GR, Reddy MM, Ryan JN (1998) Enhanced dissolution of cinnabar (mercuric sulfide) by dissolved organic matter isolated from the Florida Everglades. Environ Sci Technol 32:3305–3311

Ravichandran M, Aiken GR, Ryan JN, Reddy MM (1999) Inhibition of precipitation and aggregation of metacinnabar (mercuric sulfide) by dissolved organic matter isolated from the Florida Everglades. Environ Sci Technol 33:1418–1423

Raymond PA, Bauer JE (2001) Riverine export of aged terrestrial organic matter to the North Atlantic Ocean. Nature 409:497–500

Reche I, Pace ML, Cole JJ (2000) Modeled effects of dissolved organic carbon and solar spectra on photobleaching in lake ecosystems. Ecosystems 3:419–432

Reche I, Pulido-Villena E, Conde-Porcuna JM, Carrillo P (2001) Photoreactivity of dissolved organic matter from high-mountain lakes of Sierra Nevada, Spain. Arct Antarct Alp Res 33:426–434

Reddy MM, Aiken GR (2001) Fulvic acid-sulfide ion competition for mercury ion binding in the Florida Everglades. Water Air Soil Pollut 132:89–104

Reemtsma T, Bredow A, Gehring M (1999) The nature and kinetics of organic matter release from soil by salt solution. Eur J Soil Sci 50:53–64

Reinikainen J, Hyvärinen H (1997) Humic- and fulvic-acid stratigraphy of the Holocene sediments from a small lake in Finnish Lapland. Holocene 7:401–407

Reiss F (1977) Qualitative and quantitative investigations on the macrobenthic fauna of Central Amazon lakes. I. Lago Tupé, a black water lake on the lower Rio Negro. Amazoniana 6:203–235

Reitner B, Herndl GJ, Herzig A (1997) Role of ultraviolet-B radiation on photochemical and microbial oxygen consumption in a humic-rich shallow lake. Limnol Oceanogr 42:950–960

Renberg I, Hellberg T (1982) The pH history of lakes in south-western Sweden, as calculated from the subfossil diatom flora of the sediment. Ambio 11:30–33.

Renneberg AJ, Dudas MJ (2001) Transformation of elemental mercury to inorganic and organic forms in mercury and hydrocarbon contaminated soils. Chemosphere 45:1103–1109

Rex RW (1960) Electron paramagnetic resonance studies of stable free radicals in lignins and humic acids. Nature 188:1185–1186

Rey D, Cuany A, Pautou MP, Meyran JC (1999a) Differential sensitivity of mosquito taxa against vegetable tannins. J Chem Ecol 25:537–547

Rey D, Marigo G, Pautou MP (1996) Composés phénoliques chez *Alnus glutinosa* et contrôle des populations larvaires de Culicidae. CR Acad Sci Paris Sci Vie 319:1035–1042

Rey D, Martins D, David JP, Pautou MP, Long A, Marigo G, Meyran JC (2000) Role of vegetable tannins in habitat selection among mosquito communities from the Alpine hydrosystems. CR Acad Sci Paris Sci Vie 323:391-398

Rey D, Pautou MP, Meyran JC (1999b) Histopathological effects of tannic acid on the midgut epithelium of some aquatic diptera larvae. J Invertebr Pathol 73:173–181

Ribas G, Carbonell E, Creus A, Xamena N, Marcos R (1997) Genotoxicity of humic acid in cultured human lymphocytes and its interaction with the herbicides alachlor and maleic hydrazide. Environ Mol Mutagen 29:272–276

Rice JA, MacCarthy P (1991) Statistical evaluation of the elemental composition of humic substances. Org Geochem 17:635–648

Richards JG, Burnison BK, Playle RC (1999) Natural and commercial dissolved organic matter protects against the physiological effects of a combined cadmium and copper exposure on rainbow trout. Can J Fish Aquat Sci 56:407–418

Richards JG, Curtis PJ, Burnison BK, Playle RC (2001) Effects of natural organic matter source on reducing metal toxicity to rainbow trout (*Oncorhynchus mykiss*) and on metal binding to their gills. Environ Toxicol Chem 20:1159–1166

Ridge I, Pillinger JM (1996) Towards understanding the nature of algal inhibitors from barley straw. Hydrobiologia 340:301–305

Ridge I, Walters J, Street M (1999) Algal growth control by terrestrial leaf litter: a realistic tool? Hydrobiologia 395/396:173–180

Riede UN, Zeck-Knapp G, Freudenberg N, Keller HK, Seubert B (1991) Humate induced activation of human granulocytes. Virchows Archiv B Cell Pathol 60:27–34

Riemann B, Christoffersen K (1993) Microbial trophodynamics in temperate lake. Mar Microb Food Webs 7:69–100

Rittschof D (1990) Peptide mediated behaviors in marine organisms: evidence for a common theme. J Chem Ecol 16:261–272

Rodhe H, Rood MJ (1986) Temporal evaluation of nitrogen compounds in Swedish precipitation since 1955. Nature 321:762–764

Roila T, Kortelainen P, David MB, Mäkinen I (1994a) Acid-base characteristics of DOC in Finnish lakes. In: Senesi N, Miano TM (eds) Humic substances in the global environment and implications for human health. Elsevier, Amsterdam, pp 863–868

Roila T, Kortelainen P, David MB, Mäkinen I (1994b) Effect of organic anions on acid neutralizing capacity in surface waters. Environ Intern 20:369–372

Rook JJ (1974) Formation of haloforms during chlorination of natural waters. Water Treat Exam 23:234–243

Rose-Janes NG, Playle RC (2000) Protection by two complexing agents, thiosulphate and dissolved organic matter, against the physiological effects of silver nitrate to rainbow trout (*Oncorhynchus mykiss*) in ion-poor water. Aquat Toxicol 52:1–18

Rosseland BO, Blakar IA, Bulger A, Korglund F, Kvellestad A, Lydersen E, Oughton D, Salbu B, Stauernes M, Vogt R (1992) The mixing zone between limed and acid river waters: complex aluminum chemistry and extreme toxicity for salmonids. Environ Pollut 78:3–8

Rosseland BO, Eldhuset TD, Staurnes M (1990) Environmental effects of aluminum. Environ Geochem Health 12:17–27

Rosseland BO, Staurnes M (1994) Physiological mechanisms for toxic effects and resistance to acidic water: an ecophysiological and ecotoxicological approach. In: Steinberg CEW, Wright RF (eds) Acidification of freshwater ecosystems: implications for the future. Wiley, Chichester, pp 227–246

Rothhaupt KO (1992) Stimulation of phosphorus-limited phytoplankton by bacterivorous flagellates in laboratory experiments. Limnol Oceanogr 37:750–759

Rothhaupt KO (1996) Laboratory experiments with a mixotrophic chrysophyte and obligately phagotrophic and phototrophic competitors. Ecology 77:716–724

Rothhaupt KO, Güde H (1992) The influence of spatial and temporal concentration gradients on phosphate partitioning between different size fractions of plankton: further evidence and possible causes. Limnol Oceanogr 37:739–749

Rouleau C, Block M, Tjälve H (1998) Kinetics and body distribution of waterborne $^{65}Zn(II)$, $^{109}Cd(II)$, $^{203}Hg(II)$, and $CH_3{}^{203}Hg(II)$ in phantom midge larvae (*Chaoborus americanus*) and effects of complexing agents. Environ Sci Technol 32:1230–1236

Roy S, Ihantola R, Hanninen O (1992) Peroxidase activity in lake macrophytes and its relation to pollution tolerance. Environ Exp Bot 32:457–464

Rozan TF, Benoit G, Marsh H, Chin YP (1999) Intercomparison of DPASV and ISE for the measurement of Cu complexation characteristics of NOM in freshwater. Environ Sci Technol 33:1766–1770

Ryhänen R (1968) Die Bedeutung der Humussubstanzen im Stoffhaushalt der Gewässer Finnlands. Mitt Intern Verein Limnol 14:168–178

Saber PA, Dunson WA (1978) Toxicity of bog water to embryonic and larval anuran amphibians. J Exper Zool 204:33–42

Sabljic A, Güsten H, Schönherr J, Riederer M (1990) Modeling plant uptake of airborne organic chemicals. 1. Plant cuticle/water partitioning and molecular connectivity. Environ Sci Technol 24:1321–1326

Sachse A, Babenzien D, Ginzel G, Gelbrecht J, Steinberg CEW (2001b) Characterization of dissolved organic carbon (DOC) of a dystrophic lake and an adjacent fen. Biogeochemistry 54:279–296

Sachse A, Gelbrecht J, Steinberg CEW (2001a) Composition of dissolved organic carbon (DOC) in the porewater of fens adjacent to surface waters. In: Swift RS, Spark KM (eds) Understanding and managing organic matter in soils, sediments and waters. Proc Intern Humic Substances Society 9, pp 21–28

Sachse A, Henrion R, Gelbrecht J, Steinberg CEW: Characterization of dissolved organic carbon (DOC) in river-systems influenced by different catchment characteristics and internal processes. (submitted)

Saint-Paul U, Zuanon J, Villacorta Correa, MA, Farcía M, Noemi Fabré N, Berger U, Junk WJ (2000) Fish communities in central Amazonian white- and blackwater floodplains. Environ Biol Fish 57:235–250

Saito K, Matsumoto M, Sekine T, Murakoshi I (1989) Inhibitory substances from *Myriophyllum brasiliense* on growth of blue-green algae. J Nat Prod 52:1221–1226

Sakkas VA, Konstantinou IK, Albanis TA (2001) Photodegradation study of the antifouling booster biocide dichlofluanid in aqueous media by gas chromatographic techniques. J Chromatogr A 930:135–144

Salonen K, Hammar T (1986) On the importance of dissolved organic matter in the nutrition of zooplankton in some lake waters. Oecologia 8:246–253

Salonen K, Jones RI, de Haan H, James M (1994) Radiotracer study of phosphorus uptake by plankton and redistribution in the water column of a small humic lake. Limnol Oceanogr 39:69–83

Salonen K, Kankaala P, Tulonen T, Hammar T, James M, Metsälä TR, Arvola L (1992) Planktonic food chains of a highly humic lake. II. A mesocosm experiment in summer during dominance of heterotrophic processes. Hydrobiologia 229:143–157

Salonen K, Tulonen T (1990) Photochemical and biological transformation of dissolved humic substances. Verh Intern Verein Limnol 24:294

Salonen K, Vähätalo A (1994) Photochemical mineralisation of dissolved organic matter in Lake Skjervatjern. Environ Intern 20:307–312

Sanchèz-Cortès S, Francioso O, Ciavatta C, Govi M, Gessa C (1997) Surface-enhance Raman spectroscopy of peat humic acid fractions. In: Drozd J, Gonet SS, Senesi N, Weber J (eds) The role of humic substances in the ecosystems and in environmental protection. Polish Soc Humic Substances, Wroclaw, pp 133–137

Santschi PH, Buo L, Baskaran M, Trumbore S, Southon J, Bianchi TS, Honeyman B, Cifuentes L (1995) Isotopic evidence for contemporary origin of high-molecular weight organic matter in oceanic environment. Geochim Cosmochim Acta 59:625–631

Sarvala J, Kankaala P, Zingel P, Arvola L (1999) Zooplankton. In: Keskitalo J, Eloranta P (eds) Limnology of humic waters. Backhuys, Leiden, pp 173–191

Saunders G (1976) Decomposition in fresh water. In: Anderson J, Madfadyen A (eds) The role of terrestrial and aquatic organisms in decomposition processes. Blackwell, Oxford, pp 341–374

Schecher WD, Driscoll CT (1987) An evaluation of uncertainty associated with aluminum equilibrium calculations. Water Resourc Res 23:525–534

Schewe C, Klöcking R, Helbig B, Schewe T (1991) Lipoxygenase inhibitory action of antiviral polymeric oxidation products of polyphenols. Biomed Biochim Acta 50:299–305

Schindler DW (1971) A hypothesis to explain differences and similarities among lakes in the Experimental Lakes Area, northwestern Ontario. J Fish Res Bd Can 28:295–301

Schindler DW (1994) Changes caused by acidification to the biodiversity, productivity, and biogeochemical cycles of lakes. In: Steinberg CEW, Wright RF (eds) Acidification of freshwater ecosystems: implications for the future. Wiley, Chichester, pp 153–164

Schindler DW, Bayley SE, Curtis PJ, Parker BR, Stainton MP, Kelly CA (1992) Natural and man-caused factors affecting the abundance and cycling of dissolved organic substances in Precambrian Shield lakes. Hydrobiologia 229:1–21

Schindler DW, Beaty KG, Fee EJ, Cruikshank DR, DeBruyn ER, Findlay DL, Linsey GA, Shearer JA, Stainton MP, Turner MA (1990) Effects of climate warming on lakes of the central boreal forest. Science 250:967–970

Schindler DW, Curtis PJ, Bayley SE, Parker BR, Beaty KG, Stainton MP (1997) Climate-induced changes in the dissolved organic carbon budgets of boreal lakes. Biogeochemistry 36:9–28

Schindler DW, Curtis PJ, Parker BR, Stainton MP (1996) Consequences of climate warming and lake acidification for UV-B penetration in North American boreal lakes. Nature 379:705–708

Schindler DW, Mills KH, Malley DF, Findlay DL, Shearer JA, Davies IJ, Turner MA, Linsey GA, Cruikshank DR (1985) Long-term ecosystem stress: the effects of years of experimental acidification on a small lake. Science 28:1395–1401

Schindler JE, Williams DJ, Zimmermann AP (1976) Investigation of extracellular electron transport by humic acids. In: Nriagu JO (ed) Environmental biogeochemistry. 1: carbon, nitrogen, phosphorus, sulfur, and selenium cycles. Ann Arbor Sci Publ, Ann Arbor, Michigan, pp 109–115

Schinner F, Sonnleitner R (1996) Bodenökologie: Mikrobiologie und Bodenenzymatik. Springer, Berlin

Schirmer C (1989) Nacheiszeitliche Versauerungsgeschichte des Großen Arbersees (Bayerischer Wald). Versuch einer Rekonstruktion mittels der subfossilen Cladoceeren. Diploma Thesis, Free University Berlin, 128 pp + appendix

Schlautman MA, Morgan JJ (1993) Effects of aqueous chemistry on the binding of polycyclic aromatic hydrocarbons by dissolved humic materials. Environ Sci Technol

27:961–969

Schlesinger WH (1991) Biogeochemistry. An analysis of global change. Academic Press, San Diego, 443 pp

Schlickewei W, Riede UN, Kuner EH (1991) Steigerung der Sehnenreißfestigkeit unter Einwirkung eines Huminates (HS 1500). Hefte Unfallkd 220:558

Schmitt P, Hertkorn N, Hoppe A, Garrison AW, Freitag D, Kettrup A (1997) Capillary electrophoresis methods: new tools in the characterization of humic substances. In: Drozd J, Gonet SS, Senesi N, Weber J (eds) The role of humic substances in the ecosystems and in environmental protection. Polish Soc Humic Substances, Wroclaw, pp 183–191

Schmitt-Kopplin P, Hertkorn N, Schulten HR, Kettrup A (1998) Structural changes in a dissolved soil humic acid during photochemical degradation process under O_2 and N_2 atmosphere. Environ Sci Technol 32:2531–2541

Schneider J, Riede U (1992) Untersuchungen zur antiviralen Aktivität von synthetischen Huminstoffen insbesondere gegen HIV-1, HCMV und HSV-1. Research Report to Sopar Pharma GmbH Mannheim (with permission)

Schneider J, Werner A, Weis R, Männer C, Riede UN (1994) HIV-Virostatic effects of humic acids. (Abstract) Pathol Res Pract 190:245

Schnitzer M (1972) Humic substances: chemistry and reactions. In: Schnitzer M, Khan SU (eds) Humic substances in the environment. Marcel Dekker, New York, pp 1–64

Schnitzer M (1982) Organic matter characterization. In: Page AL, Miller RH, Keeney DR (eds) Methods of soil analysis, part 2. Chemical and microbiological properties. Agron Monogr No 9, Madison, pp 581–593

Schnitzer M (1994) A chemical structure for humic acid. Chemical ^{13}C NMR, colloid chemical, and electron microscopic evidence. In: Senesi N, Miano TM (eds) Humic substances in the global environment and implications for human health. Elsevier, Amsterdam, pp 57–69

Schnitzer M, Hindle CA, Meglic M (1986) The analysis of organic matter in soil extracts and whole soils by pyrolysis-mass spectrometry. Advan Agron 55:167–217

Schöler HF (1998) Fluxes of trichloroacetic acid between atmosphere, biota, soil, and groundwater. Eurochlor Report, Brussels

Scholz O, Marxsen J (1996) Sediment phosphatases of the Breitenbach, a first order central European stream. Arch Hydrobiol 135:433–450

Schönfelder I (2000) Indikation der Gewässerbeschaffenheit durch Diatomeen. In: Steinberg CEW, Calmano W, Klapper H, Wilken RD (eds) Handbuch Angewandte Limnologie, ecomed-Verlag, 9th Supplement, pp 1–61

Schönfelder I, Gelbrecht J, Schönfelder J, Steinberg CEW (2002) Littoral diatoms and their chemical environment: relationships in northeastern German lakes and rivers. J Phycol 38:66–82

Schramm KW, Jüttner I, Winkler R, Steinberg CEW, Kettrup A (1994) PCDD/F in recent and historical sediment layers of two German lakes. Organohalogen Comp 20:179–182

Schramm KW, Reischl A, Hutzinger O (1987) UNITree, a multimedia compartment model to estimate the fate of lipophilic compounds in plants. Chemosphere 16:2653–2663

Schramm KW, Winkler R, Casper P, Kettrup A (1997) PCDD/F in recent and historical sediment layers of Lake Stechlin, Germany. Water Res 31:1525–1531

Schreckenbach K (1990) Immunoprophylaxe in der Fischproduktion. In: Horsch F (ed) Immunoprophylaxe bei Nutztieren. Fischer, Jena

Schreckenbach K, Heidrich S, Meinelt T, Steinberg CEW (2002) Gelöste Huminstoffe – VI: Huminstoffe in Aquakultur und Aquaristik. Wasser Boden 54/5:37-41

Schreckenbach K, Knösche R, Seubert BJ, Höke H (1994) Klinische Prüfung des Synthesehuminstoffes HS 1500 bei Eiern und Larven von Regenbogenforellen (*Oncorhynchus mykiss*). Report, Institute of Inland Fisheries, Potsdam Sacrow, Germany

Schreckenbach K, Kühnert M, Haase A, Höke H (1996) Gutachten über die Wirkung des Arzneimittelgrundstoffes HS 1500 bei Nutz- und Zierfischen in der Aquakultur und Aquaristik. Report, Institute of Inland Fisheries, Potsdam Sacrow, Germany

Schreckenbach K, Meinelt T, Spangenberg R, Staaks G, Kalettka T, Spangenberg M, Stüber A (1991) Untersuchungen zur Wirkung des Synthesehuminstoffes RHS 1500 auf Süßwasserfische der Aquakultur. Report, Institute of Inland Fisheries, Berlin, Germany

Schreiner C (1990) Entwicklung eines Indikationssystems der Versauerung von Fließgewässern mit Hilfe von Diatomeen. In: Umweltbundesamt (ed): Monitoringprogramm für versauerte Gewässer durch Luftschadstoffe in der Bundesrepublik Deutschland im Rahmen der ECE. Report No 102 04 362 Umweltbundesamt Berlin, pp 261–282

Schulten HR (1999a) Interactions of dissolved organic matter with xenobiotic compounds: molecular modeling in water. Environ Toxicol Chem 18:1643–1655

Schulten HR (1999b) Analytical pyrolysis and computational chemistry of aquatic humic substances and dissolved organic matter. J Analyt Appl Pyrol 49:385–415

Schulten HR, Leinweber P (2000) New insights into organic-mineral particles: composition, properties and models of molecular structure. Bio Fertil Soils 30:399–432

Schulten HR, Leinweber P, Jandl G (2002) Analytical pyrolysis of humc substances and dissolved organic matter in water. In: Frimmel FH, Abbt-Braun G, Heumann KG, Hock B, Lüdemann HD, Spiteller M (eds) Refractory organic substances in the environment. Wiley-VCH, Weinheim, pp 163–187

Schulten HR, Thomsen M, Carlsen L (2001) Humic complexes of diethyl phthalate: molecular modelling of the sorption process. Chemosphere 45:357–369

Schultz H (1962) Die virucide Wirkung der Huminsäuren im Torfmull auf das Virus der Maul- und Klauenseuche. Dtsch tierärztl Wschr 69:613–616

Schultz JC (1989) Tannin-insect interactions. In: Hemingway RW, Karchesy JJ (eds) Chemistry and significance of condensed tannins. Plenum Press, New York, pp 417–433

Schultz ML (2002) Influence of natural organic matter quality on metal toxicity and accumulation on the gills of rainbow trout (*Oncorhynchus mykiss*). MSc Thesis, University of Waterloo, Ontario, 181 pp

Schwarzenbach RP, Gschwend PM, Imboden DM (1993) Environmental organic chemistry. Wiley, New York, 681 pp

Scott DT, McKnight DM, Blunt-Harris EL, Kolesar SE, Lovley DR (1998) Quinone moieties act as electron acceptors in the reduction of humic substances by humic-reducing microorganisms. Environ Sci Technol 32:2984–2989

Scott DT, McKnight DM, Hrncir D, Voelker B (2001) Effects of dissolved organic carbon on manganese photoreduction in mountain streams. Meeting of the Amer Soc Limnol Oceanogr, Albuquerque, New Mexico, Abstract SS 02–16

Scully NM, Lean DRS (1994) The attenuation of ultraviolet radiation in temperate lakes. Arch Hydrobiol Beih Ergebn Limnol 43:135–144

Scully NM, McQueen DJ, Lean DRS, Cooper WJ (1996) Hydrogen peroxide formation: the interaction of ultraviolet radiation and dissolved organic carbon in lake waters along a 43–75°N gradient. Limnol Oceanogr 41:540–548

Scully NM, Vincent WF, Lean DRS, Cooper WJ (1997) Implication of ozone depletion for surface-water photochemistry: sensitivity of clear lakes. Aquat Sci 59:260–274

Scully NM, Vincent WF, Lean DRS, MacIntyre S (1998) Hydrogen peroxide as a natural trace of mixing in surface layers. Aquat Sci 60:169—186

Seitzinger SP, Sander RW, Styles R (2002) Bioavailability of DON from natural and anthropogenic sources to estuarine plankton. Limnol Oceanogr 47:353–366

Sell A, Overbeck J (1992) Exudates: phytoplankton-bacterioplankton interactions in Plusssee. J Plankton Res 14:1199–1215

Semenov AD (1972) Organic matter in surface waters of USSR. Dissert Rostov/Don, USSR (in Russian)

Semenov AD, Semenova IM, Goncharova IA, Starodomskaja AG, Datsko VG (1964) Infrared spectra of aquatic humic substances. Gidrochim Mater 37:157–161 (in Russian)

Senesi N, Loffredo E, D'Orazio V, Brunetti G, Miano TM, La Cava P (2001) Adsorption of pesticides by humic acids from organic amendments and soils. In: Clapp CE, Hayes MHB, Senesi N, Bloom PR, Jardine PM (eds) Humic substances and chemical contaminants. Soil Science Society of America, Madison, pp 125–153

Senesi N, Steelink C (1989) Application of ESR spectroscopy to the study of humic substances. In: Hayes MHB, MacCarthy P, Malcolm RL, Swift RS (eds) Humic substances II: in search of structure. Wiley, New York, pp 374–408

Seppä H, Weckström J (1999) Holocene vegetational and limnological changes in the Fennoscandian tree-line area as documented by pollen and diatom records from Lake Tsuolbmajavri, Finland. Écoscience 6:621–635

Servos MR, Muir DCG, Webster GR (1989) The effect of dissolved organic matter on the bioavailability of polychlorinated dibenzo-*p*-dioxins. Aquat Toxicol 14:169–184

Shapiro J (1957) Chemical and biological studies on the yellow organic acids of lake water. Limnol Oceanogr 2:161–179

Shaw LJ, Beaton Y, Glover LA, Killham K, Osborn D, Mehard EA (2000) Bioavailability of 2,4-dichlorophenol associated with soil water-soluble humic material. Environ Sci Technol 34:4721–4726

Shaw PJ (1994) The effect of pH, dissolved humic substances, and ionic composition on the transfer of iron and phosphate to particulate size fractions in epilimnetic lake water. Limnol Oceanogr 39:1734–1743

Shaw PJ, de Haan H, Jones RI (1992) The effect of acidification on abiotic interactions of dissolved humic substances, iron and phosphate in epilimnetic water from the HUMEX lake Skjervatjern. Environ Internat 18:577–588

Shaw PJ, Jones RI, de Haan H (2000) The influence of humic substances on the molecular weight distribution of phosphate and iron in epilimnetic lake water. Freshwater Biol 45:383–393

Sherr EB (1988) Direct use of high molecular weight polysaccharide by heterotrophic flagellates. Nature 335:348–351

Shevchenko MA, Taran PN (1963) Investigations of elemental composition of humic matter. Gidrohim Mater 35:149–156 (in Russian)

Shindo H, Huang PM (1984) Significance of Mn(IV) oxide in abiotic formation of organic nitrogen complexes in natural environment. Nature 308:57–58

Shuman MS (1990) Carboxylic acidity of aquatic organic matter: possible systematic errors introduced by XAD extraction. In: Perdue EM, Gjessing ET (eds) Organic acids in aquatic ecosystems. Wiley, Chichester, pp 97–109

Siegfried CA, Bloomfield JA, Sutherland JW (1989) Planktonic rotifer community structure in Adirondack, New York, USA lakes in relation to acidity, trophic status and related water quality characteristics. Hydrobiologia 175:33–48

Sigg L, Stumm W (1991) Aquatische Chemie. Verlag der Fachvereine Zürich/Teubner Stuttgart, 2nd edn, pp 388

Sigg L, Xue H, Kistler D, Schönenberger R (2000) Size fractionation (dissolved, colloidal and particulate) of trace metals in the Thur River, Switzerland. Aquat Geochem 6:413–434

Šimek K, Babenzien D, Bittl T, Koschel R, Macek M, Nedoma J, Vrba J (1998) Microbial food webs in an artificially divided acidic bog lake. Internat Rev Hydrobiol 83:3–18

Sinsabaugh RL, Foreman CM (2001) Activity profiles of bacterioplankton in a eutrophic river. Freshwater Biol 46:1239–1249

Sinsabaugh RL, Linkins AE (1987) Inhibition of the *Trichoderma viride* cellulase complex by leaf litter extracts. Soil Biol Biochem 19:719–725

Siuda W, Chróst RJ (2001) Utilization of selected dissolved organic phosphorus compounds by bacteria in lake water under non-limiting orthophosphate conditions. Pol J Environ Stud 10:475–483

Sjöblom Å, Meili M, Sundbom M (2000) The influence of humic substances on the speciation and bioavailability of dissolved mercury and methylmercury, measured as uptake by *Chaoborus* larvae and loss by volatilization. Sci Total Environ 261:115–124

Skiba U, Cresser MS (1986) Effects of precipitation acidity on the chemistry and microbiology of Sitka spruce litter leachate. Environ Poll A 42:65–78

Skopintsev BA (1934) Analysis of organic matter in waters with high content of chlorides. Zh Prikl Him 7:376–382 (in Russian)

Skopintsev BA (1947) About oxygen equivalent of organic matter content in natural waters. DAN SSSR 2:212–218 (in Russian)

Skopintsev BA (1979) Organic matter. In: Mordukhai-Boltovskoi (ed) The River Volga and its life. Monographiae biologicae 33, Dr W Junk, The Hague

Skopintsev BA (1982) The humus of the world ocean waters and the humus of the Earth Soil In: Nikanorov A M, Valyashko M G (eds) Geochemistry of natural waters. Gidrometeoizdat, Leningrad (in Russian), pp 180–189

Skuja H (1948) Taxonomie des Phytoplanktons einiger Seen in Uppland, Schweden. Symb Bot Ups 9:1–399

Skuja H (1956) Taxonomische und biologische Studien über das Phytoplankton schwedischer Binnengewässer. Nova Acta Reg Soc Sci Uppsal 16:1–404

Skyllberg U, Xia K, Bloom PR, Nater EA, Bleam WF (2000) Binding of Mercury(II) to reduced sulfur in soil organic matter along upland-peat soil transects. J Environ Qual 29:855–865

Smock L, Roeding C (1986) The trophic basis of production of a southeastern USA blackwater stream. Holarc Ecol 9:165–174

Smock LA, Gilinsky E (1992) Coastal plain blackwater streams. In: Hackney CT, Adams MS, Martin WH (eds) Biodiversity of the southeastern United States – aquatic communities. Wiley, New York, pp 271–301

Smol JP, Charles DF, Whitehead DR (1984) Mallomonadacean (Chrysophyceae) assemblages and their relationships with limnological characteristics in 38 Adirondack (New York) lakes. Can J Bot 62:911–923

Sokoloff N (1989) The history of Kashin-Beck disease. New York State J Med 89:343–351

Sonesten L (2001) Mercury content in roach (*Rutilus rutilus* L.) in circumneutral lakes – effects of catchment area and water chemistry. Environ Poll 112:471–481

Sörensen M, Schindelin A J, Frimmel FH (1995) Aufbereitungsorientierte Aspekte des photochemischen Abbaus natürlicher Wasserinhaltsstoffe. Wasser-Abwasser 4:194–199

Specht CH, Kumke MU, Frimmel FH (2000) Characterization of NOM absorption to clay minerals by size exclusion chromatography. Water Res 34:4063–4069

Spieser OH, Scholz W (1992) German patent P 4224750.0

Spokes LJ, Liss PS (1995) Photochemically induced redox reactions in seawater. 1. Cations. Mar Chem 49:201–213

Sprengel C (1831-1832) Chemie für Landwirthe, Forstmänner und Cameralisten. Göttingen

Stabel HH (1977) Bound carbohydrates as stable components in Lake Schöhsee and in *Scenedesmus* cultures. Arch Hydrobiol Suppl 53:159–254 (in German)

Stabel HH, Moaledj K, Overbeck J (1979) On the degradation of dissolved organic molecules from Plusssee by oligocarbophilic bacteria. Arch Hydrobiol Beih Ergebn Limnol 12:95–104

Stabel HH, Steinberg CEW (1976) Cleavage of macromolecular allochthonous soluble organic matter. Naturwissenschaften 63:533

Steelink C, Tollin G (1962) Stable free radicals in soil humic acid. Biochem Biophys Acta 59:25–34

Steinberg CEW (1976) Über die Kopräzipitation ninhydrinpositiver Stoffe. Vom Wasser 47:275–280

Steinberg CEW (1977) Schwer abbaubare, stickstoffhaltige gelöste organische Substanzen im Schöhsee und in Algenkulturen. Arch Hydrobiol Suppl 53:48–148

Steinberg CEW (1980) Species of dissolved metals derived from oligotrophic hard water. Water Res 14:1239–1250

Steinberg CEW (1991) Fate of organic matter during natural and anthropogenic lake acidification. Water Res 25:1453–1458

Steinberg CEW, Bach S (1996) Growth promotion by a groundwater fulvic acid in a bacteria/algae system. Acta hydrochim hydrobiol 24:98–100

Steinberg CEW, Baltes GF (1984) Influence of metal compounds on fulvic acid/molybdenum blue reactive phosphate associations. Arch Hydrobiol 100:61–71

Steinberg CEW, Brüggemann R (2001) Ambiguous ecological control by dissolved humic matter (DHM) and natural organic matter (NOM): trade-offs between specific and non-specific effects. Acta hydrochim hydrobiol 29:399–411

Steinberg CEW, Burkert U (2002) Ökologische Regulation in Binnengewässern: Gelöste Huminstoffe -Teil IV: Energie für das aquatische Nahrungsnetz ohne Einwirkung von Licht. Wasser Boden 54/3:36–42

Steinberg CEW, Fyson A, Nixdorf B (1999) Extrem saure Seen in Deutschland. Biologie in unserer Zeit 29:98–109

Steinberg CEW, Haitzer M, Brüggemann R, Perminova IV, Yashchenko NYu, Petrosyan VS (2000) Towards a quantitative structure activity relationship (QSAR) of dissolved humic substances as detoxifying agents in freshwaters. Internat Rev Hydrobiol 85:253–266

Steinberg CEW, Haitzer M, Hoess S, Pflugmacher S, Welker M (2000) Regulatory impact of humic substances in freshwaters. Verhandl Internat Verein Limnol 27:2488–2491

Steinberg CEW, Herrmann A (1981) Utilization of dissolved metal organic compounds by freshwater microorganisms. Verh Internat Verein Limnol 21:231–235

Steinberg CEW, Hoppe A, Hertkorn N, Jüttner I, Bruckmeier B (2001) Changes of humic substance constituents in Großer Arbersee during acidification. Acta hydrochim hy-

drobiol 29:78–87
Steinberg CEW, Höss S, Brüggemann R (2002) Further evidence that humic substances have the potential to modulate the fertility of the nematode *Caenorhabditis elegans*. Intern Rev Hydrobiol 87:121–133
Steinberg CEW, Kühnel W (1987) Influence of cation acids on dissolved humic substances under acidified conditions. Water Res 21:95–98
Steinberg CEW, Mayr C, Lorenz R, Spieser OH, Kettrup A (1994) Dissolved humic material amplifies irritant effects of terbutylazine (triazine herbicide) on fish. Naturwissenschaften 81:225–227
Steinberg CEW, Münster U (1985) Geochemistry and ecological role of humic substances in lakewater. In: Aiken GR, McKnight DM, Wershaw RL, MacCarthy P (eds) Humic substances in soil, sediment, and water. Wiley, New York, pp 105–145
Steinberg CEW, Schönfelder I (2001) Ökologische Regulation in Binnengewässern: Gelöste Huminstoffe – Teil II: Klimaeinflüsse. Wasser Boden 53/12:37–41
Steinberg CEW, Sturm A, Kelbel J, Lee SK, Hertkorn N, Freitag D, Kettrup A (1992) Changes of acute toxicity of organic chemicals to *Daphnia magna* in the presence of dissolved humic material (DHM). Acta hydrochim hydrobiol 20:326–332
Steinberg CEW, Xu Y, Lee SK, Freitag D, Kettrup A (1993) Effect of dissolved humic material (DHM) on bioavailability of some organic xenobiotics to *Daphnia magna*. Chem Spec Bioavail 5:1–9
Stepanauskas R, Leonardson L, Tranvik LJ (1999) Bioavailability of wetland-derived DON to freshwater and marine bacterioplankton. Limnol Oceanogr 44:1477–1485
Stewart AJ (1984) Interactions between dissolved humic material and organic toxicants. In: Cowser KE (ed) Synthetic fossil fuel technologies. Results of health and environmental studies. Butterworth, Boston, pp 505–521
Stewart AJ, Wetzel RG (1982) Influence of dissolved humic materials on carbon assimilation and alkaline phosphatase activity in natural algal-bacterial assemblages. Freshwater Biol 12:369–380
Stites DL, Benke AC, Gillespie DM (1995) Population dynamics, growth, and production of the Asiatic clam, *Cobicula fluminea*, in a blackwater river. Can J Fish Aquat Sci 52:425–437
Stone AT, Morgan JJ (1984) Reduction and dissolution of manganese(III) and manganese(IV) oxides by organics. 2. Survey of the reactivity of organics. Environ Sci Tech 18:617–624
Stordal MC, Santschi PH, Gill GA (1996) Colloidal pumping: evidence for the coagulation process using natural colloids tagged with ^{203}Hg. Environ Sci Technol 30:3335–3340
Strauss EA, Lamberti GA (2002) Effect of dissolved organic carbon quality on microbial decomposition and nitrification rates in stream sediments. Freshwater Biol 47:65–74
Strohal P, Huljev D (1971) Investigation of mercury pollutant interaction with humic acid by means of radiotracers. In: Nuclear techniques in environmental pollution, IAEA symposium proceedings, International Atomic Energy Agency, Vienna pp 439–446
Strome DJ, Miller MC (1978) Photolytic changes in dissolved humic substances. Verh Internat Verein Limnol 20:1248–1254
Struyk Z, Sposito G (2001) Redox properties of standard humic acids. Geoderma 102:329–346
Stuermer DH (1975) The characterization of humic substances in seawater. Ph.D. Thesis, Woods Hole Oceanographic Institution, Woods Hole, Massachusetts, 187 pp
Stuermer DH, Harvey GR (1974) Humic substances from seawater. Nature 250:480–481

Stumm W, Morgan JJ (1981) Aquatic Chemistry. Wiley, New York

Suffet IH, Jafvert CT, Kukkonen J, Servos MR, Spacie A, Williams LL, Noblet JA (1994) Synopsis of discussion session: influences of particulate and dissolved material on the bioavailability of organic compounds. In: Hamelink JL, Landrum PF, Bergman HL, Benson WH (eds) Bioavailability: physical, chemical and biological interactions. CRC Press, Boca Raton, pp 93–108

Sun L, Perdue EM, Meyer JL, Weis J (1997) Use of elemental composition to predict bioavailability of dissolved organic matter in a Georgia river. Limnol Oceanogr 42:714–721

Sunda W, Guilliard RRL (1976) The relationship between cupric ion activity and the toxicity of copper to phytoplankton. J Mar Res 34:511–529

Sunda WG, Huntsman SA (1994) Photoreduction of manganese oxides in seawater. Mar Chem 46:133–152

Sütfeld R (1998) Polymerization of resorcinol by an cryptophycean exoenzyme. Phytochemistry 49:451–159

Szymczak W, Wolf M, Wittmaack K (2000) Characterisation of fulvic acids and glycyrhizic acid by time-of-flight secondary ion mass spectrometry. Acta hydrochim hydrobiol 28:350–358

Tan KH, Tantiwiramanond D (1983) Effect of humic acid on modulation and dry matter production of soybean, peanut, and clover. Soil Sci Soc Amer J 47:1121–1124

Tarr MA, Wang W, Bianchi TS, Engelhaupt E (2001) Mechanisms of ammonia and amino acid photoproduction from aquatic humic and colloidal matter. Water Res 35:3688–3696

Teunissen PJM, Field JA (1998) 2-chloro-1,4-dimethoxybenzene as a mediator of lignin peroxidase catalyzed oxidations. FEBS Lett 439:219–223

Thaer AD (ed) (1808) Grundriss der Chemie für Landwirte. Realschulbuchhandlung, Berlin

Tham J, Jansen W, Rahmann H (1997) Effect of humic material on aquatic invertebrates in streams of a raised bog complex. In: Drozd J, Gonet SS, Senesi N, Weber J (eds) The role of humic substances in the ecosystems and in environmental protection. Polish Soc Humic Substances, Wroclaw, pp 929–935

Thieme J, Schmidt C, Abbt-Braun G, Specht C, Frimmel FH (2002) X-ray microscopy studies of refractory organic substances. In: Frimmel FH, Abbt-Braun G, Heumann KG, Hock B, Lüdemann HD, Spiteller M (eds) Refractory organic substances in the environment. Wiley-VCH, Weinheim, 239–248

Thienemann A (1925) Die Binnengewässer Mitteleuropas – Eine limnologische Einführung., Schweizerbart, Stuttgart, pp 200–207

Thomas JD (1997) The role of dissolved organic matter, particularly free amino acids and humic substances, in freshwater ecosystems. Freshwater Biol 38:1–36

Thurman EM, Malcolm RL (1981) Preparative isolation of aquatic humic substances. Environ Sci Technol 15:463–466

Tilman D, Wedin D, Knops J (1996) Productivity and sustainability influenced by biodiversity in grassland ecosystems. Nature 379:718–720

Tipping E (1981) The adsorption of aquatic humic substances by iron oxides. Geochim Cosmochim Acta 45:191–199

Tipping E (1994) WHAM – A chemical equilibrium model and computer code for waters, sediments and soils incorporating a discrete site/electrostatic model of ion-binding by humic substances. Comp Geocsci 20:973–1023

Tipping E (1998) Humic ion-binding model VI: an improved description of the interaction of protons and metal ions with humic substances. Aquat Geochem 4:3–48

Tipping E, Backes CA, Hurley MA (1988) The complexation of protons, aluminium and calcium by aquatic humic substances: a model incorporating binding-site heterogeneity and macroionic effects. Water Res 22:597–611

Tipping E, Reddy MM, Hurley MA (1990) Modeling electrostatic and heterogeneity effects on proton dissociation from humic substances. Environ Sci Technol 24:1700–1705

Tombácz E, Rice JA, Ren S (1997) Effect of conformational changes on aggregation processes in humic acid solution. In: Drozd J, Gonet SS, Senesi N, Weber J (eds) The role of humic substances in the ecosystems and in environmental protection. Polish Soc Humic Substances, Wroclaw, pp 43–50

Town RM, Powell HK (1993) Ion-selective electrode potentiometric studies on the complexation of copper (II) by soil-derived humic and fulvic acids. Anal Chim Acta 279:221–233

Traina SJ, McAvoy DC, Versteeg DJ (1996) Association of linear alkylbenzene sulfonates with dissolved humic substances and its effect on bioavailability. Environ Sci Technol 30:1300–1309

Tranvik LJ (1988) Availability of dissolved organic carbon for planktonic bacteria in oligotrophic lakes of differing humic content. Microb Ecol 16:311–322

Tranvik LJ (1989) Bacterioplankton growth, grazing mortality and quantitative relationship to primary production in a humic and clearwater lake. J Plankton Res 11:985–1000

Tranvik LJ (1990) Bacterioplankton growth on fractions of dissolved organic carbon of different molecular weights from humic and clear waters. Appl Environ Microbiol 56:1672–1677

Tranvik LJ (1992a) Allochthonous dissolved organic matter as energy source for pelagic bacteria and the concept of the microbial loop. Hydrobiologia 229:107–114

Tranvik LJ (1992b) Rapid microbial production and degradation of humic-like substances in lake water. Arch Hydrobiol Ergebn Limnol 37:43–50

Tranvik LJ (1993) Microbial transformation of labile dissolved organic matter into humic–like matter in seawater. FEMS – Microbiol Ecol 12:177–183

Tranvik LJ (1998) Degradation of dissolved organic matter in humic waters by bacteria. In: Hessen DO, Tranvik LJ (eds) Aquatic humic substances – ecology and biogeochemistry. Springer, Berlin, pp 259–283

Tranvik LJ, Bertilsson S (2001) Contrasting effects of solar UV radiation on dissolved organic sources for bacterial growth. Ecol Lett 4:458–463

Tranvik LJ, Höfle MG (1987) Bacterial growth in mixed cultures on dissolved organic carbon from humic and clear waters. Appl Environ Microbiol 53:482–488

Tranvik LJ, Sieburth JM (1989) Effects of flocculated humic matter on free and attached pelagic microorganisms. Limnol Oceanogr 34:688–699

Traunspurger W, Haitzer M, Höss S, Beier S, Ahlf W, Steinberg CEW (1997) Ecotoxicological assessment of aquatic sediments with *Caenorhabditis elegans* (Nematoda) – a method for testing liquid medium and whole sediment samples. Environ Toxicol Chem 16:245–250

Tsuji K, Naito S, Kondo F, Ishikawa N, Watanabe MF, Suzuki M, Harada KI (1994) Stability of microcystin from cyanobacteria: effect of light on decomposition and isomerization. Environ Sci Technol 28:173–177

Tulonen T, Salonen K, Arvola L (1992) Effects of different molecular weight fractions of dissolved organic matter on the growth of bacteria, algae, and protozoa from a highly humic lake. Hydrobiologia 229:239–252

Twiss MR, Granier L, Lafrance P, Campbell PGC (1999) Bioaccumulation of 2,2',5,5'-tetrachlorobiphenyl and pyrene by picoplankton (*Synechococcus leopoliensis*, Cyanophyceae): influence of variable humic acid concentrations and pH. Environ Toxicol Chem 18:2063–2069

Uhle ME, Chin YP, Aiken GR, McKnight DM (1999) Binding of polychlorinated biphenyls to aquatic humic substances: the role of substrate and sorbate properties on partitioning. Environ Sci Technol 33:2715–2718

Vähätalo A, Salonen K (1997) Photochemical degradation of chromophoric dissolved organic matter and its contribution to bacterial respiration in a humic lake. Humus – Nord Humus Newslett 4:14

Vähätalo AV, Salkinoja-Salonen M, Taala P, Salonen K (2000) Spectrum of the quantum yield for photochemical mineralization of dissolved organic carbon in a humic lake. Limnol Oceanogr 45:664–676

Valentine RL, Zepp RG (1993) Formation of carbon monoxide from the photodegradation of terrestrial dissolved organic matter to simple substrates for rapid bacterial metabolism. Limnol Oceanogr 40:1369–1380

Valli K, Gold MH (1991) Degradation of 2,4-dichlorophenol by the lignin-degrading fungus *Phanerochaete chrysosporium*. J Bacteriol 173:345–352

van Bergen PF, Nott CJ, Bull ID, Poulton PR, Evershed RP (1998) Organic geochemical studies of soils from the Rothamsted Classical Experiments – IV. Preliminary results from a study of the effect of soil pH on organic matter decay. Org Geochem 29:1779–1795

Vance GF, David MB (1991) Forest soil response to acid and salt additions of sulphate: III. Solubilization and composition of dissolved organic carbon. Soil Sci 151:297–305

Vaughan D, Malcolm RE, Ord BG (1985) Influence of humic substances on biochemical processes in plants. In: Vaughan D, Malcolm RE (eds) Soil organic matter and biological activity. Martinus Nijhoff, Dordrecht, pp 77–108

Vaughan PP, Blough NV (1998) Photochemical formation of hydroxyl radical by constituents of natural waters. Environ Sci Technol 32:2947–2953

Vereecken H, Nitzsche O, Schulze M (2001) Analysis of the transport of hydrophobic organic xenobiotics in the presence of dissolved organic carbon using soil column experiments. In: Clapp CE, Hayes MHB, Senesi N, Bloom PR, Jardine PM (eds) Humic substances and chemical contaminants. Soil Science Society of America, Madison, pp 449–470

Vialaton D, Pilichowski JF, Baglio D, Paya-Perez A, Larsen B, Richard C (2001) Phototransformation of propiconazole in aqueous media. J Agr Food Chem 49:5377–5382

Vigneault B, Percot A, Lafleur M, Campbell PGC (2000) Permeability changes in model and phytoplankton membranes in the presence of aquatic humic substances. Environ Sci Technol 34:3907–3913

Vincent WF, Laurion I, Pienitz R (1998) Arctic and Antarctic lakes as optical indicators of global change. Ann Glaciol 27:691–696

Visser SA (1964) A physico-chemical study of the properties of humic acids and their changes during humification. J Soil Sci 15:202–219

Visser SA (1984) Seasonal changes in the concentration and colour of humic substances in some aquatic environments. Freshwater Biol 14:79–87

Visser SA (1985) Physiological action of humic substances on microbial cells. Soil Biol Biochem 17:457–462

Voelker BM (1994) Iron redox cycling in surface waters: effects of humic substances and light. Dissertation No. 10901, Eidgenössische Technische Hochschule, Zürich, Switzerland 121 pp

Voelker BM, Morel FMM, Sulzberger B (1997) Iron redox cycling in surface waters: effects of humic substances and light. Environ Sci Technol 31:1004–1011

Vogl J, Heumann KG (1997) Determination of heavy metal complexes with humic substances by HPLC/ICP-MS coupling using on-line isotope dilution technique. Fresenius J Anal Chem. 359:438–441

Vogt RD, Ranneklev SB, Mykkelbost TC (1994) The impact of acid treatment on soilwater chemistry at the HUMEX site. Environ Intern 20:277–286

Vogt RD, Seip HM, Ranneklev S (1992) Soil and soil water studies at the HUMEX site. Environ Intern 18:555–564

Volk CJ, Volk CB, Kaplan LA (1997) Chemical composition of biodegradable dissolved organic matter in streamwater. Limnol Oceanogr 42:39–44

Vörös L, Kovács A, Borgulya É, K-Balogh K (2000) Effect of ultraviolet radiation on unicellular algae in presence of humic substances. Proc 10th Meet Intern Humic Subst Soc, Toulouse, pp 883–887

Vuori KM, Muotka T (1999) Benthic communities in humic streams. In: Keskitalo J, Eloranta P (eds) Limnology of humic waters. Backhuys, Leiden, pp 193–207

Wagoner DB, Christman RF, Cauchon G, Paulson R (1997) Molar mass and size of Suwannee river natural organic matter using multi-angle laser light scattering. Environ Sci Technol 31:937–941

Waiser MJ, Robarts RD (2000) Changes in composition and reactivity of allochthonous DOM in a prairie saline lake. Limnol Oceanogr 45:763–774

Waite TD, Wrigley IC, Szymczak R (1988) Photoassisted dissolution of a colloidal manganese oxide in the presence of fulvic acid. Environ Sci Technol 22:778–785

Walenciak O, Zwisler W, Gross EM (2002) Influence of *Myriophyllum spicatum* derived tannins on gut microbiota of its herbivore *Acentria ephemerella* (Lepidoptera: Pyralidae). J Chem Ecol (in press)

Walker I (1992) Life history traits of shrimps (Decapoda, Palaemonidae) of Amazonian inland waters and their phylogenetic interpretation. Stud Neotrop Fauna Environ 27:131–143

Wallschläger D, Desai MVM, Spengler M, Windmöller CC, Wilken RD (1998) How humic substances dominate mercury geochemistry in contaminated floodplain soils and sediments. J Environ Qual 27:1044–1054

Wallschläger D, Desai MVM, Wilken RD (1996) The role of humic substances in the aqueous mobilization of mercury from contaminated floodplain soils. Water Air Soil Pollut 90:507–520

Wang MC, Huang PM (1997) Catalytic power of birnessite in abiotic formation of humic polycondensates from glycine and pyrogallol. In: Drozd J, Gonet SS, Senesi N, Weber J (eds) The role of humic substances in the ecosystems and in environmental protection. Polish Soc Humic Substances, Wroclaw, pp 59–65

Wang W, Tarr MA, Bianchi TS, Engelhaupt E (2000) Ammonium photoproduction from aquatic humic and colloidal matter. Aquat Geochem 6:275–292

Wang WH, Bray CM, Jones MN (1999) The fate of ^{14}C-labelled humic substances in rice cells in cultures. J Plant Physiol 154:203–211

Watras CJ, Bloom NS (1992) Mercury and methylmercury in individual zooplankton: implications for bioaccumulation. Limnol Oceanogr 37:1313–1318

Watt BE, Hayes TM, Hayes MHB, Price RT, Malcolm RL, Jakeman P (1996) Sugar and amino acids in humic substances isolated from British and Irish waters. In: Clapp CE, Hayes MHB, Senesi N, Griffith SM (eds) Humic substances and organic matter in soil and water environments. Intern Humic Subst Soc, St. Paul, pp 81–91

Weber EJ, Colon D, Baughman GL (2001) Sediment-associated reactions of aromatic amines. 1. Elucidation of sorption mechanisms. Environ Sci Technol 35:2470–2475

Weber JH (1993) Review of possible paths for abiotic methylation of mercury(II) in the aquatic environment. Chemosphere 26:2063–2077

Weckström J (2001) Assessment of diatoms as markers of environmental changes in northern Fennoscandia. Dissert Univ Helsinki, Finland

Weete JD (1976) Algal and fungal waxes. In: Kolattukudy PE (ed) Chemistry and biochemistry of natural waxes. Elsevier, New York, pp 364–378

Wehr JD, Brown LM, O'Grady K (1985) Physiological ecology of the bloom-forming alga *Chrysochromulina breviturrita* Nich. (Prymnesiophyceae) from lakes influenced by acid precipitation. Can J Bot 67:2231–2239

Wehr JD, Brown LM, O'Grady K (1987) Highly specialized nitrogen metabolism in a freshwater phytoplankter, *Chrysochromulina breviturrita.* Can J Fish Aquat Sci 44:736–742

Welch IM, Barrett PRF, Gibson MT, Ridge I (1990) Barley straw as an inhibitor of algal growth I: studies in the Chesterfield Canal. J Appl Phycol 2:231–239

Welhouse GJ, Bleam WF (1993a) Atrazine hydrogen-bonding potentials. Environ Sci Technol 27:494–500

Welhouse GJ, Bleam WF (1993b) Cooperative hydrogen bonding to atrazine. Environ Sci Technol 27:500–505

Welker M, Hoeg S, Steinberg CEW (1999) Hepatotoxic cyanobacteria in the shallow lake Müggelsee. Hydrobiologia 408/409:263–268

Welker M, Jones GJ, Steinberg CEW (2001) Release and persistence of microcystins in natural waters. In: Chorus I (ed) Cyanotoxins – occurrence, effects, controlling factors. Springer, Heidelberg, pp 83–101

Welker M, Steinberg CEW (1999) Indirect photolysis of cyanotoxins: one possible mechanism of their low persistence. Water Res 33:1159–1164

Welker M, Steinberg CEW (2000) Rates of humic substances photosensitized degradation of microcystin-LR in natural waters. Environ Sci Technol 34:3415–3419

Welsh PG, Skidmore JF, Sprey DJ, Dixon DG, Hodson PV, Hutchinson NJ, Hickie BE (1993) Effect of pH and dissolved organic carbon on the toxicity of copper to larval fathead minnows (*Pimephales promelas*) in natural lake water of low alkalinity. Can J Fish Aquat Sci 50:1356–1362

Wershaw RL (1986) A new model for humic materials and their interactions with hydrophobic organic chemicals in soil-water and sediment-water systems. J Contam Hydrol 1:29–45

Wershaw RL (1989) Molecular aggregate structure. In: Humic Substances in the Suwannee River, Georgia: interactions, properties, and proposed structures. US Geol Survey Open File Report 87–557, pp 354–356

Wershaw RL, Burcar PJ, Goldberg MC (1969) Interaction of pesticides with natural organic material. Environ Sci Technol 3:271–273

Wershaw RL, Pinckney DJ, Llaguno EC, Vincente-Beckett V (1990) NMR characterization of humic acid fractions from different Philippine soils and sediments. Anal Chim Acta 232:31–42

Wershaw RL, Thorn KA, Pinckney DJ, MacCarthy P, Rice JA, Hemond HF (1986) Application of a membrane model to the secondary structure of humic materials in peat. In: Fuchsman CH (ed) Peat and water. Elsevier, New York, pp 137–157

West CC (1984) Dissolved organic carbon facilitated transport of neutral organic compounds in subsurface systems. Ph.D. Diss Rice University, Houston TX

Westcott K, Kalff J (1996) Environmental factors affecting methyl mercury accumulation in zooplankton. Can J Fish Aquat Sci 53:2221–2228

Westerhoff P, Aiken GR, Amy G, Debroux J (1999) Relationships between the structure of natural organic matter and its reactivity toward molecular ozone and hydroxyl radicals. Water Res 33:2265–2276

Wetzel RG (1990) Land-water interfaces: metablic and limnological regulators. Verh Internat Verein Limnol 24:6–24

Wetzel RG (1991) Extracelluar enzymatic interactions in aquatic ecosystems: storage, redistribution, interspecific communication. In: Chróst RJ (ed) Microbial enzymes in aquatic environments. Brock/Springer Ser Contempor Biosci, Springer Berlin, pp 6–28

Wetzel RG (1993) Humic compounds from wetlands: complexation, inactivation, and reactivation of surface-bound and extracellular enzymes. Verh Intern Verein Limnol 25:122–128

Wetzel RG (1995) Death, detritus, and energy flow in aquatic ecosystems. Freshwater Biol 33:83–89

Wetzel RG (2001) Limnology. Lake and River Ecosystems. 3rd edn. Academic Press, San Diego

Wetzel RG, Hatcher PG, Bianchi TS (1995) Natural photolysis by ultraviolet irradiance of recalcitrant dissolved organic matter to simple substrates for rapid bacterial metabolism. Limnol Oceanogr 40:1369–1380

Wetzel RG, Howe MJ (1999) High production in a herbaceous perennial plant achieved by continuous growth and synchronized population dynamics. Aquat Bot 64:111–129

Wiegand C, Meems N, Timoveyev M, Steinberg CEW, Pflugmacher S: More evidence for humic substances acting as biogeochemicals on organisms. In: Ghabbour EA, Davies G (eds) Humic substances: nature's most versatile materials. Taylor & Francis, New York (in press)

Wiegand C, Pflugmacher S, Giese M, Frank H, Steinberg CEW (1999a) Uptake, toxicity and effects on detoxication enzymes of atrazine and trifluoracetate in embryos of zebrafish. Ecotoxicol Environ Safe 45:122–131

Wiegand C, Pflugmacher S, Oberemm A, Meems N, Beattie KA, Steinberg CEW, Codd GA (1999b) Uptake and effects of microcystin-LR on detoxication enzymes of early life stages of the zebra fish (*Danio rerio*). Environ Toxicol 14:89–95

Wiegand C, Pflugmacher S, Oberemm A, Steinberg CEW (2000) Activity development of selected detoxication enzymes during the ontogenesis of the zebrafish (*Danio rerio*). Intern Rev Hydrobiol 85:413–422

Wilkinson AE, Hesketh N, Higgo JJW, Tipping E (1993) The determination of the molecular mass of humic substances from natural waters by analytical ultracentrifugation. Colloids Surfaces A 73:19–28

Wilkinson KJ, Jones HG, Campbell PGC, Lachance M (1992) Estimating organic acid contributions to surface water acidity in Quebec (Canada). Water Air Soil Poll 61:57–74

Wilkinson KJ, Lead JR, Hosse M, Balnois E (2000) Multi-method structural characterization of freshwater humic substances. Proc 10th Meet Intern Humic Subst Soc, 24–28 July 2000, Toulouse, pp 691–694

Wilkinson KJ, Nègre JC, Buffle J (1997) Coagulation of colloidal material in surface waters: the role of natural organic matter. J Contam Hydrol 26:229–243

Williams MW, Losleben M, Caine N, Greenland D (1996) Changes in climate and hydrochemical responses in a high-elevation catchment in the Rocky Mountains, USA. Limnol Oceanogr 41:939–946

Williams PM, Druffel ERM (1987) Radiocarbon in dissolved organic matter in the central North Pacific Ocean. Nature 330:246–248

Winner RW (1984) The toxicity and bioaccumulation of cadmium and copper as affected by humic acid. Aquat Toxicol 5:264–274

Winner RW (1986) Interactive effects of water hardness and humic acid on the chronic toxicity of cadmium to *Daphnia pulex*. Aquat Toxicol 8:281–293

Wojciechowski I, Górniak A (1990) Influence of the brown humic and fulvic acids originating from nearby peat bogs on phytoplankton activity in the littoral of two lakes in Mid-Eastern Poland. Verh Intern Verein Limnol. 24:295–297.

Wolfe MF, Schwarzbach S, Sulaiman RA (1998) Effects of mercury on wildlife: a comprehensive review. Environ Toxicol Chem 17:146–160

Wolff CJM, Halmans MTH, van der Heijde HB (1981) The formation of singlet oxygen in surface waters. Chemosphere 10:59–62

Wood CM, Hogstrand C, Galves F, Munger RS (1996) The physiology of waterborne silver toxicity in freshwater rainbow trout (*Oncorhynchus mykiss*) 1. The effects of ion Ag^+. Aquat Toxicol 35:93–109

Wood CM, McDonald DG (1987) The physiology of acid/aluminum stress in trout. Ann Soc R Zool Belg 117:399–410

Wood CM, Playle RC, Hogstand C (1999) Physiological and modeling of mechanisms of silver uptake and toxicity in fish. Environ Toxicol Chem 18:71–83

Woodward J (1699) Thoughts and experiments on vegetation. Phil Trans R Soc London B 21:382–398

Wotton RS (1977) The size of particles ingested by moorland stream blackfly larvae (Simuliidae). Oikos 29:332–335

Wren CD, Scheider WA, Wales DL, Muncaster BW, Gray IM (1991) Relation between mercury concentrations in walleye (*Stizostedion vitreum vitreum*) and northern pike (*Esox lucius*) in Ontario lakes and influence of environmental factors. Can J Fish Aquat Sci 48:132–139

Wright RF, Lotse E, Semb A (1988) Reversibility of acidification shown by whole-catchment experiments. Nature 334:670–675

Xenopoulos MA, Bird DF (1997) Effect of acute exposure to hydrogen peroxide on the production of phytoplankton and bacterioplankton in a mesohumic lake. Photochem Photobiol 66:471–478

Xia K, Skyllberg UL, Bleam WF, Bloom PR, Nater EA, Helmke PA (1999) X-ray absorption spectroscopic evidence for the complexation of Hg(II) by reduced sulfur in soil humic substances. Environ Sci Technol 33:257–261

Xia K, Weesner F, Bleam WF, Bloom PR, Skyllberg UL, Helmke PA (1998) XANES study of oxidation states of sulfur in aquatic and soil humic substances. Soil Sci Soc Am J 62:1240–1246

Xing BS (2001) Sorption of anthropogenic organic compounds by soil organic matter: a mechanistic consideration. Can J Soil Sci 81:317–323

Xue HB, Jansen S, Prasch A, Sigg L (2001) Nickel speciation and complexation kinetics in freshwater by ligand exchange and DPCSV. Environ Sci Technol 35:539–546

Xue HB, Kistler D, Sigg L (1995) Competition of copper and zinc for strong ligands in a eutrophic lake. Limnol Oceanogr 40:1142–1152

Xue HB, Sigg L (1999) Comparison of the complexation of Cu and Cd by humic or fulvic acids and by ligands observed in lake waters. Aquat Geochem 5:313–335

Yamamoto S, Ishiwatari R (1989) A study of the formation mechanism of sedimentary humic substances. II. Protein-based melanoidin model. Org Geochem 14:479–489

Yamamoto S, Ishiwatari R (1992) A study of the formation mechanism of sedimentary humic substances. III. Evidence for the protein-based melanoidin model. Sci Total Environ 117/118:279–292

Yan ND (1983) Effects of changes in pH on transparency and thermal regimes of Lohi Lake, near Sudbury, Ontario. Can J Fish Aquat Sci 40:621–623

Yan ND (1986) Empirical prediction of crustacean zooplankton biomass in nutrient-poor Canadian Shield lakes. Can J Fish Aquat Sci 43:788–796

Yan ND, Keller W, Scully NM, Lean DRS, Dillon P (1996) Increased UV-B penetration in a lake owing to drought-induced acidification. Nature 381:141–143

Yang HL, Tu SC, Lu FJ, Chiu HC (1994) Plasma protein C activity is enhanced by arsenic but inhibited by fluorescent humic acid associated with Blackfoot disease. Amer J Hematol 46:264–269

Yin Y, Allen HE, Huang CP, Sanders PF (1997) Interaction of Hg(II) with soil-derived humic substances. Anal Chim Acta 341:73–82

Zak D, Gelbrecht J, Steinberg CEW: Phosphorus retention at the redox interface of peatlands adjacent to surface waters in Northeast Germany. (submitted)

Zeck-Kapp G, Nauck M, Riede UN, Block L, Freudenberg N, Seubert B (1991) Niedermolekulare Huminstoffe als proinflammatorische Zellsignale. (Abstract) Verh Dtsch Ges Path 75:504

Zepp RG, Baughman GL, Schlotzhauer PF (1981) Comparison of photochemical behavior of various humic substances in water. I. Sunlight induced reactions of aquatic pollutants photosensitized by humic substances. Chemosphere 10:109–117

Zepp RG, Braun AM, Hoigné J, Leenheer JA (1987a) Photoproduction of hydrated electrons from natural organic solutes in aquatic environments. Environ Sci Technol 21:485–490

Zepp RG, Callaghan TV, Erickson DJ (1995) Effects of increased solar ultraviolet radiation on biogeochemical cycles. Ambio 24:181–187

Zepp RG, Hoigné J, Bader H (1987b) Nitrate-induced photooxidation of trace organic chemicals in water. Environ Sci Technol 21:443–450

Zepp RG, Wolfe NL, Baughman GL, Hollis RC (1977) Singlet oxygen in natural waters. Nature 207:421–423

Zhilin DM (1998) Reactivity and detoxifying properties of humic substances in relation to Hg(II). Ph.D. Thesis, Lomonosov Moscow State University, Moscow

Zsolnay A, Görlitz H (1994) Water extractable organic matter in arable soils: effect of drought and long-term fertilization. Soil Biol Biochem 26:1257–1261

Index

MIX
Papier aus verantwortungsvollen Quellen
Paper from responsible sources
FSC® C105338

If you have any concerns about our products,
you can contact us on
ProductSafety@springernature.com

In case Publisher is established outside the EU,
the EU authorized representative is:
Springer Nature Customer Service Center GmbH
Europaplatz 3, 69115 Heidelberg, Germany

Printed by Libri Plureos GmbH
in Hamburg, Germany